SI UNITS USED IN MECHANICS

Quantity	Unit	SI Symbol
(*Base Units*)		
Length	meter°	m
Mass	kilogram	kg
Time	second	s
(*Derived Units*)		
Acceleration, linear	meter/second^2	m/s^2
Acceleration, angular	radian/second^2	rad/s^2
Area	meter^2	m^2
Density	kilogram/meter^3	kg/m^3
Force	newton	N $(= \text{kg}\cdot\text{m/s}^2)$
Frequency	hertz	Hz $(= 1/\text{s})$
Impulse, linear	newton-second	N·s
Impulse, angular	newton-meter-second	N·m·s
Moment of force	newton-meter	N·m
Moment of inertia, area	meter^4	m^4
Moment of inertia, mass	kilogram-meter^2	kg·m^2
Momentum, linear	kilogram-meter/second	kg·m/s $(= \text{N·s})$
Momentum, angular	kilogram-meter^2/second	kg·m^2/s $(= \text{N·m·s})$
Power	watt	W $(= \text{J/s} = \text{N·m/s})$
Pressure, stress	pascal	Pa $(= \text{N/m}^2)$
Product of inertia, area	meter^4	m^4
Product of inertia, mass	kilogram-meter^2	kg·m^2
Spring constant	newton/meter	N/m
Velocity, linear	meter/second	m/s
Velocity, angular	radian/second	rad/s
Volume	meter^3	m^3
Work, energy	joule	J $(= \text{N·m})$
(*Supplementary and Other Acceptable Units*)		
Distance (navigation)	nautical mile	$(= 1.852 \text{ km})$
Mass	ton (metric)	t $(= 1000 \text{ kg})$
Plane angle	degrees (decimal)	°
Plane angle	radian	—
Speed	knot	(1.852 km/h)
Time	day	d
Time	hour	h
Time	minute	min

° Also spelled *metre*.

SI UNIT PREFIXES

Multiplication Factor	Prefix	Symbol
1 000 000 000 000 $= 10^{12}$	terra	T
1 000 000 000 $= 10^{9}$	giga	G
1 000 000 $= 10^{6}$	mega	M
1 000 $= 10^{3}$	kilo	k
100 $= 10^{2}$	hecto	h
10 $= 10$	deka	da
0.1 $= 10^{-1}$	deci	d
0.01 $= 10^{-2}$	centi	c
0.001 $= 10^{-3}$	milli	m
0.000 001 $= 10^{-6}$	micro	μ
0.000 000 001 $= 10^{-9}$	nano	n
0.000 000 000 001 $= 10^{-12}$	pico	p

SELECTED RULES FOR WRITING METRIC QUANTITIES

1. (a) Use prefixes to keep numerical values generally between 0.1 and 1000.
   (b) Use of the prefixes hecto, decka, deci, and centi should be generally avoided except for certain areas or volumes where the numbers would be otherwise awkward.
   (c) Use prefixes only in the numerator of unit combinations. The one exception is the base unit kilogram. (*Example:* write kN/m not N/mm; J/kg not mJ/g)
   (d) Avoid double prefixes. (*Example:* write GN not kMN)
2. Unit designations
   (a) Use a dot for multiplication of units. (*Example:* write N·m not Nm)
   (b) Avoid ambiguous double solidus. (*Example:* write N/m^2 not N/m/m)
   (c) Exponents refer to entire unit. (*Example:* mm^2 means (mm)^2)
3. Number grouping
   Use a space rather than a comma to separate numbers in groups of three, counting from the decimal point in both directions. (*Example:* 4 607 321.048 72) Space may be omitted for numbers of four digits. (*Example:* 4296 or 0.0476)

ENGINEERING MECHANICS

DYNAMICS

ENGINEERING MECHANICS

VOLUME 2

DYNAMICS

SECOND EDITION

J.L. MERIAM
University of California
Santa Barbara

L.G. KRAIGE
Virginia Polytechnic Institute and
State University

JOHN WILEY AND SONS
New York • Chichester • Brisbane • Toronto • Singapore

SI VERSION

Sponsoring Editor Charity Robey
Production Supervisor Pamela A. Pelton
Cover design by Sheila Granda
Illustrations by John Balbalis with the assistance of the Wiley
    Illustration Department
Manuscript Editor John Thomas
    under the supervision of Bruce Safford

Library of Congress Cataloging-in-Publication Data
(Revised for volume 2)

Meriam, J. L. (James L.)
    Engineering mechanics.

    Bibliography: v. 2, p.
    Includes indexes.
    Contents: v. 1. Statics — v. 2. Dynamics.
    1. Mechanics, Applied. I. Kraige, L. G. (L. Glenn) II. Title.
TA350.M458   1987   620.1   86-13272

Printed in the United States of America
10 9 8 7 6 5 4 3 2 1

FOREWORD

The innovations and contributions of Dr. James L. Meriam to the field of engineering mechanics cannot be overstated. He has undoubtedly had as much influence on instruction in mechanics during the last thirty years as any one individual. His first books on mechanics in 1951 literally reconstructed undergraduate mechanics and became the definitive textbooks for the next decade. His texts were logically organized, easy to read, directed to the average engineering undergraduate, and were packed with exciting examples of real-life engineering problems superbly illustrated. These books became the model for other engineering mechanics texts in the 1950s and 1960s.

Dr. Meriam began his work in mechanical engineering at Yale University where he earned his B.E., M. Eng., and Ph.D. degrees. He had early industrial experience with Pratt and Whitney Aircraft and the General Electric Company, which stimulated his first contributions to mechanics in mathematical and experimental stress analysis. In the Second World War he served in the U.S. Coast Guard.

Dr. Meriam was a member of the faculty of the University of California, Berkeley, for twenty-one years where he served as Professor of Engineering Mechanics, Assistant Dean of Graduate Studies, and Chairman of the Division of Mechanics and Design. In 1963 he became Dean of Engineering at Duke University where he devoted his full energies to the development of its School of Engineering. In 1972 Professor Meriam followed his desire to return to full-time teaching and served as Professor of Mechanical Engineering at California Polytechnic State University, and more recently as a visiting professor at the University of California, Santa Barbara. Professor Meriam has always placed great emphasis on teaching, and this trait has been recognized by his students wherever he has taught. At Berkeley in 1963 he was the first recipient of the Outstanding Faculty Award of Tau Beta Pi, given primarily for excellence in teaching, and in 1978 received the Distinguished Educator Award for Outstanding Service to Engineering Mechanics Education from the American Society for Engineering Education.

Professor Meriam was the first author to show clearly how the method of virtual work in statics can be employed to solve a class of problems largely neglected by previous authors. In dynamics, plane motion became understandable, and in his later editions, three-dimensional kinematics and kinetics received the same treatment. He is credited with original developments in the theory of variable-mass dynamics, which are contained in his *Dynamics, 2nd Edition*. Professor Meriam has also been a leader in promoting the use of SI units, and his *SI Versions* of *Statics* and *Dynamics* published in 1975 were the first mechanics textbooks in SI units in this country.

Welcomed as a coauthor of this second edition of ENGINEERING MECHANICS is Dr. L. Glenn Kraige whose interests and qualifications in mechanics place him in a position of singular excellence to make significant contributions to mechanics education. Dr. Kraige earned his B.S., M.S., and Ph.D. degrees at the University of Virginia, principally in aerospace engineering, and he currently serves as Associate Professor of Engineering Science and Mechanics at Virginia Polytechnic Institute and State University.

In addition to his recognized research and publications in the field of spacecraft dynamics, Professor Kraige has devoted his attention to the teaching of mechanics at both introductory and advanced levels. His teaching earned him the Outstanding Educator Award from his department in 1977, the Phillip J. Sporn Award for outstanding teaching of engineering subjects in 1978, and the William E. Wine Award for outstanding university-level teaching in 1984. In his teaching, Professor Kraige stresses the development of analytical capabilities and at the same time the strengthening of physical insight and engineering judgment. More recently, he has also become a leader in the development of motion simulation software for use on personal computers.

The second edition of ENGINEERING MECHANICS promises to exceed the high standards set by the previous edition. The new edition contains one of the most outstanding collections of instructive and interesting problems yet assembled. With increased emphasis on the analytical rigor and unity of mechanics, and with continued emphasis on challenging engineering applications, the new text will appeal to a wide audience of students, teachers, and engineers, and will further extend the authors' contributions to mechanics.

R. F. Steidel
Professor of Mechanical Engineering
University of California, Berkeley

PREFACE
To the Student

As you undertake the study of engineering mechanics, first statics and then dynamics, you will be building a foundation of analytical capability for the solution of a great variety of engineering problems. Modern engineering practice demands a high level of analytical capability, and you will find that your study of mechanics will help you immensely in developing this capacity.

In engineering mechanics we learn to construct and solve mathematical models which describe the effects of force and motion on a variety of structures and machines that are of concern to engineers. In applying our principles of mechanics we formulate these models by incorporating appropriate physical assumptions and mathematical approximations. Both in the formulation and solution of mechanics problems you will have frequent occasion to use your background in plane and solid geometry, scalar and vector algebra, trigonometry, analytic geometry, and calculus. Indeed, you are likely to discover new significance to these mathematical tools as you make them work for you in mechanics.

Your success in mechanics (and throughout engineering) will be highly contingent upon developing a well-disciplined method of attack from hypothesis to conclusion in which the applicable principles are applied rigorously. Years of experience in teaching and engineering disclose the importance of developing the ability to represent one's work in a clear, logical, and concise manner. Mechanics is an excellent place in which to develop these habits of logical thinking and effective communication.

ENGINEERING MECHANICS contains a large number of sample problems in which the solutions are presented in detail. Also included in these examples are helpful observations that mention common errors and pitfalls to be avoided. In addition, the book contains a large selection of simple, introductory problems and problems of intermediate difficulty to help you gain initial confidence and understanding of each new topic. Also included are many problems which illustrate significant and contemporary engineering situations to stimulate your interest and help you to develop an appreciation for the many applications of mechanics in engineering.

We are pleased to extend our encouragement to you as a student of mechanics. We hope this book will provide both help and stimulation as you develop your background in engineering.

J. L. Meriam

Santa Barbara, California
January 1987

L. Glenn Kraige

Blacksburg, Virginia
January 1987

PREFACE
To the Instructor

The primary purpose of the study of engineering mechanics is to develop the capacity to predict the effects of force and motion in the course of carrying out the creative design function of engineering. Successful prediction requires more than a mere knowledge of the physical and mathematical principles of mechanics. Prediction also requires the ability to visualize physical configurations in terms of real materials, actual constraints, and the practical limitations which govern the behavior of machines and structures. One of our primary objectives in teaching mechanics should be to help the student develop this ability to visualize, which is so vital to problem formulation. Indeed, the construction of a meaningful mathematical model is often a more important experience than its solution. Maximum progress is made when the principles and their limitations are learned together within the context of engineering application.

Courses in mechanics are often regarded by students as a difficult requirement and frequently as an uninteresting academic hurdle as well. The difficulty stems from the extent to which reasoning from fundamentals, as distinguished from rote learning, is required. The lack of interest that is frequently experienced is due primarily to the extent to which mechanics is presented as an academic discipline often lacking in engineering purpose and challenge. This attitude is traceable to the frequent tendency in the presentation of mechanics to use problems mainly as a vehicle to illustrate theory rather than to develop theory for the purpose of solving problems. When the first view is allowed to predominate, problems tend to become overly idealized and unrelated to engineering with the result that the exercise becomes dull, academic, and uninteresting. This approach deprives the student of much of the valuable experience in formulating problems and thus of discovering the need for and meaning of theory. The second view provides by far the stronger motive for learning theory and leads to a better balance between theory and application. The crucial role of interest and purpose in providing the strongest possible motive for learning cannot be overemphasized.

Furthermore, we should stress the view that, at best, theory can only approximate the real world of mechanics rather than the view that the real world approximates the theory. This difference in philosophy is indeed basic and distinguishes the *engineering* of mechanics from the *science* of mechanics.

During the past twenty years there has been a strong trend in engineering education to increase the extent and level of theory in the engineering-science courses. Nowhere has this trend been felt more than in mechanics courses. To the extent that students are prepared to handle the accelerated treatment, the trend is beneficial. There is evidence and justifiable concern, however, that a significant disparity has more recently appeared between coverage and comprehension. Among the contributing factors we note three trends. First, emphasis on the geometric and physical meanings of prerequisite mathematics appears to have diminished. Second, there has been a significant reduction and even elimination of instruction in graphics, which in the past served to enhance the visualization and representation of mechanics problems. Third, in advancing the mathematical level of our treatment of mechanics there has been a tendency to allow the notational manipulation of vector operations to mask or replace geometric visualization. Mechanics is inherently a subject which depends on geometric and physical perception, and we should increase our efforts to develop this ability.

One of our responsibilities as teachers of mechanics is to use the mathematics which is most appropriate for the problem at hand. The use of vector notation for one-dimensional problems is usually trivial; for two-dimensional problems it is often optional; but for three-dimensional problems it is quite essential. As we introduce vector operations in two-dimensional problems, it is especially important that their geometric meaning be emphasized. A vector equation is brought to life by a sketch of the corresponding vector polygon, which often discloses through its geometry the shortest solution. There are, of course, many mechanics problems where the complexity of variable interdependence is beyond the normal powers of visualization and physical perception, and reliance on analysis is essential. Nevertheless, our students become better engineers when their abilities to perceive, visualize, and represent are developed to the fullest.

As teachers of engineering mechanics we have the strongest obligation to the engineering profession to set reasonable standards of performance and to uphold them. In addition, we have a serious responsibility to encourage our students to think for themselves. Too much help with details that students should be reasonably able to handle from prerequisite subjects can be as bad as too little help and can easily condition them to becoming overly dependent on others rather than to exercise their own initiative and ability. Also, when mechanics is subdivided into an excessive number of small compartments, each with detailed and repetitious instructions, students can have difficulty seeing the "forest for the trees" and, consequently, fail to perceive the unity of mechanics and the far-reaching applicability of its few basic principles and methods.

This second edition of ENGINEERING MECHANICS, as with the first edition, is written with the foregoing philosophy in mind. It is intended primarily for the first engineering course in mechanics, generally taught in the second year of study. The book omits a number of the more advanced topics contained in the senior author's more extensive treatments, *Statics* and *Dynamics, 2nd Edition* (1971) and *SI Version* (1975), and is designed especially to facilitate self-study. To this end a major feature of the book is the extensive treatment of sample problems, which are presented in a single-page format for more convenient study. In addition to presenting the solution in detail, each sample problem also contains comments and cautions keyed to salient points in the solution and printed in colored type. These comments alert students to common pitfalls and should provide a valuable aid to their self-study efforts.

A special feature of the second edition is the inclusion of selected problems that are especially appropriate for computer solution. These problems are separately identified and are placed at the end of the Problem Review articles that conclude most of the chapters. With the increasing availability of personal computers, attention to their role in mechanics education must be addressed. The authors wish to emphasize that the experience in formulating problems, where reason and judgment are developed, is vastly more important for the student than is the manipulative exercise in carrying out the solution. For this reason the time devoted to computer usage in mechanics should be carefully limited. On the other hand, the computer can be used to reveal results in special problems where conventional solutions are awkward or inadequate. The computer-oriented problems in the second edition are included to help guide the student into the type of problem for which machine solution is especially beneficial in revealing some critical conclusion not otherwise readily discernable. To conserve adequate time for problem formulation, it is suggested that the student be encouraged or required to develop and carry out an appropriate computer program for only a limited few of these special problems.

In the second edition increased attention has been given to identifying the more complete mathematical solutions to selected problems which illustrate the importance of matching the number of dependent variables with the number of available independent equations.

Volume 2, Dynamics, contains 114 sample problems and 1313 unsolved problems from which a wide choice of assignments can be made. Of these problems over 50 percent are totally new with the balance selected from the preceding editions. Each problem set begins with relatively simple, uncomplicated problems to help students gain confidence with the new topic. Many practical problems and examples of interesting engineering situations drawn from a wide range of applications are represented in the problem collection. Simple numerical values have been used throughout, however, so as not to complicate the solutions and divert attention from the principles. The problems are arranged generally in order of increasing difficulty, and the answers to most of them are given. The more difficult problems are identified by the symbol ▶ and may often be used to provide a comprehensive classroom experience when solved by the instructor. All numerical solutions have been carried out and checked with an electronic calculator without rounding intermediate values. Consequently, the final answers should be correct to within the number of significant figures cited. The authors are confident that the book is exceptionally free from error.

ENGINEERING MECHANICS is written in a style which is both concise and friendly. The major emphasis is focused on basic principles and methods rather than on a multitude of special cases. Strong effort has been made to show both the cohesiveness of the relatively few fundamental ideas and the great variety of problems which these few ideas will solve.

The logical division between particle dynamics and rigid-body dynamics, with each part treating the kinematics prior to the kinetics, has been followed in Chapters 2 and 3. This arrangement greatly facilitates a more thorough and rapid excursion in rigid-body dynamics with the prior benefit of a comprehensive introduction to particle dynamics.

Chapter 3 on particle kinetics focuses on the three basic methods, force-mass-acceleration, work-energy, and impulse-momentum. The special topics of impact, central-force motion, and relative motion are grouped together in Section D of Chapter 3 on special applications and serve as optional material to be assigned according to instructor preference and available time. With this arrangement the attention of the student is focused more strongly on the three basic approaches to kinetics which are developed in Sections A, B, and C of the chapter.

Chapter 4 on systems of particles is an extension of the principles of motion for a single particle and develops the general relationships which are so basic to modern comprehension of dynamics. The chapter also includes the topics of steady mass flow and variable mass, which may be considered as optional material depending on the time available.

In Chapter 5 on the kinematics of rigid bodies in plane motion, emphasis is placed jointly on geometrical visualization and analysis and on rigorous mathematical solution of the vector equations of relative velocity and relative acceleration. This dual approach serves the purpose of reinforcing the meaning of vector mathematics.

Chapter 6 on the kinetics of rigid bodies has been redesigned to place greater emphasis on the basic equations which govern all categories of plane motion. Strong dependence is placed on the identification of knowns and unknowns and on the necessary and sufficient motion equations which guarantee solution. Special emphasis is also placed on forming the direct equivalence between the actual applied forces and couples and their $m\bar{a}$ and $\bar{I}\alpha$ resultants. In this way the versatility of the moment principle is emphasized, and the student is encouraged to think directly in terms of resultant dynamics effects.

Chapter 7, which may be treated as optional, provides a basic introduction to three-dimensional dynamics which is sufficient to solve many of the more common space-motion problems. For students who later pursue more advanced work in dynamics, Chapter 7 will provide a solid foundation. Gyroscopic motion with steady precession is treated in two ways. The first approach makes use of the analogy between the relation of force and linear-momentum vectors and the relation of moment and angular-momentum vectors. With this treatment the student can understand the gyroscopic phenomenon of steady precession and handle most of the engineering problems on gyros without a detailed study of three-dimensional dynamics. The second approach makes use of the more general momentum equations for three-dimensional rotation where all components of momentum are accounted for.

Chapter 8 is devoted to the topic of vibrations, which has been expanded to a separate chapter. This increased emphasis will be especially useful for engineering students whose only exposure to vibrations is acquired in the basic dynamics course.

Moments and products of inertia of mass are included in Appendix B, and Appendix C contains a summary review of selected topics of elementary mathematics that the student should be prepared to use in mechanics.

Special recognition is due Dr. A. L. Hale of the Bell Telephone Laboratories for his continuing contribution in the form of invaluable suggestions and accurate checking of the manuscript. Dr. Hale has rendered similar service to all previous versions of the senior author's books on mechanics, and his input has been a great asset. In addition, appreciation is expressed to Professor J. M. Henderson of the University of California, Davis, for helpful comments and suggestions of selected problems. Contribution by the staff of John Wiley & Sons, especially editor Bill Stenquist and illustrator John Balbalis, during the planning and production of the book reflects a high degree of professional competence and is duly recognized. The contribution of Virginia Polytechnic Institute and State University through the generous support of Dr. Paul Torgersen, Dean of the College of Engineering, and Dr. Daniel Frederick, Head of the Department of Engineering Science and Mechanics, is especially acknowledged. Finally, we wish to acknowledge the patience and forebearance of our wives, Julia and Dale, during the many hours required to prepare this manuscript.

J. L. Meriam

Santa Barbara, California
January 1987

L. Glenn Kraige

Blacksburg, Virginia
January 1987

CONTENTS

ENGINEERING MECHANICS

DYNAMICS

PART I

DYNAMICS OF PARTICLES

INTRODUCTION TO DYNAMICS

1

1/1 HISTORY AND MODERN APPLICATIONS

Dynamics is that branch of mechanics which deals with the motion of bodies under the action of forces. The study of dynamics in engineering usually follows the study of statics, which deals with the action of forces on bodies at rest. Dynamics has two distinct parts—*kinematics*, which is the study of motion without reference to the forces which cause motion, and *kinetics*, which relates the action of forces on bodies to their resulting motions. The student of engineering will find that a thorough comprehension of dynamics will provide one of his most useful and powerful tools for analysis in engineering.

Historically, dynamics is a relatively recent subject compared with statics. The beginning of a rational understanding of dynamics is credited to Galileo (1564–1642), who made careful observations concerning bodies in free fall, motion on an inclined plane, and motion of the pendulum. He was largely responsible for bringing a scientific approach to the investigation of physical problems. Galileo was continually under severe criticism for refusing to accept the established beliefs of his day, such as the philosophies of Aristotle which held, for example, that heavy bodies fall more rapidly than light bodies. The lack of accurate means for the measurement of time was a severe handicap to Galileo, and further significant development in dynamics awaited the invention of the pendulum clock by Huygens in 1657. Newton (1642–1727), guided by Galileo's work, was able to make an accurate formulation of the laws of motion and, hence, to place dynamics on a sound basis. Newton's famous work was published in the first edition of his *Principia*,* which is generally recognized as one of the greatest of all recorded contributions to knowledge. In addition to stating the laws governing the motion of a particle, Newton

---

* The original formulations of Sir Isaac Newton may be found in the translation of his *Principia* (1687), revised by F. Cajori, University of California Press, 1934.

was the first to correctly formulate the law of universal gravitation. Although his mathematical description was accurate, he felt that the concept of remote transmission of gravitational force without a supporting medium was an absurd notion. Following Newton's time, important contributions to mechanics were made by Euler, D'Alembert, Lagrange, Laplace, Poinsot, Coriolis, Einstein, and others.

In terms of engineering application, dynamics is an even more recent science. Only since machines and structures have operated with high speeds and appreciable accelerations has it been necessary to make calculations based on the principles of dynamics rather than on the principles of statics. The rapid technological developments of the present day require increasing application of the principles of mechanics, particularly dynamics. These principles are basic to the analysis and design of moving structures, to fixed structures subject to shock loads, to robotic devices, to automatic control systems, to rockets, missiles, and spacecraft, to ground and air transportation vehicles, to electron ballistics of electrical devices, and to machinery of all types such as turbines, pumps, reciprocating engines, hoists, machine tools, etc. Students whose interests lead them into one or more of these and many other activities will find a constant need for applying the fundamentals of dynamics.

1/2 BASIC CONCEPTS

The concepts basic to mechanics were set forth in Art. 1/2 of *Vol. 1 Statics*. They are summarized here along with additional comments of special relevance to the study of dynamics.

Space is the geometric region occupied by bodies. Position in space is determined relative to some geometric reference system by means of linear and angular measurements. The basic frame of reference for the laws of Newtonian mechanics is the *primary inertial system* or *astronomical frame of reference*, which is an imaginary set of rectangular axes assumed to have no translation or rotation in space. Measurements show that the laws of Newtonian mechanics are valid for this reference system as long as any velocities involved are negligible compared with the speed of light, which is 300 000 km/s or 186,000 mi/sec. Measurements made with respect to this reference are said to be *absolute*, and this reference system may be considered "fixed" in space. A reference frame attached to the surface of the earth has a somewhat complicated motion in the primary system, and a correction to the basic equations of mechanics must be applied for measurements made relative to the earth's reference frame. In the calculation of rocket and space flight trajectories, for example, the absolute motion of the earth becomes an important parameter. For most engineering problems of machines and structures which remain on the earth's surface, the corrections are extremely small and may be neglected. For these problems the laws of mechanics may be applied directly for measurements made relative to the earth, and in a practical sense such measurements will be referred to as *absolute*.

Time is a measure of the succession of events and is considered an absolute quantity in Newtonian mechanics.

Mass is the quantitative measure of the inertia or resistance to change in motion of a body. Mass is also the property which gives rise to gravitational attraction.

Force is the vector action of one body on another. The properties of forces have been thoroughly treated in *Vol. 1 Statics*.

A *particle* is a body of negligible dimensions. Also, when the dimensions of a body are irrelevant to the description of its motion or the action of forces upon it, the body may be treated as a particle. An airplane, for example, may be treated as a particle for the description of its flight path.

A *rigid body* is a body whose changes in shape are negligible compared with the overall dimensions of the body or with the changes in position of the body as a whole. As an example of the assumption of rigidity, the small flexural movement of the wing tip of an airplane flying through turbulent air is clearly of no consequence to the description of the motion of the airplane as a whole along its flight path. For this purpose, then, the treatment of the airplane as a rigid body offers no complication. On the other hand, if the problem is one of examining the internal stresses in the wing structure due to changing dynamic loads, then the deformable characteristics of the structure would have to be examined, and for this purpose the airplane could no longer be considered a rigid body.

Vector and *scalar* quantities have been treated extensively in *Vol. 1 Statics*, and their distinction should be perfectly clear by now. Scalar quantities are printed in lightface italic type, and vectors are shown in boldface type. Thus V is the scalar magnitude of the vector $\mathbf{V}$. It is important that we use an identifying mark, such as an underline $\underline{V}$, for all handwritten vectors to take the place of the boldface designation in print. For two nonparallel vectors recall, for example, that $\mathbf{V}_1 + \mathbf{V}_2$ and $V_1 + V_2$ have two entirely different meanings.

It is assumed that the reader is familiar with the geometry and algebra of vectors through previous study of statics and mathematics. Students who need to review these topics will find a brief summary of them in Appendix C along with other mathematical relations which find frequent use in mechanics. Experience has shown that the geometry of mechanics is often a source of difficulty. Mechanics by its very nature is geometrical, and students should bear this in mind as they review their mathematics. In addition to vector algebra, dynamics requires the use of vector calculus, and the essentials of this topic will be developed in the text as they are needed.

Dynamics involves the frequent use of time derivatives of both vectors and scalars. As a notational shorthand a dot over a quantity will frequently be used to indicate a derivative with respect to time. Thus $\dot{x}$ means dx/dt and $\ddot{x}$ stands for d^2x/dt^2.

1/3 NEWTON'S LAWS

Newton's three laws of motion, stated in Art. 1/4 of *Vol. 1 Statics*, are restated here because of their special significance to dynamics. In modern terminology they are

Law I.   A particle remains at rest or continues to move in a straight line with a constant velocity if there is no unbalanced force acting on it.

Law II.   The acceleration of a particle is proportional to the resultant force acting on it and is in the direction of this force.*

Law III.   The forces of action and reaction between interacting bodies are equal in magnitude, opposite in direction, and collinear.

These laws have been verified by countless physical measurements. The first two laws hold for measurements made in an absolute frame of reference but are subject to some correction when the motion is measured relative to a reference system having acceleration, such as one attached to the earth's surface.

Newton's second law forms the basis for most of the analysis in dynamics. For a particle of mass m subjected to a resultant force **F** the law may be stated as

$$\boxed{\mathbf{F} = m\mathbf{a}} \tag{1/1}$$

where **a** is the resulting acceleration measured in a nonaccelerating frame of reference. Newton's first law is a consequence of the second law since there is no acceleration when the force is zero, and the particle is either at rest or is moving with constant velocity. The third law constitutes the principle of action and reaction with which we should be thoroughly familiar from our work in statics.

---

* To some it is preferable to interpret Newton's second law as meaning that the resultant force acting on a particle is proportional to the time rate of change of momentum of the particle and that this change is in the direction of the force. Both formulations are equally correct when applied to a particle of constant mass.

1/4 UNITS

The International System of metric units (SI) is defined and used in *Vol. 2 Dynamics*. To become familiar with this system it is necessary to think directly in SI units. Although in engineering practice numerical conversion from the U.S. customary or British system to SI must be accomplished for some years, it would be a mistake to assume that familiarity with SI units can be achieved simply by the conversion of numerical results from the old system. Tables defining the SI units and giving numerical conversions between U.S. customary and SI units are included inside the front cover of the book. Charts comparing selected quantities in SI and U.S. customary units are included inside the back cover of the book to facilitate conversion and to help establish a feel for the relative size of units in both systems.

The four fundamental quantities of mechanics and their units and symbols for the two systems are summarized in the following table.

QUANTITY	DIMENSIONAL SYMBOL	SI UNITS		U.S. CUSTOMARY UNITS	
		UNIT	SYMBOL	UNIT	SYMBOL
Mass	M	Base units { kilogram	kg	slug	—
Length	L	meter*	m	Base units { foot	ft
Time	T	second	s	second	sec
Force	F	newton	N	pound	lb

* Also spelled *metre*.

In SI the units for mass, length, and time are taken as base units and the units for force are derived from Newton's second law of motion, Eq. 1/1. In the U.S. customary system the units for force, length, and time are base units and the units for mass are derived from the second law. The SI system is termed an *absolute* system since mass is taken to be an absolute or base quantity. The U.S. customary system is termed a *gravitational* system since force (as measured from gravitational pull) is taken as a base quantity. This distinction is a fundamental difference between the two systems.

In SI units by definition one newton is that force which will give a one-kilogram mass an acceleration of one meter per second squared. In the U.S. customary system a 32.1740-pound mass (1 slug) will have an acceleration of one foot per second squared when acted upon by a force of one pound. Thus for each system we have from Eq. 1/1

SI UNITS	U.S. CUSTOMARY UNITS
$(1 \text{ N}) = (1 \text{ kg})(1 \text{ m/s}^2)$	$(1 \text{ lb}) = (1 \text{ slug})(1 \text{ ft/sec}^2)$
$\text{N} = \text{kg} \cdot \text{m/s}^2$	$\text{slug} = \text{lb-sec}^2/\text{ft}$

In SI units each unit and symbol stands for only one quantity. The kilogram is to be used *only* as a unit of mass and *never* as a unit of force. On the other hand in U.S. customary units the *pound* is used both as a unit of force (lbf) and a unit of mass (lbm). The distinction between the meaning as force or mass is usually clear from the situation at hand. The symbol for pound force is most frequently written merely lb.

Additional quantities used in mechanics and their equivalent base units will be defined as they are introduced in the chapters which follow. However, for convenient reference these quantities along with their SI base units are listed in one place in the first table inside the front cover of the book.

Detailed guidelines have been established for the consistent use of SI units, and these guidelines have been followed throughout this book. The most essential ones are summarized inside the front cover, and the student should observe these rules carefully.

1/5 GRAVITATION

Newton's law of gravitation which governs the mutual attraction between bodies is

$$F = G\frac{m_1 m_2}{r^2} \tag{1/2}$$

where $F =$ the mutual force of attraction between two particles
$G =$ a universal constant called the constant of gravitation
$m_1, m_2 =$ the masses of the two particles
$r =$ the distance between the centers of the particles

The value of the gravitational constant from experimental data is $G = 6.673(10^{-11}) \text{ m}^3/(\text{kg} \cdot \text{s}^2)$. The only gravitational force of appreciable magnitude is the force due to the attraction of the earth. It was shown in *Vol. 1 Statics*, for example, that each of two iron spheres 100 mm in diameter is attracted to the earth with a gravitational force of 37.1 N, which is called its *weight*, but the force of mutual attraction between them if they are just touching is only 0.000 000 095 1 N.

Since the gravitational attraction or weight of a body is a force, it should always be expressed in force units, newtons (N) in SI units and pounds force (lb) in U.S. customary units. Unfortunately the mass unit kilogram (kg) has been used widely in common practice as a measure of weight. When expressed in kilograms the word "weight" technically means mass. To avoid confusion the word "weight" in this book shall be restricted to mean the force of gravitational attraction, and it will always be expressed in newtons or pounds force.

The force of gravitational attraction of the earth on a body depends on the position of the body relative to the earth. If the earth were a perfect sphere of the same volume, a body with a mass of exactly 1 kg would be attracted to the earth by a force of 9.825 N on the earth's surface, 9.822 N at an altitude of 1 km, 9.523 N at an altitude of 100 km, 7.340 N at an altitude of 1000 km, and 2.456 N at an altitude equal to the mean radius 6371 km of the earth. It is at once apparent that the variation in gravitational attraction of high-altitude rockets and spacecraft becomes a major consideration.

Every object which is allowed to fall in a vacuum at a given position on the earth's surface will have the same acceleration g as can be seen by combining Eqs. 1/1 and 1/2 and canceling the term representing the mass of the falling object. This combination gives

$$g = \frac{Gm_e}{R^2}$$

where m_e is the mass of the earth and R is the radius of the earth.* The mass m_e and the mean radius R of the earth have been found through experimental measurements to be $5.976(10^{24})$ kg and $6.371(10^6)$ m, respectively. These values, together with the value of G already cited, when substituted into the expression for g, give a mean value

$$g = 9.825 \text{ m/s}^2$$

The acceleration due to gravity as determined from the gravitational law is the acceleration which would be measured from a set of axes whose origin is at the center of the earth but which does not rotate with the earth. With respect to these "fixed" axes, then, this value may be termed the absolute value of g. Because of the fact that the earth rotates, the acceleration of a freely falling body as measured from a position attached to the earth's surface is slightly less than the absolute value. Accurate values of the gravitational acceleration as measured relative to the earth's surface account for the fact that the earth is a rotating oblate spheroid with flattening at the poles. These values may be calculated to a high degree of accuracy from the 1980 International Gravity Formula, which is

$$g = 9.780\ 327(1 + 0.005\ 279 \sin^2 \gamma + 0.000\ 023 \sin^4 \gamma + \cdots)$$

---

* It can be proved that the earth, when taken as a sphere with a symmetrical distribution of mass about its center, may be considered a particle with its entire mass concentrated at its center.

where γ is the latitude and g is expressed in meters per second squared. The formula is based upon an ellipsoidal model of the earth and also accounts for the effect of the earth's rotation. The absolute acceleration due to gravity as determined for a nonrotating earth may be computed from the relative values to a close approximation by adding $3.382(10^{-2}) \cos^2 \gamma$ m/s^2, which removes the effect of the earth's rotation. The variation of both the absolute and the relative values of g with latitude is shown in Fig. 1/1 for sea-level conditions.*

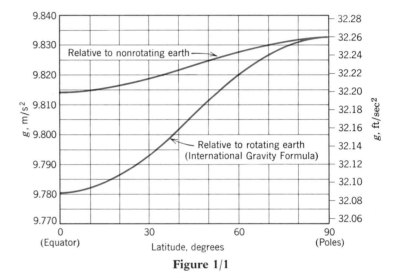

Figure 1/1

The standard value which has been adopted internationally for the gravitational acceleration relative to the rotating earth at sea level and at a latitude of 45° is 9.806 65 m/s^2 or 32.1740 ft/sec^2. This value differs very slightly from that obtained by evaluating the International Gravity Formula for $\gamma = 45°$. The reason for the small difference is that the earth is not exactly ellipsoidal as assumed in the formulation of the International Gravity Formula.

The proximity of large land masses and the variations in the density of the earth's crust also influence the local value of g by a small but detectable amount. In almost all engineering problems where measurements are made on the surface of the earth the difference between the absolute and relative values of the gravitational acceleration and the effect of local variations are neglected, and 9.81 m/s^2 in SI units and 32.2 ft/sec^2 in U.S. customary units are used for the sea-level value of g.

---

* Students will be able to derive these relations for a spherical earth following their study of relative motion in Chapter 3.

The variation of g with altitude is easily determined by the gravitational law. If g_0 represents the absolute acceleration due to gravity at sea level, the absolute value at an altitude h is

$$g = g_0 \frac{R^2}{(R + h)^2}$$

where R is the radius of the earth.

The earth's gravitational attraction on a body may be calculated from the results of the simple gravitational experiment. If the gravitational force of attraction or true weight of a body is **W**, then, since the body will fall with an absolute acceleration **g** in a vacuum, Eq. 1/1 gives

$$\boxed{\mathbf{W} = m\mathbf{g}} \tag{1/3}$$

The apparent weight of a body as determined by a spring balance, calibrated to read the correct force and attached to the surface of the earth, will be slightly less than its true weight. The difference is due to the rotation of the earth. The ratio of the apparent weight to the apparent or relative acceleration due to gravity still gives the correct value of mass. The apparent weight and the relative acceleration due to gravity are, of course, the quantities which are measured in experiments conducted on the surface of the earth.

1/6 DIMENSIONS

A given dimension such as length can be expressed in a number of different units such as meters, millimeters, or kilometers. Thus the word *dimension* is distinguished from the word *unit*. Physical relations must always be dimensionally homogeneous, that is, the dimensions of all terms in an equation must be the same. It is customary to use the symbols L, M, T, and F to stand for length, mass, time, and force, respectively. In SI units force is a derived quantity and from Eq. 1/1 has the dimensions of mass times acceleration or

$$F = ML/T^2$$

One important use of the theory of dimensions is found in checking the dimensional correctness of some derived physical relation. The following expression for the velocity v of a body of mass m which is moved from rest a horizontal distance x by a force F may be derived:

$$Fx = \tfrac{1}{2}mv^2$$

where the $\tfrac{1}{2}$ is a dimensionless coefficient resulting from integration. This equation is dimensionally correct since substitution of *L*, *M*, and *T* gives

$$[MLT^{-2}][L] = [M][LT^{-1}]^2$$

Dimensional homogeneity is a necessary condition for correctness, but it is not sufficient since the correctness of dimensionless coefficients cannot be checked in this way.

1/7 FORMULATION AND SOLUTION OF DYNAMICS PROBLEMS

The study of dynamics is directed toward the understanding and description of the various quantities involved in the motions of bodies. This description, which is largely mathematical, enables predictions to be made of dynamical behavior. A dual thought process is necessary in formulating this description. It is necessary to think both in terms of the physical situation and in terms of the corresponding mathematical description. Analysis of every problem will require this repeated transition of thought between the physical and the mathematical. Without question, one of the greatest difficulties encountered by students is the inability to make this transition of thought freely. They should recognize that the mathematical formulation of a physical problem represents an ideal and limiting description, or model, which approximates but never quite matches the actual physical situation.

In the course of constructing the idealized mathematical model for any given engineering problem, certain approximations will always be involved. Some of these approximations may be mathematical, whereas others will be physical. For instance, it is often necessary to neglect small distances, angles, or forces compared with large distances, angles, or forces. If the change in velocity of a body with time is nearly uniform, then an assumption of constant acceleration may be justified. An interval of motion which cannot be easily described in its entirety is often divided into small increments each of which can be approximated. The retarding effect of bearing friction on the motion acquired by a machine as the result of applied forces or moments may often be neglected if the friction forces are small. However, these same friction forces cannot be neglected if the purpose of the inquiry is a determination of the drop in efficiency of the machine due to the friction process. Thus the degree of assumption involved depends on what information is desired and on the accuracy required. The student should be constantly alert to the various assumptions called for in the formulation of real problems. The ability to understand and make use of the appropriate assumptions in the course of the formulation and solution of engineering problems is, certainly, one of the most important characteristics of a successful engineer. Along with the development of the principles and analytical tools needed for modern dynamics, one of the major aims of this book is to provide a maximum of opportunity to develop ability in formulating good mathematical models. Strong emphasis is placed on a wide range of practical problems which not only require the full exercise of theory but also force consideration of the decisions which must be made concerning relevant assumptions.

An effective method of attack on dynamics problems, as in all engineering problems, is essential. The development of good habits in formulating problems and in representing their solutions will prove to be an invaluable asset. Each solution should proceed with a logical sequence of steps from hypothesis to conclusion, and its representation should include a clear statement of the following parts, each clearly identified:

1. Given data
2. Results desired
3. Necessary diagrams
4. Calculations
5. Answers and conclusions

In addition it is well to incorporate a series of checks on the calculations at intermediate points in the solution. The reasonableness of numerical magnitudes should be observed, and the accuracy and dimensional homogeneity of terms should be checked frequently. It is also important that the arrangement of work be neat and orderly. Careless solutions which cannot be read easily by others are of little or no value. It will be found that the discipline involved in adherence to good form will in itself be an invaluable aid to the development of the abilities for formulation and analysis. Many problems which at first may seem difficult and complicated become clear and straightforward once they are begun with a logical and disciplined method of attack.

The subject of dynamics is based upon a surprisingly few fundamental concepts and principles which, however, are extended and applied over an exceedingly wide range of conditions. One of the most valuable aspects of the study of dynamics is the experience afforded in reasoning from fundamentals. This experience cannot be obtained merely by memorizing the kinematical and dynamical equations which describe various motions. It must be obtained through exposure to a wide variety of problem situations which force the choice, use, and extension of basic principles to meet the given conditions.

In describing the relations between forces and the motions they produce, it is essential that the system to which a principle is applied be clearly defined. At times a single particle or a rigid body is the system to be isolated, whereas at other times two or more bodies taken together constitute the system. The definition of the system to be analyzed is made clear by constructing its *free-body diagram*. This diagram consists of a closed outline of the external boundary of the system defined. All bodies which contact and exert forces on the system but are not a part of it are removed and replaced by vectors representing the forces they exert *on* the system isolated. In this way we make a clear distinction between the action and reaction of each force, and account is taken of *all* forces on and external to the system. It is assumed that all students are familiar with the technique of drawing free-body diagrams from their prior work in statics.

In applying the laws of dynamics, numerical values of the quantities may be used directly in proceeding toward the solution, or algebraic symbols may be used to represent the quantities involved and the answer left as a formula. With numerical substitution the magnitudes of all quantities expressed in their particular units are evident at each stage of the calculation. This approach offers advantage when the practical significance of the magnitude of each term is important. The symbolic solution, however, has several advantages over the numerical solution. First, the abbreviation achieved by the use of symbols aids in focusing attention on the interconnection between the physical situation and its related mathematical description. Second, a symbolic solution permits a dimensional check to be made at every step, whereas dimensional homogeneity may not be checked when only numerical values are used. Third, a symbolic solution may be used repeatedly for obtaining answers to the same problem when different sets and sizes of units are used. Facility with both forms of solution is essential, and ample practice with each should be sought in the problem work.

Students will find that solutions to the various equations of dynamics may be obtained in one of three ways. First, a direct mathematical solution by hand calculation may be carried out with answers appearing either as algebraic symbols or as numerical results. The large majority of the problems come under this category. Second, certain problems may be readily handled by graphical solutions, such as with the determination of velocities and accelerations in two-dimensional relative motion of rigid bodies. Third, there are a number of problems in *Vol. 2 Dynamics* which are designated as *computer-oriented problems*. They appear at the end of the Review Problem sets and are selected to illustrate the type of problem for which solution by computer offers a distinct advantage. The choice of the most expedient method of solution is an important aspect of the experience to be gained from the problem work. We emphasize, however, that the most important experience in learning mechanics lies in the formulation of problems as distinct from their solution per se.

PROBLEMS

(Refer to Table D/2 in Appendix D for relevant solar-system values.)

1/1 Determine the weight in newtons of a person whose mass is 85 kg. In addition, convert the person's weight to pounds and calculate the corresponding mass in lb-sec^2/ft (or slugs).

$$Ans.\ W = 834\ N$$
$$W = 187.5\ lb$$
$$m = 5.82\ lb\text{-}sec^2/ft$$

1/2 At what altitude h above the north pole is the weight of an object reduced to one-half of its earth-surface value? Assume a spherical earth of radius R and express h in terms of R.

1/3 A space shuttle is in a circular orbit at an altitude of 250 kilometers. Calculate the absolute value of g at this altitude and determine the corresponding weight of a shuttle passenger who weighs 880 N when standing on the surface of the earth at a latitude of 45°. Are the terms "zero-g" and "weightless," which are sometimes used to describe conditions aboard orbiting spacecraft, correct in the absolute sense?

$$Ans.\ g = 9.10\ m/s^2$$
$$W = 816\ N,\ No$$

1/4 Calculate the force F_s exerted by the sun on an 80-kg man as he stands on the surface of the earth at 45° latitude and compare F_s to the force exerted on him by the earth.

1/5 Consider two iron spheres, each of diameter 100 mm, which are just touching. At what distance r from the center of the earth will the force of mutual attraction between the contacting spheres be equal to the force exerted by the earth on one of the spheres? $Ans.\ r = 1.258(10^8)\ km$

1/6 Determine the absolute weight and the weight relative to the rotating earth of an 80-kg man if he is standing on the surface of the earth at a latitude of 45°. (Refer to Fig. 1/1.)

Ans. $W_{\text{abs}} = 785.8$ N
$W_{\text{rel}} = 784.5$ N

Problem 1/7

1/7 Determine the ratio R_A of the force exerted by the sun on the moon to that exerted by the earth on the moon for position A of the moon. Repeat for moon position B.

Ans. $R_A = 2.19$
$R_B = 2.21$

1/8 Check the following equation for dimensional homogeneity:

$$mv = \int_{t_1}^{t_2} F\,dt$$

where m is mass, v is velocity, F is force, and t is time.

Problem 1/9

1/9 Calculate the distance d from the center of the earth at which a particle experiences equal attractions from the earth and moon. The particle is restricted to the line which joins the centers of the earth and moon. Justify the two solutions physically.

Ans. $d = 346\ 022$ km or $432\ 348$ km

1/10 At what latitude is the value of g relative to the rotating earth equal to the equatorial value of g relative to the nonrotating earth? Approximate your answer by inspecting Fig. 1/1, then find a more precise answer by using the International Gravity Formula.

KINEMATICS OF PARTICLES

2

2/1 INTRODUCTION

Kinematics is that branch of dynamics which describes the motion of bodies without reference to the forces which either cause the motion or are generated as a result of the motion. Kinematics is often referred to as the "geometry of motion." The design of cams, gears, linkages, and other machine elements to control or produce certain desired motions and the calculation of flight trajectories for aircraft, rockets, and spacecraft are a few examples of kinematic problems which engage the attention of engineers. A thorough working knowledge of kinematics is an absolute prerequisite to kinetics, which is the study of the relationships between motion and the corresponding forces which cause or accompany the motion.

We start our study of kinematics by first discussing in this chapter the motions of points or particles. A particle is a body whose physical dimensions are so small compared with the radius of curvature of its path that we can treat the motion of the particle as that of a point. For example, the wingspan of a jet transport flying between Los Angeles and New York is of no consequence compared with the radius of curvature of its flight path, and the treatment of the airplane as a particle or point should raise no question.

There are a number of ways in which the motion of a particle can be described, and the choice of the most convenient or appropriate method depends a great deal on experience and on how the data are given. Let us get an overview of the several methods developed in this chapter by referring to Fig. 2/1 which shows a particle P moving along some general path in space. If the particle is confined to a specified path, as with a bead sliding along a fixed wire, its motion is said to be *constrained*. If there are no physical guides, the motion is said to be *unconstrained*. A small rock tied to the end of a string and whirled in a circle undergoes constrained motion until the string breaks, after which instant its motion is unconstrained.

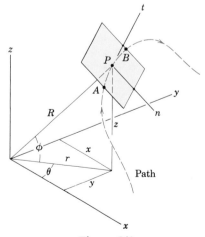

Figure 2/1

The position of particle *P* at any time *t* can be described by specifying its rectangular coordinates* *x*, *y*, *z*, its cylindrical coordinates *r*, *θ*, *z*, or its spherical coordinates *R*, *θ*, *φ*. The motion of *P* may also be described by measurements along the tangent *t* and normal *n* to the curve. The direction of *n* lies in the local plane of the curve.[†] These last measurements are known as *path variables*.

The motion of particles (or rigid bodies) may be described by using coordinates measured from fixed reference axes (*absolute-motion* analysis) or by using coordinates measured from moving reference axes (*relative-motion* analysis). Both descriptions will be developed and applied in the articles which follow.

With this conceptual picture of the description of particle motion in mind, we will now restrict our attention in the first part of this chapter to the case of *plane motion* where all movement occurs in or can be represented as occurring in a single plane. A large proportion of the motions of machines and structures in engineering can be represented as plane motion. Later, in Chapter 7 an introduction to three-dimensional motion is presented. We will begin our discussion of plane motion with *rectilinear motion*, which is motion along a straight line, and follow it with a description of motion along a plane curve.

2/2 RECTILINEAR MOTION

Figure 2/2

Consider a particle *P* moving along a straight line, Fig. 2/2. The position of *P* at any instant of time *t* may be specified by its distance *s* measured from some convenient reference point *O* fixed on the line. At time *t* + Δ*t* the particle has moved to *P'* and its coordinate becomes *s* + Δ*s*. The change in the position coordinate during the interval Δ*t* is called the *displacement* Δ*s* of the particle. The displacement would be negative if the particle moved in the negative *s*-direction.

The average velocity of the particle during the interval Δ*t* is the displacement divided by the time interval or $v_{av} = \Delta s / \Delta t$. As Δ*t* becomes smaller and approaches zero in the limit, the average velocity approaches the *instantaneous velocity* of the particle, which is $v = \lim\limits_{\Delta t \to 0} \dfrac{\Delta s}{\Delta t}$ or

$$v = \frac{ds}{dt} = \dot{s} \qquad (2/1)$$

* Often called *Cartesian* coordinates, named after René Descartes (1596–1650), a French mathematician who was one of the inventors of analytic geometry.
[†] This plane is known as the *osculating* plane, which comes from the Latin word *osculari* meaning "to kiss." The plane which contains *P* and the two points *A* and *B*, one on either side of *P*, becomes the osculating plane as the distances between the points approach zero.

Thus the velocity is the time rate of change of the position coordinate s. The velocity is positive or negative depending on whether the corresponding displacement is positive or negative.

The average acceleration of the particle during the interval Δt is the change in its velocity divided by the time interval or $a_{av} = \Delta v/\Delta t$. As Δt becomes smaller and approaches zero in the limit, the average acceleration approaches the *instantaneous acceleration* of the particle, which is $a = \lim\limits_{\Delta t \to 0} \dfrac{\Delta v}{\Delta t}$ or

$$a = \frac{dv}{dt} = \dot{v} \qquad \text{or} \qquad a = \frac{d^2 s}{dt^2} = \ddot{s} \qquad (2/2)$$

The acceleration is positive or negative depending on whether the velocity is increasing or decreasing. Note that the acceleration would be positive if the particle had a negative velocity which was becoming less negative. If the particle is slowing down, the particle is said to be *decelerating*.

Velocity and acceleration are actually vector quantities, as we will see for curvilinear motion beginning with Art. 2/3. For rectilinear motion in the present article, where the direction of the motion is that of the given straight-line path, the sense of the vector along the path is described by a plus or minus sign. In our treatment of curvilinear motion we will account for the change in direction of the velocity and acceleration vectors as well as their change in magnitude.

By eliminating the time dt between Eq. 2/1 and the first of Eqs. 2/2 a differential equation relating displacement, velocity, and acceleration results* which is

$$v \, dv = a \, ds \qquad \text{or} \qquad \dot{s} \, d\dot{s} = \ddot{s} \, ds \qquad (2/3)$$

Equations 2/1, 2/2, and 2/3 are the differential equations for the rectilinear motion of a particle. Problems in rectilinear motion involving finite changes in the motion variables are solved by integration of these basic differential relations. The position coordinate s, the velocity v, and the acceleration a are algebraic quantities, so that their signs, positive or negative, must be carefully observed. Note that the positive directions for v and a are the same as the positive direction for s.

Interpretation of the differential equations governing rectilinear motion is considerably clarified by representing the relationships among s, v, a, and t graphically. Figure 2/3a represents a schematic plot of the variation of s with t from time t_1 to time

* Differential quantities can be multiplied and divided in exactly the same way as other algebraic quantities.

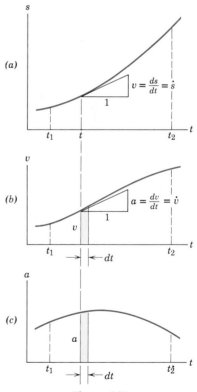

Figure 2/3

t_2 for some given rectilinear motion. By constructing the tangent to the curve at any time t we obtain the slope, which is the velocity $v = ds/dt$. Thus the velocity may be determined at all points on the curve and plotted against the corresponding time as shown in Fig. 2/3b. Similarly, the slope dv/dt of the v-t curve at any instant gives the acceleration at that instant, and the a-t curve can therefore be plotted as in Fig. 2/3c.

We now see from Fig. 2/3b that the area under the v-t curve during time dt is $v\,dt$ which from Eq. 2/1 is the displacement ds. Consequently the net displacement of the particle during the interval from t_1 to t_2 is the corresponding area under the curve which is

$$\int_{s_1}^{s_2} ds = \int_{t_1}^{t_2} v\,dt \qquad \text{or} \qquad s_2 - s_1 = (\text{area under } v\text{-}t \text{ curve})$$

Similarly, from Fig. 2/3c we see that the area under the a-t curve during time dt is $a\,dt$ which, from the first of Eqs. 2/2, is dv. Thus the net change in velocity between t_1 and t_2 is the corresponding area under the curve, which is

$$\int_{v_1}^{v_2} dv = \int_{t_1}^{t_2} a\,dt \qquad \text{or} \qquad v_2 - v_1 = (\text{area under } a\text{-}t \text{ curve})$$

Two additional graphical relations are noted. When the acceleration a is plotted as a function of the position coordinate s, Fig. 2/4a, the area under the curve during a displacement ds is $a\,ds$ which, from Eq. 2/3, is $v\,dv = d(v^2/2)$. Thus the net area under the curve between position coordinates s_1 and s_2 is

$$\int_{v_1}^{v_2} v\,dv = \int_{s_1}^{s_2} a\,ds \qquad \text{or} \qquad \tfrac{1}{2}(v_2{}^2 - v_1{}^2) = (\text{area under } a\text{-}s \text{ curve})$$

When the velocity v is plotted as a function of the position coordinate s, Fig. 2/4b, the slope of the curve at any point A is dv/ds. By constructing the normal AB to the curve at this point, we see from the similar triangles that $\overline{CB}/v = dv/ds$. Thus, from Eq. 2/3, $\overline{CB} = v(dv/ds) = a$, the acceleration. It is necessary that the velocity and position coordinate axes have the same numerical scales so that the acceleration read on the position coordinate scale in meters, say, will represent the actual acceleration in the units meters per second squared.

The graphical representations described are useful not only in visualizing the relationships among the several motion quantities but also in approximating results by graphical integration or differentiation when a lack of knowledge of the mathematical relationship prevents its expression as an explicit mathematical function. Experimental data and motions which involve discontinuous relationships between the variables are frequently analyzed graphically.

If the position coordinate s is known for all values of the

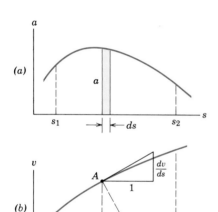

Figure 2/4

time t, then successive mathematical or graphical differentiation with respect to t gives the velocity v and acceleration a. In many problems, however, the functional relationship between position coordinate and time is unknown, and we must determine it by successive integration from the acceleration. Acceleration is determined by the forces which act on moving bodies and is computed from the equations of kinetics discussed in subsequent chapters. Depending on the nature of the forces, the acceleration may be specified as a function of time, velocity, position coordinate, or as a combined function of these quantities. The procedure for integrating the differential equation in each case is indicated as follows:

(*a*) *Constant acceleration.* When a is constant, the first of Eqs. 2/2 and 2/3 may be integrated directly. For simplicity with $s = s_0$, $v = v_0$, and $t = 0$ designated at the beginning of the interval, then for a lapse of time t the integrated equations become

$$\int_{v_0}^{v} dv = a \int_{0}^{t} dt \qquad \text{or} \qquad v = v_0 + at$$

$$\int_{v_0}^{v} v\, dv = a \int_{s_0}^{s} ds \qquad \text{or} \qquad v^2 = v_0^2 + 2a(s - s_0)$$

Substitution of the integrated expression for v into Eq. 2/1 and integration with respect to t give

$$\int_{s_0}^{s} ds = \int_{0}^{t} (v_0 + at)\, dt \qquad \text{or} \qquad s = s_0 + v_0 t + \tfrac{1}{2}at^2$$

These relations are necessarily restricted to the special case where the acceleration is constant. The integration limits depend on the initial and final conditions and for a given problem may be different from those used here. It may be more convenient, for instance, to begin the integration at some specified time t_1 rather than at time $t = 0$.

Caution: One of the most frequent mistakes made by students is the attempt to use the foregoing equations for problems of variable acceleration where they do not apply since they have been integrated for constant acceleration only.

(*b*) *Acceleration given as a function of time, a = f(t).* Substitution of the function into the first of Eqs. 2/2 gives $f(t) = dv/dt$. Multiplying by dt separates the variables and permits integration. Thus

$$\int_{v_0}^{v} dv = \int_{0}^{t} f(t)\, dt \qquad \text{or} \qquad v = v_0 + \int_{0}^{t} f(t)\, dt$$

From this integrated expression for v as a function of t, the position coordinate s is obtained by integrating Eq. 2/1 which, in form, would be

$$\int_{s_0}^{s} ds = \int_{0}^{t} v\, dt \qquad \text{or} \qquad s = s_0 + \int_{0}^{t} v\, dt$$

If the indefinite integral is employed, the end conditions are used to establish the constants of integration with results which are identical with those obtained by using the definite integral.

If desired, the displacement s may be obtained by a direct solution of the second-order differential equation $\ddot{s} = f(t)$ obtained by substitution of $f(t)$ into the second of Eqs. 2/2.

(*c*) *Acceleration given as a function of velocity*, $a = f(v)$. Substitution of the function into the first of Eqs. 2/2 gives $f(v) = dv/dt$, which permits separating the variables and integrating. Thus

$$t = \int_{0}^{t} dt = \int_{v_0}^{v} \frac{dv}{f(v)}$$

This result gives t as a function of v. Then it would be necessary to solve for v as a function of t so that Eq. 2/1 can be integrated to obtain the position coordinate s as a function of t.

Alternatively, the function $a = f(v)$ can be substituted into the first of Eqs. 2/3 giving $v\, dv = f(v)\, ds$. The variables are now separated and the equation integrated in the form

$$\int_{v_0}^{v} \frac{v\, dv}{f(v)} = \int_{s_0}^{s} ds \qquad \text{or} \qquad s = s_0 + \int_{v_0}^{v} \frac{v\, dv}{f(v)}$$

Note that this equation gives s in terms of v without explicit reference to t.

(*d*) *Acceleration given as a function of displacement*, $a = f(s)$. Substitution of the function into Eq. 2/3 and integrating gives the form

$$\int_{v_0}^{v} v\, dv = \int_{s_0}^{s} f(s)\, ds \qquad \text{or} \qquad v^2 = v_0{}^2 + 2\int_{s_0}^{s} f(s)\, ds$$

Next we solve for v to give $v = g(s)$, a function of s. Now we can substitute ds/dt for v, separate variables, and integrate in the form

$$\int_{s_0}^{s} \frac{ds}{g(s)} = \int_{0}^{t} dt \qquad \text{or} \qquad t = \int_{s_0}^{s} \frac{ds}{g(s)}$$

which gives t as a function of s. Lastly we can rearrange to obtain s as a function of t.

In each of the foregoing cases when the acceleration varies according to some functional relationship, the ability to solve the equations by direct mathematical integration will depend on the form of the function. In cases where the integration is excessively awkward or difficult, integration by graphical, numerical, or computer methods may be utilized.

Sample Problem   2/1

The position coordinate of a particle which is confined to move along a straight line is given by $s = 2t^3 - 24t + 6$ where s is measured in meters from a convenient origin and where t is in seconds. Determine (a) the time required for the particle to reach a velocity of 72 m/s from its initial condition at $t = 0$, (b) the acceleration of the particle when $v = 30$ m/s, and (c) the net displacement of the particle during the interval from $t = 1$ s to $t = 4$ s.

Solution. The velocity and acceleration are obtained by successive differentiation of s with respect to the time. Thus

$$[v = \dot{s}] \qquad\qquad v = 6t^2 - 24 \text{ m/s}$$
$$[a = \dot{v}] \qquad\qquad a = 12t \text{ m/s}^2$$

(a) Substituting $v = 72$ m/s into the expression for v gives us $72 = 6t^2 - 24$ from which $t = \pm 4$ s. The negative root describes a ① mathematical solution for t before the initiation of motion, so this root is of no physical interest. Thus the desired result is

$$t = 4 \text{ s} \qquad\qquad\qquad Ans.$$

(b) Substituting $v = 30$ m/s into the expression for v gives $30 = 6t^2 - 24$ from which the positive root is $t = 3$ s, and the corresponding acceleration is

$$a = 12(3) = 36 \text{ m/s}^2 \qquad\qquad Ans.$$

(c) The net displacement during the specified interval is

$$\Delta s = s_4 - s_1 \qquad \text{or}$$
$$\Delta s = [2(4^3) - 24(4) + 6] - [2(1^3) - 24(1) + 6]$$
$$= 54 \text{ m} \qquad\qquad\qquad Ans.$$

② which represents the net advancement of the particle along the s-axis from the position it occupied at $t = 1$ s to its position at $t = 4$ s.

To help visualize the motion the values of s, v, and a are plotted against the time t as shown. Because the area under the v-t curve rep-③ resents displacement, we see that the net displacement from $t = 1$ s to $t = 4$ s is the positive area Δs_{2-4} less the negative area Δs_{1-2}.

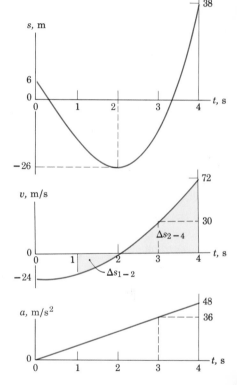

① Be alert to the proper choice of sign when taking a square root. When the situation calls for only one answer, the positive root is not always the one you may need.

② Note carefully the distinction between italic s for the position coordinate and the vertical s for seconds.

③ Note from the graphs that the values for v are the slopes ($\dot{s}$) of the s-t curve and that the values for a are the slopes ($\dot{v}$) of the v-t curve. *Suggestion:* Integrate $v\, dt$ for each of the two intervals and check the answer for Δs. Show that the total distance traveled during the interval $t = 1$ s to $t = 4$ s is 74 m.

Sample Problem 2/2

A particle moves along the x-axis with an initial velocity $v_x = 50$ m/s at the origin when $t = 0$. For the first 4 s it has no acceleration, and thereafter it is acted upon by a retarding force which gives it a constant acceleration $a_x = -10$ m/s². Calculate the velocity and the x-coordinate of the particle for the conditions of $t = 8$ s and $t = 12$ s
① and find the maximum positive x-coordinate reached by the particle.

① Learn to be flexible with symbols. The position coordinate x is just as valid as s.

Solution. The velocity of the particle after $t = 4$ s is computed from

② $\left[\int dv = \int a\, dt\right]$ $\quad \int_{50}^{v_x} dv_x = -10 \int_{4}^{t} dt \quad v_x = 90 - 10t$ m/s

② Note that we integrate to a general time t and then substitute specific values.

and is plotted as shown. At the specified times the velocities are

$$t = 8\ \text{s}, \qquad v_x = 90 - 10(8) = 10\ \text{m/s}$$
$$t = 12\ \text{s}, \qquad v_x = 90 - 10(12) = -30\ \text{m/s} \qquad Ans.$$

The x-coordinate of the particle at any time greater than 4 s is the distance traveled during the first 4 s plus the distance traveled after the discontinuity in acceleration occurred. Thus

$$\left[\int ds = \int v\, dt\right] \quad x = 50(4) + \int_{4}^{t} (90 - 10t)\, dt = -5t^2 + 90t - 80\ \text{m}$$

For the two specified times

$$t = 8\ \text{s}, \qquad x = -5(8^2) + 90(8) - 80 = 320\ \text{m}$$
$$t = 12\ \text{s}, \qquad x = -5(12^2) + 90(12) - 80 = 280\ \text{m} \qquad Ans.$$

The x-coordinate for $t = 12$ s is less than that for $t = 8$ s since the motion is in the negative x-direction after $t = 9$ s. The maximum positive x-coordinate is, then, the value of x for $t = 9$ s which is

$$x_{\text{max}} = -5(9^2) + 90(9) - 80 = 325\ \text{m} \qquad Ans.$$

③ These displacements are seen to be the net positive areas under the v-t graph up to the value of t in question.

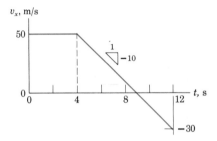

③ Show that the total distance traveled by the particle in the 12 seconds is 370 m.

Sample Problem 2/3

The spring-mounted slider moves in the horizontal guide with negligible friction and has a velocity v_0 in the s-direction as it crosses the mid-position where $s = 0$ and $t = 0$. The two springs together exert a retarding force to the motion of the slider which gives it an acceleration proportional to the displacement but oppositely directed and equal to $a = -k^2 s$ where k is constant. (The constant is arbitrarily squared for later convenience in the form of the expressions.) Determine the expressions for the displacement s and velocity v as functions of the time t.

Solution I. Since the acceleration is specified in terms of the displacement, the differential relation $v \, dv = a \, ds$ may be integrated. Thus

① $$\int v \, dv = \int -k^2 s \, ds + C_1 \text{ a constant,} \qquad \text{or} \qquad \frac{v^2}{2} = -\frac{k^2 s^2}{2} + C_1$$

① We have used an indefinite integral here and evaluated the constant of integration. For practice obtain the same results by using the definite integral with the appropriate limits.

When $s = 0$, $v = v_0$, so that $C_1 = v_0^2/2$, and the velocity becomes

$$v = +\sqrt{v_0^2 - k^2 s^2}$$

The plus sign of the radical is taken when v is positive (in the plus s-direction). This last expression may be integrated by substituting $v = ds/dt$. Thus,

② $$\int \frac{ds}{\sqrt{v_0^2 - k^2 s^2}} = \int dt + C_2 \text{ a constant,} \qquad \text{or} \qquad \frac{1}{k} \sin^{-1} \frac{ks}{v_0} = t + C_2$$

② Again try the definite integral here as above.

With the requirement of $t = 0$ when $s = 0$, the constant of integration becomes $C_2 = 0$, and we may solve the equation for s so that

$$s = \frac{v_0}{k} \sin kt \qquad\qquad\qquad Ans.$$

The velocity is $v = \dot{s}$ which gives

$$v = v_0 \cos kt \qquad\qquad\qquad Ans.$$

Solution II. Since $a = \ddot{s}$, the given relation may be written at once as

$$\ddot{s} + k^2 s = 0$$

This is an ordinary linear differential equation of second order for which the solution is well known and is

$$s = A \sin Kt + B \cos Kt$$

where A, B, and K are constants. Substitution of this expression into the differential equation shows that it satisfies the equation provided that $K = k$. The velocity is $v = \dot{s}$ which becomes

$$v = Ak \cos kt - Bk \sin kt$$

③ This motion is called *simple harmonic motion* and is characteristic of all oscillations where the restoring force, and hence the acceleration, is proportional to the displacement but opposite in sign.

The boundary condition $v = v_0$ when $t = 0$ requires that $A = v_0/k$, and the condition $s = 0$ when $t = 0$ gives $B = 0$. Thus the solution is

③ $$s = \frac{v_0}{k} \sin kt \qquad \text{and} \qquad v = v_0 \cos kt \qquad\qquad Ans.$$

Sample Problem   2/4

A freighter is moving at a speed of 8 knots when its engines are suddenly stopped. If it takes 10 minutes for the freighter to reduce
① its speed to 4 knots, determine and plot the distance s in nautical miles moved by the ship and its speed v in knots as functions of the time t during this interval. The deceleration of the ship is proportional to the square of its speed, so that $a = -kv^2$.

① Recall that one knot is the speed of one nautical mile (1852 m) per hour. Work directly in the units of nautical miles and hours.

Solution.   The speeds and the time are given, so we may substitute the expression for acceleration directly into the basic definition $a = dv/dt$ and integrate. Thus

$$-kv^2 = \frac{dv}{dt} \qquad \frac{dv}{v^2} = -k\,dt \qquad \int_8^v \frac{dv}{v^2} = -k \int_0^t dt$$

② $$-\frac{1}{v} + \frac{1}{8} = -kt \qquad v = \frac{8}{1 + 8kt}$$

② We choose to integrate to a general value of v and its corresponding time t so that we may obtain the variation of v with t.

Now we substitute the end limits of $v = 4$ knots and $t = \frac{10}{60} = \frac{1}{6}$ hour and get

$$4 = \frac{8}{1 + 8k(1/6)} \qquad k = \frac{3}{4}\,\text{mi}^{-1} \qquad v = \frac{8}{1 + 6t} \qquad Ans.$$

The speed is plotted against the time as shown.

The distance is obtained by substituting the expression for v into the definition $v = ds/dt$ and integrating. Thus

$$\frac{8}{1 + 6t} = \frac{ds}{dt} \qquad \int_0^t \frac{8\,dt}{1 + 6t} = \int_0^s ds \qquad s = \frac{4}{3}\ln(1 + 6t) \qquad Ans.$$

The distance s is also plotted against the time as shown, and we see that the ship has moved through a distance $s = \frac{4}{3}\ln(1 + \frac{6}{6}) = \frac{4}{3}\ln 2 = 0.924$ mi (nautical) during the 10 minutes.

PROBLEMS

2/1 The motion of a particle along a straight line is governed by the equation $s = 3 + 2t^2 - (2/15)t^3$ where s is the position coordinate in meters and t is the time in seconds. Plot the velocity v of the particle against the time t for the first 10 s of motion and find the acceleration a for $t = 0$, 5, and 10 s.

$$Ans.\ a_0 = 4\ m/s^2,\ a_5 = 0,\ a_{10} = -4\ m/s^2$$

2/2 The position of a certain particle is described by $s = 4t^3 + 3t^2 - 18t + 5$ where s is in meters and t is in seconds. Determine the time t and the acceleration a when $v = 0$. Consider positive values of t only. $\qquad Ans.\ t = 1\ s,\ a = 30\ m/s^2$

2/3 In the final stages of a moon landing the lunar module descends under retro-thrust of its descent engine to within $h = 6$ m of the lunar surface where it has a downward velocity of 3 m/s. If the descent engine is cut off abruptly at this point, compute the impact velocity of the landing gear with the moon. Lunar gravity is $\frac{1}{6}$ of the earth's gravity. $\qquad Ans.\ v = 5.35\ m/s$

Problem 2/3

2/4 The velocity of a particle which moves along the x-axis is given by $\dot{x} = 40 - 3t^2$ m/s where t is in seconds. Calculate the displacement Δx of the particle during the interval from $t = 2$ s to $t = 4$ s.

2/5 The graph shows the displacement-time history for the rectilinear motion of a particle during an 8-second interval. Determine the average velocity v_{av} during the interval and, to within reasonable limits of accuracy, find the instantaneous velocity v when $t = 4$ s.

Problem 2/5

2/6 During an 8-second interval the velocity of a particle moving in a straight line varies with the time as shown. Within reasonable limits of accuracy, determine the amount Δa by which the acceleration at $t = 4$ s exceeds the average acceleration during the interval. What is the displacement Δs during the interval?

$$Ans.\ \Delta a = 0.50\ m/s^2,\ \Delta s = 64\ m$$

Problem 2/6

Problem 2/7

Problem 2/9

"Zero-g" test facility

Problem 2/10

2/7 Experimental data for the motion of a particle along a straight line yield measured values of the velocity v for various position coordinates s. A smooth curve is drawn through the points as shown in the graph. Determine the acceleration of the particle when $s = 20$ m.

Ans. $a = 1.2$ m/s^2

2/8 A projectile is fired vertically with an initial velocity of 250 m/s. Calculate the maximum altitude h reached by the projectile and the time t after firing for it to return to the ground. Neglect air resistance and take the gravitational acceleration to be constant at 9.81 m/s^2.

Ans. $h = 3190$ m, $t = 51.0$ s

2/9 A ball is thrown vertically up with a velocity of 24 m/s at the edge of a 60-m cliff. Calculate the height h to which the ball rises and the total time t after release for the ball to reach the bottom of the cliff. Neglect air resistance and take the downward acceleration to be 9.81 m/s^2.

2/10 To test the effects of "weightlessness" for short periods of time a test facility is designed which accelerates a test package vertically up from A to B by means of a gas-activated piston and allows it to ascend and descend from B to C to B under free-fall conditions. The test chamber consists of a deep well and is evacuated to eliminate any appreciable air resistance. If a constant acceleration of $40g$ from A to B is provided by the piston and if the total test time for the "weightless" condition from B to C to B is 10 s, calculate the required working height h of the chamber. Upon returning to B the test package is recovered in a basket filled with polystyrene pellets inserted in the line of fall.

Ans. $h = 125.7$ m

2/11 A sprinter reaches his top speed of 10.6 m/s in t seconds from rest with essentially constant acceleration. If he maintains his speed and covers the 100-m distance in 10.5 s, find the acceleration interval t and his average starting acceleration a.

2/12 The pilot of a jet transport brings the engines to full takeoff power before releasing the brakes as the aircraft is standing on the runway. The jet thrust remains constant, and the aircraft has a near-constant acceleration of $0.6g$. If the takeoff speed is 200 km/h, calculate the distance s and time t from rest to takeoff.   *Ans.* $s = 262$ m, $t = 9.44$ s

2/13 A jet aircraft with a landing speed of 200 km/h has a maximum of 600 m of available runway after touchdown in which to reduce its ground speed to 30 km/h. Compute the average acceleration a required of the aircraft during braking.

2/14 A particle moves along a straight line with a velocity in millimeters per second given by $v = 400 - 16t^2$ where t is in seconds. Calculate the net displacement Δs and the total distance D traveled during the first 6 seconds of motion.
$$Ans. \ \Delta s = 1.248 \text{ m}, D = 1.419 \text{ m}$$

2/15 A retarding force acts on a particle moving initially with a velocity of 100 m/s and gives it a deceleration as recorded by the oscilloscope record shown. Approximate the velocity of the particle at $t = 4$ s and at $t = 8$ s.

Problem 2/15

2/16 A vacuum-propelled capsule for a high-speed tube transportation system of the future is being designed for operation between two stations 10 km apart. If maximum acceleration and deceleration are to have a limiting magnitude of $0.6g$ and if velocities are to be limited to 400 km/h, determine the minimum time t for the capsule to make the 10-km trip.   *Ans.* $t = 1.81$ min

2/17 The position of a particle in millimeters is given by $s = 27 - 12t + t^2$ where t is in seconds. Plot the s-t and v-t relationships for the first 9 seconds. Determine the net displacement Δs during that interval and the total distance D traveled. By inspection of the s-t relationship, what conclusion can you reach regarding the acceleration?

Problem 2/18

Problem 2/19

Problem 2/21

Problem 2/22

2/18 Packages enter the 3-m chute at A with a velocity of 1.2 m/s and have an acceleration of $0.3g$ from A to B. If the packages come to a stop at C, 2.8 seconds after leaving A, calculate the acceleration a of the packages from B to C. Also find the time required to go from B to C.

Ans. $a = -2.65$ m/s², $\Delta t = 1.723$ s

2/19 A motorcycle patrolman starts from rest at A two seconds after a car, speeding at the constant rate of 120 km/h, passes point A. If the patrolman accelerates at the rate of 6 m/s² until he reaches his maximum permissible speed of 150 km/h, which he maintains, calculate the distance s from point A to the point at which he overtakes the car.

2/20 A single-stage rocket is launched vertically from rest, and its thrust is programmed to give the rocket a constant upward acceleration of 6 m/s². If the fuel is exhausted 20 s after launch, calculate the maximum velocity v_m and the subsequent maximum altitude h reached by the rocket.

Ans. $v_m = 120$ m/s, $h = 1.934$ km

2/21 The spring-loaded body oscillates in the vertical direction with a y-coordinate specified by $y = y_0 \sin 2\pi nt$ where y is measured from the equilibrium position. The maximum value of y is y_0, and n is the constant number of complete cycles per unit of time t. Determine the velocity $v_{1/2}$ and acceleration $a_{1/2}$ of the body when $t = 1/(2n)$ which is the time for $\frac{1}{2}$ of a complete oscillation.

2/22 A particle moves along the positive x-axis with an acceleration a_x in m/s² which increases linearly with x expressed in millimeters, as shown on the graph for an interval of its motion. If the velocity of the particle at $x = 40$ mm is 0.4 m/s, determine the velocity at $x = 120$ mm.   *Ans.* $v = 0.8$ m/s

2/23 The acceleration of a particle which moves in the positive x-direction varies with its position as shown. If the velocity of the particle is 0.8 m/s when $x = 0$, determine the velocity v of the particle when $x = 1.4$ m.

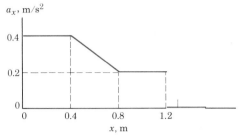

Problem 2/23

2/24 A particle moves along the x-axis with a constant acceleration. When $t = 0$, $x = 4$ m and $\dot{x} = 3$ m/s. Also, when $t = 4$ s, a maximum value of x is observed. Determine x_{max} and the value of x when $t = 12$ s. Plot x versus t.

$$\textit{Ans. } x_{\text{max}} = 10 \text{ m}, \ x_{12} = -14 \text{ m}$$

2/25 A ship reduces speed as it approaches the entrance to a harbor in still water. The speed is reduced from 10 knots as the ship passes the first entrance buoy to 4 knots 6 minutes later as it passes the second buoy. Approximate the distance Δs between the buoys in nautical miles by using the graph directly. (Recall that 1 knot equals one nautical mile per hour.)

$$\textit{Ans. } \Delta s = 0.60 \text{ mi (nautical)}$$

Problem 2/25

2/26 In a test of air resistance under conditions of deceleration a small test vehicle is given an initial velocity against the air flow in a wind tunnel, and its position x from the start is measured as a function of time and tabulated. Approximate the velocity v and acceleration a in the x-direction at $t = 2$ s. Draw a smooth curve through the plotted points and solve graphically.

t, s	x, mm	t, s	x, mm
0	0	3.5	2000
0.5	475	4.0	2100
1.0	800	4.5	2075
1.5	1225	5.0	2050
2.0	1425	5.5	1975
2.5	1700	6.0	1725
3.0	1875		

$$\textit{Ans. } v \cong 480 \text{ mm/s}, \ a \cong -210 \text{ mm/s}^2$$

Problem 2/27

Problem 2/28

Problem 2/29

Problem 2/30

2/27 We are given the *s-t* curve for the motion of a particle. We are also given that the slope of the curve at $t = 2$ min and $t = 6$ min is 0.33 km/min. Construct the corresponding *v-t* curve to within the accuracy of the given information using km/h for *v*.

2/28 A particle starts from rest at $x = -2$ m and moves along the *x*-axis with the velocity history shown. Plot the corresponding acceleration and displacement histories for the two seconds. Find the time *t* when the particle crosses the origin.

2/29 A subway train travels between two of its station stops with the acceleration schedule shown. Determine the time interval Δt during which the train brakes to a stop with a deceleration of 2 m/s² and find the distance *s* between stations.

Ans. $\Delta t = 10$ s, $s = 416$ m

2/30 Prove that the acceleration of a particle in rectilinear motion is constant when the square of its velocity varies linearly with the displacement *s*. For the case illustrated, determine the acceleration a_1 during the first interval and a_2 during the second interval.

Ans. $a_1 = 0.141$ m/s², $a_2 = -0.208$ m/s²

2/31 A body which moves in a straight line between two points A and B a distance of 50 m apart has a velocity whose square increases linearly with the distance traveled, as shown on the graph. Determine the displacement Δs of the body during the last 2 seconds before arrival at B.

Problem 2/31

2/32 A projectile is fired horizontally into a resisting medium with a velocity v_0, and the resulting deceleration is equal to cv^n, where c and n are constants and v is the velocity within the medium. Find the expression for the velocity v of the projectile in terms of the time t of penetration.

$$Ans. \ v = [v_0{}^{1-n} + c(n-1)t]^{1/(1-n)}$$

2/33 A retarding force is applied to a body moving in a straight line so that, during an interval of its motion, its speed v decreases with increased position coordinate s according to the relation $v^2 = k/s$, where k is a constant. If the body has a forward speed of 50 mm/s and its position coordinate is 225 mm at time $t = 0$, determine the speed v at $t = 3$ s.

2/34 The aerodynamic resistance to motion of a car is nearly proportional to the square of its velocity. Additional frictional resistance is constant, so that the acceleration of the car when coasting may be written $a = -C_1 - C_2 v^2$ where C_1 and C_2 are constants that depend on the mechanical configuration of the car. If the car has an initial velocity v_0 when the engine is disengaged, derive an expression for the distance D required for the car to coast to a stop.

Problem 2/34

$$Ans. \ D = \frac{1}{2C_2} \ln \left(1 + \frac{C_2}{C_1} v_0{}^2 \right)$$

2/35 The steel ball A of diameter D slides freely on the horizontal rod which leads to the pole face of the electromagnet. The force of attraction obeys an inverse-square law, and the resulting acceleration of the ball is $a = K/(L-x)^2$ where K is a measure of the strength of the magnetic field. If the ball is released from rest at $x = 0$, determine the velocity v with which it strikes the pole face.

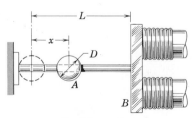

Problem 2/35

2/36 The velocity of a particle moving along a straight line decreases linearly with its position coordinate s from 9 m/s at $s = 0$ to a value approaching zero as s approaches 18 m. Show that the particle never reaches $s = 18$ m, and find the acceleration a when $s = 12$ m. *Ans.* $a = -1.5$ m/s^2

2/37 A self-propelled vehicle of mass m whose engine delivers constant power P has an acceleration $a = P/(mv)$ where all frictional resistance is neglected. Determine expressions for the distance s traveled and the corresponding time t required by the vehicle to increase its speed from v_1 to v_2.

Problem 2/38

2/38 To a close approximation the pressure behind a rifle bullet varies inversely with the position x of the bullet along the barrel. Thus the acceleration of the bullet may be written as $a = k/x$ where k is a constant. If the bullet starts from rest at $x = 7.5$ mm and if the muzzle velocity of the bullet is 600 m/s at the end of the 750-mm barrel, compute the acceleration of the bullet as it passes the midpoint of the barrel at $x = 375$ mm.
Ans. $a = 104.2$ km/s^2

2/39 A certain lake is proposed as a landing area for large jet aircraft. The touchdown speed of 160 km/h upon contact with the water is to be reduced to 30 km/h in a distance of 400 m. If the deceleration is proportional to the square of the velocity of the aircraft through the water, $a = -Kv^2$, find the value of the design parameter K, which would be a measure of the size and shape of the landing gear vanes which plow through the water. Also find the time t elapsed during the specified interval.

Problem 2/40

2/40 A test projectile is fired horizontally into a viscous liquid with a velocity v_0. The retarding force is proportional to the square of the velocity, so that the acceleration becomes $a = -kv^2$. Derive expressions for the distance D traveled in the liquid and the corresponding time t required to reduce the velocity to $v_0/2$. Neglect any vertical motion.
Ans. $D = 0.693/k$, $t = 1/(kv_0)$

2/41 The horizontal motion of the plunger and shaft is arrested by the resistance of the attached disk which moves through the oil bath. If the velocity of the plunger is v_0 in the position A where $x = 0$ and $t = 0$, and if the deceleration is proportional to v so that $a = -kv$, derive expressions for the velocity v and position coordinate x in terms of the time t. Also express v in terms of x.

Problem 2/41

2/42 A small object is released from rest in a tank of oil. The downward acceleration of the object is $g - kv$ where g is the constant acceleration due to gravity, k is a constant which depends on the viscosity of the oil and shape of the object, and v is the downward velocity of the object. Derive expressions for the velocity v and vertical drop y as functions of the time t after release.

$$Ans. \ v = \frac{g}{k}(1 - e^{-kt}), \ y = \frac{g}{k}\left[t - \frac{1}{k}(1 - e^{-kt})\right]$$

2/43 In a test drop in still air, a small steel ball is released from rest at a considerable height. Its initial acceleration g is reduced by the decrement kv^2, where k is a constant and v is the downward velocity. Determine the maximum velocity attainable by the ball and find the vertical drop y in terms of the time t from release.

2/44 A particle starts from rest and has an acceleration given by $a = 3.2 - 0.2v^2$ m/s^2 where v is its velocity in meters per second. Calculate the time t required to reach a velocity of 2 m/s and find the corresponding displacement s of the particle.

Ans. $t = 0.687$ s, $s = 719$ mm

2/45 The acceleration of a certain machine part in rectilinear motion varies with time as shown. If the part has a velocity of 8 mm/s in the negative s-direction when $t = 0$, determine its velocity when $t = 8$ s and its net displacement Δs during the first 4 seconds.

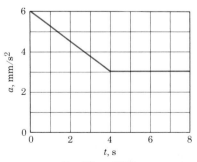

Problem 2/45

2/46 A car starts from rest with an acceleration in meters per second squared given by $a = 4 - \frac{3}{4}t$ where t is in seconds. When the car reaches a maximum velocity, it maintains that velocity until it has gone a total distance of 200 m. Calculate the time t required by the car to go 100 m from its starting position. *Ans.* $t = 11.15$ s

Problem 2/47

Problem 2/49

Problem 2/50

Problem 2/51

2/47 An earth satellite traveling in a circular orbit a distance h above the surface of the earth is suddenly deprived of its orbital velocity at A and falls to the earth's surface. Use the expression for the varying gravitational acceleration g, explained in Art. 1/5, where g_0 is the value of g at the surface of the earth, and derive an expression for the velocity v with which the satellite would strike the surface of the earth if there were no atmosphere. The radius of the earth is R.

2/48 A car starts from rest and accelerates at a constant rate until it reaches 100 km/h in a distance of 60 m, at which time the clutch is disengaged. The car then slows down to a velocity of 50 km/h in an additional distance of 120 m with a deceleration which is proportional to its velocity. Find the time t for the car to travel the 180 m.

Ans. $t = 10.31$ s

2/49 The acceleration of a particle in the x-direction changes with its position coordinate x as shown. If the particle has a velocity $\dot{x} = 3$ m/s when $x = 0$, determine its velocity v when $x = 10$ m. Also find the time t required to move the 10 meters.

2/50 A bumper, consisting of a nest of three springs, is used to arrest the horizontal motion of a large mass which is traveling at 40 m/s as it contacts the bumper. The two outer springs cause a deceleration proportional to the spring deformation. The center spring increases the deceleration rate when the compression exceeds 0.5 m as shown on the graph. Determine the maximum compression x of the outer springs. *Ans.* $x = 0.831$ m

2/51 The forward acceleration a of a certain test vehicle is measured experimentally during an interval of its motion and plotted against its position coordinate s as shown. At the 125-m position the driving mechanism shifts abruptly, and a discontinuity in the acceleration occurs. If the velocity of the vehicle is 30 km/h at $s = 50$ m, determine to within the accuracy of the plot the velocity v of the vehicle for each 25-m increment in displacement. Plot v against s and find the velocity at $s = 175$ m.

Ans. $v_{175} = 34.3$ m/s

2/52 Approximate the v-t data for the decelerating ship of Prob. 2/25 by using the exponential relation $v = v_0 e^{-kt}$ where v_0 and k are constants determined by the initial and final conditions. Then calculate the distance s between the buoys.

Ans. $s = 0.65$ mi (nautical)

2/53 Approximate the v-t data for the decelerating ship of Prob. 2/25 by the form of the expression derived in Sample Problem 2/4, namely, $v = v_0/(1 + v_0 kt)$ and calculate the distance s between the two buoys.

Ans. $s = 0.61$ mi (nautical)

▶**2/54** A rocket propels a space probe for measuring micrometeorite density vertically upward from the north pole and separates from the probe at burnout conditions. The probe continues to move upward subject to the influence of the earth's gravity and has an upward velocity of 16 000 km/h at an altitude of 160 km above the earth's surface at the pole. Calculate the maximum distance h from the earth's surface reached by the probe. The polar radius of the earth is 6357 km, and the value of g at the pole is 9.833 m/s². *Ans.* $h = 1420$ km

▶**2/55** A particle which is constrained to move in a straight line is subjected to an accelerating force which increases with time and a retarding force which increases directly with the position coordinate x. The resulting acceleration is $a = Kt - k^2 x$ where K and k are positive constants and where both x and $\dot{x}$ are zero when $t = 0$. Determine x as a function of t.

$$Ans. \ x = \frac{K}{k^3}(kt - \sin kt)$$

▶**2/56** The vertical acceleration of a certain solid-fuel rocket is given by $a = ke^{-bt} - cv - g$, where k, b, and c are constants, v is the vertical velocity acquired, and g is the gravitational acceleration, essentially constant for atmospheric flight. The exponential term represents the effect of a decaying thrust as fuel is burned, and the term $-cv$ approximates the retardation due to atmospheric resistance. Determine the expression for the vertical velocity of the rocket t seconds after firing.

$$Ans. \ v = \frac{g}{c}(e^{-ct} - 1) + \frac{k}{c-b}(e^{-bt} - e^{-ct})$$

2/3 PLANE CURVILINEAR MOTION

We now begin our description of the motion of a particle along a curved path which lies in a single plane. This motion is seen to be a special case of the more general three-dimensional motion introduced in Art. 2/1 and illustrated in Fig. 2/1. If we let the plane of motion be the *x-y* plane, for instance, then the coordinates *z* and ϕ of Fig. 2/1 are both zero and *R* becomes the same as *r*. As mentioned previously, the vast majority of the motions of points or particles encountered in engineering practice may be represented as plane motion.

Before pursuing the description of plane curvilinear motion in any specific set of coordinates, we will first describe the motion with the aid of vector analysis, the results of which are independent of any particular coordinate system. What follows in this article will constitute one of the most basic of all treatments in dynamics, namely, development of the *time derivative of a vector*. A large fraction of the entire subject of dynamics hinges directly on the time rates of change of vector quantities, and all students of dynamics are well advised to master this topic at the outset because they will have frequent occasion to use it.

Consider now the continuous motion of a particle along a plane curve as represented in Fig. 2/5. At time *t* the particle is at position *A*, which is located by the *position vector* **r** measured from some convenient fixed origin *O*. If both the magnitude and direction of **r** are known at time *t*, then the position of the particle is completely specified. At time $t + \Delta t$ the particle is at *A'*, located by the position vector $\mathbf{r} + \Delta \mathbf{r}$. We note, of course, that this combination is vector addition and not scalar addition. The *displacement* of the particle during time Δt is the vector $\Delta \mathbf{r}$ which represents the vector change of position and is clearly independent of the choice of origin. If an origin were chosen at some different location, the position vector **r** would be changed, but $\Delta \mathbf{r}$ would be unchanged. The *distance* actually traveled by the particle as it

Figure 2/5

moves along the path from A to A' is the scalar length Δs measured along the path. Thus we distinguish between the vector displacement $\Delta \mathbf{r}$ and the scalar distance Δs.

The *average velocity* of the particle between A and A' is defined as $\mathbf{v}_{av} = \Delta \mathbf{r}/\Delta t$, which is a vector whose direction is that of $\Delta \mathbf{r}$ and whose magnitude is the magnitude of $\Delta \mathbf{r}$ divided by Δt. The average *speed* of the particle between A and A' is the scalar quotient $\Delta s/\Delta t$. Clearly the magnitude of the average velocity and the speed approach one another as the interval Δt decreases and A and A' become closer together.

The *instantaneous velocity* $\mathbf{v}$ of the particle is defined as the limiting value of the average velocity as the time interval approaches zero. Thus

$$\mathbf{v} = \lim_{\Delta t \to 0} \frac{\Delta \mathbf{r}}{\Delta t}$$

We observe that the direction of $\Delta \mathbf{r}$ approaches that of the tangent to the path as Δt approaches zero and, hence, the velocity $\mathbf{v}$ is always a vector tangent to the path. We now extend the basic definition of the derivative of a scalar quantity to include a vector quantity and write

$$\mathbf{v} = \frac{d\mathbf{r}}{dt} = \dot{\mathbf{r}} \qquad (2/4)$$

The derivative of a vector is itself a vector having both a magnitude and a direction. The magnitude of $\mathbf{v}$ is called the *speed* and is the scalar

$$v = |\mathbf{v}| = \frac{ds}{dt} = \dot{s}$$

At this point we make a careful distinction between the magnitude of the derivative and the derivative of the magnitude. The magnitude of the derivative may be written in any one of the several ways $|d\mathbf{r}/dt| = |\dot{\mathbf{r}}| = \dot{s} = |\mathbf{v}| = v$ and represents the magnitude of the velocity or the speed of the particle. On the other hand the derivative of the magnitude is written $d|\mathbf{r}|/dt = dr/dt = \dot{r}$, and represents the rate at which the length of the position vector $\mathbf{r}$ is changing. Thus these two derivatives have two entirely different meanings, and we must be extremely careful to distinguish between them in our thinking and in our notation. For this and other reasons students are urged to adopt a consistent notation for their handwritten work for all vector quantities to distinguish them from scalar quantities. For simplicity the underline $\underline{v}$ is recommended. Other handwritten symbols such as $\vec{v}$, $\underline{v}$, and $\hat{v}$ are sometimes used.

With the concept of velocity as a vector established, we return to Fig. 2/5 and denote the velocity of the particle at A by the tangent vector $\mathbf{v}$ and the velocity at A' by the tangent vector $\mathbf{v}'$. Clearly there is a vector change in the velocity during the time Δt. The velocity $\mathbf{v}$ at A plus (vectorially) the change $\Delta \mathbf{v}$ must equal the velocity at A', so we may write $\mathbf{v}' - \mathbf{v} = \Delta \mathbf{v}$. Inspection of the vector diagram shows that $\Delta \mathbf{v}$ depends both on the change in magnitude (length) of $\mathbf{v}$ and on the change in direction of $\mathbf{v}$. These two changes are fundamental characteristics of the derivative of a vector.

The *average acceleration* of the particle between A and A' is defined as $\Delta \mathbf{v}/\Delta t$, which is a vector whose direction is that of $\Delta \mathbf{v}$. The magnitude of this average acceleration is the magnitude of $\Delta \mathbf{v}$ divided by Δt.

The *instantaneous acceleration* $\mathbf{a}$ of the particle is defined as the limiting value of the average acceleration as the time interval approaches zero. Thus

$$\mathbf{a} = \lim_{\Delta t \to 0} \frac{\Delta \mathbf{v}}{\Delta t}$$

By definition of the derivative, then, we write

$$\mathbf{a} = \frac{d\mathbf{v}}{dt} = \dot{\mathbf{v}} \tag{2/5}$$

As the interval Δt becomes smaller and approaches zero, the direction of the change $\Delta \mathbf{v}$ approaches that of the differential change $d\mathbf{v}$ and, hence, $\mathbf{a}$. The acceleration $\mathbf{a}$, then, includes the effect of change in magnitude of $\mathbf{v}$ and the change of direction of $\mathbf{v}$. It is apparent, in general, that the direction of the acceleration of a particle in curvilinear motion is neither tangent to the path nor normal to the path. We do observe, however, that the acceleration component which is normal to the path points toward the center of curvature of the path.

A further approach to the visualization of acceleration is shown in Fig. 2/6, where the position vectors to three arbitrary positions on the path of the particle are shown for illustrative pur-

Figure 2/6

pose. There is a velocity vector tangent to the path corresponding to each position vector, and the relation is $\mathbf{v} = \dot{\mathbf{r}}$. Now if these velocity vectors are plotted from some arbitrary point *C*, then a curve, known as the *hodograph*, is formed. The derivatives of these velocity vectors will be the acceleration vectors $\mathbf{a} = \dot{\mathbf{v}}$ which are tangent to the hodograph. We see that the acceleration bears the same relation to the velocity as the velocity bears to the position vector.

The geometric portrayal of the derivatives of the position vector **r** and velocity vector **v** in Fig. 2/5 can be used to describe the derivative of any vector quantity with respect to *t* or with respect to any other scalar variable. With the introduction of the derivative of a vector in the definitions of velocity and acceleration it is important at this point to establish the rules under which the differentiation of vector quantities may be carried out. These rules are the same as for the differentiation of scalar quantities except for the case of the cross product where the order of the terms must be preserved. These rules are covered in Art. C/7 of Appendix C and should be reviewed at this point.

For curvilinear motion of a particle in a plane there are three different coordinate systems which are in common use to describe the vector relationships developed in this article. An important lesson to be learned from the study of these coordinate systems is the proper choice of reference system for a given problem. This choice is usually revealed by the manner in which the motion is generated or by the form in which the data are specified. Each of the three coordinate systems will now be developed and illustrated.

2/4 *RECTANGULAR COORDINATES* (*x-y*)

This system of coordinates is particularly useful for describing motions where the *x*- and *y*-components of acceleration are independently generated or determined. The resulting curvilinear motion is, then, obtained by a vector combination of the *x*- and *y*-components of the position vector, the velocity, and the acceleration.

The particle path of Fig. 2/5 is shown again in Fig. 2/7 along with *x*- and *y*-axes. The position vector **r**, the velocity **v**, and the acceleration **a** of the particle as developed in Art. 2/3 are represented in Fig. 2/7 together with their *x*- and *y*-components. With the aid of the unit vectors **i** and **j** we may write the vectors **r**, **v**, and **a** in terms of their *x*- and *y*-components. Thus

$$\boxed{\begin{aligned} \mathbf{r} &= x\mathbf{i} + y\mathbf{j} \\ \mathbf{v} = \dot{\mathbf{r}} &= \dot{x}\mathbf{i} + \dot{y}\mathbf{j} \\ \mathbf{a} = \dot{\mathbf{v}} = \ddot{\mathbf{r}} &= \ddot{x}\mathbf{i} + \ddot{y}\mathbf{j} \end{aligned}} \qquad (2/6)$$

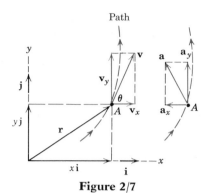

Figure 2/7

As we differentiate with respect to time we observe that the unit vectors have no time derivatives since their magnitudes and directions remain constant. The scalar magnitudes of the components of $\mathbf{v}$ and $\mathbf{a}$ are merely $v_x = \dot{x}$, $v_y = \dot{y}$ and $a_x = \dot{v}_x = \ddot{x}$, $a_y = \dot{v}_y = \ddot{y}$. (As drawn in Fig. 2/7, a_x is in the negative x-direction, so that $\ddot{x}$ would be a negative number.)

As observed previously the direction of the velocity is always tangent to the path, and from the figure it is clear that

$$v^2 = v_x{}^2 + v_y{}^2 \qquad v = \sqrt{v_x{}^2 + v_y{}^2} \qquad \tan \theta = \frac{v_y}{v_x}$$

$$a^2 = a_x{}^2 + a_y{}^2 \qquad a = \sqrt{a_x{}^2 + a_y{}^2}$$

If the angle θ is measured counterclockwise from the x-axis to $\mathbf{v}$ for the configuration of axes shown, then we may also observe that $dy/dx = \tan \theta = v_y/v_x$.

If the coordinates x and y are each known independently as a function of time, $x = f_1(t)$ and $y = f_2(t)$, then for any value of the time we may combine them to obtain $\mathbf{r}$. Similarly we combine their first derivatives $\dot{x}$ and $\dot{y}$ to obtain $\mathbf{v}$ and their second derivatives $\ddot{x}$ and $\ddot{y}$ to obtain $\mathbf{a}$. Or, if the acceleration components a_x and a_y are given as functions of the time, we may integrate each one separately with respect to time, once to obtain v_x and v_y and again to obtain $x = f_1(t)$ and $y = f_2(t)$. Elimination of the time t between these last two parametric equations gives the equation of the curved path $y = f(x)$.

From the foregoing discussion it should be recognized that the rectangular-coordinate representation of curvilinear motion is merely the superposition of the components of two simultaneous rectilinear motions in the x- and y-directions. Therefore, everything covered in Art. 2/2 on rectilinear motion may be applied separately to the x-motion and to the y-motion.

Sample Problem 2/5

The curvilinear motion of a particle is defined by $v_x = 50 - 16t$ and $y = 100 - 4t^2$ where v_x is in meters per second, y is in meters, and t is in seconds. It is also known that $x = 0$ when $t = 0$. Plot the path of the particle and determine its velocity and acceleration when the position $y = 0$ is reached.

Solution. The x-coordinate is obtained by integrating the expression for v_x, and the x-component of the acceleration is obtained by differentiating v_x. Thus

$$\left[\int dx = \int v_x\, dt \right] \qquad \int_0^x dx = \int_0^t (50 - 16t)\, dt \qquad x = 50t - 8t^2 \text{ m}$$

$$[a_x = \dot{v}_x] \qquad a_x = \frac{d}{dt}(50 - 16t) \qquad a_x = -16 \text{ m/s}^2$$

The y-components of velocity and acceleration are

$$[v_y = \dot{y}] \qquad v_y = \frac{d}{dt}(100 - 4t^2) \qquad v_y = -8t \text{ m/s}$$

$$[a_y = \dot{v}_y] \qquad a_y = \frac{d}{dt}(-8t) \qquad a_y = -8 \text{ m/s}^2$$

We now calculate corresponding values of x and y for various values of t and plot x against y to obtain the path as shown.

When $y = 0$, $0 = 100 - 4t^2$, so $t = 5$ s. For this value of the time we have

$$v_x = 50 - 16(5) = -30 \text{ m/s}$$
$$v_y = -8(5) = -40 \text{ m/s}$$
$$v = \sqrt{(-30)^2 + (-40)^2} = 50 \text{ m/s}$$
$$a = \sqrt{(-16)^2 + (-8)^2} = 17.9 \text{ m/s}^2$$

The velocity and acceleration components and their resultants are shown on the separate diagrams for point A where $y = 0$. Thus for this condition we may write

$$\mathbf{v} = -30\mathbf{i} - 40\mathbf{j} \text{ m/s} \qquad \text{Ans.}$$
$$\mathbf{a} = -16\mathbf{i} - 8\mathbf{j} \text{ m/s}^2 \qquad \text{Ans.}$$

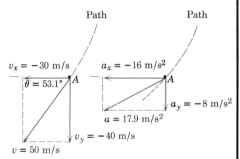

We observe that the velocity vector lies along the tangent to the path as it should but that the acceleration vector is not tangent to the path. Note especially that the acceleration vector has a component which points toward the inside of the curved path. We concluded from our diagram in Fig. 2/5 that it is impossible for the acceleration to have a component which points toward the outside of the curve.

Sample Problem 2/6

A rocket has expended all its fuel when it reaches position A where it has a velocity $\mathbf{u}$ at an angle θ with respect to the horizontal. It then begins unpowered flight and attains a maximum added height h at position B after traveling a horizontal distance s from A. Determine the expressions for h and s, the time t of flight from A to B, and the equation of the path. For the interval concerned, assume a flat earth with a constant gravitational acceleration g and neglect any atmospheric resistance.

Solution. Since all motion components are directly expressible in terms of horizontal and vertical coordinates, a rectangular set of
① axes x-y will be employed. With the neglect of atmospheric resistance, $a_x = 0$ and $a_y = -g$, and the resulting motion is a direct superposition of two rectilinear motions with constant acceleration. Thus

$$[dx = v_x \, dt] \qquad x = \int_0^t u \cos \theta \, dt \qquad x = ut \cos \theta$$

$$[dv_y = a_y \, dt] \qquad \int_{u \sin \theta}^{v_y} dv_y = \int_0^t (-g) \, dt \qquad v_y = u \sin \theta - gt$$

$$[dy = v_y \, dt] \qquad y = \int_0^t (u \sin \theta - gt) \, dt \qquad y = ut \sin \theta - \tfrac{1}{2}gt^2$$

Position B is reached when $v_y = 0$, which occurs for $0 = u \sin \theta - gt$ or

$$t = (u \sin \theta)/g \qquad\qquad \textit{Ans.}$$

Substitution of this value for the time into the expression for y gives the maximum added altitude

$$h = u\left(\frac{u \sin \theta}{g}\right) \sin \theta - \frac{1}{2} g \left(\frac{u \sin \theta}{g}\right)^2 \qquad h = \frac{u^2 \sin^2 \theta}{2g} \qquad \textit{Ans.}$$

The horizontal distance is seen to be

② $$s = u\left(\frac{u \sin \theta}{g}\right) \cos \theta \qquad s = \frac{u^2 \sin 2\theta}{2g} \qquad \textit{Ans.}$$

which is clearly a maximum when $\theta = 45°$. The equation of the path is obtained by eliminating t from the expressions for x and y, which gives

$$y = x \tan \theta - \frac{gx^2}{2u^2} \sec^2 \theta \qquad\qquad \textit{Ans.}$$

③ This equation describes a vertical parabola as indicated in the figure.

① Note that this problem is simply the description of projectile motion neglecting atmospheric resistance.

② We see that the total range and time of flight for a projectile fired above a horizontal plane would be twice the respective values of s and t given here.

③ If atmospheric resistance were to be accounted for, the dependency of the acceleration components on the velocity would have to be established before an integration of the equations could be carried out. This becomes a much more difficult problem.

PROBLEMS

(In the following problems where motion as a projectile in air is involved, neglect air resistance unless otherwise stated and use $g = 9.81$ m/s^2.)

2/57 A particle which moves with curvilinear motion has coordinates in millimeters which vary with the time t in seconds according to $x = 3t^2 - 4t$ and $y = 4t^2 - \frac{1}{3}t^3$. Determine the magnitudes of the velocity **v** and acceleration **a** and the angles which these vectors make with the x-axis when $t = 2$ s.

> *Ans.* $v = 14.42$ mm/s with $\theta_x = 56.3°$
> $a = 7.21$ mm/s^2 with $\theta_x = 33.7°$

2/58 The position vector of a point which moves in the x-y plane is given by

$$\mathbf{r} = \left(\frac{2}{3}\, t^3 - \frac{3}{2}\, t^2 \right)\mathbf{i} + \frac{t^4}{12}\, \mathbf{j}$$

where **r** is in meters and t is in seconds. Determine the velocity **v** and the acceleration **a** when $t = 3$ s. What do the results tell about the curvature of the path at this instant?

> *Ans.* $\mathbf{v} = 9(\mathbf{i} + \mathbf{j})$ m/s, $\mathbf{a} = 9(\mathbf{i} + \mathbf{j})$ m/s^2

2/59 The x- and y-motions of guides A and B with right-angle slots control the curvilinear motion of the connecting pin P, which slides in both slots. For a short interval the motions are governed by $x = 20 + \frac{1}{4}t^2$ and $y = 15 - \frac{1}{6}t^3$ where x and y are in millimeters and t is in seconds. Calculate the magnitudes of the velocity **v** and acceleration **a** of the pin for $t = 2$ s. Sketch the direction of the path and indicate its curvature for this instant.

> *Ans.* $v = 2.24$ mm/s, $a = 2.06$ mm/s^2

Problem 2/59

2/60 If the positioning screws of Prob. 2/59 revolve periodically and transmit motions to their respective slotted guides given by $x = 30 + 24 \sin(\pi t/2)$ and $y = 40 + 32 \cos(\pi t/2)$, where x and y are in millimeters and t is in seconds, determine the magnitudes of the velocity **v** and acceleration **a** of the pin P for $t = 2.5$ s. Sketch the direction of the path and indicate its curvature at this instant.

2/61 The y-coordinate of a particle in curvilinear motion is given by $y = 4t^3 - 3t$ where y is in meters and t is in seconds. Also, the particle has an acceleration in the x-direction given by $a_x = 12t$ m/s^2. If the velocity of the particle in the x-direction is 4 m/s when $t = 0$, calculate the magnitudes of the velocity $\mathbf{v}$ and acceleration $\mathbf{a}$ of the particle when $t = 1$ s. Construct $\mathbf{v}$ and $\mathbf{a}$ in your solution.

$$\textit{Ans. } v = 13.45 \text{ m/s with } \theta_x = 42.0°$$
$$a = 26.8 \text{ m/s}^2 \text{ with } \theta_x = 63.4°$$

2/62 Calculate the minimum possible magnitude of the muzzle velocity $\mathbf{u}$ which a projectile must have when fired from point A to reach a target B on the same horizontal plane 12 km away.

Problem 2/62

2/63 A broadjumper approaches his takeoff board A with a horizontal velocity of 10 m/s. Determine the vertical component v_y of the velocity of his center of gravity at takeoff for him to make the jump shown. What is the vertical rise h of his center of gravity?

$$\textit{Ans. } v_y = 3.68 \text{ m/s}, \ h = 0.690 \text{ m}$$

Problem 2/63

2/64 A rock is thrown horizontally from a tower at A and hits the ground 3.5 s later at B. The line of sight from A to B makes an angle of 50° with the horizontal. Compute the magnitude of the initial velocity $\mathbf{u}$ of the rock. *Ans. $u = 14.41$ m/s*

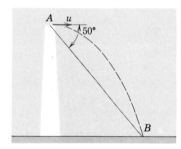

Problem 2/64

2/65 A particle P moves on the circular path of radius r as shown. If $\dot{\theta} = \omega$, a constant, prove that the acceleration of P is always directed toward the center.

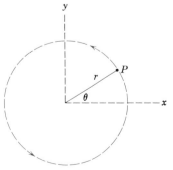

Problem 2/65

2/66 A ball is dropped from rest at the top of a 60-m tower. A horizontal wind gives it a constant horizontal acceleration, and the ball hits the ground 3 m west of its release point. Neglect vertical air resistance and calculate the horizontal acceleration a_x and the horizontal velocity v_x of the ball as it hits the ground. Neglect any effects associated with the rotation of the earth.

Ans. $a_x = 0.490$ m/s^2, $v_x = 1.716$ m/s

Problem 2/66

2/67 With what minimum horizontal velocity u can a boy throw a rock at A and have it just clear the obstruction at B?

Problem 2/67

2/68 The pilot of an airplane carrying a package of mail to a remote outpost wishes to release the package at the right moment to hit the recovery location A. What angle θ with the horizontal should the pilot's line of sight to the target make at the instant of release? The airplane is flying horizontally at an altitude of 100 m with a velocity of 200 km/h.

Ans. $\theta = 21.7°$

Problem 2/68

2/69 A particle moves in the x-y plane with a y-component of velocity in meters per second given by $v_y = 8t$ with t in seconds. The acceleration of the particle in the x-direction in meters per second squared is given by $a_x = 4t$ with t in seconds. When $t = 0$, $y = 2$ m, $x = 0$, and $v_x = 0$. Find the equation of the path of the particle and calculate the magnitude of the velocity $\mathbf{v}$ of the particle for the instant when its x-coordinate reaches 18 m.

Problem 2/70

Problem 2/72

Problem 2/73

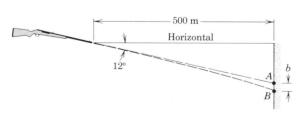

Problem 2/74

2/70 For a certain interval of motion the pin P is forced to move in the fixed parabolic slot by the vertical slotted guide, which moves in the x-direction at the constant rate of 20 mm/s. All measurements are in millimeters and seconds. Calculate the magnitudes of the velocity **v** and acceleration **a** of pin P when $x = 60$ mm.

Ans. $v = 25$ mm/s, $a = 5$ mm/s^2

2/71 If the vertical slotted guide in Prob. 2/70 is given an oscillatory motion $x = 60 \sin \pi t$, where x and y are in millimeters and t is in seconds, calculate the magnitudes of the velocity **v** and acceleration **a** of the pin P for $t = \frac{1}{2}$ s.

2/72 A bomber flying with a horizontal velocity of 500 km/h at an altitude of 5 km is to score a direct hit on a train moving with a constant velocity of 100 km/h in the same direction and in the same vertical plane. Calculate the angle θ between the line of sight to the target and the horizontal at the instant the bomb should be released.

Ans. $\theta = 54.6°$

2/73 A rocket runs out of fuel in the position shown and continues in unpowered flight above the atmosphere. If its velocity in this position was 1000 km/h, calculate the maximum additional altitude h acquired and the corresponding time t to reach it. The gravitational acceleration during this phase of its flight is 9.39 m/s^2.

2/74 The rifle is aimed at point A and fired. Calculate the distance b below A to the point B where the bullet strikes. The muzzle velocity is 800 m/s.

Ans. $b = 2.00$ m

2/75 By what angle δ should the pilot of a dive bomber lead his target if he is to release his bomb at an altitude $h = 1.8$ km at a speed $u = 1000$ km/h while diving at the angle $\theta = 45°$?

Problem 2/75

2/76 Ball bearings leave the horizontal trough with a velocity of magnitude u and fall through the 70-mm-diameter hole as shown. Calculate the permissible range of u which will enable the balls to enter the hole. Take the dotted positions to represent the limiting conditions.

Ans. $u_{max} = 1.135$ m/s, $u_{min} = 0.744$ m/s

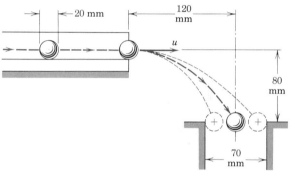

Problem 2/76

2/77 In the cathode-ray tube electrons traveling horizontally from their source with the velocity v_0 are deflected by an electric field E due to the voltage gradient across the plates P. The deflecting force causes an acceleration in the vertical direction on the sketch equal to eE/m where e is the electron charge and m is its mass. When clear of the plates the electrons travel in straight lines. Determine the expression for the deflection δ for the tube and plate dimensions shown.

Ans. $\delta = \dfrac{eEl}{mv_0{}^2}\left(\dfrac{l}{2} + b\right)$

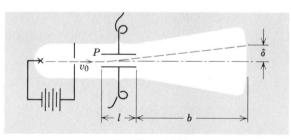

Problem 2/77

2/78 To pass inspection small ball bearings must bounce through an opening of limited size at the top of their trajectory when rebounding from a heavy plate as shown. Calculate the angle θ made by the rebound velocity with the horizontal and the velocity v of the balls as they pass through the opening.

Problem 2/78

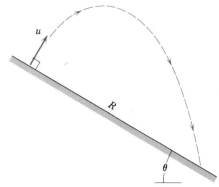

Problem 2/79

2/79 A projectile is fired with a velocity u at right angles to the slope, which is inclined at an angle θ with the horizontal. Derive an expression for the distance R to the point of impact.   *Ans.* $R = \dfrac{2u^2}{g} \tan \theta \sec \theta$

Problem 2/80

2/80 A projectile is fired with a velocity **u** at the entrance A to a horizontal tunnel of length L and height H. Determine the minimum value of u and the corresponding value of the angle θ for which the projectile will reach B at the other end of the tunnel without touching the top of the tunnel.

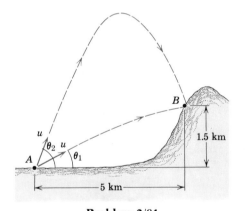

Problem 2/81

2/81 The muzzle velocity of a long-range rifle at A is $u = 400$ m/s. Determine the two angles of elevation θ which will permit the projectile to hit the mountain target B.   *Ans.* $\theta_1 = 26.1°, \theta_2 = 80.6°$

Problem 2/82

2/82 A football player attempts a 30-m field goal. If he is able to impart a velocity u of 30 m/s to the ball, compute the minimum angle θ for which the ball will clear the crossbar of the goal. *Hint:* Let $m = \tan \theta$.

2/83 A long-range artillery rifle at *A* is aimed at an angle of 45° with the horizontal, and its shell is just able to clear the mountain peak at the top of its trajectory. Determine the magnitude *u* of the muzzle velocity, the height *H* of the mountain above sea level, and the range *R* to the sea.

   Ans. $u = 396$ m/s, $H = 4600$ m, $R = 16.58$ km

Problem 2/83

2/84 The pilot of an airplane pulls into a steep 45° climb at 300 km/h and releases a package at position *A*. Calculate the horizontal distance *s* and the time *t* from the point of release to the point at which the package strikes the ground.

Problem 2/84

2/85 Calculate the firing angle θ of the anti-aircraft gun with a muzzle velocity of 600 m/s to score a direct hit on an aircraft flying horizontally at 1000 km/h. The gun is fired at the instant when the aircraft is directly overhead. If a near miss occurs when the shell is on its way up but a direct hit is scored when the shell is on its way down, calculate the time *t* after firing for the hit to be scored if the altitude of the aircraft is 6000 m.

   Ans. $\theta = 62.4°$, $t = 95.6$ s

Problem 2/85

2/86 An object which is released from rest relative to the earth at *A* at a distance *h* above a horizontal surface will appear not to fall straight down by virtue of the effect of the earth's rotation. It may be shown that the object has an eastward acceleration relative to the horizontal surface on the earth equal to $2v_y\omega \cos \gamma$, where v_y is the free-fall downward velocity, ω is the angular velocity of the earth, and γ is the north latitude. Determine the expression for the distance δ to the east of the vertical line through *A* at which the object strikes the ground.

   Ans. $\delta = \dfrac{2\sqrt{2}}{3}\, \omega h \sqrt{\dfrac{h}{g}} \cos \gamma$

Problem 2/86

Problem 2/87

2/87 A rocket is released at point A from a jet aircraft flying horizontally at 1000 km/h at a 1500-m altitude. If the thrust of the rocket gives it a constant horizontal acceleration of 0.5g, determine the angle θ between the horizontal and the line of sight to the target for a direct hit. *Ans.* $\theta = 15.0°$

Problem 2/88

▶**2/88** A point P is located by its position vector $\mathbf{r} = (b_1 \cos \theta)\mathbf{i} + (b_2 \sin \theta)\mathbf{j}$ where b_1 and b_2 are constants and θ is the angle between $\mathbf{r}$ and the x-axis. If θ increases at the constant rate $\dot{\theta}$, show that P moves on an elliptical path with an acceleration proportional to r and directed along $\mathbf{r}$ toward the origin.

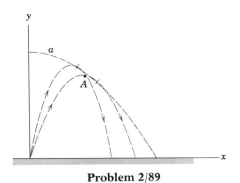

Problem 2/89

▶**2/89** Determine the equation for the envelope a of the parabolic trajectories of a projectile fired at any angle but with a fixed muzzle velocity u. (*Hint:* Substitute $m = \tan \theta$, where θ is the firing angle with the horizontal, into the equation of the trajectory. The two roots m_1 and m_2 of the equation written as a quadratic in m give the two firing angles for the two trajectories shown such that the shells pass through the same point A. Point A will approach the envelope a as the two roots approach equality.) Neglect air resistance and assume g is constant.

$$Ans. \ y = \frac{u^2}{2g} - \frac{gx^2}{2u^2}$$

2/5 *NORMAL AND TANGENTIAL COORDINATES* (*n-t*)

In the introduction to this chapter, Art. 2/1, it was mentioned that one of the common descriptions of curvilinear motion involves the use of path variables, which are measurements made along the tangent *t* and normal *n* to the path of the particle. These coordinates provide a very natural description for curvilinear motion and are frequently the most direct and convenient coordinates to use. The *n*- and *t*-coordinates are considered to move along the path with the particle as seen in Fig. 2/8 where the particle advances from *A* to *B* to *C*. The positive direction for *n* at any position is always taken toward the center of curvature of the path. As seen from Fig. 2/8 the positive *n*-direction will shift from one side of the curve to the other side if the curvature changes direction.

The coordinates *n* and *t* will now be used to describe the velocity **v** and acceleration **a** which were introduced in Art. 2/3 for the curvilinear motion of a particle. For this purpose we introduce unit vectors $\mathbf{e}_n$ in the *n*-direction and $\mathbf{e}_t$ in the *t*-direction, as shown in Fig. 2/9*a* for the position of the particle at point *A* on its path. During a differential increment of time *dt* the particle moves a differential distance *ds* along the curve from *A* to *A'*. With the radius of curvature of the path at this position designated by *ρ*, we see that $ds = \rho\, d\beta$, where *β* is in radians. We see that it is unnecessary to consider the differential change in *ρ* between *A* and *A'* since a higher-order term would be introduced which disappears in the limit. Thus the magnitude of the velocity may be written $v = ds/dt = \rho\, d\beta/dt$, and we may write the velocity as the vector

$$\mathbf{v} = v\mathbf{e}_t = \rho\dot{\beta}\mathbf{e}_t \qquad (2/7)$$

The acceleration **a** of the particle was defined in Art. 2/3 as $\mathbf{a} = d\mathbf{v}/dt$, and we observed from Fig. 2/5 that the acceleration is a vector which reflects both the change in magnitude and the change in direction of **v**. We now differentiate **v** in Eq. 2/7 by applying the ordinary rule for the differentiation of the product of a scalar and a vector* and get

$$\mathbf{a} = \frac{d\mathbf{v}}{dt} = \frac{d(v\mathbf{e}_t)}{dt} = v\dot{\mathbf{e}}_t + \dot{v}\mathbf{e}_t \qquad (2/8)$$

where the unit vector $\mathbf{e}_t$ now has a derivative because its direction changes. To find $\dot{\mathbf{e}}_t$ we analyze the change in $\mathbf{e}_t$ during a differential increment of motion as the particle moves from *A* to *A'* in Fig. 2/9*a*. The unit vector $\mathbf{e}_t$ correspondingly changes to $\mathbf{e}_t'$, and the vector difference $d\mathbf{e}_t$ is shown in the *b*-part of the figure. The

* See Art. C/7 of Appendix C.

Figure 2/8

(a)

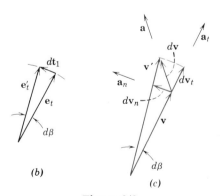

(b)

(c)

Figure 2/9

vector $d\mathbf{e}_t$ in the limit has a magnitude equal to the length of the arc $|\mathbf{e}_t| \, d\beta = d\beta$ obtained by swinging the unit vector $\mathbf{e}_t$ through the angle $d\beta$ expressed in radians. The direction of $d\mathbf{e}_t$ is given by $\mathbf{e}_n$. Thus we may write $d\mathbf{e}_t = \mathbf{e}_n \, d\beta$. Dividing by $d\beta$ gives

$$\frac{d\mathbf{e}_t}{d\beta} = \mathbf{e}_n$$

or dividing by dt gives $d\mathbf{e}_t/dt = (d\beta/dt)\mathbf{e}_n$ which may be written

$$\dot{\mathbf{e}}_t = \dot{\beta}\mathbf{e}_n \tag{2/9}$$

With the substitution of Eq. 2/9 and $\dot{\beta}$ from the relation $v = \rho\dot{\beta}$, Eq. 2/8 for the acceleration becomes

$$\mathbf{a} = \frac{v^2}{\rho}\mathbf{e}_n + \dot{v}\mathbf{e}_t \tag{2/10}$$

where

$$a_n = \frac{v^2}{\rho} = \rho\dot{\beta}^2 = v\dot{\beta}$$

$$a_t = \dot{v} = \ddot{s}$$

$$a = \sqrt{a_n^{\,2} + a_t^{\,2}}$$

We may also note that $a_t = \dot{v} = d(\rho\dot{\beta})/dt = \rho\ddot{\beta} + \dot{\rho}\dot{\beta}$. However this relation finds little use since we seldom have reason to compute $\dot{\rho}$.

A clear understanding of Eq. 2/10 comes only when the geometry of the physical changes which it describes can be clearly seen. Figure 2/9c shows the velocity vector $\mathbf{v}$ when the particle is at A and $\mathbf{v}'$ when it is at A'. The vector change in the velocity is $d\mathbf{v}$, which establishes the direction of the acceleration $\mathbf{a}$. The n-component of $d\mathbf{v}$ is labeled $d\mathbf{v}_n$, and in the limit its magnitude equals the length of the arc generated by swinging the vector $\mathbf{v}$ as a radius through the angle $d\beta$. Thus $|d\mathbf{v}_n| = v \, d\beta$ and the n-component of acceleration is $a_n = |d\mathbf{v}_n|/dt = v(d\beta/dt) = v\dot{\beta}$ as before. The t-component of $d\mathbf{v}$ is labeled $d\mathbf{v}_t$, and its magnitude is simply the change dv in the magnitude or length of the velocity vector. Therefore the t-component of acceleration is $a_t = dv/dt = \dot{v} = \ddot{s}$ as before. The acceleration vectors resulting from their corresponding vector changes in velocity are shown in Fig. 2/9c.

It is especially important to observe that the normal component of acceleration a_n is *always directed toward the center of curvature C*. The tangential component of acceleration, on the other hand, will be in the positive t-direction of motion if the speed v is increasing and in the negative t-direction if the speed is decreasing. In Fig. 2/10 are shown schematic representations of the variation in the acceleration vector for a particle moving from A

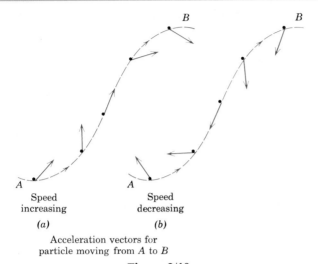

Speed increasing

(a)

Speed decreasing

(b)

Acceleration vectors for particle moving from *A* to *B*

Figure 2/10

to *B* with (*a*) increasing speed and (*b*) decreasing speed. At an inflection point on the curve the normal acceleration v^2/ρ goes to zero since ρ becomes infinite.

Circular motion is an important special case of plane curvilinear motion where the radius of curvature ρ becomes the constant radius r of the circle and the angle β is replaced by the angle θ measured from any convenient radial reference to *OP*, Fig. 2/11. The velocity and the acceleration components for the circular motion of the particle *P* become

$$v = r\dot{\theta}$$
$$a_n = v^2/r = r\dot{\theta}^2 = v\dot{\theta}$$
$$a_t = \dot{v} = r\ddot{\theta}$$

(2/11)

We find repeated use for Eqs. 2/10 and 2/11 in dynamics, and these relations and the principles behind them should be *mastered thoroughly.*

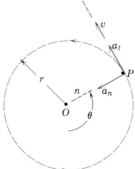

Figure 2/11

Sample Problem   2/7

To anticipate the dip and hump in the road the driver of a car applies her brakes to produce a uniform deceleration. Her speed is 100 km/h at the bottom A of the dip and 50 km/h at the top C of the hump, which is 120 m along the road from A. If the passengers experience a total acceleration of 3 m/s² at A and if the radius of curvature of the hump at C is 150 m, calculate (a) the radius of curvature ρ at A, (b) the acceleration at the inflection point B, and (c) the total acceleration at C.

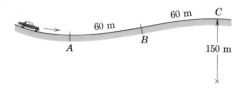

① **Solution.**   The dimensions of the car are small compared with those of the path, so we will treat the car as a particle. The velocities are

$$v_A = \left(100 \frac{\text{km}}{\text{h}}\right)\left(\frac{1 \text{ h}}{3600 \text{ s}}\right)\left(1000 \frac{\text{m}}{\text{km}}\right) = 27.78 \text{ m/s}$$

$$v_C = 50 \frac{1000}{3600} = 13.89 \text{ m/s}$$

We find the constant deceleration along the path from

$$\left[\int v \, dv = \int a_t \, ds\right] \qquad \int_{v_A}^{v_C} v \, dv = a_t \int_0^s ds$$

$$a_t = \frac{1}{2s}(v_C{}^2 - v_A{}^2) = \frac{(13.89)^2 - (27.78)^2}{2(120)} = -2.41 \text{ m/s}^2$$

(a) **Condition at A.**   With the total acceleration given and a_t determined we can easily compute a_n and hence ρ from

$[a^2 = a_n{}^2 + a_t{}^2]$ $a_n{}^2 = 3^2 - (2.41)^2 = 3.19$ $a_n = 1.78 \text{ m/s}^2$

$[a_n = v^2/\rho]$ $\rho = v^2/a_n = (27.78)^2/1.78 = 432 \text{ m}$ *Ans.*

(b) **Condition at B.**   Since the radius of curvature is infinite at the inflection point, $a_n = 0$ and

$$a = a_t = -2.41 \text{ m/s}^2 \qquad\qquad Ans.$$

(c) **Condition at C.**   The normal and total acceleration become

$[a_n = v^2/\rho]$ $a_n = (13.89)^2/150 = 1.29 \text{ m/s}^2$

With unit vectors $\mathbf{e}_n$ and $\mathbf{e}_t$ in the n- and t-directions the acceleration may be written

$$\mathbf{a} = 1.29\mathbf{e}_n - 2.41\mathbf{e}_t \text{ m/s}^2$$

where the magnitude of $\mathbf{a}$ is

$[a = \sqrt{a_n{}^2 + a_t{}^2}]$ $a = \sqrt{(1.29)^2 + (-2.41)^2} = 2.73 \text{ m/s}^2$ *Ans.*

The acceleration vectors representing the conditions at each of the three points are shown for clarification.

① Actually the radius of curvature to the road differs by about one meter from that to the path followed by the center of mass of the passengers, but we have neglected this relatively small difference.

Sample Problem 2/8

A certain rocket maintains a horizontal attitude of its axis during the powered phase of its flight at high altitude. The thrust imparts a horizontal component of acceleration of 6 m/s², and the downward acceleration component is the acceleration due to gravity at that altitude, which is $g = 9$ m/s². At the instant represented the velocity of the mass center G of the rocket along the 15-deg direction of its trajectory is $20(10^3)$ km/h. For this position determine (a) the radius of curvature of the flight trajectory, (b) the rate at which the speed v is increasing, (c) the angular rate $\dot{\beta}$ of the radial line from G to the center of curvature C, and (d) the vector expression for the total acceleration **a** of the rocket.

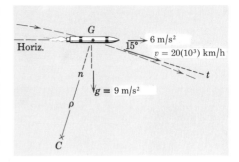

Solution. We observe that the radius of curvature appears in the expression for the normal component of acceleration, so we use n- and t-coordinates to describe the motion of G. The n- and t-components of the total acceleration are obtained by resolving the given horizontal and vertical accelerations into their n- and t-components and then combining. From the figure we get

$$a_n = 9 \cos 15° - 6 \sin 15° = 7.14 \text{ m/s}^2$$
$$a_t = 9 \sin 15° + 6 \cos 15° = 8.12 \text{ m/s}^2$$

(a) We may now compute the radius of curvature from

$$[a_n = v^2/\rho] \qquad \rho = \frac{v^2}{a_n} = \frac{[20(10^3)/3.6]^2}{7.14} = 4.32(10^6) \text{ m} \qquad Ans.$$

(b) The rate at which v is increasing is simply the t-component of acceleration.

$$[\dot{v} = a_t] \qquad\qquad \dot{v} = 8.12 \text{ m/s}^2 \qquad\qquad Ans.$$

(c) The angular rate $\dot{\beta}$ of line GC depends on v and ρ and is given by

$$[v = \rho\dot{\beta}] \qquad \dot{\beta} = v/\rho = \frac{20(10^3)/3.6}{4.32(10^6)} = 12.85(10^{-4}) \text{ rad/s} \qquad Ans.$$

(d) With unit vectors $\mathbf{e}_n$ and $\mathbf{e}_t$ for the n- and t-directions, respectively, the total acceleration becomes

$$\mathbf{a} = 7.14\mathbf{e}_n + 8.12\mathbf{e}_t \text{ m/s}^2 \qquad\qquad Ans.$$

① Alternatively we could find the resultant acceleration and then resolve it into n- and t-components.

② To convert from km/h to m/s multiply by $\dfrac{1000 \text{ m/km}}{3600 \text{ s/h}}$ or divide by 3.6, which is easily remembered.

PROBLEMS

2/90 A particle moves in a circular path of 0.3-m radius. Calculate the magnitude a of the acceleration of the particle (a) if its speed is constant at 0.6 m/s and (b) if its speed is 0.6 m/s but is increasing at the rate of 0.9 m/s each second.

Ans. (a) $a = 1.2$ m/s^2, (b) $a = 1.5$ m/s^2

2/91 A particle P moves in a circular path of 2-m radius. At the instant considered, the speed of the particle is increasing at the rate of 6 m/s^2, and the magnitude of its total acceleration is 10 m/s^2. Determine the speed v of the particle at this instant.

Ans. $v = 4$ m/s

2/92 A racing car traveling at a constant speed negotiates a horizontal turn where the radius of curvature is 300 m. What is the maximum speed v of the car if its normal acceleration cannot exceed $0.8g$ without the car sliding?

2/93 The car passes through a dip in the road at A with a constant speed which gives its mass center G an acceleration equal to $0.5g$. If the radius of curvature of the road at A is 100 m, and if the distance from the road to the mass center G of the car is 0.6 m, determine the speed v of the car.

Ans. $v = 79.5$ km/h

0.6 m

Problem 2/93

2/94 A ship which moves at a steady 20-knot speed (1 knot = 1.852 km/h) executes a turn to port by changing its compass heading at a constant counterclockwise rate. If it requires 60 s to alter course 90°, calculate the magnitude of the acceleration **a** of the ship during the turn.

2/95 Consider the polar axis of the earth to be fixed in space and compute the magnitude of the acceleration **a** of a point P on the earth's surface at latitude 40° north. The mean diameter of the earth is 12 742 km and its angular velocity is $0.729(10^{-4})$ rad/s.

Ans. $a = 0.0259$ m/s^2

Problem 2/95

2/96 To simulate a condition of "weightlessness" in its cabin, a jet transport traveling at 800 km/h moves on a sustained vertical curve as shown. At what rate $\dot{\beta}$ in degrees per second should the pilot drop his longitudinal line of sight to effect the desired condition? The maneuver takes place at a mean altitude of 8 km, and the gravitational acceleration may be taken as 9.79 m/s². *Ans.* $\dot{\beta} = 2.52$ deg/s

Problem 2/96

2/97 The velocity of a particle in plane curvilinear motion at a certain instant is given by $\mathbf{v} = 3\mathbf{i} + 4\mathbf{j}$ m/s. At this instant the radius of curvature of the path of the particle is 6.25 m, and the speed of the particle is decreasing at the rate of 2 m/s each second. Write the expressions for the velocity and acceleration of the particle using the path-variable unit vectors $\mathbf{e}_n$ and $\mathbf{e}_t$.

2/98 Write the vector expression for the acceleration $\mathbf{a}$ of the mass center G of the simple pendulum in both n-t and x-y coordinates for the instant when $\theta = 60°$ if $\dot{\theta} = 2.00$ rad/s and $\ddot{\theta} = 2.45$ rad/s².

Problem 2/98

2/99 Pin P in the crank PO engages the horizontal slot in the guide C and controls its motion on the fixed vertical rod. Determine the velocity $\dot{y}$ and the acceleration $\ddot{y}$ of guide C for a given value of θ if (a) $\dot{\theta} = \omega$ and $\ddot{\theta} = 0$ and (b) if $\dot{\theta} = 0$ and $\ddot{\theta} = \alpha$.
Ans. (a) $\dot{y} = r\omega \sin \theta$, $\ddot{y} = r\omega^2 \cos \theta$
(b) $\dot{y} = 0$, $\ddot{y} = r\alpha \sin \theta$

2/100 If the vertical rod of Prob. 2/99 is replaced by an elevating screw which raises the horizontal slotted guide C at a constant rate $\dot{y} = v_0$, determine the n- and t-components of the acceleration of pin P in its circular motion for a given value of θ.

Problem 2/99

Problem 2/101

2/101 The car *C* increases its speed at the constant rate of 1.5 m/s² as it rounds the curve shown. If the magnitude of the total acceleration of the car is 2.5 m/s² at point *A* where the radius of curvature is 200 m, compute the speed *v* of the car at this point.   *Ans. v = 72 km/h*

2/102 A car is traveling around a circular track of 240-m radius. If the magnitude of its total acceleration is 3 m/s² at the instant when its speed is 75 km/h, determine the rate at which the car is changing its speed.

Problem 2/103

2/103 At the bottom *A* of the vertical inside loop the magnitude of the total acceleration of the airplane is 3*g*. If the airspeed is 800 km/h and is increasing at the rate of 20 km/h per second, calculate the radius of curvature *ρ* of the path at *A*.
                      Ans. ρ = 1709 m

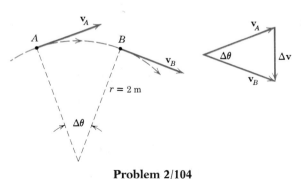

Problem 2/104

2/104 A particle moves on a circular path of radius *r* = 2 m with a constant speed of 10 m/s. During the interval from *A* to *B* the velocity undergoes a vector change **Δv**. Divide this change by the time interval between the two points to obtain the average normal acceleration for (*a*) Δθ = 30°, (*b*) Δθ = 15°, and (*c*) Δθ = 5°. Compare the results with the instantaneous normal acceleration.

2/105 Magnetic tape is being transferred from reel A to reel B and passes around idler pulleys C and D. At a certain instant point P_1 on the tape is in contact with pulley C and point P_2 is in contact with pulley D. If the normal component of acceleration of P_1 is 40 m/s^2 and the tangential component of acceleration of P_2 is 30 m/s^2 at this instant, compute the corresponding speed v of the tape, the magnitude of the total acceleration of P_1, and the magnitude of the total acceleration of P_2.

$\qquad$ *Ans.* $v = 2$ m/s, $a_1 = 50$ m/s^2, $a_2 = 85.4$ m/s^2

Problem 2/105

2/106 The command module of a lunar mission is orbiting the moon in a circular path at an altitude of 200 km above the moon's surface. Consult Table D/2, Appendix D, for any needed information about the moon and compute the magnitude of the orbital velocity $\mathbf{v}$ of the module with respect to the moon.

2/107 A satellite which moves in a circular orbit around the earth at a height $h = 319$ km above the surface of the earth must have a velocity of 27 790 km/h. Calculate the gravitational acceleration g for that altitude. The mean radius R of the earth is 6371 km. (Check your answer by computing g from the gravitational law, which gives $g = g_0(R/[R + h])^2$ where $g_0 = 9.821$ m/s^2 from Table D/2 in Appendix D.) $\qquad$ *Ans.* $g = 8.91$ m/s^2

2/108 A scientific satellite is injected into an elliptical orbit about the earth and has a velocity of 32 000 km/h at its point of nearest approach (perigee) 400 km above the earth. Calculate the radius of curvature ρ of the orbit at this point. The absolute gravitational acceleration at the surface of the earth is 9.821 m/s^2, and the mean diameter of the earth is 12 742 km.

2/109 An earth satellite which moves in the elliptical equatorial orbit shown has a velocity v in space of 17 970 km/h when it passes the end of the semiminor axis at A. The earth has an absolute surface value of g of 9.821 m/s^2 and has a radius of 6371 km. Determine the radius of curvature ρ of the orbit at A. $\qquad$ *Ans.* $\rho = 18\,480$ km

Problem 2/109

$u = 600$ m/s

60°

Horizontal

Problem 2/110

a_1

30°

g

Problem 2/111

Vertical

100 km/h

1
⎕
5

Problem 2/112

2/110 A bullet is fired from the gun with a muzzle velocity of 600 m/s in a direction 60° above the horizontal. The acceleration of the bullet has two components: one in the direction opposite to the velocity is due to air resistance and the other, gravitational acceleration $g = 9.81$ m/s², is due to the weight of the bullet. Compute the radius of curvature ρ of the path of the bullet just after it leaves the gun.

Ans. $\rho = 73.4$ km

2/111 A rocket traveling above the atmosphere at an altitude of 500 km would have a free-fall acceleration $g = 8.43$ m/s² in the absence of forces other than gravitational attraction. Because of thrust, however, the rocket has an additional acceleration component a_1 of 8.80 m/s² tangent to its trajectory, which makes an angle of 30° with the vertical at the instant considered. If the velocity v of the rocket is 30 000 km/h at this position, compute the radius of curvature ρ of the trajectory and the rate at which v is changing with time.

2/112 The muzzle velocity of a certain rifle is 600 m/s. If the rifle is fired with the barrel pointed vertically up from a vehicle traveling down a 20-percent grade at a speed of 100 km/h, determine the radius of curvature ρ of the path of the bullet at its maximum altitude. Neglect air resistance.

Ans. $\rho = 75.6$ m

2/113 Magnetic tape runs over the idler pulley in a computer as shown. If the total acceleration of a point *P* on the tape in contact with the pulley makes an angle of 4° with the tangent to the tape at time $t = 0$ when the velocity v of the tape is 4 m/s, determine the time t required to bring the pulley to a stop with constant deceleration. Assume no slipping between the pulley and the tape.

Problem 2/113

2/114 A particle moving in the *x-y* plane has a position vector given by $\mathbf{r} = \frac{3}{2}t^2\mathbf{i} + \frac{2}{3}t^3\mathbf{j}$ where $\mathbf{r}$ is in meters and t is in seconds. Calculate the radius of curvature ρ of the path for the position of the particle when $t = 2$ s. Sketch the velocity $\mathbf{v}$ and the curvature of the path for this particular instant.

Ans. $\rho = 41.7$ m

2/115 A race driver traveling at a speed of 250 km/h on the straightaway applies his brakes at point *A* and reduces his speed at a uniform rate to 200 km/h at *C* in a distance of $150 + 150 = 300$ m. Calculate the magnitude of the total acceleration of the race car an instant after it passes point *B*.

Ans. $a = 8.42$ m/s²

Problem 2/115

2/116 A projectile is fired at an angle of 30° above the horizontal with a muzzle velocity of 460 m/s. Determine the radius of curvature ρ of the trajectory at the position of the particle 10 seconds after firing. Neglect atmospheric resistance so that the gravitational attraction is the only force to be considered. The acceleration of the projectile, consequently, is *g* down.

2/117 For an interval of its motion the slotted guide of Prob. 2/99 is given an upward acceleration $\ddot{y} = 25$ m/s². At the instant when $\theta = 60°$, the velocity of the guide is $\dot{y} = 2$ m/s. Determine the tangential acceleration a_t of the pin *P* at this instant in its circular motion of radius $r = 160$ mm.

Ans. $a_t = 9.62$ m/s²

Problem 2/118

Problem 2/119

Problem 2/121

2/118 The mine skip is being hauled to the surface over the curved track by the cable around the 750-mm drum which turns at the constant clockwise speed of 120 rev/min. The shape of the track is defined by $y = x^2/16$ where x and y are in meters. Calculate the magnitude of the total acceleration of the skip as it reaches a level of 1 m below the top. Neglect the dimensions of the skip compared with those of the path. (Recall that the radius of curvature is given by $\rho = [1 + (dy/dx)^2]^{3/2}/(d^2y/dx^2)$.)

Ans. $a = 1.986$ m/s^2

2/119 The pin P is constrained to move in the slotted guides which move at right angles to one another. At the instant represented, A has a velocity to the right of 0.2 m/s which is decreasing at the rate of 0.75 m/s each second. At the same time B is moving down with a velocity of 0.15 m/s which is decreasing at the rate of 0.5 m/s each second. For this instant determine the radius of curvature ρ of the path followed by P. Is it possible to determine also the time rate of change of ρ?

Ans. $\rho = 1.25$ m

▶**2/120** A particle starts from rest at the origin and moves along the positive branch of the curve $y = 2x^{3/2}$ so that the distance s measured from the origin along the curve varies with the time t according to $s = 2t^3$, where x, y, and s are in millimeters and t is in seconds. Find the magnitude of the total acceleration **a** of the particle when $t = 1$ s. (Refer to Art. C/10 of Appendix C for an expression for ρ.)

Ans. $a = 12.17$ mm/s^2

▶**2/121** The horizontal plunger A, which operates the 70° bell crank BOC, has a velocity to the right of 75 mm/s and is speeding up at the rate of 100 mm/s per second at the position for which $\theta = 30°$. Compute the angular acceleration $\ddot{\theta}$ of the bell crank at this instant.

Ans. $\ddot{\theta} = -0.399$ rad/s^2

2/6 POLAR COORDINATES (*r-θ*)

We now consider the third description of plane curvilinear motion, namely, polar coordinates where the particle is located by the radial distance *r* from a fixed pole and by an angular measurement *θ* to the radial line. Polar coordinates are particularly useful when a motion is constrained through the control of a radial distance and an angular position or when an unconstrained motion is observed by measurements of a radial distance and an angular position.

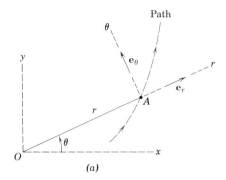

(a)

Figure 2/12*a* shows the polar coordinates *r* and *θ* which locate a particle traveling on a curved path. An arbitrary fixed line, such as the *x*-axis, is used as a reference for the measurement of *θ*. Unit vectors $\mathbf{e}_r$ and $\mathbf{e}_\theta$ are established in the positive *r*- and *θ*-directions, respectively. The position vector **r** to the particle at *A* has a magnitude equal to the radial distance *r* and a direction specified by the unit vector $\mathbf{e}_r$. Thus we express the location of the particle at *A* by the vector

$$\mathbf{r} = r\mathbf{e}_r$$

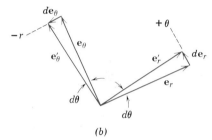

(b)

Figure 2/12

As we differentiate this relation with respect to time to obtain $\mathbf{v} = \dot{\mathbf{r}}$ and $\mathbf{a} = \dot{\mathbf{v}}$, we will need expressions for the time derivatives of both unit vectors $\mathbf{e}_r$ and $\mathbf{e}_\theta$. We obtain $\dot{\mathbf{e}}_r$ and $\dot{\mathbf{e}}_\theta$ in exactly the same way that we derived $\dot{\mathbf{e}}_t$ in the preceding article. During time *dt* the coordinate directions rotate through the angle *dθ*, and the unit vectors also rotate through the same angle from $\mathbf{e}_r$ and $\mathbf{e}_\theta$ to $\mathbf{e}_r'$ and $\mathbf{e}_\theta'$ as shown in Fig. 2/12*b*. We note that the vector change $d\mathbf{e}_r$ is in the plus *θ*-direction and that $d\mathbf{e}_\theta$ is in the minus *r*-direction. Since their magnitudes in the limit are equal to the unit vector as radius times the angle *dθ* in radians, we may write them as $d\mathbf{e}_r = \mathbf{e}_\theta\, d\theta$ and $d\mathbf{e}_\theta = -\mathbf{e}_r\, d\theta$. If we divide these equations by *dθ* we have

$$\frac{d\mathbf{e}_r}{d\theta} = \mathbf{e}_\theta \qquad \text{and} \qquad \frac{d\mathbf{e}_\theta}{d\theta} = -\mathbf{e}_r$$

Or, if we divide them by *dt*, we have $d\mathbf{e}_r/dt = (d\theta/dt)\mathbf{e}_\theta$ and $d\mathbf{e}_\theta/dt = -(d\theta/dt)\mathbf{e}_r$ or simply

$$\dot{\mathbf{e}}_r = \dot{\theta}\mathbf{e}_\theta \qquad \text{and} \qquad \dot{\mathbf{e}}_\theta = -\dot{\theta}\mathbf{e}_r \qquad\qquad \textbf{(2/12)}$$

We are now ready to differentiate $\mathbf{r} = r\mathbf{e}_r$ with respect to time. Using the rule for differentiating the product of a scalar and a vector gives

$$\mathbf{v} = \dot{\mathbf{r}} = \dot{r}\mathbf{e}_r + r\dot{\mathbf{e}}_r$$

With the substitution of $\dot{\mathbf{e}}_r$ from Eq. 2/12 the vector expression for the velocity becomes

$$\mathbf{v} = \dot{r}\mathbf{e}_r + r\dot{\theta}\mathbf{e}_\theta \qquad (2/13)$$

where

$$v_r = \dot{r}$$

$$v_\theta = r\dot{\theta}$$

$$v = \sqrt{v_r{}^2 + v_\theta{}^2}$$

The r-component of $\mathbf{v}$ is merely the rate at which the vector $\mathbf{r}$ stretches. The θ-component of $\mathbf{v}$ is due to the rotation of $\mathbf{r}$.

We now differentiate the expression for $\mathbf{v}$ to obtain the acceleration $\mathbf{a} = \dot{\mathbf{v}}$. Note that the derivative of $r\dot{\theta}\mathbf{e}_\theta$ will produce three terms since all three factors are variable. Thus

$$\mathbf{a} = \dot{\mathbf{v}} = (\ddot{r}\mathbf{e}_r + \dot{r}\dot{\mathbf{e}}_r) + (\dot{r}\dot{\theta}\mathbf{e}_\theta + r\ddot{\theta}\mathbf{e}_\theta + r\dot{\theta}\dot{\mathbf{e}}_\theta)$$

Substitution of $\dot{\mathbf{e}}_r$ and $\dot{\mathbf{e}}_\theta$ from Eq. 2/12 and collecting terms give

$$\mathbf{a} = (\ddot{r} - r\dot{\theta}^2)\mathbf{e}_r + (r\ddot{\theta} + 2\dot{r}\dot{\theta})\mathbf{e}_\theta \qquad (2/14)$$

where

$$a_r = \ddot{r} - r\dot{\theta}^2$$

$$a_\theta = r\ddot{\theta} + 2\dot{r}\dot{\theta}$$

$$a = \sqrt{a_r{}^2 + a_\theta{}^2}$$

We may write the θ-component alternatively as

$$a_\theta = \frac{1}{r}\frac{d}{dt}(r^2\dot{\theta})$$

which can be verified easily by carrying out the differentiation. This form for a_θ will be found useful when we treat the angular momentum of particles in the next chapter.

An adequate appreciation of the terms in Eq. 2/14 comes only when the geometry of the physical changes can be clearly seen. For this purpose Fig. 2/13*a* is developed to show the velocity vectors and their r- and θ-components at position A and at position A' after an infinitesimal movement. Each of these components undergoes a change in magnitude and direction as shown in Fig. 2/13*b*. In this figure we see the following changes:

Magnitude change of v$_r$. This change is simply the increase in length of v_r or $dv_r = d\dot{r}$, and the corresponding acceleration term is $d\dot{r}/dt = \ddot{r}$ in the positive r-direction.

Direction change of v$_r$. The magnitude of this change is seen from the figure to be $v_r \, d\theta = \dot{r} \, d\theta$, and its contribution to the acceleration becomes $\dot{r} \, d\theta/dt = \dot{r}\dot{\theta}$ which is in the positive θ-direction.

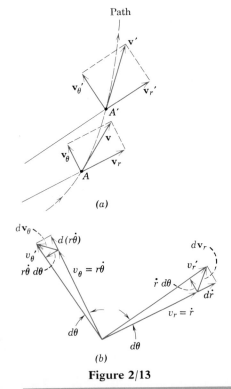

(a)

(b)

Figure 2/13

Magnitude change of $\mathbf{v}_\theta$. This term is the change in length of $\mathbf{v}_\theta$ or $d(r\dot\theta)$, and its contribution to the acceleration is $d(r\dot\theta)/dt = r\ddot\theta + \dot r\dot\theta$ and is in the positive θ-direction.

Direction change of $\mathbf{v}_\theta$. The magnitude of this change is $v_\theta\,d\theta = r\dot\theta\,d\theta$, and the corresponding acceleration term is seen to be $r\dot\theta(d\theta/dt) = r\dot\theta^2$ in the negative r-direction.

Collecting terms gives $a_r = \ddot r - r\dot\theta^2$ and $a_\theta = r\ddot\theta + 2\dot r\dot\theta$ as obtained previously. We see that the term $\ddot r$ is the acceleration which the particle would have along the radius in the absence of a change in θ. The term $-r\dot\theta^2$ is the normal component of acceleration if r were constant as in circular motion. The term $r\ddot\theta$ is the tangential acceleration which the particle would have if r were constant but is only a part of the acceleration due to the change in magnitude of $\mathbf{v}_\theta$ when r is variable. Lastly, the term $2\dot r\dot\theta$ is composed of two effects. The first effect comes from that portion of the change in magnitude $d(r\dot\theta)$ of v_θ due to the change in r, and the second effect comes from the change in direction of $\mathbf{v}_r$. The term $2\dot r\dot\theta$ represents, therefore, a combination of changes and is not so easily perceived as are the other acceleration terms.

We note carefully the difference between the vector change $d\mathbf{v}_r$ in $\mathbf{v}_r$ and the change dv_r in the magnitude of v_r. Similarly, the vector change $d\mathbf{v}_\theta$ is not the same as the change dv_θ in the magnitude of v_θ. When we divide these changes by dt to obtain expressions for the derivatives, we see clearly that the magnitude of the derivative $|d\mathbf{v}_r/dt|$ and the derivative of the magnitude dv_r/dt are *not* the same. We also note carefully that a_r is not $\dot v_r$ and that a_θ is not $\dot v_\theta$.

The total acceleration $\mathbf{a}$ and its components are represented in Fig. 2/14. If $\mathbf{a}$ has a component normal to the path, we know from our analysis of n- and t-components in Art. 2/5 that the sense of the n-component *must* be toward the center of curvature.

For motion in a circular path with r constant the components of Eqs. 2/13 and 2/14 become simply

$$v_r = 0 \qquad\qquad v_\theta = r\dot\theta$$
$$a_r = -r\dot\theta^2 \qquad a_\theta = r\ddot\theta$$

This description is the same as that obtained with n- and t-components where the θ- and t-directions coincide but where the positive r-direction is in the negative n-direction. Hence $a_r = -a_n$ for circular motion.

The expressions for a_r and a_θ in scalar form may also be obtained by direct differentiation of the coordinate relations $x = r\cos\theta$ and $y = r\sin\theta$ to get $a_x = \ddot x$ and $a_y = \ddot y$. Each of these rectangular components of acceleration may then be resolved into r- and θ-components which, when combined, will yield the expressions of Eq. 2/14.

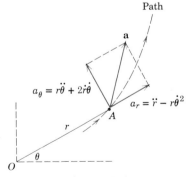

Figure 2/14

Sample Problem 2/9

Rotation of the radially slotted arm is governed by $\theta = 0.2t + 0.02t^3$ where θ is in radians and t is in seconds. Simultaneously the power screw in the arm engages the slider B and controls its distance from O according to $r = 0.2 + 0.04t^2$ where r is in meters and t is in seconds. Calculate the magnitudes of the velocity and acceleration of the slider for the instant when $t = 3$ s.

Solution. The coordinates and their time derivatives which appear in the expressions for velocity and acceleration in polar coordinates are obtained first and evaluated for $t = 3$ s.

$r = 0.2 + 0.04t^2 \qquad r_3 = 0.2 + 0.04(3^2) = 0.56$ m

$\dot{r} = 0.08t \qquad \dot{r}_3 = 0.08(3) = 0.24$ m/s

$\ddot{r} = 0.08 \qquad \ddot{r}_3 = 0.08$ m/s^2

$\theta = 0.2t + 0.02t^3 \quad \theta_3 = 0.2(3) + 0.02(3^3) = 1.14$ rad

$\qquad\qquad\qquad$ or $\theta_3 = 1.14(180/\pi) = 65.3°$

$\dot{\theta} = 0.2 + 0.06t^2 \quad \dot{\theta}_3 = 0.2 + 0.06(3^2) = 0.74$ rad/s

$\ddot{\theta} = 0.12t \qquad \ddot{\theta}_3 = 0.12(3) = 0.36$ rad/s^2

The velocity components are obtained from Eq. 2/13 and for $t = 3$ s are

$[v_r = \dot{r}] \qquad\qquad v_r = 0.24$ m/s

$[v_\theta = r\dot{\theta}] \qquad\qquad v_\theta = 0.56(0.74) = 0.414$ m/s

$[v = \sqrt{v_r^2 + v_\theta^2}] \qquad v = \sqrt{(0.24)^2 + (0.414)^2} = 0.479$ m/s $\qquad$ *Ans.*

The velocity and its components are shown for the specified position of the arm.

The acceleration components are obtained from Eq. 2/14 and for $t = 3$ s are

$[a_r = \ddot{r} - r\dot{\theta}^2] \qquad a_r = 0.08 - 0.56(0.74)^2 = -0.227$ m/s^2

$[a_\theta = r\ddot{\theta} + 2\dot{r}\dot{\theta}] \qquad a_\theta = 0.56(0.36) + 2(0.24)(0.74) = 0.557$ m/s^2

$[a = \sqrt{a_r^2 + a_\theta^2}] \qquad a = \sqrt{(-0.227)^2 + (0.557)^2} = 0.601$ m/s^2 $\quad$ *Ans.*

The acceleration and its components are also shown for the 65.3° position of the arm.

① We see that this problem is an example of constrained motion where the center B of the slider is mechanically constrained by the rotation of the slotted arm and by engagement with the turning screw.

Sample Problem 2/10

A tracking radar lies in the vertical plane of the path of a rocket which is coasting in unpowered flight above the atmosphere. For the instant when $\theta = 30°$ the tracking data give $r = 8(10^4)$ m, $\dot{r} = 1200$ m/s, and $\dot{\theta} = 0.80$ °/s. The acceleration of the rocket is due only to gravitational attraction and for its particular altitude is 9.20 m/s² vertically down. For these conditions determine the velocity v of the rocket and the values of $\ddot{r}$ and $\ddot{\theta}$.

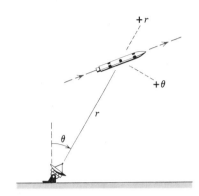

Solution. The components of velocity from Eq. 2/13 are

$$[v_r = \dot{r}] \qquad v_r = 1200 \text{ m/s}$$

① $[v_\theta = r\dot{\theta}] \qquad v_\theta = 8(10^4)(0.80)\left(\frac{\pi}{180}\right) = 1117 \text{ m/s}$

$[v = \sqrt{v_r^2 + v_\theta^2}] \qquad v = \sqrt{(1200)^2 + (1117)^2} = 1639 \text{ m/s}$ *Ans.*

Since the total acceleration of the rocket is $g = 9.20$ m/s² down, we can easily find its r- and θ-components for the given position. As shown in the figure they are

② $$a_r = -9.20 \cos 30° = -7.97 \text{ m/s}^2$$
$$a_\theta = 9.20 \sin 30° = 4.60 \text{ m/s}^2$$

We now equate these values to the polar-coordinate expressions for a_r and a_θ which contain the unknowns $\ddot{r}$ and $\ddot{\theta}$. Thus from Eq. 2/14

③ $[a_r = \ddot{r} - r\dot{\theta}^2] \qquad -7.97 = \ddot{r} - 8(10^4)\left(0.80 \frac{\pi}{180}\right)^2$

$$\ddot{r} = 7.63 \text{ m/s}^2$$

$[a_\theta = r\ddot{\theta} + 2\dot{r}\dot{\theta}] \qquad 4.60 = 8(10^4)\ddot{\theta} + 2(1200)\left(0.80 \frac{\pi}{180}\right)$

$$\ddot{\theta} = -3.61(10^{-4}) \text{ rad/s}^2 \qquad Ans.$$

① We observe that the angle θ in polar coordinates need not always be taken positive in a counterclockwise sense.

② Note that the r-component of acceleration is in the negative r-direction, so it carries a minus sign.

③ We must be careful to convert $\dot{\theta}$ from °/s to rad/s.

Problem 2/122

PROBLEMS

2/122 The position of the slider P in the rotating slotted arm OA is controlled by a power screw as shown. At the instant represented $\dot{\theta} = 8$ rad/s and $\ddot{\theta} = -20$ rad/s². Also at this same instant $r = 200$ mm, $\dot{r} = -300$ mm/s, and $\ddot{r} = 0$. For this instant determine the r- and θ-components of the acceleration of P.   *Ans.* $a_r = -12.8$ m/s², $a_\theta = -8.80$ m/s²

2/123 The polar coordinates of a particle are given by $r = 2 + 5t^2/2$ and $\theta = 0.5 + 3t^2/4$ where r is in millimeters, θ is in radians, and t is in seconds. Determine the magnitude of the velocity **v** and the r- and θ-components of the acceleration **a** of the particle when $t = 2$ s.
$$\text{*Ans.* } v = 37.4 \text{ mm/s}, \quad a_r = -103 \text{ mm/s}^2$$
$$a_\theta = 78 \text{ mm/s}^2$$

2/124 A pin which is constrained by two machine parts which it joins has plane curvilinear motion governed by its polar coordinates $r = r_0 e^{kt}$ and $\theta = Ct$ where r_0, k, and C are constants and θ is in radians. Determine the expressions for the magnitudes of the velocity **v** and acceleration **a** of the pin in terms of the time t.
$$\text{*Ans.* } v = r_0 e^{kt} \sqrt{k^2 + C^2}, \quad a = r_0(k^2 + C^2)e^{kt}$$

2/125 A particle P moves on a curved path defined by polar coordinates. At a certain instant when $\theta = 5\pi/4$, the θ-component of acceleration is 16 mm/s². Also $r = 5$ mm, $\ddot{r} = 6$ mm/s², and $\dot{\theta} = 0.5$ rad/s and is increasing at the rate of 4 rad/s each second. Use whatever information you need and compute the magnitude of the velocity of the particle.

2/126 At a certain instant the polar coordinates of a particle are $r = 60$ mm and $\theta = 30°$. At this same instant $\dot{r} = 20\sqrt{3}$ mm/s and $\dot{x} = 20$ mm/s. Determine the magnitude of the velocity **v** of the particle and the value of $\dot{\theta}$ for this position.
$$\text{*Ans.* } v = 40 \text{ mm/s}, \quad \dot{\theta} = 0.333 \text{ rad/s}$$

2/127 The curvilinear motion of a particle is governed by the polar coordinates $r = t^3/3$ and $\theta = 2 \cos (\pi t/6)$ where r is in meters, θ is in radians, and t is in seconds. Specify the velocity $\mathbf{v}$ and acceleration $\mathbf{a}$ of the particle when $t = 2$ s.

2/128 For a certain curvilinear motion of a particle expressed in polar coordinates the product $r^2\dot{\theta}$ in m²/s varies with the time t in seconds measured over a short period as shown. Approximate the θ-component of the acceleration of the particle at the instant when $t = 5$ s at which time $r = 1/3$ m.

Ans. $a_\theta = -8$ m/s²

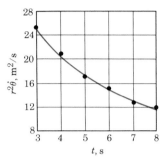

Problem 2/128

2/129 A satellite m moves in an elliptical orbit around the earth. There is no force on the satellite in the θ-direction, so that $a_\theta = 0$. Prove Kepler's second law of planetary motion which says that the radial line r sweeps through equal areas in equal times. The area dA swept by the radial line during time dt is shaded in the figure.

Problem 2/129

2/130 A particle P moves along a path, given by $r = f(\theta)$, which is symmetrical about the line $\theta = 0$. As the particle passes the position $\theta = 0$ where the radius of curvature of the path is ρ, the velocity of P is v. Derive an expression for $\ddot{r}$ in terms of v, r, and ρ for the motion of the particle at this point.

Ans. $\ddot{r} = -v^2\left(\dfrac{1}{\rho} - \dfrac{1}{r}\right)$

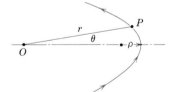

Problem 2/130

2/131 The cam has a shape such that the center of the roller A which follows the contour moves on a limaçon defined by $r = b - c \cos \theta$, where $b > c$. If the cam does not rotate, determine the magnitude of the total acceleration of A in terms of θ if the slotted arm revolves with a constant counterclockwise angular rate $\dot{\theta} = \omega$.

Problem 2/131

Problem 2/132

Problem 2/133

Problem 2/134

Problem 2/135

2/132 The rocket is fired vertically and tracked by the radar shown. When θ reaches 60°, other corresponding measurements give the values $r = 9$ km, $\ddot{r} = 21$ m/s², and $\dot{\theta} = 0.02$ rad/s. Calculate the magnitudes of the velocity and acceleration of the rocket at this position.

Ans. $v = 360$ m/s
$a = 20.1$ m/s²

2/133 A jet plane flying at a constant speed v at an altitude $h = 10$ km is being tracked by radar located at O directly below the line of flight. If the angle θ is decreasing at the rate of 0.020 rad/s when $\theta = 60°$, determine the value of $\ddot{r}$ at this instant and the magnitude of the velocity **v** of the plane.

2/134 The path of a fluid particle P in a certain centrifugal pump with radial vanes is to be approximated by the spiral $r = r_0 e^{n\theta}$, where n is a dimensionless constant. If the pump turns at a constant rate $\dot{\theta} = K$, determine the expression for the total acceleration of the particle just prior to leaving the vane in terms of R, K, and n.   *Ans.* $a = RK^2(n^2 + 1)$

2/135 The piston of the hydraulic cylinder gives pin A a constant velocity $v = 0.9$ m/s in the direction shown for an interval of its motion. For the instant when $\theta = 60°$, determine $\dot{r}$, $\ddot{r}$, $\dot{\theta}$, and $\ddot{\theta}$ where $r = \overline{OA}$.

2/136 If the piston of the hydraulic cylinder of Prob. 2/135 starts from rest with an acceleration $\dot{v} = 1.2$ m/s² when $\theta = 60°$, determine $\dot{r}$, $\ddot{r}$, $\dot{\theta}$, and $\ddot{\theta}$ for this instant.

      Ans. $\dot{r} = 0$, $\ddot{r} = -0.6$ m/s², $\dot{\theta} = 0$, $\ddot{\theta} = 6$ rad/s²

2/137 The fixed horizontal guide carries a slider and pin P whose motion is controlled by the rotating slotted arm OA. If the arm is revolving about O at the constant rate $\dot{\theta} = 2$ rad/s for an interval of its motion, determine the magnitudes of the velocity and acceleration of the slider in the slot for the instant when $\theta = 60°$. Also find the r-components of the velocity and acceleration.

Problem 2/137

2/138 The slotted arm OA forces the small pin to move in the fixed spiral guide defined by $r = K\theta$. Arm OA starts from rest at $\theta = \pi/4$ and has a constant counterclockwise angular acceleration $\ddot{\theta} = \alpha$. Determine the magnitude of the acceleration of the pin when $\theta = 3\pi/4$.

      Ans. $a = 10.76K\alpha$

Problem 2/138

2/139 The circular disk rotates about its center O with a constant angular velocity $\omega = \dot{\theta}$ and carries the two spring-loaded plungers shown. The distance b which each plunger protrudes from the rim of the disk varies according to $b = b_0 \sin 2\pi n t$ where b_0 is the maximum protrusion, n is the constant frequency of oscillation of the plungers in the radial slots, and t is the time. Determine the maximum magnitudes of the r- and θ-components of the acceleration of the ends A of the plungers during their motion.

      Ans. $|a_r|_{\max} = (4\pi^2 n^2 + \omega^2)b_0 + r_0\omega^2$
                 $|a_\theta|_{\max} = 4\pi b_0 n\omega$

Problem 2/139

Problem 2/140

Problem 2/141

2/140 The earth satellite of Prob. 2/109, shown again here, has a velocity $v = 17\,970$ km/h as it passes the end of the semiminor axis at A. Gravitational attraction produces an acceleration $a = a_r = -1.556$ m/s² as calculated from the gravitational law. For this position calculate the rate $\dot{v}$ at which the speed of the satellite is changing and the quantity $\ddot{r}$. *Ans.* $\dot{v} = -0.778$ m/s², $\ddot{r} = -0.388$ m/s²

2/141 A rocket follows a trajectory in the vertical plane and is tracked by radar from point A. At a certain instant the radar measurements give $r = 10.5$ km, $\dot{r} = 480$ m/s, $\dot{\theta} = 0$, and $\ddot{\theta} = -0.00720$ rad/s². Sketch the position of the rocket for this instant and determine the radius of curvature ρ of the trajectory at this position of the rocket.
Ans. $\rho = 3.05$ km

2/142 If the radar-tracked rocket of Prob. 2/141 reaches a velocity of 360 m/s when $\theta = 45°$ at which time $r = 5.4$ km, $\dot{r} = 312$ m/s, $\ddot{r} = 16.86$ m/s², $a_\theta = -4.8$ m/s², and θ is increasing, determine a_r and $\ddot{\theta}$.

2/143 At a certain instant a particle which moves on a curved path has coordinates $x = 8$ m and $y = 6$ m and velocity components $\dot{x} = -3$ m/s and $\dot{y} = 4$ m/s. Also at this instant $a_r = 10$ m/s² and $a_\theta = 15$ m/s². For this instant determine $\dot{r}$, $\ddot{r}$, $\dot{\theta}$, $\ddot{\theta}$, and the radius of curvature ρ of the path. (*Suggestion:* Sketch the geometry of your solution as you proceed.)
Ans. $\dot{r} = 0$, $\ddot{r} = 12.5$ m/s², $\dot{\theta} = 0.5$ rad/s, $\ddot{\theta} = 1.5$ rad/s², $\rho = 2.5$ m

2/144 A particle moves along a curved path defined by polar coordinates r and θ. For a certain position of the particle $r = 0.6$ m, $\dot{r} = 1.2$ m/s, $\ddot{r} = 4.8$ m/s², $\theta = \pi/3$ rad, $\dot{\theta} = -2$ rad/s, and $\ddot{\theta} = 12$ rad/s². Determine the magnitude of the velocity **v** and find the r- and θ-components of acceleration. Also specify the acceleration a_t tangent to the path. Sketch the velocity and acceleration vectors.

2/145 A rocket is tracked by radar from its launching point *A*. When it is 10 seconds into its flight, the following radar measurements are recorded: $r = 2200$ m, $\dot{r} = 500$ m/s, $\ddot{r} = 4.66$ m/s^2, $\theta = 22°$, $\dot{\theta} = 0.0788$ rad/s, and $\ddot{\theta} = -0.0341$ rad/s^2. For this instant determine the angle β between the horizontal and the direction of the trajectory of the rocket and find the magnitudes of its velocity **v** and acceleration **a**.

Ans. $\beta = 48.9°$, $v = 529$ m/s, $a = 9.76$ m/s^2

Problem 2/145

2/146 For an interval of motion the drum of radius *b* turns clockwise at a constant rate ω in radians per second and causes the carriage *P* to move to the right as the unwound length of the connecting cable is shortened. Use polar coordinates *r* and θ and derive expressions for the velocity *v* and acceleration *a* of *P* in the horizontal guide in terms of the angle θ. Check your solution by a direct differentiation with time of the relation $x^2 + h^2 = r^2$.

Problem 2/146

2/147 The slotted arm *OA* oscillates about *O* within the limits shown and drives the crank *CP* through the pin *P*. For an interval of the motion $\dot{\theta} = K$, a constant. Determine the magnitude of the corresponding total acceleration of *P* for any value of θ within the range for which $\dot{\theta} = K$. Use polar coordinates *r* and θ. Show that the magnitudes of the velocity and acceleration of *P* in its circular path are constant. Ans. $v = 2bK$, $a = 4bK^2$

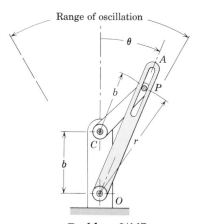

Range of oscillation

Problem 2/147

Problem 2/148

2/148 During a portion of a vertical loop an airplane flies in an arc of radius $\rho = 600$ m with a constant speed $v = 400$ km/h. When the airplane is at A, the angle made by **v** with the horizontal is $\beta = 30°$, and radar tracking gives $r = 800$ m and $\theta = 30°$. Calculate v_r, v_θ, a_r, and $\ddot\theta$ for this instant.

2/149 At a certain instant the coordinates and their time derivatives for a particle moving in plane curvilinear motion are $x = 100$ mm, $y = 75$ mm, $\dot x = 75$ mm/s, and $\dot y = 100$ mm/s. At this same instant the particle has a radial acceleration $a_r = 250$ mm/s^2, and its θ-acceleration is zero. For this instant determine the magnitude of its velocity **v**, the values of $\ddot r$ and $\ddot\theta$, the radius of curvature ρ of the path, and the acceleration component a_t tangent to the path. (*Suggestion:* Sketch the geometry of your solution as you proceed.)

> *Ans.* $v = 125$ mm/s, $\ddot r = 260$ mm/s^2
> $\ddot\theta = -0.538$ rad/s^2, $\rho = 223$ mm
> $a_t = 240$ mm/s^2

Problem 2/150

2/150 Pin A moves in a circle of 90-mm radius as crank AC revolves at the constant rate $\dot\beta = 60$ rad/s. The slotted link rotates about point O as the rod attached to A moves in and out of the slot. For the position $\beta = 30°$, determine $\dot r$, $\ddot r$, $\dot\theta$, and $\ddot\theta$.

> *Ans.* $\dot r = 3.58$ m/s, $\ddot r = 315$ m/s^2, $\dot\theta = 17.86$ rad/s
> $\ddot\theta = -1510$ rad/s^2

2/151 If crank CA of Prob. 2/150 starts from rest at $\beta = 30°$ with an angular acceleration $\ddot\beta = 8$ rad/s^2, determine the values of $\ddot r$ and $\ddot\theta$ at this instant.

2/152 For the mechanism of Prob. 2/147 the slotted arm OA is given a periodic oscillation governed by $\theta = \theta_0 \sin(2\pi t/\tau)$ where τ is the period (time for one complete oscillation) and t is the time. The total range of oscillation is $2\theta_0$. Determine the expression for the magnitude of the acceleration of the pin P when (*a*) $t = \tau/4$ and (*b*) $t = \tau/2$.

> *Ans.* (*a*) $a = 8\pi^2\theta_0 b/\tau^2$
> (*b*) $a = -a_r = 16\pi^2\theta_0^2 b/\tau^2$

2/153 During reentry a space capsule A is tracked by radar station B located in the vertical plane of the trajectory. The values of r and θ are read against time and are recorded in the following table. Determine the velocity v of the capsule when $t = 40$ s. Explain how the acceleration of the capsule could be found from the given data.

t, s	r, km	θ, deg	t, s	r, km	θ, deg
0	36.4	110.5	60	19.0	45.0
5	29.9	100.0	70	18.8	38.5
10	26.2	91.0	80	18.7	32.8
15	24.1	83.7	90	18.7	27.0
20	22.7	77.7	100	18.7	21.6
30	20.9	67.7	110	18.8	16.8
40	20.1	58.6	120	19.0	12.0
50	19.3	52.0			

Ans. $v = 1020$ km/h

Problem 2/153

▶**2/154** A centrifugal pump with radial vanes, similar to that illustrated with Prob. 2/134, rotates at the constant rate $\dot{\theta} = K$. An element of the fluid being pumped will be considered here as a smooth particle P which is introduced at the radius r_0 without radial velocity and which moves outward along the vane without friction. With no force on the particle in the radial direction, its acceleration in this direction is zero. Under these conditions, determine the equation of the path of the particle if the time t is zero for $r = r_0$. *Ans.* $r = r_0 \cosh Kt$

▶**2/155** If the slotted arm (Prob. 2/131) is revolving counterclockwise at the constant rate of 40 rev/min and the cam is revolving clockwise at the constant rate of 30 rev/min, determine the magnitude of the acceleration of the center of the roller A when the cam and arm are in the relative position for which $\theta = 30°$. The limaçon has the dimensions $b = 100$ mm and $c = 75$ mm. (*Caution:* Redefine the coordinates as necessary after noting that the θ in the expression $r = b - c \cos \theta$ is not the absolute angle appearing in Eq. 2/14.) *Ans.* $a = 3.68$ m/s^2

40 rev/min

30 rev/min

Problem 2/155

2/7 *SPACE CURVILINEAR MOTION*

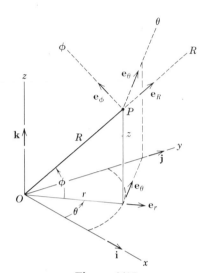

Figure 2/15

The general case of three-dimensional motion of a particle along a space curve was introduced in Art. 2/1 and illustrated in Fig. 2/1. Mention was made of the three coordinate systems, rectangular (x-y-z), cylindrical (r-θ-z), and spherical (R-θ-ϕ), which are commonly used to describe this motion. These systems are indicated in Fig. 2/15, which also shows the unit vectors for the three coordinate systems.

Before describing the use of these coordinate systems we may observe that a path-variable description, using n- and t-coordinates, which we developed in Art. 2/5, may be applied in the osculating plane shown in Fig. 2/1. We defined this plane earlier as the plane which contains the curve at the location in question. We see that the velocity $\mathbf{v}$, which is along the tangent t to the curve, lies in the osculating plane. The acceleration $\mathbf{a}$ also lies in the osculating plane and, as in the case of plane motion, has a component $a_t = \dot{v}$ tangent to the path due to the change in magnitude of the velocity and a component $a_n = v^2/\rho$ normal to the curve due to the change in direction of the velocity. As before ρ is the radius of curvature of the path at the point in question and would be measured in the osculating plane. This description of motion, which we found to be natural and direct for many plane-motion problems, finds little use for space motion because the osculating plane continually shifts its orientation, thus making its use as a reference awkward. We shall confine our attention, therefore, to the three fixed coordinate systems shown in Fig. 2/15.

(*a*) *Rectangular coordinates* (*x-y-z*).   The extension from two to three dimensions offers no particular difficulty. We merely add the z-coordinate and its two time derivatives to the two-dimensional expressions of Eqs. 2/6 so that the position vector $\mathbf{R}$, the velocity $\mathbf{v}$, and the acceleration $\mathbf{a}$ become

$$
\begin{aligned}
\mathbf{R} &= x\mathbf{i} + y\mathbf{j} + z\mathbf{k} \\
\mathbf{v} = \dot{\mathbf{R}} &= \dot{x}\mathbf{i} + \dot{y}\mathbf{j} + \dot{z}\mathbf{k} \\
\mathbf{a} = \dot{\mathbf{v}} = \ddot{\mathbf{R}} &= \ddot{x}\mathbf{i} + \ddot{y}\mathbf{j} + \ddot{z}\mathbf{k}
\end{aligned}
\qquad (2/15)
$$

Note that in three dimensions we are using $\mathbf{R}$ in place of $\mathbf{r}$ for the position vector.

(*b*) *Cylindrical coordinates* (*r-θ-z*).   If we understand the polar-coordinate description of plane motion, then there should be no difficulty with cylindrical coordinates because all that is required is the addition of the z-coordinate and its two time derivatives. The position vector **R** to the particle for cylindrical coordinates is simply

$$\mathbf{R} = r\mathbf{e}_r + z\mathbf{k}$$

In place of Eq. 2/13 for plane motion we may write the velocity as

$$\mathbf{v} = \dot{r}\mathbf{e}_r + r\dot{\theta}\mathbf{e}_\theta + \dot{z}\mathbf{k} \qquad (2/16)$$

where

$$v_r = \dot{r}$$

$$v_\theta = r\dot{\theta}$$

$$v_z = \dot{z}$$

$$v = \sqrt{v_r^2 + v_\theta^2 + v_z^2}$$

Similarly, the acceleration is written by adding the *z*-component to Eq. 2/14, which gives us

$$\mathbf{a} = (\ddot{r} - r\dot{\theta}^2)\mathbf{e}_r + (r\ddot{\theta} + 2\dot{r}\dot{\theta})\mathbf{e}_\theta + \ddot{z}\mathbf{k} \qquad (2/17)$$

where

$$a_r = \ddot{r} - r\dot{\theta}^2$$

$$a_\theta = r\ddot{\theta} + 2\dot{r}\dot{\theta} = \frac{1}{r}\frac{d}{dt}(r^2\dot{\theta})$$

$$a_z = \ddot{z}$$

$$a = \sqrt{a_r^2 + a_\theta^2 + a_z^2}$$

Whereas the unit vectors $\mathbf{e}_r$ and $\mathbf{e}_\theta$ have time derivatives due to the change in their directions, we note that the unit vector **k** in the *z*-direction remains fixed in direction and therefore has no time derivative.

(*c*) *Spherical coordinates* (*R-θ-φ*).   When a radial distance and two angles are used to specify the position of a particle, as in the case of radar measurements, for example, spherical coordinates R, θ, ϕ are used. Derivation of the expression for the velocity **v** is easily obtained, but the expression for the acceleration **a** is more complex because of the added geometry. Consequently only the results will be cited here.* First we designate unit vectors $\mathbf{e}_R$, $\mathbf{e}_\theta$, $\mathbf{e}_\phi$ as shown in Fig. 2/15. Note that the unit vector $\mathbf{e}_R$ is in the direction which the particle P would move if R increases but θ and ϕ are held constant. Also the unit vector $\mathbf{e}_\theta$ is in the

* For a complete derivation of **v** and **a** in spherical coordinates see the senior author's book *Dynamics*, 2nd Edition 1971 or SI Version 1975 (John Wiley & Sons, Inc.)

direction which P would move if θ increases while R and ϕ are held constant. Lastly the unit vector $\mathbf{e}_\phi$ is in the direction which P would move if ϕ increases while R and θ are held constant. The resulting expressions for $\mathbf{v}$ and $\mathbf{a}$ are

$$\boxed{\mathbf{v} = v_R\mathbf{e}_R + v_\theta\mathbf{e}_\theta + v_\phi\mathbf{e}_\phi} \tag{2/18}$$

where
$$v_R = \dot{R}$$
$$v_\theta = R\dot{\theta}\cos\phi$$
$$v_\phi = R\dot{\phi}$$

and
$$\boxed{\mathbf{a} = a_R\mathbf{e}_R + a_\theta\mathbf{e}_\theta + a_\phi\mathbf{e}_\phi} \tag{2/19}$$

where
$$a_R = \ddot{R} - R\dot{\phi}^2 - R\dot{\theta}^2\cos^2\phi$$

$$a_\theta = \frac{\cos\phi}{R}\frac{d}{dt}(R^2\dot{\theta}) - 2R\dot{\theta}\dot{\phi}\sin\phi$$

$$a_\phi = \frac{1}{R}\frac{d}{dt}(R^2\dot{\phi}) + R\dot{\theta}^2\sin\phi\cos\phi$$

We should mention that linear algebraic transformations between any two of the three coordinate-system expressions for velocity or acceleration may be developed. These transformations make it possible to express the motion components in rectangular coordinates, for example, if the components are known in spherical coordinates, or vice versa.* These transformations are easily handled with the aid of matrix algebra and a simple computer program.

---

* These coordinate transformations are developed and illustrated in the senior author's book *Dynamics*, 2nd Edition 1971 or SI Version 1975 (John Wiley & Sons, Inc.)

Sample Problem 2/11

The power screw starts from rest and is given a rotational speed $\dot{\theta}$ which increases uniformly with time t according to $\dot{\theta} = kt$ where k is a constant. Determine the expressions for the velocity v and acceleration a of the center of ball A when the screw has turned through one complete revolution from rest. The lead of the screw (advancement per revolution) is L.

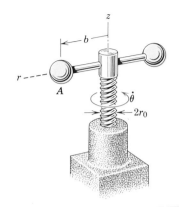

Solution. The center of ball A moves in a helix on the cylindrical surface of radius b, and the cylindrical coordinates r, θ, z are clearly indicated.

Integrating the given relation for $\dot{\theta}$ gives $\theta = \Delta\theta = \int \dot{\theta}\,dt = \frac{1}{2}kt^2$. For one revolution from rest we have

$$2\pi = \tfrac{1}{2}kt^2 \qquad \text{giving} \qquad t = 2\sqrt{\pi/k}$$

Thus the angular rate at one revolution is

$$\dot{\theta} = kt = k(2\sqrt{\pi/k}) = 2\sqrt{\pi k}$$

① The helix angle γ of the path followed by the center of the ball governs the relation between the θ- and the z-components of velocity and is given by $\tan\gamma = L/(2\pi b)$. Now from the figure we see that $v_\theta = v\cos\gamma$. Substituting $v_\theta = r\dot{\theta} = b\dot{\theta}$ from Eq. 2/16 gives $v = v_\theta/\cos\gamma = b\dot{\theta}/\cos\gamma$. With $\cos\gamma$ obtained from $\tan\gamma$ and with $\dot{\theta} = 2\sqrt{\pi k}$ we have for the one-revolution position

$$v = 2b\sqrt{\pi k}\,\frac{\sqrt{L^2 + 4\pi^2 b^2}}{2\pi b} = \sqrt{\frac{k}{\pi}}\,\sqrt{L^2 + 4\pi^2 b^2} \qquad Ans.$$

The acceleration components from Eq. 2/17 become

③ $[a_r = \ddot{r} - r\dot{\theta}^2]$ $a_r = 0 - b(2\sqrt{\pi k})^2 = -4b\pi k$

   $[a_\theta = r\ddot{\theta} + 2\dot{r}\dot{\theta}]$ $a_\theta = bk + 2(0)(2\sqrt{\pi k}) = bk$

   $[a_z = \ddot{z} = \dot{v}_z]$ $a_z = \dfrac{d}{dt}(v_z) = \dfrac{d}{dt}(v_\theta \tan\gamma) = \dfrac{d}{dt}(b\dot{\theta}\tan\gamma)$

$$= (b\tan\gamma)\ddot{\theta} = b\frac{L}{2\pi b}k = \frac{kL}{2\pi}$$

Now we combine the components to give the magnitude of the total acceleration, which becomes

$$a = \sqrt{(-4b\pi k)^2 + (bk)^2 + \left(\frac{kL}{2\pi}\right)^2}$$

$$= bk\sqrt{(1 + 16\pi^2) + L^2/(4\pi^2 b^2)} \qquad Ans.$$

① Unless we recognize that the helix angle of the path governs the z-motion in relation to the rotation of the screw, we cannot complete the problem. We must also be careful to divide the lead L by the circumference $2\pi b$ and not the diameter $2b$ to obtain $\tan\gamma$. If in doubt unwrap one turn of the helix traced by the center of the ball.

② Sketch a right triangle and recall that for $\tan\beta = a/b$ the cosine of β becomes $b/\sqrt{a^2 + b^2}$.

③ The negative sign for a_r is consistent with our previous knowledge that the normal component of acceleration is directed toward the center of curvature.

PROBLEMS

2/156 Consider the power screw of Sample Problem 2/11 with a lead $L = 30$ mm. If $b = 150$ mm and if the screw turns at a constant rate of 4 rev/s, calculate the magnitudes of the velocity and acceleration of the center of ball A.

Ans. $v = 3.77$ m/s
$a = |a_r| = 94.7$ m/s^2

2/157 The velocity and acceleration of a particle moving along a space curve are given for a certain instant by $\mathbf{v} = 6\mathbf{i} - 2\mathbf{j} + \mathbf{k}$ m/s and $\mathbf{a} = 2\mathbf{i} + 3\mathbf{j} - 6\mathbf{k}$ m/s^2. Show that $\mathbf{v}$ and $\mathbf{a}$ are perpendicular and use this fact to determine $\dot{v}$. Also determine the corresponding radius of curvature ρ in the osculating plane. *Ans.* $\rho = 5.86$ m

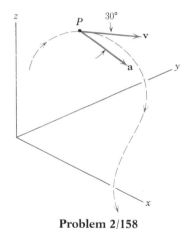

Problem 2/158

2/158 The particle P moves along a space curve and has a velocity $\mathbf{v} = 4\mathbf{i} + 3\mathbf{j}$ m/s at the position represented. At this same instant the total acceleration $\mathbf{a}$ of P has a magnitude of 10 m/s^2 and makes an angle of 30° with $\mathbf{v}$. Calculate the radius of curvature ρ measured in the osculating plane for this instant and determine the acceleration of P tangent to its path.

Ans. $\rho = 5$ m, $a_t = 8.66$ m/s^2

2/159 A particle moves on a space curve governed by $x = 60 \cos \omega t$, $y = 40 \sin \omega t$, and $z = 30t^2$ where x, y, and z are in millimeters, t is in seconds, and $\omega = 2$ rad/s. Describe the path of the particle and compute the magnitude of its acceleration $\mathbf{a}$ when $t = 4$ s.

2/160 An amusement ride called the "corkscrew" takes the passengers through the upside-down curve of a horizontal cylindrical helix. The velocity of the cars as they pass position A is 15 m/s, and the component of their acceleration measured along the tangent to the path is $g \cos \gamma$ at this point. The effective radius of the cylindrical helix is 5 m, and the helix angle is $\gamma = 40°$. Compute the magnitude of the acceleration of the passengers as they pass position A. *Ans.* $a = 27.5$ m/s^2

Problem 2/160

2/161 The rotating element in a mixing chamber is given a periodic axial movement $z = z_0 \sin 2\pi nt$ while it is rotating at the constant angular velocity $\dot{\theta} = \omega$. Determine the expression for the maximum magnitude of the acceleration of a point A on the rim of radius r. The frequency n of vertical oscillation is constant.   *Ans.* $a_{max} = \sqrt{r^2\omega^4 + 16n^4\pi^4 z_0{}^2}$

Problem 2/161

2/162 The rotating nozzle sprays a large circular area and turns with the constant angular rate $\dot{\theta} = K$. Particles of water move along the tube at the constant rate $\dot{l} = c$ relative to the tube. Write expressions for the magnitudes of the velocity and acceleration of a water particle P for a given position l in the rotating tube.

$$Ans.\ v = \sqrt{c^2 + K^2 l^2 \sin^2 \beta}$$
$$a = K \sin \beta \sqrt{K^2 l^2 + 4c^2}$$

Problem 2/162

2/163 The power screw shown in Sample Problem 2/11 is given a rotational oscillation $\theta = \pi \sin 2\pi nt$ where the angle θ and the time t are measured from the mid position and the ball travels between C and D as shown in the separate figure here. The constant frequency (number of complete oscillations per unit time) is n. Determine the magnitude of the acceleration of the center of ball A when it reaches D at $\theta = \pi$. The lead of the screw is L.

Problem 2/163

2/164 An industrial robot is being used to position a small part P. Calculate the magnitude of the acceleration **a** of P for the instant when $\beta = 30°$ if $\dot{\beta} = 10$ degrees per second and $\ddot{\beta} = 20$ degrees per second squared at this same instant. The base of the robot is revolving at the constant rate $\omega = 40$ degrees per second. During the motion arms AO and AP remain perpendicular.   *Ans.* $a = 218.5$ mm/s^2

Problem 2/164

Problem 2/165

▶ **2/165** The particle P moves down the spiral path which is wrapped around the surface of a right circular cone of base radius b and altitude h. The angle γ between the tangent to the curve at any point and a horizontal tangent to the cone at this point is constant. Also the motion of the particle is controlled so that $\dot{\theta}$ is constant. Determine the expression for the radial acceleration a_r of the particle for any value of θ.

$$Ans.\ a_r = b\dot{\theta}^2(\tan^2\gamma\sin^2\beta - 1)e^{-\theta\tan\gamma\sin\beta}$$
$$\text{where } \beta = \tan^{-1}(b/h)$$

▶ **2/166** Assign unit vectors $\mathbf{e}_R$, $\mathbf{e}_\theta$, and $\mathbf{e}_\phi$ along the spherical coordinate directions, Fig. 2/15, and determine $\dot{\mathbf{e}}_R$, $\dot{\mathbf{e}}_\theta$, and $\dot{\mathbf{e}}_\phi$.

$$Ans.\ \dot{\mathbf{e}}_R = \dot{\phi}\mathbf{e}_\phi + (\dot{\theta}\cos\phi)\mathbf{e}_\theta$$
$$\dot{\mathbf{e}}_\theta = -(\dot{\theta}\cos\phi)\mathbf{e}_R + (\dot{\theta}\sin\phi)\mathbf{e}_\phi$$
$$\dot{\mathbf{e}}_\phi = -\dot{\phi}\mathbf{e}_R - (\dot{\theta}\sin\phi)\mathbf{e}_\theta$$

▶ **2/167** Use the results of Prob. 2/166 to obtain the acceleration components in spherical coordinates, Eqs. 2/19, by a direct differentiation of the position vector $\mathbf{R} = R\mathbf{e}_R$.

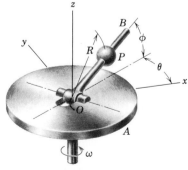

Problem 2/168

▶ **2/168** The disk A rotates about the vertical z-axis with a constant speed $\omega = \dot{\theta} = \pi/3$ rad/s. Simultaneously the hinged arm OB is elevated at the constant rate $\dot{\phi} = 2\pi/3$ rad/s. At time $t = 0$, both $\theta = 0$ and $\phi = 0$. The angle θ is measured from the fixed reference x-axis. The small sphere P slides out along the rod according to $R = 50 + 200t^2$, where R is in millimeters and t is in seconds. Determine the magnitude of the total acceleration $\mathbf{a}$ of P when $t = \frac{1}{2}$ s.   *Ans.* $a = 0.904$ m/s^2

▶ **2/169** The cars of an amusement-park ride are attached to arms of length R which are hinged to a central rotating collar that drives the assembly about the vertical axis with a constant angular rate $\omega = \dot{\theta}$. The cars rise and fall with the track according to the relation $z = (h/2)(1 - \cos 2\theta)$. Determine the expressions for the R-, θ-, and ϕ-components of the velocity **v** of each car as it passes the position $\theta = \pi/4$ rad.

$$Ans.\ v_R = 0,\ v_\theta = R\omega \sqrt{1 - \left(\frac{h}{2R}\right)^2}$$

$$v_\phi = \frac{h\omega}{\sqrt{1 - \left(\frac{h}{2R}\right)^2}}$$

Problem 2/169

▶ **2/170** The revolving crane has a boom of length $OP = 24$ m and is turning about its vertical axis at a constant rate of 2 rev/min. At the same time, the boom is being lowered at the constant rate $\dot{\beta} = 0.10$ rad/s. Calculate the magnitude of the velocity and acceleration of the end P of the boom at the instant when the position $\beta = 30°$ is passed.

$$Ans.\ v = 3.48 \text{ m/s},\ a = 1.104 \text{ m/s}^2$$

▶ **2/171** In a test of the actuating mechanism for a telescoping antenna on a spacecraft the supporting shaft rotates about the fixed z-axis with an angular rate $\dot{\theta}$. Determine the R-, θ-, and ϕ-components of the acceleration **a** of the end of the antenna at the instant when $L = 1.2$ m and $\beta = 45°$ if the rates $\dot{\theta} = 2$ rad/s, $\dot{\beta} = \frac{3}{2}$ rad/s, and $\dot{L} = 0.9$ m/s are constant during the motion.

$$Ans.\ a_R = -5.10 \text{ m/s}^2$$
$$a_\theta = 7.64 \text{ m/s}^2$$
$$a_\phi = -0.3 \text{ m/s}^2$$

Problem 2/170

▶ **2/172** Determine the expressions for the R-, θ-, and ϕ-components of acceleration of each car of the amusement-park ride of Prob. 2/169 as it reaches the position $\theta = \omega t = \pi/2$ rad.

$$Ans.\ a_R = -R\omega^2\left(1 - \frac{h^2}{R^2}\right),\ a_\theta = 0$$

$$a_\phi = -h\omega^2 \frac{1 + (h/R)^2}{\sqrt{1 - (h/R)^2}}$$

Problem 2/171

2/8 RELATIVE MOTION (TRANSLATING AXES)

In the previous articles of this chapter we have described particle motion using coordinates which were referred to fixed reference axes. The displacements, velocities, and accelerations so determined are termed *absolute*. But it is not always possible or convenient to use a fixed set of axes for the observation of motion, and there are many engineering problems for which the analysis of motion is simplified by using measurements made with respect to a moving reference system. These measurements, when combined with the absolute motion of the moving coordinate system, permit us to determine the absolute motion in question. This approach is known as a *relative-motion* analysis.

The motion of the moving coordinate system is specified with respect to a fixed coordinate system. Strictly speaking, this fixed system in Newtonian mechanics is the primary inertial system which is assumed to have no motion in space. For engineering purposes the fixed system may be taken as any system whose absolute motion is negligible for the problem at hand. For most earth-bound engineering problems it is sufficiently precise to take for the fixed reference system a set of axes attached to the earth, in which case we neglect the motion of the earth. For the motion of satellites around the earth, a nonrotating coordinate system with origin on the earth's axis of rotation is chosen. For interplanetary travel, a nonrotating coordinate system fixed to the sun would be used. Hence the choice of the fixed system depends on the type of problem involved.

We will confine our attention in this article to moving reference systems which translate but do not rotate. Motion measured in rotating systems will be discussed in Art. 5/7 of Chapter 5 on rigid-body kinematics where this approach finds special but important application. We will also confine our attention here to relative motion analysis for plane motion.

Consider now two particles A and B which may have separate curvilinear motions in a given plane or in parallel planes, Fig. 2/16. We will arbitrarily attach the origin of a set of translating (nonrotating) axes x-y to particle B and observe the motion of A from our moving position on B. The position vector of A as measured relative to the frame x-y is $\mathbf{r}_{A/B} = x\mathbf{i} + y\mathbf{j}$ where the subscript notation "A/B" means "A relative to B" or "A with respect to B." The unit vectors along the x- and y-axes are $\mathbf{i}$ and $\mathbf{j}$, and x and y are the coordinates of A measured in x-y. The absolute position of B is defined by the vector $\mathbf{r}_B$ measured from the origin of the fixed axes X-Y. The absolute position of A is seen, therefore, to be determined by the vector

$$\mathbf{r}_A = \mathbf{r}_B + \mathbf{r}_{A/B}$$

We now differentiate this vector equation once with respect to

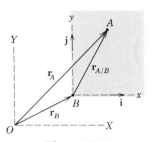

Figure 2/16

time to obtain velocities and twice to obtain accelerations. Thus

$$\dot{\mathbf{r}}_A = \dot{\mathbf{r}}_B + \dot{\mathbf{r}}_{A/B} \qquad \text{or} \qquad \boxed{\mathbf{v}_A = \mathbf{v}_B + \mathbf{v}_{A/B}} \qquad (2/20)$$

$$\ddot{\mathbf{r}}_A = \ddot{\mathbf{r}}_B + \ddot{\mathbf{r}}_{A/B} \qquad \text{or} \qquad \boxed{\mathbf{a}_A = \mathbf{a}_B + \mathbf{a}_{A/B}} \qquad (2/21)$$

In Eq. 2/20 the velocity which we observe A to have from our position at B attached to the moving axes x-y is $\dot{\mathbf{r}}_{A/B} = \mathbf{v}_{A/B} = \dot{x}\mathbf{i} + \dot{y}\mathbf{j}$. This term is the velocity of A with respect to B. Similarly in Eq. 2/21 the acceleration which we observe A to have from our nonrotating position on B is $\ddot{\mathbf{r}}_{A/B} = \dot{\mathbf{v}}_{A/B} = \ddot{x}\mathbf{i} + \ddot{y}\mathbf{j}$. This term is the acceleration of A with respect to B. We note carefully that the unit vectors $\mathbf{i}$ and $\mathbf{j}$ have no derivatives because their directions as well as their magnitudes remain unchanged. (Later when we discuss rotating reference axes we will be obliged to account for the derivatives of the unit vectors when they change direction.)

In words, Eq. 2/20 (or 2/21) states that the absolute velocity (or acceleration) of A equals the absolute velocity (or acceleration) of B plus, vectorially, the velocity (or acceleration) of A relative to B. The relative term is the velocity (or acceleration) measurement which an observer attached to the moving coordinate system x-y would make. We may express the relative motion terms in whatever coordinate system is convenient—rectangular, normal and tangential, or polar—and the formulations in the preceding articles may be used for this purpose. The appropriate fixed system of the previous articles becomes the moving system in the present article.

The selection of the moving point B for attachment of the reference coordinate system is arbitrary. As shown in Fig. 2/17 point A could be used just as well for the attachment of the moving system, in which case the three corresponding relative motion equations for position, velocity, and acceleration are

$$\mathbf{r}_B = \mathbf{r}_A + \mathbf{r}_{B/A} \qquad \mathbf{v}_B = \mathbf{v}_A + \mathbf{v}_{B/A} \qquad \mathbf{a}_B = \mathbf{a}_A + \mathbf{a}_{B/A}$$

It is seen, therefore, that $\mathbf{r}_{B/A} = -\mathbf{r}_{A/B}$, $\mathbf{v}_{B/A} = -\mathbf{v}_{A/B}$, and $\mathbf{a}_{B/A} = -\mathbf{a}_{A/B}$.

An important observation to be made in relative motion analysis is that the acceleration of a particle as observed in a translating system x-y is the same as that observed in a fixed system X-Y if the moving system has a constant velocity. This conclusion broadens the application of Newton's second law of motion, treated in Chapter 3. We conclude, consequently, that a set of axes which has a constant absolute velocity may be used in place of a "fixed" system for the determination of accelerations. A translating reference system which has no acceleration is known as an *inertial system*.

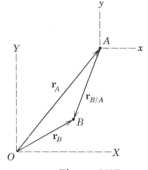

Figure 2/17

Sample Problem   2/12

Passengers in the jet transport A flying east at a speed of 800 km/h observe a second jet plane B which passes under the transport in horizontal flight. Although the nose of B is pointed in the 45° northeast direction, plane B appears to the passengers in A to be moving away from the transport at the 60° angle as shown. Determine the true velocity of B.

Solution.   The moving reference axes x-y are attached to A from which the relative observations are made. We write, therefore,

①
$$\mathbf{v}_B = \mathbf{v}_A + \mathbf{v}_{B/A}$$

Next we identify the knowns and unknowns. The velocity $\mathbf{v}_A$ is given both in magnitude and direction. The 60° direction of $\mathbf{v}_{B/A}$, the velocity which B appears to have to the moving observers in A, is known, ② and the true velocity of B is in the 45° direction in which it is heading. The two remaining unknowns are the magnitudes of $\mathbf{v}_B$ and $\mathbf{v}_{B/A}$. We ③ may solve the vector equation in any one of three ways.

(*I*) *Graphical.*   We start the vector sum at some point P by drawing $\mathbf{v}_A$ to a convenient scale and then construct a line through the tip of $\mathbf{v}_A$ with the known direction of $\mathbf{v}_{B/A}$. The known direction of $\mathbf{v}_B$ is then drawn through P, and the intersection C yields the unique solution enabling us to complete the vector triangle and scale off the unknown magnitudes, which are found to be

$$v_{B/A} = 586 \text{ km/h} \quad \text{and} \quad v_B = 717 \text{ km/h} \qquad Ans.$$

(*II*) *Trigonometric.*   A sketch of the vector triangle is made to reveal the trigonometry, which gives

④
$$\frac{v_B}{\sin 60°} = \frac{v_A}{\sin 75°} \qquad v_B = 800\,\frac{\sin 60°}{\sin 75°} = 717 \text{ km/h} \qquad Ans.$$

(*III*) *Vector algebra.*   Using unit vectors $\mathbf{i}$ and $\mathbf{j}$ we express each of the velocities in vector form as

$$\mathbf{v}_A = 800\mathbf{i} \text{ km/h} \qquad \mathbf{v}_B = (v_B \cos 45°)\mathbf{i} + (v_B \sin 45°)\mathbf{j}$$

$$\mathbf{v}_{B/A} = (v_{B/A} \cos 60°)(-\mathbf{i}) + (v_{B/A} \sin 60°)\mathbf{j}$$

Substituting these relations into the relative-velocity equation and solving separately for the $\mathbf{i}$ and $\mathbf{j}$ terms give

$$\text{(\textbf{i}-terms)} \qquad v_B \cos 45° = 800 - v_{B/A} \cos 60°$$

$$\text{(\textbf{j}-terms)} \qquad v_B \sin 45° = v_{B/A} \sin 60°$$

⑤ Solving simultaneously yields the unknown velocity magnitudes

$$v_{B/A} = 586 \text{ km/h} \quad \text{and} \quad v_B = 717 \text{ km/h} \qquad Ans.$$

It is worth noting the solution of this problem from the viewpoint of an observer in B. With reference axes attached to B we would write $\mathbf{v}_A = \mathbf{v}_B + \mathbf{v}_{A/B}$. The apparent velocity of A as observed by B is then $\mathbf{v}_{A/B}$, which is the negative of $\mathbf{v}_{B/A}$.

① We treat each airplane as a particle.

② We assume no side slip due to cross wind.

③ Students should become familiar with all three solutions.

④ We must be prepared to recognize the appropriate trigonometric relation, which here is the law of sines.

⑤ We can see that the graphical or trigonometric solution is shorter than the vector algebra solution in this particular problem.

Sample Problem 2/13

Car A is accelerating in the direction of its motion at the rate of 1.2 m/s². Car B is rounding a curve of 150-m radius at a constant speed of 54 km/h. Determine the velocity and acceleration which car B appears to have to an observer in car A if car A has reached a speed of 72 km/h for the positions represented.

Solution. We choose nonrotating reference axes attached to car A since the motion of B with respect to A is desired.

Velocity. The relative-velocity equation is

$$\mathbf{v}_B = \mathbf{v}_A + \mathbf{v}_{B/A}$$

and the velocities of A and B for the position considered have the magnitudes

$$v_A = 72/3.6 = 20 \text{ m/s} \qquad v_B = 54/3.6 = 15 \text{ m/s}$$

The triangle of velocity vectors is drawn in the sequence required by the equation, and application of the law of cosines and the law of sines gives

$$v_{B/A} = 18.03 \text{ m/s} \qquad \theta = 46.1° \qquad\qquad Ans.$$

Acceleration. The relative-acceleration equation is

$$\mathbf{a}_B = \mathbf{a}_A + \mathbf{a}_{B/A}$$

The acceleration of A is given, and the acceleration of B is normal to the curve in the n-direction and has the magnitude

$$[a_n = v^2/\rho] \qquad a_B = (15)^2/150 = 1.5 \text{ m/s}^2$$

The triangle of acceleration vectors is drawn in the sequence required by the equation as illustrated. Solving for the x- and y-components of $\mathbf{a}_{B/A}$ gives us

$$(a_{B/A})_x = 1.5 \cos 30° - 1.2 = 0.099 \text{ m/s}^2$$
$$(a_{B/A})_y = 1.5 \sin 30° = 0.750 \text{ m/s}^2$$

from which $a_{B/A} = \sqrt{(0.099)^2 + (0.750)^2} = 0.757 \text{ m/s}^2$ *Ans.*

The direction of $\mathbf{a}_{B/A}$ may be specified by the angle β which, by the law of sines, becomes

$$\frac{1.5}{\sin \beta} = \frac{0.757}{\sin 30°} \qquad \beta = \sin^{-1}\left(\frac{1.5}{0.757} 0.5\right) = 97.5° \qquad Ans.$$

① Alternatively we could use either a graphical or a vector algebraic solution.

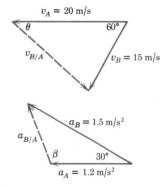

② Be careful to choose between the two values 82.5° and 180 − 82.5 = 97.5°.

Suggestion: To gain familiarity with the manipulation of vector equations, it is suggested that the student rewrite the relative-motion equations in the form $\mathbf{v}_{B/A} = \mathbf{v}_B - \mathbf{v}_A$ and $\mathbf{a}_{B/A} = \mathbf{a}_B - \mathbf{a}_A$ and redraw the vector polygons to conform with these alternative relations.

Caution: So far we are only prepared to handle motion relative to *nonrotating* axes. If we had attached the reference axes rigidly to car B, they would rotate with the car, and we would find that the velocity and acceleration terms relative to the rotating axes are *not* the negative of those measured from the nonrotating axes moving with A. Rotating axes are treated in Art. 5/7.

Problem 2/173

PROBLEMS

2/173 The jet transport B is flying north with a velocity $v_B = 600$ km/h when a smaller aircraft A passes underneath the transport headed in the 60° direction shown. To passengers in B, however, A appears to be flying sideways and moving east. Determine the actual velocity of A and the velocity which A appears to have relative to B.

Ans. $v_A = 1200$ km/h, $v_{A/B} = 1039$ km/h

Problem 2/174

2/174 The passenger aircraft B is flying east with a velocity $v_B = 800$ km/h. A military jet traveling south with a velocity $v_A = 1200$ km/h passes under B at a slightly lower altitude. What velocity does A appear to have to a passenger in B, and what is the direction of that apparent velocity?

Ans. $v_{A/B} = 1442$ km/h 33.7° west of south

2/175 A ship capable of making a speed of 16 knots through still water is to maintain a true course due west while encountering a 3-knot current running from north to south. What should be the heading of the ship (measured clockwise from the north to the nearest degree)? How long does it take the ship to proceed 24 nautical miles due west?

Problem 2/176

2/176 A sailboat moving in the direction shown is tacking to windward against a north wind. The log registers a hull speed of 6.5 knots. A "telltale" (light string tied to the rigging) indicates that the direction of the apparent wind is 35° from the centerline of the boat. What is the true wind velocity v_w?   *Ans.* $v_w = 14.4$ knots

2/177 Two cars A and B are traveling along straight roads as shown. If the time rate of increase of the distance $\overline{AB}$ separating the two cars equals the magnitude of the relative velocity between the cars, what can be said concerning the velocities of the two cars?

Problem 2/177

2/178 The destroyer moves at 30 knots (1 knot = 1.852 km/h) and fires a rocket at an angle which trails the line of sight to the fixed target by the angle α. The launching velocity is 75 m/s relative to the ship and has an angle of elevation of 30° above the horizontal. If the missile continues to move in the same vertical plane as that determined by its absolute velocity at launching, determine α for $\theta = 60°$.   *Ans.* $\alpha = 11.88°$

Problem 2/178

2/179 A small ship capable of making a speed of 6 knots through still water maintains a heading due east while being set to the south by an ocean current. The actual course of the boat is from A to B, a distance of 10 nautical miles which requires exactly 2 hours. Determine the speed v_w of the current and its direction measured clockwise from the north.

Problem 2/179

2/180 Ship A is headed west at a speed of 15 knots, and ship B is headed southwest. The relative bearing θ of B with respect to A is 80° and is unchanging. If the distance between A and B is 8 nautical miles at 3:00 P.M., when would collision occur if neither ship altered course or speed? At what speed is ship B traveling?   *Ans.* 3:37 P.M., $v_B = 18.0$ knots

Problem 2/180

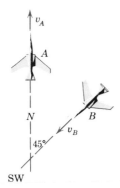

Problem 2/181

2/181 Airplane A is flying north with a constant horizontal velocity of 500 km/h. Airplane B is flying southwest at the same altitude with a velocity of 500 km/h. From the frame of reference of A determine the magnitude v_r of the apparent or relative velocity of B. Also find the magnitude of the apparent velocity v_n with which B appears to be moving sideways or normal to its centerline. Would the results be different if the two airplanes were flying at different but constant altitudes?

2/182 In Prob. 2/181 if aircraft A is accelerating in its northward direction at the rate of 3 km/h each second while aircraft B is slowing down at the rate of 4 km/h each second in its southwesterly direction, determine the acceleration in m/s² which B appears to have to an observer in A and specify its direction (β) measured clockwise from the north.

Ans. $a_{B/A} = 0.787$ m/s², $\beta = 93.5°$

Problem 2/183

2/183 Car A rounds a curve of 150-m radius at a constant speed of 54 km/h. At the instant represented, car B is moving at 81 km/h but is slowing down at the rate of 3 m/s². Determine the velocity and acceleration of car A as observed from car B.

Problem 2/184

2/184 For the instant represented car A has a speed of 100 km/h, which is increasing at the rate of 8 km/h each second. Simultaneously car B also has a speed of 100 km/h as it rounds the turn and is slowing down at the rate of 8 km/h each second. Determine the acceleration which car B appears to have to an observer in car A.

Ans. $\mathbf{a}_{B/A} = -4.44\mathbf{i} + 2.57\mathbf{j}$ m/s²

2/185 Two ships A and B are moving with constant speeds v_A and v_B, respectively, along straight intersecting courses. The navigator of ship B notes the time rates of change of the separation distance r between the ships and the bearing angle θ. Show that $\ddot{\theta} = -2\dot{r}\dot{\theta}/r$ and $\ddot{r} = r\dot{\theta}^2$.

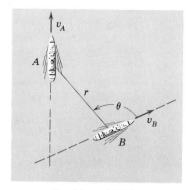

Problem 2/185

2/186 For the coplanar or parallel-plane motion of three particles A, B, and C prove that the relative velocities and relative accelerations measured from nonrotating axes obey the equations $\mathbf{v}_{A/B} = \mathbf{v}_{A/C} + \mathbf{v}_{C/B}$ and $\mathbf{a}_{A/B} = \mathbf{a}_{A/C} + \mathbf{a}_{C/B}$.

2/187 For the instant represented car A is rounding the circular curve at a constant speed of 50 km/h while car B is slowing down at the rate of 8 km/h per second. Determine the acceleration which car A appears to have an observer in car B.

Ans. $\mathbf{a}_{A/B} = 1.92\mathbf{i} + 1.11\mathbf{j} \text{ m/s}^2$

Problem 2/187

2/188 For the instant represented in the figure for Prob. 2/187, car B is accelerating in the direction of its motion, and car A has a speed of 50 km/h which is increasing. If the acceleration of A as observed from B is zero for this instant, determine the magnitude of the acceleration of B and the rate at which the speed of A is changing.

2/189 An earth satellite is put into a circular polar orbit at an altitude of 240 km, which requires an orbital velocity of 27 940 km/h with respect to the center of the earth considered fixed in space. In going from south to north, when the satellite passes over an observer on the equator, in which direction does the satellite appear to be moving? The equatorial radius of the earth is 6378 km, and the angular velocity of the earth is $0.729(10^{-4})$ rad/s.

Ans. Apparent direction 3.43° west of north

Problem 2/190

Problem 2/191

Problem 2/193

2/190 The spacecraft S approaches the planet Mars along a trajectory b-b in the orbital plane of Mars with an absolute velocity of 19 km/s. Mars has a velocity of 24.1 km/s along its trajectory a-a. Determine the angle β between the line of sight S-M and the trajectory b-b when Mars appears from the spacecraft to be approaching it head on.

Ans. $\beta = 55.6°$

2/191 Airplane A is flying horizontally with a constant speed of 200 km/h and is towing the glider B, which is gaining altitude. If the tow cable has a length $r = 60$ m and θ is increasing at the constant rate of 5 degrees per second, determine the magnitudes of the velocity $\mathbf{v}$ and acceleration $\mathbf{a}$ of the glider for the instant when $\theta = 15°$.

Ans. $v_B = 206$ km/h, $a_B = 0.457$ m/s² from B to A

2/192 If the airplane in Prob. 2/191 is increasing its speed in level flight at the rate of 5 km/h each second and is unreeling the glider tow cable at the constant rate $\dot{r} = 2$ m/s while θ remains constant, determine the magnitude of the acceleration of the glider B.

▶**2/193** The aircraft A with radar detection equipment is flying horizontally at an altitude of 12 km and is increasing its speed at the rate of 1.2 m/s each second. Its radar locks onto an aircraft flying in the same direction and in the same vertical plane at an altitude of 18 km. If A has a speed of 1000 km/h at the instant that $\theta = 30°$, determine the values of $\ddot{r}$ and $\ddot{\theta}$ at this same instant if B has a constant speed of 1500 km/h.

Ans. $\ddot{r} = -0.637$ m/s²
$\ddot{\theta} = 0.1660(10^{-3})$ rad/s²

2/9 *CONSTRAINED MOTION OF CONNECTED PARTICLES*

There are occasions when the motions of particles are interrelated by virtue of the constraints of interconnecting members, and it becomes necessary to account for these constraints in order to determine the respective motions of the particles.

Consider, first, the very simple system of two interconnected particles A and B shown in Fig. 2/18. Although it should be quite evident by inspection that the horizontal motion of A is twice the vertical motion of B, we will use this example to illustrate the method of analysis which will carry over to more complex situations where the results cannot be easily perceived by inspection. Clearly the motion of B is the same as that of the center of its pulley, so we establish position coordinates y and x measured from a convenient fixed datum. The total length of the cable is

$$L = x + \frac{\pi r_2}{2} + 2y + \pi r_1 + b$$

With L, r_2, r_1, and b all constant, the first and second time derivatives of the equation give

$$0 = \dot{x} + 2\dot{y} \quad \text{or} \quad v_A = 2v_B$$

$$0 = \ddot{x} + 2\ddot{y} \quad \text{or} \quad a_A = 2a_B$$

Figure 2/18

Since the results do not depend on the lengths or pulley radii, we should be able to analyze the motion without considering them. In the lower-left portion of Fig. 2/18 is shown an enlarged view of the horizontal diameter $A'B'C$ of the lower pulley at an instant of time. Clearly A' and A have the same motion magnitudes, as do B and B'. During an infinitesimal motion of A' it is easy to see that B' moves half as far since point C as a point on the fixed portion of the cable momentarily has no motion. Thus, with differentiation by time in mind, we can see the velocity and acceleration relationships by inspection. The pulley, in effect, is a wheel which rolls on the fixed vertical cable. (The kinematics of a rolling wheel will be treated more extensively in Chapter 5 on rigid-body motion.) The system of Fig. 2/18 is said to have *one degree of freedom* since only one variable, either x or y, is needed to specify the positions of all parts of the system.

A system with *two degrees of freedom* is shown in Fig. 2/19. Here the position of the lower cylinder and pulley C depend on the separate specifications of the two coordinates y_A and y_B. The lengths of the cables attached to cylinders A and B may be written, respectively, as

$$L_A = y_A + 2y_D + \text{constant}$$

$$L_B = y_B + y_C + (y_C - y_D) + \text{constant}$$

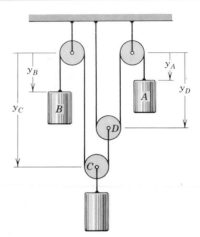

Figure 2/19

and their time derivatives are

$$0 = \dot{y}_A + 2\dot{y}_D \quad \text{and} \quad 0 = \dot{y}_B + 2\dot{y}_C - \dot{y}_D$$

$$0 = \ddot{y}_A + 2\ddot{y}_D \quad \text{and} \quad 0 = \ddot{y}_B + 2\ddot{y}_C - \ddot{y}_D$$

Eliminating the terms in $\dot{y}_D$ and $\ddot{y}_D$ gives

$$\dot{y}_A + 2\dot{y}_B + 4\dot{y}_C = 0 \quad \text{and} \quad \ddot{y}_A + 2\ddot{y}_B + 4\ddot{y}_C = 0$$

Clearly, the signs of all three terms cannot be positive physically. So, for example, if both A and B have downward velocities $v_A = +\dot{y}_A$ and $v_B = +\dot{y}_B$, then C will have an upward velocity $v_C = -\dot{y}_C = v_A/4 + v_B/2$.

These results may also be found by inspection of the motions of the two pulleys at C and D. Due to an increment dy_A (with y_B held fixed), the center of D moves up an amount $dy_A/2$ which causes an upward movement $dy_A/4$ for the center of C. Due to an increment dy_B (with y_A held fixed), the center of C moves up a distance $dy_B/2$. Combination of the two movements gives an upward movement

$$-dy_C = \frac{dy_A}{4} + \frac{dy_B}{2}$$

so that $v_C = v_A/4 + v_B/2$ as before. Visualization of the actual geometry of the motion is an important ability.

A second type of constraint where the direction of the connecting member changes with the motion is illustrated in the second of the two sample problems which follow.

Sample Problem 2/14

In the pulley configuration shown, cylinder A has a downward velocity of 0.3 m/s. Determine the upward velocity of B. Solve in two ways.

Solution (I). The centers of the pulleys at A and B are located by the coordinates y_A and y_B measured from fixed positions. The total constant length of cable in the pulley system is

$$L = 3y_B + 2y_A + \text{constants}$$

where the constants account for the fixed lengths of cable in contact

① with the circumferences of the pulleys and the constant vertical separation between the two upper left-hand pulleys. Differentiation with time gives

$$0 = 3\dot{y}_B + 2\dot{y}_A$$

But the downward velocity of A is $v_A = \dot{y}_A = 0.3$ m/s, and the upward velocity of B is $v_B = -\dot{y}_B$. Substitution gives

$$0 = 3(-v_B) + 2(0.3) \qquad \text{or} \qquad v_B = 0.2 \text{ m/s} \qquad \textit{Ans.}$$

Solution (II). An enlarged diagram of the pulleys at A, B, and C is shown. During a differential movement ds_A of the center of pulley A, the left end of its horizontal diameter has no motion since it is attached to the fixed part of the cable. Therefore, the right-hand end has a movement of $2ds_A$ as shown. This movement is transmitted to the left-hand end of the horizontal diameter of the pulley at B. Further, from pulley C with its fixed center, we see that the displacements on each side are equal and opposite. Thus for pulley B, the right-hand end of the diameter has a downward displacement equal to the up-

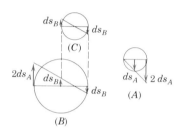

ward displacement ds_B of its center. By inspection of the geometry we conclude that

② $$2ds_A = 3ds_B \qquad \text{or} \qquad ds_B = \tfrac{2}{3}ds_A$$

Dividing by dt gives

$$v_B = \tfrac{2}{3}v_A = \tfrac{2}{3}(0.3) = 0.2 \text{ m/s} \qquad \textit{Ans.}$$

① We neglect the small angularity of the cables between B and C.

② If this ratio is not readily apparent to you, make a scale drawing of the pulley-B displacements for clarification.

Sample Problem 2/15

The tractor A is used to hoist the bale B with the pulley arrangement shown. If A has a forward velocity v_A, determine an expression for the upward velocity v_B of the bale in terms of x.

Solution. We designate the position of the tractor by the coordinate x and the position of the bale by the coordinate y, both measured from a fixed reference. The total constant length of the cable is

$$L = 2(h - y) + l = 2(h - y) + \sqrt{h^2 + x^2}$$

① Differentiation with time yields

$$0 = -2\dot{y} + \frac{x\dot{x}}{\sqrt{h^2 + x^2}}$$

Substituting $v_A = \dot{x}$ and $v_B = \dot{y}$ gives

$$v_B = \frac{1}{2}\frac{x}{\sqrt{h^2 + x^2}} \qquad \textit{Ans.}$$

① Differentiation of the relation for a right triangle occurs frequently in mechanics.

PROBLEMS

Problem 2/194

2/194 If the velocity $\dot{x}$ of block A up the incline is increasing at the rate of 0.044 m/s each second, determine the acceleration of B.

Ans. $a_B = 0.022$ m/s²

2/195 Determine the vertical rise h of the load W during 5 seconds if the hoisting drum wraps cable around it at the constant rate of 320 mm/s.

Problem 2/195

2/196 A truck equipped with a power winch on its front end pulls itself up a steep incline with the cable and pulley arrangement shown. If the cable is wound up on the drum at the constant rate of 40 mm/s, how long does it take for the truck to move 4 m up the incline?   *Ans.* $t = 3$ min 20 s

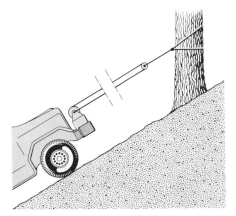

Problem 2/196

2/197 If counterweight A has a downward velocity v_A, determine the upward velocity of B by considering the relationship between differential displacements.

Problem 2/197

2/198 The vertical displacement of cylinder A in meters is given by $y = t^2/4$ where t is in seconds. Calculate the downward acceleration a_B of cylinder B. Identify the number of degrees of freedom.

  Ans. $a_B = 4$ m/s^2, one degree of freedom

Problem 2/198

2/199 Determine the relation which governs the accelerations of A, B, and C, all measured positive down. Identify the number of degrees of freedom.

Problem 2/199

Problem 2/200

2/200 Determine the relationship which governs the velocities of the four cylinders. Express all velocities as positive down. How many degrees of freedom are there?

Ans. $4v_A + 8v_B + 4v_C + v_D = 0$, three degrees of freedom

Problem 2/201

2/201 Cylinder *B* has a downward velocity in meters per second given by $v_B = t^2/2 + t^3/6$ where *t* is in seconds. Calculate the acceleration of *A* when $t = 2$ s. *Ans.* $a_A = 2$ m/s^2

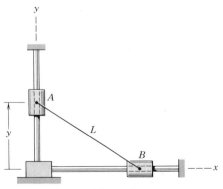

Problem 2/202

2/202 Collars *A* and *B* slide along the fixed right-angled rods and are connected by a cord of length *L*. Determine the acceleration a_x of collar *B* as a function of *y* if collar *A* is given a constant upward velocity v_A.

2/203 For a given value of y determine the upward velocity of A in terms of the downward velocity of B. Neglect the diameters of the pulleys

$$Ans. \; v_A = \frac{2y}{\sqrt{y^2 + b^2}} \, v_B$$

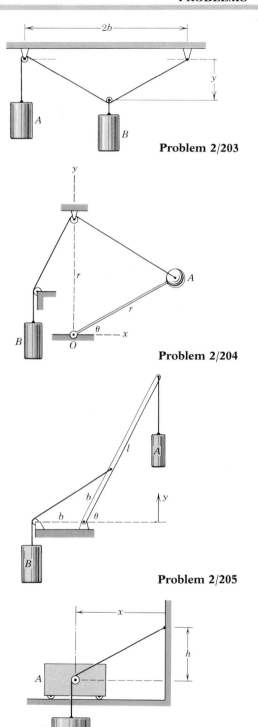

Problem 2/203

2/204 The particle A is mounted on a light rod pivoted at O and, therefore, is constrained to move in a circular arc of radius r. Determine the velocity of A in terms of the downward velocity v_B of the counterweight for any angle θ.

Problem 2/204

▶**2/205** If load B has a downward velocity v_B, determine the upward component $(v_A)_y$ of the velocity of A in terms of b, the boom length l, and the angle θ. Assume that the cable supporting A remains vertical.

$$Ans. \; (v_A)_y = \frac{l\sqrt{2(1 + \cos \theta)}}{b \tan \theta} \, v_B$$

Problem 2/205

▶**2/206** Neglect the diameter of the small pulley attached to body A and determine the magnitude of the total velocity of B in terms of the velocity v_A which body A has to the right. Assume that the cable between B and the pulley remains vertical and solve for a given value of x.

$$Ans. \; v_B = v_A \sqrt{\frac{2x^2 + h^2}{x^2 + h^2}}$$

Problem 2/206

2/10 *PROBLEM FORMULATION AND REVIEW*

In Chapter 2 we have developed and illustrated the basic elements of theory and their use for the description of particle motion. It is important for the student to review this material before proceeding to consolidate and enlarge his or her perspective of particle kinematics. The concepts and procedures developed and illustrated in this chapter form the basis for building much of the entire subject of dynamics, and satisfactory progress in the topics ahead will be contingent upon a firm grasp of Chapter 2.

By far the most important concept in Chapter 2 is the time derivative of a vector. The independent contributions of magnitude change and direction change to the time derivative are, indeed, basic. As we proceed in our study of dynamics we will have further occasion to examine the time derivatives of vectors other than position and velocity vectors, and the principles and procedures developed in Chapter 2 will be utilized directly.

When one concentrates primarily on the individual parts of a subject, the relationships among its parts are not always clearly grasped. When guided by a specific topical assignment the student automatically recognizes the problem category and the indicated method of solution. But without this guide he must choose his own method of solution. The ability to choose the most appropriate method is certainly a key to the successful formulation and solution of engineering problems. One of the best ways to acquire this ability is to compare alternative methods after each is studied.

To facilitate this comparison it will be helpful if we recognize the following breakdowns:

(a) *Type of motion.* The three conspicuous categories are
Rectilinear motion (one coordinate)
Plane curvilinear motion (two coordinates)
Space curvilinear motion (three coordinates)
Generally the geometry of a given problem permits this identification to be made readily. One exception to this categorization is encountered when only the magnitudes of the motion quantities measured along the path are of interest. In this event we may use the single distance coordinate measured along the curved path together with its scalar time derivatives giving the speed $\dot{s}$ and the tangential acceleration $\ddot{s}$.

Plane motion is easier to generate and control, particularly in machinery, than is space motion, so that a large fraction of our motion problems come under the plane-curvilinear or rectilinear categories.

(b) *Reference fixity.* We commonly make motion measurements with respect to fixed reference axes (absolute motion) and moving axes (relative motion). The acceptable choice of the fixed axes depends on the problem. Axes attached to the earth are sufficiently "fixed" for most engineering problems although important exceptions include earth–satellite and interplanetary motion, accurate projectile trajectories, navigation, and other problems. The equations of relative motion discussed in Chapter 2 are restricted to translating reference axes.

(*c*) *Coordinates.* Of prime importance is the choice of coordinates. We have developed the description of motion using

   Rectangular (Cartesian) coordinates (*x-y*) and (*x-y-z*)
   Normal and tangential coordinates (*n-t*)
   Polar coordinates (*r-θ*)
   Cylindrical coordinates (*r-θ-z*)
   Spherical coordinates (*R-θ-φ*)

When the coordinates are not specified, the choice usually depends on how the motion is generated or measured. Thus for a particle which slides radially along a rotating rod, polar coordinates are the natural ones to use. Radar tracking calls for polar or spherical coordinates. When measurements are made along a curved path, normal and tangential coordinates are indicated. An *x-y* plotter clearly involves rectangular coordinates.

Figure 2/20 is a composite representation of the *x-y*, *n-t*, and *r-θ* coordinate descriptions of the velocity **v** and acceleration **a** for curvilinear motion. It is frequently essential to transfer motion description from one coordinate description to another, and Fig. 2/20 contains the information necessary for that transition.

(*d*) *Approximations.* One of the most important abilities to acquire is that of making appropriate approximations. The assumption of constant acceleration is valid when the forces which cause the acceleration do not vary appreciably. When motion data are acquired experimentally, we must utilize the nonexact data to acquire the best possible description, often with the aid of graphical or numerical approximations.

(*e*) *Mathematical method.* We frequently have a choice of solution using scalar algebra, vector algebra, trigonometric geometry, or graphical geometry. All of these methods have been illustrated and all are important to learn. The choice of method will depend on the geometry of the problem, how the motion data are given, and the accuracy desired. Inasmuch as mechanics by its very nature is geometrical, the student is encouraged to develop his facility to sketch vector relationships both as an aid to the disclosure of appropriate geometrical and trigonometrical relations and as a means of solving vector equations graphically. Geometric portrayal is the most direct representation of the vast majority of mechanics problems.

The problems which follow are included to facilitate a review of Chapter 2 and have a random arrangement of motion category.

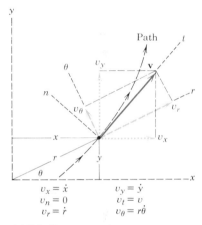

$$v_x = \dot{x} \qquad\qquad v_y = \dot{y}$$
$$v_n = 0 \qquad\qquad v_t = v$$
$$v_r = \dot{r} \qquad\qquad v_\theta = r\dot{\theta}$$

(a) Velocity components

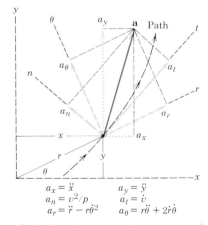

$$a_x = \ddot{x} \qquad\qquad a_y = \ddot{y}$$
$$a_n = v^2/p \qquad\quad a_t = \dot{v}$$
$$a_r = \ddot{r} - r\dot{\theta}^2 \qquad a_\theta = r\ddot{\theta} + 2\dot{r}\dot{\theta}$$

(b) Acceleration components

Figure 2/20

REVIEW PROBLEMS

2/207 At a certain instant the time derivatives of the co-ordinates of a particle moving along a curvilinear path are $\dot{x} = 60$ mm/s, $\ddot{x} = 0$, $\dot{y} = 0$, $\ddot{y} = 90$ mm/s^2. Determine the radius of curvature ρ of the path at this instant.   *Ans.* $\rho = 40$ mm

2/208 The polar coordinates of a particle are given by $r = 10$ mm, constant, and $\theta = 4t$ radians, where t is in seconds. When $\theta = 60°$, determine $\ddot{x}$.
Ans. $\ddot{x} = -80$ mm/s^2

2/209 The total acceleration **a** of a simple pendulum makes an angle β with the suspension wire at a certain angle θ. Write expressions for the corresponding values of $\dot{\theta}$ and $\ddot{\theta}$.

Problem 2/209

2/210 The launching catapult of the aircraft carrier gives the jet fighter a constant acceleration of 50 m/s^2 from rest relative to the flight deck and launches the aircraft in a distance of 100 m measured along the angled takeoff ramp. If the carrier is moving at a steady 30 knots (1 knot = 1.852 km/h), determine the magnitude v of the actual velocity of the fighter when it is launched.   *Ans.* $v = 414$ km/h

Problem 2/210

2/211 Car A negotiates a curve of 60-m radius at a constant speed of 50 km/h. When A passes the position shown, car B is 30 m from the intersection and is accelerating south toward the intersection at the rate of 1.5 m/s^2. Determine the acceleration which A appears to have when observed by an occupant of B at this instant.

Problem 2/211

2/212 An inexperienced designer of a roadbed for a new high-speed train proposes to join a straight section of track to a circular section of 600-m radius as shown. For a train which would travel at a constant speed of 150 km/h plot the magnitude of its acceleration as a function of distance along the track between points *A* and *C* and explain why this design is unacceptable.

Problem 2/212

2/213 A jet aircraft pulls up into a vertical curve as shown. As it passes the position where $\theta = 30°$, its speed is 1000 km/h and is decreasing at the rate of 15 km/h per second. If the radius of curvature ρ of the flight path is 1.5 km at this point, calculate the corresponding horizontal and vertical components, $\ddot{x}$ and $\ddot{y}$, of the acceleration of the aircraft.
Ans. $\ddot{x} = -29.3$ m/s^2, $\ddot{y} = 42.5$ m/s^2

Problem 2/213

2/214 In the differential pulley hoist shown, the two upper pulleys are fastened together to form an integral unit. The cable is wrapped around the smaller pulley with its end secured to the pulley so that it cannot slip. Determine the upward acceleration a_B of cylinder *B* if cylinder *A* has a downward acceleration of 2 m/s^2. (*Suggestion:* Analyze geometrically the consequences of a differential movement of cylinder *A*.)   *Ans.* $a_B = 0.25$ m/s^2

Problem 2/214

2/215 A stone is thrown down the slope as shown. Determine the magnitude u and direction θ of its initial velocity so that the stone will rise 12 m and still have a range of 50 m down the slope.
Ans. $u = 17.74$ m/s, $\theta = 96.7°$

Problem 2/215

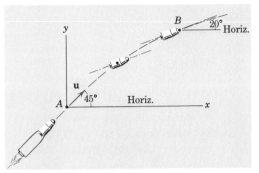

Problem 2/216

2/216 The third stage of a rocket is injected by its booster with a velocity u of 15 000 km/h at A into an unpowered coasting flight to B. At B its rocket motor is ignited when the trajectory makes an angle of 20° with the horizontal. Operation is effectively above the atmosphere, and the gravitational acceleration during this interval may be taken as 9 m/s², constant in magnitude and direction. Determine the time t to go from A to B. (This quantity is needed in the design of the ignition control system.) Also determine the corresponding increase h in altitude.

2/217 A particle P moves along a plane curve with a speed v, measured in meters per second, given by $v = 2 + 0.3t^2$ where t is the time in seconds after P passes a certain fixed point on the curve. If the total acceleration of P is 2.4 m/s² when $t = 2$ s, compute the radius of curvature ρ of the curve for the position of the particle at this instant.

Ans. $\rho = 4.93$ m

Problem 2/218

2/218 A rocket is fired vertically and tracked by the radar shown with Prob. 2/132 and repeated here. At the instant when $\theta = 60°$, measurements give $\dot{\theta} = 0.03$ rad/s and $r = 7500$ m, and the vertical acceleration of the rocket is found to be $a = 20$ m/s². For this instant determine the values of $\ddot{r}$ and $\ddot{\theta}$.

Ans. $\ddot{r} = 24.1$ m/s², $\ddot{\theta} = -1.784(10^{-3})$ rad/s²

2/219 Rotation of the arm PO is controlled by the horizontal motion of the vertical slotted link. If $\dot{x} = 1.2$ m/s and $\ddot{x} = 9$ m/s² when $x = 50$ mm, determine $\dot{\theta}$ and $\ddot{\theta}$ for this instant.

Problem 2/219

2/220 A particle P moving in plane curvilinear motion is located by the polar coordinates shown. At a particular instant $r = 2$ m, $\theta = 60°$, $v_r = 3$ m/s, $v_\theta = 4$ m/s, $a_r = -10$ m/s^2, and $a_\theta = -5$ m/s^2. For this instant calculate the radius of curvature ρ of the path of the particle. Also locate the center of curvature C graphically.

     Ans. $\rho = 5$ m, $x_C = -0.23$ m, $y_C = -3.60$ m

Problem 2/220

2/221 For the instant represented the particle P has a velocity $v = 6$ m/s in the direction shown and has acceleration components $a_x = 15$ m/s^2 and $a_\theta = -15$ m/s^2. Determine a_r, a_y, a_t, and a_n and the radius of curvature ρ of the path for this position. (*Hint:* Draw the related acceleration components of the total acceleration of the particle and take advantage of the simplified geometry for your calculations.)

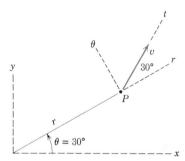

Problem 2/221

2/222 The motion of pin P is controlled by the two moving slots A and B in which the pin slides. If B has a velocity $v_B = 3$ m/s to the right while A has an upward velocity $v_A = 2$ m/s, determine the magnitude v_P of the velocity of the pin.

     Ans. $v_P = 2.72$ m/s

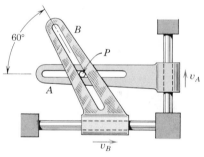

Problem 2/222

2/223 As a further test of the operation of the control mechanism for the telescoping antenna of Prob. 2/171, the vertical shaft is made to oscillate about the fixed z-axis according to $\theta = \pi/4 + 0.12 \sin 4\pi t$, where θ is in radians and t is in seconds. During the oscillation β is increasing at the rate of $\frac{3}{2}$ rad/s, and L is increasing at the rate of 0.9 m/s, both constant. Determine the magnitude of the acceleration **a** of the tip of the antenna when $\beta = 60°$ if this occurs at the position $\theta = \pi/4$ and when $L = 1.2$ m.

     Ans. $a = 7.11$ m/s^2

Problem 2/223

Problem 2/224

▶**2/224** The rod OA is held at the constant angle $\beta = 30°$ while it rotates about the vertical with a constant angular rate of 120 rev/min. Simultaneously the slider P oscillates along the rod with a variable distance from the fixed pivot O given in millimeters by $R = 400 + 100 \sin 2\pi nt$ where the frequency n of oscillation along the rod is a constant 2 cycles per second and where the time t is in seconds. Calculate the magnitude of the acceleration of the slider for an instant when its velocity $\dot{R}$ along the rod is a maximum.

Ans. $a = 35.3$ m/s^2

Problem 2/225

▶**2/225** The radar tracking antenna oscillates about its vertical axis according to $\theta = \theta_0 \cos \omega t$ where ω is the constant circular frequency and $2\theta_0$ is the double amplitude of oscillation. Simultaneously, the angle of elevation ϕ is increasing at the constant rate $\dot{\phi} = K$. Determine the expression for the magnitude a of the acceleration of the signal horn (a) as it passes position A and (b) as it passes the top position B, assuming that $\theta = 0$ at this instant.

Ans. $(a)\ a = b\sqrt{K^4 + \omega^4\theta_0^2 \cos^2 \phi}$
$(b)\ a = bK\sqrt{K^2 + 4\omega^2\theta_0^2}$

Computer-oriented problems

*2/226** The fixed cam and rotating arm of Prob. 2/131 are repeated here. The shape of the cam causes the follower A to move along a limaçon defined by $r = b - c \cos \theta$ where $b > c$. If $b = 100$ mm, $c = 50$ mm, and the arm revolves at the constant rate $\dot{\theta} = 2$ rad/s, plot the magnitudes of the velocity $\mathbf{v}$ and acceleration $\mathbf{a}$ of pin A from $\theta = 0$ to $\theta = 180°$. Justify your result for the acceleration at $\theta = 0$.

Problem 2/226

2/227 The bar OA is given an oscillatory motion about the vertical specified by $\theta = \theta_0 \sin(2\pi t/\tau)$ where θ_0 is the maximum angular displacement in radians, τ is the time (period) for one complete oscillation, and t is the time in seconds measured from an instant when OA is vertical. For $\theta_0 = \pi/2$, $\tau = 2$ s, and $\overline{OA} = l = 100$ mm, plot the magnitude a of the total acceleration of A as a function of t for the first quarter cycle ($t = 0$ to $t = 0.5$ s). Determine the maximum and minimum value of a and their corresponding values of θ.

$$\text{\textit{Ans.}} \ a_{max} = 2.44 \text{ m/s}^2 \text{ at } \theta = 0$$
$$a_{min} = 1.47 \text{ m/s}^2 \text{ at } \theta = 80.4°$$

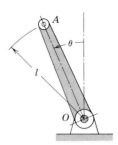

Problem 2/227

*2/228 The guide with the vertical slot is given a horizontal oscillatory motion according to $x = 100 \sin 2t$ where x is in millimeters and t is in seconds. The oscillation causes the pin P to move in the fixed parabolic slot whose shape is given by $y = x^2/100$ with y also in millimeters. Plot the magnitude v of the velocity of the pin as a function of time during the interval required for pin P to go from the center to the extremity $x = 100$ mm. Find and locate the maximum value of v and verify your results analytically.

$$\text{\textit{Ans.}} \ v_{max} = 250 \text{ mm/s at } x = 61.2 \text{ mm and}$$
$$t = 0.330 \text{ s}$$

Problem 2/228

*2/229 A ship with a total displacement of 16 000 metric tons (1 metric ton = 1000 kg) starts from rest in still water under a constant propeller thrust $T = 250$ kN. The ship develops a total resistance to motion through the water given by $R = 4.50v^2$ where R is in kilonewtons and v is in meters per second. The acceleration of the ship is $a = (T - R)/m$ where m equals the mass of the ship in metric tons. Plot the speed v of the ship in knots as a function of the distance s in nautical miles which the ship goes for the first 5 nautical miles from rest. Find the speed after the ship has gone 1 nautical mile. What is the maximum speed which the ship can reach?

$$\text{\textit{Ans.}} \ v_{1 \text{ mi}} = 11.7 \text{ knots}, \ v_{max} = 14.49 \text{ knots}$$

KINETICS OF PARTICLES

3

3/1 INTRODUCTION

According to Newton's second law, a particle will accelerate when it is subjected to unbalanced forces. Kinetics is the study of the relations between unbalanced forces and the changes in motion which they produce. In Chapter 3 we shall study the kinetics of particles. This topic requires that we combine our knowledge of two previously learned parts of mechanics, namely, the properties of forces which we developed in our earlier study of statics, and the kinematics of particle motion which we have just covered in Chapter 2. With the aid of Newton's second law of motion we are now ready to combine these two topics and prepare for the solution of engineering problems involving force, mass, and motion.

There are three general approaches to the solution of kinetics problems: (A) direct application of Newton's second law (called the force-mass-acceleration method), (B) use of work and energy principles, and (C) solution by impulse and momentum methods. Each approach has its special characteristics and advantages, and Chapter 3 is subdivided into sections according to these three methods of solution. In addition, a fourth section treats special applications and combinations of the three basic approaches. Before proceeding the student is strongly advised to review carefully the definitions and concepts of Chapter 1, as they are fundamental to the developments which follow.

SECTION A. FORCE, MASS, AND ACCELERATION

3/2 NEWTON'S SECOND LAW

The basic relation between force and acceleration is found in Newton's second law of motion, Eq. 1/1, the verification of which is entirely experimental. We will describe the fundamental meaning of this law by an ideal experiment in which force and acceleration are assumed to be measured without error. A mass particle is isolated in the primary inertial system* and is subjected to the action of a single force $\mathbf{F}_1$. The acceleration $\mathbf{a}_1$ of the particle is measured, and the ratio F_1/a_1 of the magnitudes of the force and the acceleration will be some number C_1 whose value depends on the units used for measurement of force and acceleration. The experiment is now repeated by subjecting the same particle to a different force $\mathbf{F}_2$ and measuring the corresponding acceleration $\mathbf{a}_2$. Again the ratio of the magnitudes F_2/a_2 will produce a number C_2. The experiment is repeated as many times as desired. We draw two important conclusions from the results. First, the ratios of applied force to corresponding acceleration all equal the *same* number, provided the units used for measurement are not changed in the experiments. Thus,

$$\frac{F_1}{a_1} = \frac{F_2}{a_2} = \cdots = \frac{F}{a} = C, \qquad \text{a constant}$$

We conclude that the constant C is a measure of some property of the particle which does not change. This property is the *inertia* of the particle which is its *resistance to rate of change of velocity*. For a particle of high inertia (large C) the acceleration will be small for a given force F. On the other hand, if the inertia is small, the acceleration will be large. The mass m is used as a quantitative measure of inertia, and therefore the expression $C = km$ may be written, where k is a constant to account for the units used. Thus the experimental relation becomes

$$F = kma \qquad (3/1)$$

where F is the magnitude of the resultant force acting on the particle of mass m, and a is the magnitude of the resulting acceleration of the particle.

The second conclusion we draw from the ideal experiment is that the acceleration is always in the direction of the applied force. Thus Eq. 3/1 becomes a *vector* relation and may be written

$$\mathbf{F} = km\mathbf{a} \qquad (3/2)$$

Although an actual experiment cannot be performed in the ideal manner described, the conclusions are inferred from the measurements of countless accurately performed experiments where the results are correctly predicted from the hypothesis of the ideal experiment. One of the most accurate checks lies in the precise prediction of the motions of planets based on Eq. 3/2.

(*a*) *Inertial system.* Whereas the results of the ideal experiment are obtained for measurements made relative to the

---

* The primary inertial system or astronomical frame of reference is an imaginary set of reference axes which are assumed to have no translation or rotation in space. See Art. 1/2, Chapter 1.

"fixed" primary inertial system, they are equally valid for measurements made with respect to any nonrotating reference system which translates with a constant velocity with respect to the primary system. From our study of relative motion in Art. 2/8 we saw that the acceleration measured in a system translating with no acceleration is the same as that measured in the primary system. Thus Newton's second law holds equally well in a nonaccelerating system, so that we may define an *inertial system* as any system in which Eq. 3/2 is valid.

If the ideal experiment described were performed on the surface of the earth and all measurements were made relative to a reference system attached to the earth, the measured results would show a slight discrepancy upon substitution into Eq. 3/2. This discrepancy would be due to the fact that the measured acceleration would not be the correct absolute acceleration. The discrepancy would disappear when we introduced the corrections due to the acceleration components of the earth. These corrections are negligible for most engineering problems which involve the motions of structures and machines on the surface of the earth. In such cases the accelerations measured with respect to reference axes attached to the surface of the earth may be treated as "absolute," and Eq. 3/2 may be applied with negligible error to experimental measurements made on the surface of the earth.*

There is an increasing number of problems, particularly in the fields of rocket and spacecraft design, where the acceleration components of the earth are of primary concern. For this work it is essential that the fundamental basis of Newton's law be thoroughly understood and that the appropriate absolute acceleration components be employed.

Before 1905 the laws of Newtonian mechanics had been verified by innumerable physical experiments and were considered the final description of the motion of bodies. The concept of *time*, considered an absolute quantity in the Newtonian theory, received a basically different interpretation in the theory of relativity announced by Einstein in 1905. The new concept called

---

* An example of the magnitude of the error introduced by neglect of the motion of the earth may be cited for the case of a particle which is allowed to fall from rest (relative to the earth) at a height h above the ground. We can show that the rotation of the earth gives rise to an eastward acceleration (Coriolis acceleration) relative to the earth and, neglecting air resistance, that the particle falls to the ground a distance

$$x = \frac{2}{3} \omega \sqrt{\frac{2h^3}{g}} \cos \gamma$$

east of the point on the ground directly under that from which it was dropped. The angular velocity of the earth is $\omega = 0.729(10^{-4})$ rad/s, and the latitude, north or south, is γ. At a latitude of 45° and from a height of 200 m, this eastward deflection would be $x = 43.9$ mm.

for a complete reformulation of the accepted laws of mechanics. The theory of relativity was subjected to early ridicule but has had experimental check and is now universally accepted by scientists the world over. Although the difference between the mechanics of Newton and that of Einstein is basic, there is a practical difference in the results given by the two theories only when velocities of the order of the speed of light (300×10^6 m/s) are encountered.* Important problems dealing with atomic and nuclear particles, for example, involve calculations based on the theory of relativity and are of basic concern to both scientists and engineers.

(*b*) *Units.* It is customary to take k equal to unity in Eq. 3/2, thus putting the relation in the usual form of Newton's second law

$$\boxed{\mathbf{F} = m\mathbf{a}} \qquad [1/1]$$

A system of units for which k is unity is known as a *kinetic* system. Thus for a kinetic system the units of force, mass, and acceleration are not independent. In SI units, as explained in Art. 1/4, the units of force (newtons, N) are derived by Newton's second law from the base units of mass (kilograms, kg) times acceleration (meters per second squared, m/s^2). Thus N = kg·m/s^2. This system is known as an *absolute* system since the unit for force is dependent on the absolute value of mass.

In U.S. customary units, on the other hand, the units of mass (slugs) are derived from the units of force (pounds force, lb) divided by acceleration (feet per second squared, ft/sec^2). Thus the mass units are slugs = lb-sec^2/ft. This system is known as a *gravitational* system since mass is derived from force as determined from gravitational attraction.

For measurements made relative to the rotating earth the relative value of g should be used. The internationally accepted value of g relative to the earth at sea level and at a latitude of 45° is 9.806 65 m/s^2. Except where greater precision is required the value of 9.81 m/s^2 will be used for g. For measurements relative to a nonrotating earth the absolute value of g should be used. At a latitude of 45° and at sea level the absolute value is 9.8236 m/s^2. The sea-level variation in both the absolute and relative values of

---

* The theory of relativity demonstrates that there is no such thing as a preferred primary inertial system and that measurements of time which are made in two coordinate systems which have a velocity relative to one another are different. On this basis, for example, the principles of relativity show that a clock carried by the pilot of a spacecraft traveling around the earth in a circular polar orbit of 644 km altitude at a velocity of 27 080 km/h would be slow compared with a clock at the pole by 0.000 001 85 s for each orbit.

g with latitude is shown in Fig. 1/1 of Art. 1/5.

In the U.S. customary system the standard value of g relative to the rotating earth at sea level and at a latitude of 45° is 32.1740 ft/sec². The corresponding value relative to a nonrotating earth is 32.2295 ft/sec².

Along with our primary use of SI units we will have occasional need to refer to U.S. customary units. Thus we must be absolutely sure that we have a clear understanding of the correct force and mass units in each system. These units were explained in Art. 1/4, but it will be helpful to illustrate them here using simple numbers before moving ahead to the applications of Newton's second law. Consider, first, the free-fall experiment as depicted in Fig. 3/1*a* where we release an object from rest near the earth's surface and allow it to fall freely under the influence of the force of gravitational attraction W on the body, which we call its *weight*. In SI units for a mass $m = 1$ kg the weight is $W = 9.81$ N, and the corresponding downward acceleration a is $g = 9.81$ m/s². In U.S. customary units for a mass $m = 1$ lbm (1/32.2 slug) the weight is $W = 1$ lbf and the resulting gravitational acceleration is $g = 32.2$ ft/sec². For a mass $m = 1$ slug (32.2 lbm) the weight is $W = 32.2$ lbf and the acceleration, of course, is also $g = 32.2$ ft/sec².

In Fig. 3/1*b* we illustrate the proper units with the simplest

(a) Gravitational Free-Fall

(b) Newton's Second Law

Figure 3/1

example where we accelerate an object of mass m along the horizontal with a force F. In the coherent system of SI units a force $F = 1$ N causes a mass $m = 1$ kg to accelerate at the rate $a = 1$ m/s^2. Thus 1 N $= 1$ kg·m/s^2. In the U.S. customary system (which is a gravitational, noncoherent system) a force $F = 1$ lbf causes a mass $m = 1$ lbm (1/32.2 slug) to accelerate at the rate $a = 32.2$ ft/sec^2, whereas a force $F = 1$ lbf causes a mass $m = 1$ slug (32.2 lbm) to accelerate at the rate $a = 1$ ft/sec^2.

We note that in SI units where the mass is expressed in kilograms (kg) the weight W of the body in newtons (N) is given by $W = mg$ where $g = 9.81$ m/s^2. In U.S. customary units the weight W of a body is expressed in pounds force (lbf), and the mass in slugs (lbf-sec^2/ft) is given by $m = W/g$ where $g = 32.2$ ft/sec^2. In U.S. customary units we frequently speak of the weight of a body when we really mean mass. It is entirely proper to specify the mass of a body in pounds (lbm) which must be converted to mass in slugs before substituting into Newton's second law. Unless otherwise stated, the pound (lb) is normally used as the unit of force (lbf).

3/3 EQUATION OF MOTION AND SOLUTION OF PROBLEMS

When a particle of mass m is subjected to the action of concurrent forces $\mathbf{F}_1$, $\mathbf{F}_2$, $\mathbf{F}_3$, . . . whose vector sum is $\Sigma\mathbf{F}$, Eq. 1/1 becomes

$$\boxed{\Sigma\mathbf{F} = m\mathbf{a}} \qquad (3/3)$$

In the solution of problems Eq. 3/3 is usually expressed in scalar component form using one of the coordinate systems developed in Chapter 2. The choice of the appropriate coordinate system is dictated by the type of motion involved and is a vital step in the formulation of any problem. Equation 3/3, or any one of the component forms of the force-mass-acceleration equation, is usually referred to as the *equation of motion*. The equation of motion gives the instantaneous value of the acceleration corresponding to the instantaneous values of the forces which are acting.

(*a*) *The two problems of dynamics.* We encounter two types of problems when applying Eq. 3/3. In the first type the acceleration is either specified or can be determined directly from known kinematic conditions. The corresponding forces which act on the particle whose motion is specified are then determined by direct substitution into Eq. 3/3. This problem is generally quite straightforward.

In the second type of problem the forces are specified and the resulting motion is to be determined. If the forces are constant, the acceleration is constant and is easily found from Eq. 3/3. When the forces are functions of time, position, velocity, or acceleration, Eq. 3/3 becomes a differential equation which must be integrated to determine the velocity and displacement. Problems of this second type are often more formidable, as the integration may be difficult to carry out, particularly when the force is a mixed function of two or more motion variables. In practice it is frequently necessary to resort to approximate integration techniques, either numerical or graphical, particularly when experimental data are involved. The procedures for a mathematical integration of the acceleration when it is a function of the motion variables were developed in Art. 2/2, and these same procedures apply when the force is a specified function of these same parameters, since force and acceleration differ only by the constant factor of the mass.

(*b*) *Constrained and unconstrained motion.*   There are two physically distinct types of motion, both described by Eq. 3/3. The first type is *unconstrained* motion where the particle is free of mechanical guides and follows a path determined by its initial motion and by the forces which are applied to it from external sources. An airplane or rocket in flight and an electron moving in a charged field are examples of unconstrained motion. The second type is *constrained* motion where the path of the particle is partially or totally determined by restraining guides. An ice-hockey puck moves with the partial constraint of the ice. A train moving along its track and a collar sliding along a fixed shaft are examples of more fully constrained motion. Some of the forces acting on a particle during constrained motion may be applied from outside sources and others may be the reactions on the particle from the constraining guides. *All forces*, both applied and reactive, which act *on* the particle must be accounted for in applying Eq. 3/3.

The choice of a coordinate system is frequently indicated by the number and geometry of the constraints. Thus, if a particle is free to move in space, as is the center of mass of the airplane or rocket in free flight, the particle is said to have *three degrees of freedom* since three independent coordinates are required to specify the position of the particle at any instant. All three of the scalar components of the equation of motion would have to be applied and integrated to obtain the space coordinates as a function of time. If a particle is constrained to move along a surface, as is the hockey puck or a marble sliding on the curved surface of a bowl, only two coordinates are needed to specify its position, and in this case it is said to have *two degrees of freedom*. If a particle is constrained to move along a fixed linear path, as is the collar sliding along a fixed shaft, its position may be specified by the coordinate measured along the shaft. In this case the particle would have only *one degree of freedom*.

(*c*) *Free-body diagram.*   In applying any of the force-mass-acceleration equations of motion it is absolutely necessary to account correctly for *all* forces acting on the particle. The only forces which we may neglect are those whose magnitudes are negligible compared with other forces acting, such as the forces of mutual attraction between two particles compared with their attraction to a celestial body such as the earth. The vector sum $\Sigma\mathbf{F}$ of Eq. 3/3 means the vector sum of *all* forces acting *on* the particle in question. Likewise, the corresponding scalar force summation in any one of the component directions means the sum of the components of *all* forces acting *on* the particle in that particular direction. The only reliable way to account accurately and consistently for every force is to *isolate* the particle under consideration from *all* contacting and influencing bodies and replace the bodies removed by the forces they exert on the particle isolated. The resulting *free-body diagram* is the means by which every force, known and unknown, which acts on the particle is represented and hence accounted for. Only after this vital step has been completed should the appropriate equation or equations of motion be written. The free-body diagram serves the same key purpose in dynamics as it does in statics. This purpose is simply to establish a *thoroughly reliable method* for the correct evaluation of the resultant of all actual forces acting on the particle or body in question. In statics this resultant equals zero, whereas in dynamics it is equated to the product of mass and acceleration. If we recognize that the equations of motion must be interpreted literally and exactly, and if in so doing we respect the full scalar and vector meaning of the equals sign in the motion equation, then a minimum of difficulty will be experienced. Every experienced student of engineering mechanics recognizes that careful and consistent observance of the *free-body method* is the *most important single lesson* to be learned in the study of engineering mechanics. As a part of the drawing of a free-body diagram, the coordinate axes and their positive directions should be clearly indicated. When the equations of motion are written, all force summations should be consistent with the choice of these positive directions. Also, as an aid to the identification of external forces which act on the body in question, these forces are shown as heavy red vectors in the illustrations throughout the remainder of the book. Sample Problems 3/1 through 3/5 in the next article contain five examples of free-body diagrams which may be easily reviewed as a reminder of their construction.

In solving problems the student often wonders how to get started and what sequence of steps to follow in arriving at the solution. This difficulty may be minimized by forming the habit of first recognizing some relationship between the desired unknown quantity in the problem and other quantities, known and unknown. Additional relationships between these unknowns and other quantities, known and unknown, are then perceived. Finally the dependence upon the original data is established, and the procedure for the analysis and computation is indicated. A few minutes spent organizing the plan of attack through recognition of the dependence of one quantity upon another will be time well spent and will usually prevent groping for the answer with irrelevant calculations.

3/4 RECTILINEAR MOTION

We now apply the concepts discussed in Arts. 3/2 and 3/3 to problems in particle motion starting with rectilinear motion in this article and treating curvilinear motion in Art. 3/5. In both articles we shall analyze the motions of bodies which can be treated as particles. This simplication is permitted as long as we are interested only in the motion of the mass center of the body, in which case the forces may be treated as concurrent through the mass center. When we discuss the kinetics of rigid bodies in Chapter 6, we shall account for the action of nonconcurrent forces on the motions of bodies.

If we choose the x-direction, for example, as the direction of the rectilinear motion of a particle of mass m, the acceleration in the y- and z-directions will be zero and the scalar components of Eq. 3/3 become

$$\Sigma F_x = ma_x$$
$$\Sigma F_y = 0 \qquad\qquad (3/4)$$
$$\Sigma F_z = 0$$

For cases where we are not free to choose a coordinate direction along the motion, we would have in the general case all three component equations

$$\Sigma F_x = ma_x$$
$$\Sigma F_y = ma_y \qquad\qquad (3/5)$$
$$\Sigma F_z = ma_z$$

where the acceleration and resultant force are given by

$$\mathbf{a} = a_x\mathbf{i} + a_y\mathbf{j} + a_z\mathbf{k}$$
$$a = \sqrt{a_x^{\,2} + a_y^{\,2} + a_z^{\,2}}$$
$$\Sigma\mathbf{F} = \Sigma F_x\mathbf{i} + \Sigma F_y\mathbf{j} + \Sigma F_z\mathbf{k}$$
$$|\Sigma\mathbf{F}| = \sqrt{(\Sigma F_x)^2 + (\Sigma F_y)^2 + (\Sigma F_z)^2}$$

Sample Problem 3/1

A 75-kg man stands on a spring scale in an elevator. During the first 3 s of motion from rest the tension T in the hoisting cable is 8300 N. Find the reading R of the scale in newtons during this interval and the upward velocity v of the elevator at the end of the 3 seconds. The total mass of the elevator, man, and scale is 750 kg.

Solution. The force registered by the scale and the velocity both depend on the acceleration of the elevator, which is constant during the interval for which the forces are constant. From the free-body diagram of the elevator, scale, and man taken together, the acceleration is found to be

$[\Sigma F_y = ma_y]$ $8300 - 7360 = 750a_y$ $a_y = 1.257 \text{ m/s}^2$

The scale reads the downward force exerted on it by the man's feet. The equal and opposite reaction R to this action is shown on the free-body diagram of the man alone together with his weight, and the equation of motion for him gives

① $[\Sigma F_y = ma_y]$ $R - 736 = 75(1.257)$ $R = 830 \text{ N}$ *Ans.*

The velocity reached at the end of the 3 seconds is

$\left[\Delta v = \int a\, dt\right]$ $v - 0 = \int_0^3 1.257\, dt$ $v = 3.77 \text{ m/s}$ *Ans.*

① If the scale is calibrated in kilograms it would read $830/9.81 = 84.6$ kg which, of course, is not his true mass since the measurement was made in a noninertial (accelerating) system. *Suggestion:* Rework this problem in U.S. customary units.

Sample Problem 3/2

A small inspection car with a mass of 200 kg runs along the fixed overhead cable and is controlled by the attached cable at A. Determine the acceleration of the car when the control cable is horizontal and under a tension $T = 2.4$ kN. Also find the total force P exerted by the supporting cable on the wheels.

Solution. The free-body diagram of the car and wheels taken together and treated as a particle discloses the 2.4 kN tension T, the weight $W = mg = 200(9.81) = 1962$ N, and the force P exerted on the wheel assembly by the cable.

The car is in equilibrium in the y-direction since there is no acceleration in this direction. Thus

$[\Sigma F_y = 0]$ $P - 2.4(\frac{5}{13}) - 1.962(\frac{12}{13}) = 0$ $P = 2.73 \text{ kN}$ *Ans.*

① In the x-direction the equation of motion gives

$[\Sigma F_x = ma_x]$ $2400(\frac{12}{13}) - 1962(\frac{5}{13}) = 200a$ $a = 7.30 \text{ m/s}^2$ *Ans.*

① By choosing our coordinate axes along and normal to the direction of the acceleration we are able to solve the two equations independently. Would this be so if x and y were chosen as horizontal and vertical?

Sample Problem 3/3

The 125-kg concrete block A is released from rest in the position shown and pulls the 200-kg log up the 30° ramp. If the coefficient of kinetic friction between the log and the ramp is 0.5, determine the velocity of the log as the block hits the ground at B.

Solution. The motions of the log and the block A are clearly dependent. Although by now it should be evident that the acceleration of the log up the incline is half the downward acceleration of A, we may prove it formally. The constant total length of the cable is $L = 2s_C + s_A +$ constant, where the constant accounts for the cable portions wrapped around the pulleys. Differentiating twice with respect to time gives $0 = 2\ddot{s}_C + \ddot{s}_A$. Since the acceleration of the log (and, hence, point C) *up* the incline is the negative of the *downward* measurement $\ddot{s}_C$, the acceleration of the log is $-\ddot{s}_C = \ddot{s}_A/2 = a/2$.

① We assume here that the masses of the pulleys are negligible and that they turn with negligible friction. With these assumptions the free-body diagram of the pulley C discloses force and moment equilibrium. Hence the tension in the cable attached to the log is twice that applied to the block.

Equilibrium of the log in the y-direction gives

② $[\Sigma F_y = 0]$ $N - 200(9.81)\cos 30° = 0$ $N = 1699$ N

and its equation of motion in the x-direction gives

$[\Sigma F_x = ma_x]$ $2T - 0.5(1699) - 200(9.81)\sin 30° = 200\dfrac{a}{2}$

For the block in its x-direction of motion we have

③ $[\Sigma F_x = ma_x]$ $125(9.81) - T = 125a$

Solving these two equations simultaneously gives us

$$a = 1.777 \text{ m/s}^2 \qquad T = 1004 \text{ N}$$

④ For the 6-m drop with constant acceleration the block acquires a velocity

$[v^2 = 2ax]$ $v_A = \sqrt{2(1.777)(6)} = 4.62$ m/s

Hence the velocity of the log is

$$v = \tfrac{1}{2}v_A = 2.31 \text{ m/s} \qquad\qquad \textit{Ans.}$$

① The coordinates used in expressing the final kinematic constraint relationship must be consistent with those used for the kinetic equations of motion.

② We can verify that the log will indeed move up the ramp by calculating the force in the cable necessary to initiate motion from the equilibrium condition. This force is $2T = 0.5N + 200(9.81)\sin 30° = 1831$ N   or $T = 915$ N, which is less than the 1226-N weight of block A. Hence the log will move up.

③ Note the serious error in assuming that $T = 125(9.81)$ N, in which case block A would not accelerate.

④ Because the forces on this system remain constant, the resulting accelerations also remain constant.

Sample Problem 3/4

The design model for a new ship has a mass of 10 kg and is tested in an experimental towing tank to determine its resistance to motion through the water at various speeds. The test results are plotted on the accompanying graph and the resistance R may be closely approximately by the dotted parabolic curve shown. If the model is released when it has a speed of 2 m/s, determine the time t required for it to reduce its speed to 1 m/s and the corresponding travel distance x.

Solution. We approximate the resistance–velocity relation by $R = kv^2$ and find k by substituting $R = 8$ N and $v = 2$ m/s into the equation which gives $k = 8/2^2 = 2$ N·s²/m². Thus $R = 2v^2$.

The only horizontal force on the model is R, so that

① $[\Sigma F_x = ma_x]$ $-R = ma_x$ or $-2v^2 = 10\dfrac{dv}{dt}$

We separate the variables and integrate to obtain

$$\int_0^t dt = -5 \int_2^v \frac{dv}{v^2} \qquad t = 5\left(\frac{1}{v} - \frac{1}{2}\right) \qquad \text{seconds}$$

Thus when $v = v_0/2 = 1$ m/s, the time is $t = 5(\frac{1}{1} - \frac{1}{2}) = 2.5$ s   *Ans.*

The distance traveled during the 2.5 seconds is obtained by integrating $v = dx/dt$. Thus $v = 10/(5 + 2t)$ so that

② $$\int_0^x dx = \int_0^{2.5} \frac{10}{5 + 2t}\, dt \qquad x = \frac{10}{2} \ln\,(5 + 2t)\Big]_0^{2.5} = 3.47 \text{ m} \quad Ans.$$

① Be careful to observe the minus sign for R.

② *Suggestion:* Express the distance x after release in terms of the velocity v and see if you agree with the resulting relation $x = 5 \ln\,(v_0/v)$.

Sample Problem 3/5

The collar of mass m slides up the vertical shaft under the action of a force F of constant magnitude but variable direction. If $\theta = kt$ where k is a constant and if the collar starts from rest with $\theta = 0$, determine the magnitude F of the force which will result in the collar coming to rest as θ reaches $\pi/2$. The coefficient of kinetic friction between the collar and shaft is μ_k.

Solution. After drawing the free-body diagram, we apply the equation of motion in the y-direction to get

① $[\Sigma F_y = ma_y]$ $F \cos \theta - \mu_k N - mg = m\dfrac{dv}{dt}$

where equilibrium in the horizontal direction requires that $N = F \sin \theta$. Substituting $\theta = kt$ and integrating first between general limits give

$$\int_0^t (F \cos kt - \mu_k F \sin kt - mg)\, dt = m \int_0^v dv$$

which becomes

$$\frac{F}{k}\,[\sin kt + \mu_k(\cos kt - 1)] - mgt = mv$$

For $\theta = \pi/2$ the time becomes $t = \pi/2k$ and $v = 0$ so that

② $$\frac{F}{k}\,[1 + \mu_k(0 - 1)] - \frac{mg\pi}{2k} = 0 \quad \text{and} \quad F = \frac{mg\pi}{2(1 - \mu_k)} \quad Ans.$$

① If θ were expressed as a function of the vertical displacement y instead of the time t, the acceleration would become a function of the displacement and we would use $v\, dv = a\, dy$.

② We see that the results do not depend on k, the rate at which the force changes direction.

PROBLEMS

3/1 A 300-Mg jet airliner has four engines, each of which produces a nearly constant thrust of 180 kN during the takeoff roll. Determine the length s of runway required if the takeoff speed is 220 km/h. Neglect air and rolling resistance.

Ans. $s = 778$ m

3/2 The 50-kg crate is projected along the floor with an initial speed of 8 m/s at $x = 0$. The coefficient of kinetic friction is 0.4. Calculate the time required for the crate to come to rest and the corresponding distance x traveled.

Ans. $t = 2.04$ s, $x = 8.15$ m

Problem 3/2

3/3 The 50-kg crate of Prob. 3/2 is now projected down an incline as shown with an initial speed of 8 m/s. Investigate the time t required for the crate to come to rest and the corresponding distance x traveled if (*a*) $\theta = 15°$ and (*b*) $\theta = 30°$.

Problem 3/3

3/4 The 10-Mg truck hauls the 20-Mg trailer. If the unit starts from rest on a level road with a tractive force of 20 kN between the driving wheels of the truck and the road, compute the tension T in the horizontal drawbar and the acceleration a of the rig.

Problem 3/4

3/5 As part of the design process for a restraint chair, an engineer considers the following possible set of conditions: A 12-kg child is riding in the chair which in turn is securely fastened to the seat of an automobile. The automobile is involved in a head-on collision with another vehicle. The initial speed v_0 of the car is 50 km/h and this speed is reduced to zero during the collision of 0.2-s duration. Assume a constant car deceleration during the collision and estimate the net horizontal force F which the straps of the restraint chair must exert on the child in order to keep her fixed to the chair. Treat the child as a particle and state any additional assumptions made during your analysis.

Ans. $F = 833$ N

Problem 3/5

3/6 Repeat Prob. 3/5, except assume that the automobile decelerates during the collision according to $v = v_0(b - e^{ct})$, where b and c are constants and the collision begins at time $t = 0$. Calculate the maximum horizontal force F_{max} exerted by the restraint straps on the child.

3/7 Calculate the vertical acceleration a of the 150-kg cylinder for each of the two cases illustrated. Neglect friction and the mass of the pulleys.

Ans. (a) $a = 1.40$ m/s^2
(b) $a = 3.27$ m/s^2

(a)   (b)

Problem 3/7

3/8 A 3-kg particle is moving in the horizontal y-direction with a velocity of 4 m/s when acted upon by a resultant force $F_y = 5 + 2t^2$ newtons where t is the time of application in seconds. Calculate the velocity v of the particle when $t = 3$ s.

3/9 A man pulls himself up the 15° incline by the method shown. If the combined mass of the man and cart is 100 kg, determine the acceleration of the cart while the man exerts a pull of 250 N on the rope. Neglect all friction and the mass of the rope, pulleys, and wheels.

Ans. $a = 4.96$ m/s^2

Problem 3/9

3/10 The coefficient of static friction between the flat bed of the truck and the crate it carries is 0.30. Determine the minimum stopping distance s which the truck can have from a speed of 70 km/h with constant deceleration if the crate is not to slip forward. *Ans. $s = 64.3$ m*

Problem 3/10

3/11 If the truck of Prob. 3/10 comes to a stop from an initial forward speed of 70 km/h in a distance of 50 m with uniform deceleration, determine whether or not the crate strikes the wall at the forward end of the flat bed. If the crate does strike the wall, calculate its speed relative to the truck as the impact occurs. Use the friction coefficients $\mu_s = 0.3$ and $\mu_k = 0.25$.

3/12 A car is climbing the hill of slope θ_1 at a constant speed v. If the slope decreases abruptly to θ_2 at point A, determine the acceleration a of the car just after passing point A if the driver does not change the throttle setting or shift into a different gear.   *Ans.* $a = g(\sin \theta_1 - \sin \theta_2)$

Problem 3/12

3/13 The block shown is observed to have a velocity $v_1 = 20$ m/s as it passes point A and a velocity $v_2 = 10$ m/s as it passes point B on the incline. Calculate the coefficient of kinetic friction μ_k between the block and the incline if $x = 75$ m and $\theta = 15°$.

Problem 3/13

3/14 The block-and-tackle system is released from rest with all cables taut. Neglect the mass and friction of all pulleys and determine the acceleration of each cylinder and the tensions T_1 and T_2 in the two cables.

$$\text{Ans. } a_A = 1.401 \text{ m/s}^2 \text{ up}$$
$$a_B = 5.61 \text{ m/s}^2 \text{ down}$$
$$T_1 = 16.8 \text{ N, } T_2 = 67.3 \text{ N}$$

Problem 3/14

3/15 During a reliability test, a circuit board of mass m is attached to an electromagnetic shaker and subjected to a harmonic displacement $x = X \sin \omega t$, where X is the motion amplitude, ω is the motion frequency in radians per second, and t is time. Determine the magnitude $F_{\max}$ of the maximum horizontal force which the shaker exerts on the circuit board.

Problem 3/15

Problem 3/16

Problem 3/17

Problem 3/18

Problem 3/19

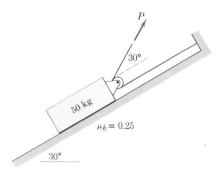

Problem 3/20

3/16 A toy train has magnetic couplers whose maximum attractive force is 0.9 N between adjacent cars as shown. What is the maximum force *P* with which a child can pull the locomotive and not break the train apart at a coupler? If *P* is slightly exceeded, which coupler fails? Neglect the mass and friction associated with all wheels.

Ans. P = 1.8 N, coupler 1

3/17 The beam and attached hoisting mechanism have a combined mass of 1200 kg with center of mass at *G*. If the initial acceleration *a* of a point *P* on the hoisting cable is 6 m/s², calculate the corresponding reaction at the support *A*.

Ans. F_A = 19.99 kN

3/18 If the coefficients of static and kinetic friction between the 20-kg block *A* and the 100-kg cart *B* are both essentially the same value of 0.50, determine the acceleration of each part for (*a*) *P* = 60 N and (*b*) *P* = 40 N.

3/19 The steel ball is suspended from the accelerating frame by the two cords *A* and *B*. Determine the acceleration *a* of the frame which will cause the tension in *A* to be twice that in *B*.

Ans. $a = \dfrac{g}{3\sqrt{3}}$

3/20 Determine the tension *P* in the cable which will give the 50-kg block a steady acceleration of 2 m/s² up the incline. *Ans. P = 227 N*

3/21 A three-car subway train is traveling down a 5-percent grade when the individual car brakes are simultaneously applied. Each 2500-kg car can generate a braking force of 0.5 times the normal force exerted on it by the tracks. Determine the train deceleration a and the forces T_1 and T_2 in the couplings 1 and 2 for the cases when (*a*) all brakes function normally, (*b*) the brakes of car *A* fail, (*c*) the brakes of car *B* fail, and (*d*) the brakes of car *C* fail.

Problem 3/21

3/22 The system is released from rest with the cable taut. Neglect the small mass and friction of the pulley and calculate the acceleration of each body and the cable tension T upon release if (*a*) $\mu_s = 0.25$, $\mu_k = 0.20$ and (*b*) $\mu_s = 0.15$, $\mu_k = 0.10$.

Problem 3/22

3/23 The pulley arrangement of Prob. 3/22 is modified as shown in the figure. For the friction coefficients $\mu_s = 0.25$ and $\mu_k = 0.20$, calculate the acceleration of each body and the tension T in the cable.

> *Ans.* $a_A = 1.450$ m/s² down
> $a_B = 0.725$ m/s² up
> $T = 105.4$ N

Problem 3/23

3/24 A cylinder of mass m rests in a supporting carriage as shown. If $\beta = 45°$ and $\theta = 30°$, calculate the maximum acceleration a which the carriage may be given up the incline so that the cylinder does not lose contact at *B*.   *Ans.* $a = 0.366g$

Problem 3/24

3/25 The total resistance R to horizontal motion of a rocket test sled is shown by the solid line in the graph. Approximate R by the dotted line and determine the distance x which the sled must travel along the track from rest to reach a velocity of 400 m/s. The forward thrust T of the rocket motors is essentially constant at 300 kN, and the total mass of the sled is also nearly constant at 2 Mg.

Problem 3/25

Problem 3/26

3/26 A player pitches a baseball horizontally toward a speed-sensing radar gun. The baseball has a mass of 0.146 kg and a circumference of 230 mm. If the speed at $x = 0$ is $v_0 = 140$ km/h, estimate the speed as a function of x. Assume that the horizontal aerodynamic drag on the baseball is given by $D = C_D(\frac{1}{2}\rho v^2)S$, where C_D is the drag coefficient, ρ is the air density, v is the speed, and S is the cross-sectional area of the baseball. Use a value of 0.4 for C_D. Neglect the y-component of the motion but comment on the validity of this assumption. Evaluate your answer for $x = 18$ m, which is the approximate distance between a pitcher's hand and home plate.

> *Ans.* $v = v_0 e^{(-\frac{1}{2}C_D\rho Sx/m)}$, $v_{18} = 123.5$ km/h

Problem 3/27

3/27 In a test to determine the crushing characteristics of styrofoam packing material, a steel cone of mass m is dropped so that it falls a distance h and then penetrates the material. The resistance R of styrofoam to penetration depends upon the cross-sectional area of the penetrating object and thus is proportional to the square of the cone penetration distance x, or $R = -kx^2$. If the cone comes to rest at a distance $x = d$, determine the constant k in terms of the test conditions and results.

> *Ans.* $k = \dfrac{3mg}{d^3}(h + d)$

Problem 3/28

3/28 A heavy chain with a mass ρ per unit length is pulled along a horizontal surface consisting of a smooth section and a rough section by the constant force P. If the chain is initially at rest on the smooth surface with $x = 0$ and if the coefficient of kinetic friction between the chain and the rough surface is μ_k, determine the velocity v of the chain when $x = L$. Assume that the chain remains taut and thus moves as a unit throughout the motion. What is the minimum value of P which will permit the chain to remain taut? (*Hint:* The acceleration must not become negative.)

> *Ans.* $v = \sqrt{\dfrac{2P}{\rho} - \mu_k gL}$
>
> $P_{min} = \mu_k \rho gL$

3/29 The chain is relased from rest with an overhanging length b. Neglect all friction and determine the velocity v of the chain when its last link leaves the edge.

Problem 3/29

3/30 A bar of length l and negligible mass connects the cart of mass M and the particle of mass m. If the cart is subjected to a constant acceleration a to the right, what is the resulting steady-state angle θ which the freely pivoting bar makes with the vertical? Determine the net force P (not shown) which must be applied to the cart to cause the specified acceleration.

$$\text{Ans. } \theta = \tan^{-1}(a/g)$$
$$P = (M + m)a$$

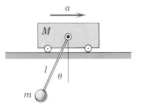

Problem 3/30

3/31 In a test of resistance to motion in an oil bath a small steel ball of mass m is released from rest at the surface ($y = 0$). If the resistance to motion is given by $R = kv$ where k is a constant, derive an expression for the depth h required for the ball to reach a velocity v.

$$\text{Ans. } h = \frac{m^2 g}{k^2} \ln\left(\frac{1}{1 - kv/(mg)}\right) - \frac{mv}{k}$$

Problem 3/31

3/32 If the steel ball of Prob. 3/31 is released from rest at the surface of a liquid in which the resistance to motion is $R = cv^2$ where c is a constant and v is the downward velocity of the ball, determine the depth h required to reach a velocity v.

3/33 The sliders A and B are connected by a light rigid bar of length $l = 0.5$ m and move with negligible friction in the horizontal slots shown. For the position when $x_A = 0.4$ m, the velocity of A is $v_A = 0.9$ m/s to the right. Determine the acceleration of each slider and the force in the bar at this instant.

Problem 3/33

Problem 3/34

Problem 3/35

Problem 3/36

Problem 3/37

3/34 For what value(s) of the angle θ will the acceleration of the 35-kg block be 9 m/s^2 to the right?

Ans. $\theta = 11.88°, 41.3°$

3/35 With the blocks initially at rest, the force P is increased slowly from zero to 260 N. Plot the accelerations of both masses as functions of P.

3/36 The inclined block A is given a constant rightward acceleration a. Determine the range of values of θ for which block B will not slip relative to block A, regardless of how large the acceleration a is. The coefficient of static friction between the blocks is μ_s.

Ans. $\tan^{-1}(1/\mu_s) \le \theta \le \dfrac{\pi}{2}$

▶**3/37** The hoisting drum A has a diameter d and turns counterclockwise with a constant angular velocity ω. Determine the tension T in the cable which attaches the mass m to the small pulley B. Express the result in terms of the variable y. The size, mass, and friction of the pulleys at C and B are negligible.

Ans. $T = mg\left[1 + \dfrac{\omega^2 b^2 d^2}{16gy^3}\right]$

▶ **3/38** The system is released from rest in the position shown. Calculate the tension T in the cord and the acceleration a of the 30-kg block. The small pulley attached to the block has negligible mass and friction. (*Suggestion:* First establish the kinematic relationship between the accelerations of the two bodies.)   *Ans. T = 138.0 N, a = 0.77 m/s²*

Problem 3/38

▶ **3/39** Two iron spheres, each of which is 100 mm in diameter, are released from rest with a center-to-center separation of 1 m. Assume an environment in space with no forces other than the force of mutual gravitational attraction and calculate the time t required for the spheres to contact each other and the absolute speed v of each sphere upon contact.

Ans. t = 13 h 33 min
v = 4.76(10⁻⁵) m/s

Problem 3/39

▶ **3/40** The lunar landing module has a mass of 17.5 Mg. As it descends onto the moon's surface, its retro-engine is shut off when it reaches a hovering condition a few meters above the moon's surface. The module then falls under the action of lunar gravity, and the coiled springs in the landing pads cushion the impact. If the stiffness of each of the three springs is 15 kN/m, determine the maximum magnitude of the "jerk" J (time rate of change of the acceleration) during compression of the springs. The acceleration due to lunar gravity is 1.62 m/s², and touchdown velocity is to be 1.5 m/s.

Ans. J = 4.65 m/s³

Problem 3/40

3/5 CURVILINEAR MOTION

We turn our attention now to the kinetics of particles which move along plane curvilinear paths. In applying Newton's second law, Eq. 3/3, we shall make use of the three coordinate descriptions of acceleration in curvilinear motion which we developed and applied in Arts. 2/4, 2/5, and 2/6.

The choice of coordinate system depends on the conditions of the problem and is one of the basic decisions to be made in solving curvilinear-motion problems. We now rewrite Eq. 3/3 in three ways the choice of which depends on which coordinate system is most appropriate.

Rectangular coordinates (Art. 2/4)

$$\Sigma F_x = ma_x$$
$$\Sigma F_y = ma_y$$

(3/6)

where $\qquad a_x = \ddot{x} \qquad$ and $\qquad a_y = \ddot{y}$

Normal and tangential coordinates (Art. 2/5)

$$\Sigma F_n = ma_n$$
$$\Sigma F_t = ma_t$$

(3/7)

where $\quad a_n = \rho\dot{\theta}^2 = v^2/\rho = v\dot{\theta}, \qquad a_t = \dot{v}, \qquad$ and $\qquad v = \rho\dot{\theta}$

Polar coordinates (Art. 2/6)

$$\Sigma F_r = ma_r$$
$$\Sigma F_\theta = ma_\theta$$

(3/8)

where $\qquad a_r = \ddot{r} - r\dot{\theta}^2 \qquad$ and $\qquad a_\theta = r\ddot{\theta} + 2\dot{r}\dot{\theta}$

In the application of these motion equations, the general procedure established in the previous article on rectilinear motion should be followed. After the motion is identified and the coordinate system is designated, the free-body diagram of the body to be treated as a particle should be drawn. The appropriate force summations are then obtained from this diagram in the usual way. The free-body diagram should be complete so that no mistakes are made in figuring the force summations. Once an assignment of reference axes is made, the expressions for both the forces and the acceleration must be consistent with that assignment. In the first of Eqs. 3/7, for example, the positive sense of the *n*-coordinate axis is *toward* the center of curvature, and so the positive sense of our force summation ΣF_n must also be *toward* the center of curvature to agree with the positive sense of the acceleration $a_n = v^2/\rho$.

Sample Problem 3/6

Determine the maximum speed v which the sliding block may have as it passes point A without losing contact with the surface.

Solution. The condition for loss of contact is that the normal force N which the surface exerts on the block goes to zero. Summing forces in the normal direction gives

$$[\Sigma F_n = ma_n] \qquad mg = m\frac{v^2}{\rho} \qquad v = \sqrt{g\rho}$$

If the speed at A were less than $\sqrt{g\rho}$, then an upward normal force exerted by the surface on the block would exist. In order for the block to have a speed at A which is greater than $\sqrt{g\rho}$, some type of constraint, such as a second curved surface above the block, would have to be introduced to provide additional downward force.

Sample Problem 3/7

Small objects are released from rest at A and slide down the smooth circular surface of radius R to a conveyor B. Determine the expression for the normal contact force N between the guide and each object in terms of θ and specify the correct angular velocity ω of the conveyor pulley of radius r to prevent any sliding on the belt as the objects transfer to the conveyor.

Solution. The free-body diagram of the object is shown together with the coordinate directions n and t. The normal force N depends on the n-component of the acceleration which, in turn, depends on the velocity. The velocity will be cumulative according to the tangential acceleration a_t. Hence we will find a_t first for any general position.

$$[\Sigma F_t = ma_t] \qquad mg \cos \theta = ma_t \qquad a_t = g \cos \theta$$

① Now we can find the velocity by integrating

$$[v\,dv = a_t\,ds] \qquad \int_0^v v\,dv = \int_0^\theta g \cos \theta\, d(R\theta) \qquad v^2 = 2gR \sin \theta$$

We obtain the normal force by summing forces in the positive n-direction, which is the direction of the n-component of acceleration.

$$[\Sigma F_n = ma_n] \qquad N - mg \sin \theta = m\frac{v^2}{R} \qquad N = 3\,mg \sin \theta \qquad Ans.$$

The conveyor pulley must turn at the rate $v = r\omega$ for $\theta = \pi/2$, so that

$$\omega = \sqrt{2gR}/r \qquad Ans.$$

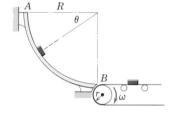

① It is essential here that we recognize the need to express the tangential acceleration as a function of position so that v may be found by integrating the kinematical relation $v\,dv = a_t\,ds$ in which all quantities are measured along the path.

Sample Problem 3/8

A 1500-kg car enters a section of curved road in the horizontal plane and slows down at a uniform rate from a speed of 100 km/h at A to a speed of 50 km/h as it passes C. The radius of curvature ρ of the road at A is 400 m and at C is 80 m. Determine the total horizontal force exerted by the road on the tires at positions A, B, and C. Point B is the inflection point where the curvature changes direction.

Solution. The car will be treated as a particle so that the effect of all forces exerted by the road on the tires will be treated as a single force. Since the motion is described along the direction of the road, normal and tangential coordinates will be used to specify the acceleration of the car. We will then determine the forces from the accelerations.

The constant tangential acceleration is in the negative t-direction and its magnitude is given by

① $[v_C^2 = v_A^2 + 2a_t\,\Delta s]$ $\quad a_t = \left|\dfrac{(50/3.6)^2 - (100/3.6)^2}{2(200)}\right| = 1.45 \text{ m/s}^2$

① Recognize the numerical value of the conversion factor from km/h to m/s as 1000/3600 or 1/3.6.

The normal components of acceleration at A, B, and C are

② $[a_n = v^2/\rho]$ At A, $a_n = \dfrac{(100/3.6)^2}{400} = 1.93 \text{ m/s}^2$

               At B, $a_n = 0$

               At C, $a_n = \dfrac{(50/3.6)^2}{80} = 2.41 \text{ m/s}^2$

② Note that a_n is always directed toward the center of curvature.

Application of Newton's second law in both the n- and t-directions to the free-body diagrams of the car gives

$[\Sigma F_t = ma_t]$ $\quad\quad\quad\quad F_t = 1500(1.45) = 2170 \text{ N}$

③ $[\Sigma F_n = ma_n]$ At A, $F_n = 1500(1.93) = 2894 \text{ N}$

               At B, $F_n = 0$

               At C, $F_n = 1500(2.41) = 3617 \text{ N}$

③ Note that the direction of F_n must agree with that of a_n.

Thus the total horizontal force acting on the tires becomes

At A, $F = \sqrt{F_n^2 + F_t^2} = \sqrt{(2894)^2 + (2170)^2} = 3617 \text{ N}$ *Ans.*

At B, $F = F_t = 2170 \text{ N}$ *Ans.*

④ At C, $F = \sqrt{F_n^2 + F_t^2} = \sqrt{(3617)^2 + (2170)^2} = 4218 \text{ N}$ *Ans.*

④ The angle made by **a** and **F** with the direction of the path can be computed if desired.

Sample Problem 3/9

Compute the magnitude v of the velocity required for the spacecraft S to maintain a circular orbit of altitude 320 kilometers above the surface of the earth.

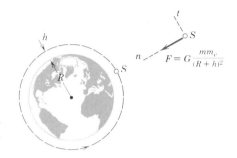

Solution. The only external force acting on the spacecraft is the
① force of gravitational attraction to the earth (i.e., its weight), as shown in the free-body diagram. Summing forces in the normal direction yields

$$[\Sigma F_n = ma_n] \quad G\frac{mm_e}{(R+h)^2} = m\frac{v^2}{(R+h)}, \quad v = \sqrt{\frac{Gm_e}{(R+h)}} = R\sqrt{\frac{g}{(R+h)}}$$

where the substitution $gR^2 = Gm_e$ has been made. Substitution of numbers gives

$$v = (6371)(1000)\sqrt{\frac{9.825}{(6371+320)(1000)}} = 7720 \text{ m/s} \quad \textit{Ans.}$$

① Note that, for observations made within an inertial frame of reference, there is no such quantity as "centrifugal force" acting in the minus n-direction. Note also that neither the spacecraft nor its occupants are "weightless," because the weight in each case is given by Newton's law of gravitation. For this altitude, the weights are only about 10% less than the earth-surface values. Finally, the term "zero-g" is also misleading. It is only when we make our observations with respect to a coordinate system which has an acceleration equal to the gravitational acceleration (such as in an orbiting spacecraft) that we appear to be in a "zero-g" environment. The quantity which does go to zero aboard orbiting spacecraft is the familiar normal force associated with, for example, an object in contact with a horizontal surface within the spacecraft.

Sample Problem 3/10

Tube A rotates about the vertical O-axis with a constant angular rate $\dot{\theta} = \omega$ and contains a small cylindrical plug B of mass m whose radial position is controlled by the cord which passes freely through the tube and shaft and is wound around the drum of radius b. Determine the tension T in the cord and the horizontal component F_θ of force exerted by the tube on the plug if the constant angular rate of rotation of the drum is ω_0 first in the direction for case (a) and second in the direction for case (b). Neglect friction.

Solution. With r a variable we use the polar-coordinate form of the equations of motion, Eqs. 3/8. The free-body diagram of B is shown in the horizontal plane and discloses only T and F_θ. The equations of motion are

$$[\Sigma F_r = ma_r] \qquad -T = m(\ddot{r} - r\dot{\theta}^2)$$
$$[\Sigma F_\theta = ma_\theta] \qquad F_\theta = m(r\ddot{\theta} + 2\dot{r}\dot{\theta})$$

Case (a) With $\dot{r} = +b\omega_0$, $\ddot{r} = 0$, and $\ddot{\theta} = 0$ the forces become

$$T = mr\omega^2 \qquad F_\theta = 2mb\omega_0\omega \qquad \textit{Ans.}$$

① **Case (b)** With $\dot{r} = -b\omega_0$, $\ddot{r} = 0$, and $\ddot{\theta} = 0$ the forces become

$$T = mr\omega^2 \qquad F_\theta = -2mb\omega_0\omega \qquad \textit{Ans.}$$

① The minus sign shows that F_θ is in the direction opposite to that shown on the free-body diagram.

PROBLEMS

Problem 3/41

Problem 3/42

Problem 3/43

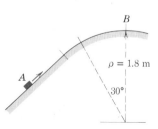

Problem 3/44

3/41 The small 2-kg block A slides down the curved path and passes the lowest point B with a speed of 3 m/s. If the radius of curvature at B is 1.5 m, determine the normal force N exerted on the block by the path at this point. Is knowledge of the friction properties necessary?

Ans. $N = 31.6$ N up, no

3/42 The 1.5-kg particle P is given an initial speed of 2 m/s at point A of the smooth guide, which is curved in the horizontal plane. If the horizontal force between the particle and the guide has a magnitude of 8 N at point B, determine the radius of curvature ρ of the path at this point.

3/43 If the 80-kg ski-jumper attains a speed of 25 m/s as he approaches the takeoff position, calculate the magnitude N of the normal force exerted by the snow on his skis just before he reaches A.

Ans. $N = 1791$ N

3/44 If the 2-kg block A passes over the top B of the circular portion of the path with a speed of 2 m/s, calculate the magnitude N of the normal force exerted by the path on the block.

3/45 The standard test to determine the maximum lateral acceleration of a car is to drive it around a 60-m-diameter circle painted on a level asphalt surface. The driver slowly increases the vehicle speed until he is no longer able to keep both wheel pairs straddling the line. If this maximum speed is 55 km/h for a 1400-kg car, determine its lateral acceleration capability a_n in g's and compute the magnitude F of the total friction force exerted by the pavement on the car tires.

Ans. $a_n = 0.793g$, $F = 10.89$ kN

Problem 3/45

3/46 A 0.8-kg slider is propelled upward at A along the fixed curved bar which lies in a vertical plane. If the slider is observed to have a speed of 4 m/s as it passes position B, determine (a) the magnitude N of the force exerted by the fixed rod on the slider and (b) the rate at which the speed of the slider is decreasing. Assume that friction is negligible.

Problem 3/46

3/47 In the design of a space station to operate outside the earth's gravitational field, it is desired to give the structure a rotational speed N which will simulate the effect of the earth's gravity for members of the crew. If the centers of the crew's quarters are to be located 12 m from the axis of rotation, calculate the necessary rotational speed N of the space station in revolutions per minute.

Ans. $N = 8.63$ rev/min

Problem 3/47

3/48 A child twirls a small 50-g ball attached to the end of a 1-m string so that the ball traces a circle in a vertical plane as shown. What is the minimum speed v which the ball must have when in position 1? If this speed is maintained throughout the circle, calculate the tension T in the string when the ball is in position 2. Neglect any small motion of the child's hand.

Problem 3/48

Problem 3/49

3/49 A jet transport plane flies in the trajectory shown in order to allow astronauts to experience the "weightless" condition similar to that aboard orbiting spacecraft. If the speed at the highest point is 900 km/h, what is the radius of curvature ρ necessary to exactly simulate the orbital "free-fall" environment?   *Ans. $\rho = 6371$ m*

3/50 Determine the altitude h (in kilometers) above the surface of the earth at which a satellite in a circular orbit has the same period, 23.9344 h, as the earth's absolute rotation. If such an orbit lies in the equatorial plane of the earth, it is said to be geosynchronous, because the satellite does not appear to move relative to an earth-fixed observer.

Problem 3/51

3/51 A small object A is held against the vertical side of the rotating cylindrical container of radius r due to centrifugal action. If the coefficient of static friction between the object and the container is μ_s, determine the expression for the minimum rotational rate $\dot{\theta} = \omega$ of the container which will keep the object from slipping down the vertical side.

$$Ans.\ \omega = \sqrt{\frac{g}{\mu_s r}}$$

3/52 Determine the height h and tension T in the cord for the conical pendulum of mass m and length l which rotates about the vertical axis at the angular rate $\dot{\theta} = \omega$.

Problem 3/52

Problem 3/53

3/53 At what angle θ should the racetrack turn of 450-m radius be banked in order that a racecar traveling at 200 km/h will have no tendency to slip sideways when rounding the turn?   *Ans. $\theta = 35.0°$*

3/54 A racetrack turn similar to that of Prob. 3/53 has a bank angle $\theta = 20°$, a radius of curvature $\rho = 300$ m, and a coefficient of static friction $\mu_s = 0.9$ between the rubber tires and asphalt surface. Determine the theoretical range of speeds over which the racecar will not slip sideways when negotiating the turn.

3/55 The hollow tube is pivoted about a horizontal axis through point O and is made to rotate in the vertical plane with a constant counterclockwise angular velocity $\dot{\theta} = 3$ rad/s. If a 0.1-kg particle is sliding in the tube toward O with a velocity of 1.2 m/s relative to the tube when the position $\theta = 30°$ is passed, calculate the magnitude N of the normal force exerted by the wall of the tube on the particle at this instant.

Ans. N = 0.1296 N

Problem 3/55

3/56 The barrel of a rifle is rotating in a horizontal plane about the vertical z-axis at the constant angular rate $\dot{\theta} = 0.5$ rad/s when a 60-g bullet is fired. If the velocity of the bullet relative to the barrel is 600 m/s just before it reaches the muzzle A, determine the resultant horizontal side thrust P exerted by the barrel on the bullet just before it emerges from A. On which side of the barrel does P act?

Problem 3/56

3/57 The slotted arm rotates about its center in a horizontal plane at the constant angular rate $\dot{\theta} = 10$ rad/s and carries a 1.5-kg spring-mounted slider which oscillates freely in the slot. If the slider has a speed of 600 mm/s relative to the slot as it crosses the center, calculate the horizontal side thrust P exerted by the slotted arm on the slider at this instant. Determine which side, A or B, of the slot is in contact with the slider.

Ans. P = 18.0 N, side A

Problem 3/57

Problem 3/58

Problem 3/59

Problem 3/61

3/58 The slotted arm revolves in the horizontal plane about the fixed vertical axis through point O. The 2-kg slider C is drawn toward O at the constant rate of 50 mm/s by pulling the cord S. At the instant for which $r = 225$ mm, the arm has a counterclockwise angular velocity $\omega = 6$ rad/s and is slowing down at the rate of 2 rad/s^2. For this instant, determine the tension T in the cord and the magnitude N of the force exerted on the slider by the sides of the smooth radial slot. Indicate which side, A or B, of the slot contacts the slider.

3/59 Calculate the tension T in a metal hoop rotating in its plane about its geometric axis with a large rim speed v. Analyze the differential element of the hoop as a particle. The mass per unit length of rim is ρ. *Ans.* $T = \rho v^2$

Problem 3/60

3/60 A 55-kg woman is standing on a scale placed on the cabin floor of a jet transport. At what rate $\dot{\theta}$ in degrees per second is the pilot raising his line of sight if the scales read 800 N when the aircraft is at the bottom of its trajectory and moving at 960 km/h? Also calculate the radius of curvature ρ of the flight path at this same condition.

3/61 The small object is placed on the inner surface of the conical dish at the radius shown. If the coefficient of static friction between the object and the conical surface is 0.30, for what range of angular velocities ω about the vertical axis will the block remain on the dish without slipping? Assume that speed changes are made slowly so that any angular acceleration may be neglected. *Ans.* $3.41 < \omega < 7.21$ rad/s

3/62 The small object of mass m is placed on the rotating conical surface at the radius shown. If the coefficient of static friction between the object and the rotating surface is 0.8, calculate the maximum angular velocity ω of the cone about the vertical axis for which the object will not slip. Assume very gradual angular velocity changes.

Problem 3/62

3/63 The small ball of mass m and its supporting wire become a simple pendulum when the horizontal cord is severed. Determine the ratio k of the tension T in the supporting wire immediately after the cord is cut to that in the wire before the cord is cut.   *Ans.* $k = \cos^2 \theta$

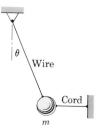

Problem 3/63

3/64 A small bead of mass m is carried by a circular hoop of radius r which rotates about a fixed vertical axis. Show how one might determine the angular speed ω of the hoop by observing the angle θ which locates the bead. Neglect friction in your analysis, but assume that a small amount of friction is present to damp out any motion of the bead relative to the hoop once a constant angular speed has been established. Note any restrictions on your solution.

Problem 3/64

Problem 3/65

Problem 3/66

Problem 3/67

3/65 A 2-kg sphere *S* is being moved in a vertical plane by a robotic arm. When the arm angle θ is 30°, its angular velocity about a horizontal axis through *O* is 50 deg/s clockwise and its angular acceleration is 200 deg/s^2 counterclockwise. In addition, the hydraulic element is being shortened at the constant rate of 500 mm/s. Determine the necessary minimum gripping force *P* if the coefficient of static friction between the sphere and the gripping surfaces is 0.5. Compare *P* to the minimum gripping force P_s required to hold the sphere in static equilibrium in the 30° position.

Ans. $P = 27.0$ N, $P_s = 19.6$ N

3/66 A tire with a diameter of 700 mm is spun up to 4000 rev/min on an off-the-car tire balance machine. A small round pebble is flung from a tread groove at this rotational speed. Estimate the magnitude *N* of the normal forces which were being exerted on the 10-g pebble by the sides of the groove if the coefficient of static friction between the pebble and the rubber is 0.95. Assume a rigid tread structure and neglect the effect of the weight of the pebble.

3/67 A flatbed truck starts from rest on a road whose constant radius of curvature is 30 m and whose bank angle is 10°. If the constant forward acceleration of the truck is 2 m/s^2, determine the time *t* after the start of motion at which the crate on the bed begins to slide. The coefficient of static friction between the crate and truck bed is $\mu_s = 0.3$, and the truck motion occurs in a horizontal plane.

Ans. $t = 5.58$ s

3/68 The 3-kg slider *A* fits loosely in the smooth 45° slot in the disk, which rotates in a horizontal plane about its center *O*. If *A* is held in position by a cord secured to point *B*, determine the tension *T* in the cord for a constant rotational velocity $\dot{\theta} =$ 300 rev/min. Would the direction of the velocity make any difference?   *Ans. T* = 314 N, no

150 mm

Problem 3/68

3/69 The 2-kg slider fits loosely in the smooth slot of the disk, which rotates about a vertical axis through point *O*. The slider is free to move slightly along the slot before one of the wires becomes taut. If the disk starts from rest at time *t* = 0 and has a constant clockwise angular acceleration of 0.5 rad/s², plot the tensions in wires 1 and 2 and the magnitude *N* of the force normal to the slot as functions of time *t* for the interval $0 \leq t \leq 5$ s.

100 mm

Problem 3/69

3/70 The rocket moves in a vertical plane and is being propelled by a thrust *T* of 32 kN. It is also subjected to an atmospheric resistance *R* of 9.6 kN. If the rocket has a velocity of 3 km/s and if the gravitational acceleration is 6 m/s² at the altitude of the rocket, calculate the radius of curvature ρ of its path for the position described and the time rate of change of the magnitude *v* of the velocity of the rocket. The mass of the rocket at the instant considered is 2000 kg.
        Ans. ρ = 3000 km, $\dot{v}$ = 6.00 m/s²

Vertical

−30°

Problem 3/70

3/71 A small coin is placed on the horizontal surface of the rotating disk. If the disk starts from rest and is given a constant angular acceleration $\ddot{\theta} = \alpha$, determine an expression for the number of revolutions *N* through which the disk turns before the coin slips. The coefficient of static friction between the coin and the disk is μ_s.

Vertical

r

Problem 3/71

Problem 3/72

Problem 3/74

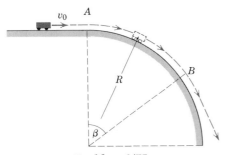

Problem 3/75

3/72 The disk with the circular groove rotates about the vertical axis with a constant speed of 30 rev/min and carries the two 4-kg spheres. Calculate the larger of the two forces of contact between the disk and each sphere. (Can this result be reached by using only one force equation?)

Ans. $R = 32.5$ N, yes

3/73 A body at rest relative to the surface of the earth rotates with the earth and therefore moves in a circular path about the polar axis of the earth considered as fixed. Derive an expression for the ratio k of the apparent weight of such a body as measured by a spring scale at the equator (calibrated to read the actual force applied) to the true weight of the body, which is the absolute gravitational attraction to the earth. The absolute acceleration due to gravity at the equator is $g = 9.815$ m/s². The radius of the earth at the equator is $R = 6378$ km, and the angular velocity of the earth is $\omega = 0.729(10^{-4})$ rad/s. If the true weight is 100 N, what is the apparent measured weight W'?

3/74 The slotted arm revolves with a constant angular velocity $\dot{\theta} = \omega$ about a vertical axis through the center O of the fixed cam. The cam is cut so that the radius to the path of the center of the pin A varies according to $r = r_0 + b \sin N\omega t$ where N equals the number of lobes—six in this case. If $\omega = 12$ rad/s, $r_0 = 100$ mm, and $b = 10$ mm, and if the spring compression varies from 11.5 N to 19.1 N from valley to crest, calculate the magnitude R of the force between the cam and the 100-g pin A as it passes over the top of the lobe in the position shown. Neglect friction.

Ans. $R = 12.33$ N

3/75 A small vehicle enters the top A of the circular path with a horizontal velocity v_0 and gathers speed as it moves down the path. Determine the expression for the angle β to the position where the vehicle leaves the path and becomes a projectile. Neglect friction and treat the vehicle as a particle.

Ans. $\beta = \cos^{-1}\left(\dfrac{2}{3} + \dfrac{v_0^2}{3gR}\right)$

3/76 A stretch of highway includes a succession of evenly spaced dips and humps, the contour of which may be represented by the relation $y = b \sin (2\pi x/L)$. What is the maximum speed at which the car A can go over a hump and still maintain contact with the road? If the car maintains this critical speed, what is the total reaction N under its wheels at the bottom of a dip? The mass of the car is m.

Problem 3/76

3/77 The car has a speed of 70 km/h at the bottom of the dip when the driver applies the brakes, causing a deceleration of 0.5g. What is the minimum seat cushion angle θ for which a package will not slide forward? The coefficient of static friction between the package and seat cushion is (a) 0.2, (b) 0.4.

Ans. (a) $\theta = 7.34°$
(b) $\theta = -3.16°$

Problem 3/77

3/78 Repeat Prob. 3/77, except let the car be at the top of a hump as shown in the figure.

Problem 3/78

3/79 A small rocket-propelled vehicle of mass m travels down the circular path of effective radius r under the action of its weight and a constant thrust T from its rocket motor. If the vehicle starts from rest at A, determine its speed v when it reaches B and the magnitude N of the force exerted by the guide on the wheels just prior to reaching B. Neglect any friction and any loss of mass of the rocket.

$$Ans.\ v = \sqrt{r\left(\frac{\pi T}{m} + 2g\right)}$$

$$N = T\pi + 3mg$$

Problem 3/79

Problem 3/80

Problem 3/81

Problem 3/83

3/80 A satellite S, which has a mass of 150 kg, is put into an elliptical orbit around the earth. At the instant represented the satellite is a distance $r = $ 10 729 km from the center of the earth and has a velocity $v = 26\ 730$ km/h making an angle $\beta = 60°$ with the radial line through S. The only force acting on the satellite is the gravitational attraction of the earth, which amounts to 519 N for these conditions. Calculate the value of $\ddot{r}$ for the condition described.   *Ans.* $\ddot{r} = 0.39$ m/s^2

3/81 The particle P is released at time $t = 0$ from the position $r = r_0$ inside the smooth tube with no velocity relative to the tube, which is driven at the constant angular velocity ω_0 about a vertical axis. Determine the radial velocity v_r, the radial position r, and the transverse velocity v_θ as functions of time t. Explain why the radial velocity increases with time in the absence of radial forces. Plot the absolute path of the particle during the time it is inside the tube for $r_0 = 0.1$ m, $l = 1$ m, and $\omega_0 = 1$ rad/s.

$$Ans.\ v_r = \frac{r_0\omega_0}{2}\left[e^{\omega_0 t} - e^{-\omega_0 t}\right] = r_0\omega_0 \sinh \omega_0 t$$

$$r = \frac{r_0}{2}\left[e^{-\omega_0 t} + e^{\omega_0 t}\right] = r_0 \cosh \omega_0 t$$

$$v_\theta = \frac{r_0\omega_0}{2}\left[e^{-\omega_0 t} + e^{\omega_0 t}\right] = r_0\omega_0 \cosh \omega_0 t$$

3/82 Remove the assumption of smooth surfaces as stated in Prob. 3/81 and assume a coefficient of kinetic friction μ_k between the particle and rotating tube. Determine the radial position r of the particle as a function of time t if it is released with no relative velocity at $r = r_0$ when $t = 0$. Assume that static friction is overcome.

3/83 The motion of the 0.5-kg pin A in the circular slot is controlled by the guide B which is being elevated by its lead screw with a constant upward velocity v_0 of 2 m/s for an interval of its motion. Calculate the magnitude P of the force exerted on the pin by the circular guide as the pin passes the position for which $\theta = 30°$. Friction is negligible.
          Ans. $P = 14.22$ N

3/84 The centrifugal pump with smooth radial vanes rotates about its vertical axis with a constant angular velocity $\dot{\theta} = \omega$. Find the magnitude N of the force exerted by a vane on a particle of mass m as it moves out along the vane. The particle is introduced at $r = r_0$ without radial velocity. Assume that the particle contacts the side of the vane only.

$$Ans. \ N = 2m\omega^2\sqrt{r^2 - r_0^2}$$

Problem 3/84

▶**3/85** Each tire on the 1350-kg car can support a maximum friction force parallel to the road surface of 2500 N. This force limit is nearly constant over all possible rectilinear and curvilinear car motions and is attainable only if the car does not skid. Under this maximum braking, determine the total stopping distance s if the brakes are first applied at point A when the car speed is 25 m/s and if the car follows the centerline of the road.

$$Ans. \ s = 47.4 \ m$$

Problem 3/85

▶**3/86** A hollow tube rotates about the horizontal axis through point O with constant angular velocity ω_0. A particle of mass m is introduced with zero relative velocity at $r = 0$ when $\theta = 0$ and slides outward through the smooth tube. Determine r as a function of θ.

Problem 3/86

Problem 3/87

Problem 3/88

Problem 3/89

Problem 3/90

▶ 3/87 A small collar of mass m is given an initial velocity of magnitude v_0 on the horizontal circular track fabricated from a slender rod. If the coefficient of kinetic friction is μ_k, determine the distance traveled before the collar comes to rest. (*Hint:* Recognize that the friction force depends upon the net normal force.)

$$Ans. \ s = \frac{r}{2\mu_k} \ln \left[\frac{v_0{}^2 + \sqrt{v_0{}^4 + r^2 g^2}}{rg} \right]$$

▶ 3/88 A flexible bicycle-type chain of length $\pi r/2$ has a mass ρ per unit length and is released after being held by its upper end in an initial rest condition in the smooth circular channel. Determine the acceleration a_t which all links experience just after release. Also find the expression for the tension T in the chain as a function of θ for the condition immediately following release. Isolate a differential element of the chain as a free body and apply the appropriate motion equation.

$$Ans. \ a_t = \frac{2g}{\pi}, \ T = \rho g r \left[\frac{2\theta}{\pi} - \sin \theta \right]$$

▶ 3/89 The coefficient of kinetic friction between the particle of mass m and the U-shaped horizontal guide tube is μ_k. For what range of initial speeds v_A at point A will the particle stop between points B and C?

$$Ans. \ \sqrt{2\mu_k g l} \le v_A \le \sqrt{rg \sinh (2\mu_k \pi) + 2\mu_k g l}$$

▶ 3/90 The slotted arm of Prob. 2/131, repeated here, rotates in a horizontal plane around the fixed cam with a constant counterclockwise velocity $\omega = 20$ rad/s. The spring has a stiffness of 5.4 kN/m and is uncompressed with $\theta = 0$. The cam has the shape $r = b - c \cos \theta$. If $b = 100$ mm, $c = 75$ mm, and the smooth roller A has a mass of 0.5 kg, find the force P exerted on A by the smooth sides of the slot for the position in which $\theta = 60°$.

$$Ans. \ P = 231 \text{ N}$$

SECTION B. WORK AND ENERGY

3/6 WORK AND KINETIC ENERGY

In the previous two articles we applied Newton's second law $\mathbf{F} = m\mathbf{a}$ to various problems of particle motion to establish the instantaneous relationship between the net force acting on a particle and the resulting acceleration of the particle. When intervals of motion were involved where the change in velocity or the corresponding displacement of the particle was required, we integrated the computed acceleration over the interval by using the appropriate kinematical equations.

There are two general classes of problems where the cumulative effects of unbalanced forces acting on a particle over an interval of motion are of interest to us. These cases involve, respectively, integration of the forces with respect to the displacement of the particle and integration of the forces with respect to the time they are applied. We may incorporate the results of these integrations directly into the governing equations of motion so that it becomes unnecessary to solve directly for the acceleration. Integration with respect to displacement leads to the equations of work and energy, which are the subject of this article. Integration with respect to time leads to the equations of impulse and momentum, discussed in Art. 3/8.

(*a*) *Work.*   The quantitative meaning of the term work will now be developed.* The work done by a force $\mathbf{F}$ during a differential displacement $d\mathbf{s}$ of its point of application O, Fig. 3/2a, is by definition the scalar quantity

$$dU = \mathbf{F} \cdot d\mathbf{s}$$

The magnitude of this dot product of the vector force and the vector displacement is $dU = F\,ds\cos\alpha$, where α is the angle between $\mathbf{F}$ and $d\mathbf{s}$. This expression may be interpreted as the displacement multiplied by the force component $F_t = F\cos\alpha$ in the direction of the displacement, as represented by the dotted lines in Fig. 3/2b. Alternatively the work dU may be interpreted as the force multiplied by the displacement component $ds\cos\alpha$ in the direction of the force, as represented by the full lines in Fig. 3/2b. With this definition of work, it should be noted that the component $F_n = F\sin\alpha$ normal to the displacement does no work. Work is positive if the working component F_t is in the direction of the displacement and negative if it is in the opposite direction.

(a)

(b)

Figure 3/2

---

* The concept of work was also developed in the study of virtual work in Chapter 7 of *Vol. 1 Statics.*

In SI units work has the units of force (N) times displacement (m) or N·m. This unit is given the special name *joule* (J) which is defined as the work done by a force of 1 N moving through a distance of 1 m in the direction of the force. Consistent use of the joule for work (and energy) rather than the units N·m will avoid possible ambiguity with the units of moment of a force or torque which are also written N·m.

It should be noted that work is a scalar as given by the dot product and involves the product of a force and a distance, both measured along the same line. Moment, on the other hand, is a vector as given by the cross product and involves the product of force and distance measured at right angles to the force.

During a finite movement s of the point of application of a force the force does an amount of work equal to

$$U = \int \mathbf{F} \cdot d\mathbf{s} = \int (F_x \, dx + F_y \, dy + F_z \, dz)$$

or

$$U = \int F_t \, ds$$

Figure 3/3

In order to carry out this integration it is necessary to know the relation between the force components and their respective coordinates or the relations between F_t and s. If the functional relationship is not known as a mathematical expression which can be integrated but is specified in the form of approximate or experimental data, then the work may be evaluated by carrying out a numerical or graphical integration which would be represented by the area under the curve of F_t versus s, as shown in Fig. 3/3.

A common example of the work done on a body by a variable force is found in the action of a spring on a movable body to which it is attached. We consider here the common linear spring of stiffness k where the force F in the spring, tension or compression, is proportional to its deformation x, so that $F = kx$. Figure 3/4 shows the two cases where the body is moved (by forces not shown) so as to stretch the spring a distance x or to compress the spring a distance x. The force exerted by the spring on the body in each case is in the sense *opposite* to the displacement so does *negative* work on the body. Thus, for both stretching and compressing the spring, the work done *on* the body is *negative* and is given by

$$U_{1\text{-}2} = -\int_{x_1}^{x_2} F \, dx = -\int_{x_1}^{x_2} kx \, dx = -\tfrac{1}{2}k(x_2{}^2 - x_1{}^2)$$

When a spring under tension is being relaxed or when a spring under compression is being relaxed, then we see from both examples in Fig. 3/4 that the deformation changes from x_2 to a

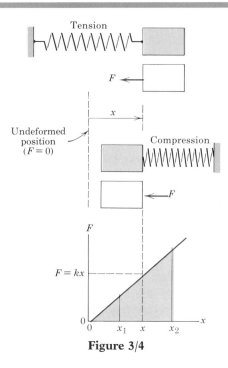

Figure 3/4

lesser deformation x_1. For this condition the force acting on the body in both cases is in the *same* sense as the displacement, and, therefore, the work done *on* the body is *positive*.

The magnitude of the work, positive or negative, is seen to be equal to the shaded trapezoidal area shown for both cases in Fig. 3/4. In calculating the work done by a spring force care must be taken to see that the units for k and x are consistent. Thus if x is in meters, k must be in N/m.

The expression $F = kx$ is actually a static relationship which is true only when elements of the spring have no acceleration. Dynamic behavior of a spring when its mass is accounted for is a fairly complex problem which will not be treated here. We shall assume that the mass of the spring is small compared with the masses of other accelerating parts of the system, in which case the linear static relationship will not involve appreciable error.

We now consider the work done on a particle of mass m, Fig. 3/5, moving along a curved path under the action of the force **F** which stands for the resultant Σ**F** of all forces acting on the particle. The position of m is established by the position vector **r**, and its displacement along its path during time dt is represented by the change $d\mathbf{r}$ in its position vector. The work dU done by **F**

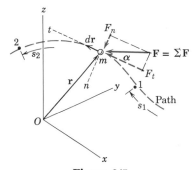

Figure 3/5

during this displacement is defined by the dot product

$$dU = \mathbf{F} \cdot d\mathbf{r}$$

This expression represents the scalar magnitude $F_t|d\mathbf{r}| = F_t\,ds = (F\cos\alpha)\,ds$ where s is the scalar distance measured along the curve.

If F_t is in the direction of $d\mathbf{r}$ as pictured, the work is positive. If F_t is in the direction opposite to $d\mathbf{r}$, the work is negative. The component F_n of the force normal to the direction of the path can do no work since its dot product with the displacement $d\mathbf{r}$ is zero.

The work done by $\mathbf{F}$ during a finite movement of the particle from point 1 to point 2 is

$$U = \int \mathbf{F} \cdot d\mathbf{r} = \int_{s_1}^{s_2} F_t\,ds$$

where the limits specify the initial and final end points of the interval of motion involved. When we substitute Newton's second law $\mathbf{F} = m\mathbf{a}$, the expression for the work of all forces becomes

$$U = \int \mathbf{F} \cdot d\mathbf{r} = \int m\mathbf{a} \cdot d\mathbf{r}$$

But $\mathbf{a} \cdot d\mathbf{r} = a_t\,ds$ where a_t is the tangential component of the acceleration of m and ds is the magnitude of $d\mathbf{r}$. In terms of the velocity v of the particle, Eq. 2/3 gives $a_t\,ds = v\,dv$. Thus the expression for the work of $\mathbf{F}$ becomes

$$U = \int \mathbf{F} \cdot d\mathbf{r} = \int mv\,dv$$

or

$$U = \int \mathbf{F} \cdot d\mathbf{r} = \tfrac{1}{2}m(v_2{}^2 - v_1{}^2) \tag{3/9}$$

where the integration is carried out between points 1 and 2 along the curve at which points the velocities have the magnitudes v_1 and v_2, respectively.

(b) Kinetic energy. The *kinetic energy* T of the particle is defined as

$$\boxed{T = \tfrac{1}{2}mv^2} \tag{3/10}$$

and is the total work which must be done on the particle to bring it from a state of rest to a velocity v. Kinetic energy T is a scalar quantity with the units of $N \cdot m$ or *joules* (J). Kinetic energy is *always* positive regardless of the direction of the velocity. Equation 3/9 may be restated simply as

$$U = \Delta T$$ (3/11)

which is the *work-energy equation* for a particle.

The equation states that the *total work done* by all forces acting on a particle during an interval of its motion equals the corresponding *change in kinetic energy* of the particle. Although T is always positive, the change ΔT may be positive, negative, or zero.

Alternatively, the work-energy relation may be thought of as the initial kinetic energy T_1 plus the work done U equals the final kinetic energy T_2 or

$$T_1 + U = T_2$$ (3/11a)

We now see from Eq. 3/11 that a major advantage of the method of work and energy is that it avoids the necessity of computing the acceleration and leads directly to the velocity changes as functions of the forces which do work. Further, the work-energy equation involves only those forces which do work and thus give rise to changes in the magnitude of the velocities.

We consider now two particles joined together by a connection which is frictionless and incapable of any deformation. The forces in the connection constitute a pair of equal and opposite forces, and the points of application of these forces necessarily have identical displacement components in the direction of the forces. Hence the net work done by these internal forces is zero during any movement of the system of the two connected particles. Thus Eq. 3/11 is applicable to the entire system, where U is the total or net work done on the system by forces external to it and ΔT is the change, $T_2 - T_1$, in the total kinetic energy of the system. The total kinetic energy is the sum of the kinetic energies of both elements of the system. It may now be observed that a further advantage of the work-energy method is that it permits the analysis of a system of particles joined in the manner described without dismembering the system.

Application of the work-energy method calls for an isolation of the particle or system under consideration. For a single particle a *free-body diagram* showing all externally applied forces should be drawn. For a system of particles rigidly connected without springs an *active-force diagram* which shows only those external forces which do work (active forces) on the entire system may be drawn.*

---

* The active-force diagram was introduced in the method of virtual work in statics. See Chapter 7 of *Vol. 1 Statics.*

(*c*) *Power.*   The capacity of a machine is measured by the time rate at which it can do work or deliver energy. The total work or energy output is not a measure of this capacity since a motor, no matter how small, can deliver a large amount of energy if given sufficient time. On the other hand a large and powerful machine is required to deliver a large amount of energy in a short period of time. Thus the capacity of a machine is rated by its *power*, which is defined as the *time rate of doing work*.

Accordingly, the power P developed by a force $\mathbf{F}$ which does an amount of work U is $P = dU/dt = \mathbf{F} \cdot d\mathbf{r}/dt$. Since $d\mathbf{r}/dt$ is the velocity $\mathbf{v}$ of the point of application of the force, we have

$$\boxed{P = \mathbf{F} \cdot \mathbf{v}} \tag{3/12}$$

Power is clearly a scalar quantity, and in SI units it has the units of $N \cdot m/s = J/s$. The special unit for power is the *watt* (W), which equals one joule per second (J/s).

(*d*) *Efficiency.*   The ratio of the work done *by* a machine to the work done *on* the machine during the same time interval is called the *mechanical efficiency* e_m of the machine. This definition assumes that the machine operates uniformly so that there is no accumulation or depletion of energy within it. Efficiency is always less than unity since every device operates with some loss of energy and since energy cannot be created within the machine. In mechanical devices which involve moving parts there will always be some loss of energy due to the negative work of kinetic friction forces. This work is converted to heat energy which in turn is dissipated to the surroundings. The mechanical efficiency at any instant of time may be expressed in terms of mechanical power P by

$$e_m = \frac{P_{\text{output}}}{P_{\text{input}}} \tag{3/13}$$

In addition to energy loss by mechanical friction there may also be electrical and thermal energy loss in which case the *electrical efficiency* e_e and *thermal efficiency* e_t are also involved. The *overall efficiency* e in such instances would be

$$e = e_m e_e e_t$$

Sample Problem 3/11

Calculate the velocity v of the 50-kg crate when it reaches the bottom of the chute at B if it is given an initial velocity of 4 m/s down the chute at A. The coefficient of kinetic friction is 0.30.

Solution. The free-body diagram of the crate is drawn and includes the normal force R and the kinetic friction force F calculated in the usual manner. The work done by the component of the weight down the plane is positive, whereas that done by the friction force is negative. The total work done on the crate during the motion is

① $[U = Fs]$ $U = [50(9.81) \sin 15° - 142]10 = -152$ J

The change in kinetic energy is $T_2 - T_1 = \Delta T$

$[T = \frac{1}{2}mv^2]$ $\Delta T = \frac{1}{2}(50)(v^2 - 4^2)$

The work-energy equation gives

$[U = \Delta T]$ $-152 = 25(v^2 - 16)$

$$v^2 = 9.93 \text{ (m/s)}^2   v = 3.15 \text{ m/s}   \textit{Ans.}$$

50(9.81) N

$\mu_k R = 142$ N

$R = 474$ N

① Since the net work done is negative, we obtain a decrease in the kinetic energy.

Sample Problem 3/12

The flatbed truck, which carries an 80-kg crate, starts from rest and attains a speed of 72 km/h in a distance of 75 m on a level road with constant acceleration. Calculate the work done by the friction force acting on the crate during this interval if the static and kinetic coefficients of friction between the crate and the truck bed are (a) 0.30 and 0.28, respectively, or (b) 0.25 and 0.20, respectively.

Solution. If the crate does not slip on the bed, its acceleration will be that of the truck, which is

$[v^2 = 2as]$ $a = \dfrac{v^2}{2s} = \dfrac{(72/3.6)^2}{2(75)} = 2.67$ m/s^2

Case (a). This acceleration requires a friction force on the block of

$[F = ma]$ $F = 80(2.67) = 213$ N

which is less than the maximum possible value of $\mu_s N = 0.30(80)(9.81) = 235$ N. Therefore the crate does not slip and the work done by the actual static friction force of 213 N is

① $[U = Fs]$ $U = 213(75) = 16\,000$ J   or   16.0 kJ

Case (b). For $\mu_s = 0.25$ the maximum possible friction force is $0.25(80)(9.81) = 196$ N, which is slightly less than the value of 213 N required for no slipping. Therefore we conclude that the crate slips, and the friction force is governed by the kinetic coefficient and is $F = 0.20(80)(9.81) = 157$ N. The acceleration becomes

$[F = ma]$ $a = F/m = 157/80 = 1.96$ m/s^2

The distances traveled by the crate and the truck are in proportion to their accelerations. Thus the crate has a displacement of $(1.96/2.67)75 = 55.2$ m, and the work done by kinetic friction is

② $[U = Fs]$ $U = 157(55.2) = 8660$ J   or   8.66 kJ   *Ans.*

80(9.81) N

F a

80(9.81) N

① We note that static friction forces do no work when the contacting surfaces are both at rest. When they are in motion, however, as in this problem, the static friction force acting on the crate does positive work and that acting on the truck bed does negative work.

② This problem shows that a kinetic friction force can do positive work when the surface which supports the object and generates the friction force is in motion. If the supporting surface is at rest, then the kinetic friction force acting on the moving part always does negative work.

Sample Problem 3/13

The 50-kg block at A is mounted on rollers so that it moves along the fixed horizontal rail with negligible friction under the action of the constant 300-N force in the cable. The block is released from rest at A with the spring to which it is attached extended an initial amount $x_1 = 0.233$ m. The spring has a stiffness $k = 80$ N/m. Calculate the velocity v of the block as it reaches position B.

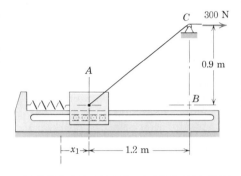

Solution. It will be assumed initially that the stiffness of the spring is small enough to allow the block to reach position B. The active-force diagram for the system composed of both block and cable is shown for a general position. The spring force $80x$ and the 300-N tension are the only forces external to this system which do work on the system. The force exerted on the block by the rail, the weight of the block, and the reaction of the small pulley on the cable do no work on the system and are not included on the active-force diagram.

As the block moves from $x = 0.233$ m to $x = 0.233 + 1.2 = 1.433$ m, the work done by the spring force acting on the block is negative and equals

① $[U = \int F \, dx]$ $\qquad U = -\int_{0.233}^{1.433} 80x \, dx = -40x^2 \Big]_{0.233}^{1.433} = -80.0$ J

① If the variable x had been measured from the starting position A, the spring force would be $80(0.233 + x)$, and the limits of integration would be 0 and 1.2 m.

The work done on the system by the constant 300-N force in the cable is the force times the net horizontal movement of the cable over pulley C, which is $\sqrt{(1.2)^2 + (0.9)^2} - 0.9 = 0.6$ m. Thus the work done is $300(0.6) = 180$ J. We now apply the work-energy equation to the system and get

$[U = \Delta T] \qquad -80.0 + 180 = \tfrac{1}{2}(50)(v^2 - 0) \qquad v = 2.0$ m/s $\qquad$ *Ans.*

We take special note of the advantage to our choice of system. If the block alone had constituted the system, the horizontal component of the 300-N cable tension acting on the block would have to be integrated over the 1.2-m displacement. This step would require considerably more effort than was needed in the solution as presented. If there had been appreciable friction between the block and its guiding rail, we would have found it necessary to isolate the block alone in order to compute the variable normal force and, hence, the variable friction force. Integration of the friction force over the displacement would then be required to evaluate the negative work which it would do.

Sample Problem 3/14

The power winch A hoists the 360-kg log up the 30° incline at a constant speed of 1.2 m/s. If the power output of the winch is 4 kW, compute the coefficient of kinetic friction μ_k between the log and the incline. If the power is suddenly increased to 6 kW, what is the corresponding instantaneous acceleration a of the log?

Solution. From the free-body diagram of the log we get $N = 360(9.81) \cos 30° = 3058$ N, and the kinetic friction force becomes $3058\mu_k$. For constant speed the forces are in equilibrium so that

$[\Sigma F_x = 0]$ $T - 3058\mu_k - 360(9.81) \sin 30° = 0$ $T = 3058\mu_k + 1766$

The power output of the winch gives the tension in the cable

① $[P = Tv]$ $T = P/v = 4000/1.2 = 3333$ N

Substituting T gives $3333 = 3058\mu_k + 1766$ $\mu_k = 0.513$ *Ans.*

When the power is increased, the tension momentarily becomes

$[P = Tv]$ $T = P/v = 6000/1.2 = 5000$ N

and the corresponding acceleration is given by

$[\Sigma F_x = ma_x]$ $5000 - 3058(0.513) - 360(9.81) \sin 30° = 360a$

②
$$a = 4.63 \text{ m/s}^2 \qquad\qquad Ans.$$

① Note the conversion from kilowatts to watts. Also remember to use J/s rather than N·m/s.

② As the speed increases, the acceleration will drop until the speed stabilizes at a value higher than 1.2 m/s.

Sample Problem 3/15

A satellite of mass m is put into an elliptical orbit around the earth. At point A its distance from the earth is $h_1 = 500$ km and it has a velocity $v_1 = 30\ 000$ km/h. Determine the velocity v_2 of the satellite as it reaches point B, a distance $h_2 = 1200$ km from the earth.

Solution. The satellite is moving outside of the earth's atmosphere so that the only force acting on it is the gravitational attraction of the earth. With the mass and radius of the earth expressed by m_e and R, respectively, the gravitational law of Eq. 1/2 gives $F = Gmm_e/r^2 = gR^2m/r^2$ when the substitution $Gm_e = gR^2$ is made for the surface values $F = mg$ and $r = R$. The work done by F is due only to the radial component of motion along the line of action of F and is negative for increasing r.

$$U = -\int_{r_1}^{r_2} F\,dr = -mgR^2 \int_{r_1}^{r_2} \frac{dr}{r^2} = mgR^2 \left(\frac{1}{r_2} - \frac{1}{r_1} \right)$$

The work-energy equation $U = \Delta T$ gives

① $mgR^2 \left(\dfrac{1}{r_2} - \dfrac{1}{r_1} \right) = \tfrac{1}{2} m(v_2{}^2 - v_1{}^2)$ $v_2{}^2 = v_1{}^2 + 2gR^2 \left(\dfrac{1}{r_2} - \dfrac{1}{r_1} \right)$

② Substituting the numerical values gives

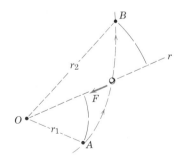

$$2(9.81)[(6371)(10^3)]^2 \left(\frac{10^{-3}}{6371 + 1200} - \frac{10^{-3}}{6371 + 500} \right)$$

$= 69.44(10^6) - 10.72(10^6) = 58.73(10^6)$ (m/s)^2

$v_2 = 7663$ m/s   or $v_2 = 7663(3.6) = 27\ 590$ km/h   *Ans.*

① Note that the result is independent of the mass of the satellite.

② Consult Table D/2, Appendix D, to find the radius R of the earth.

PROBLEMS

Problem 3/91

3/91 In the design of a spring bumper for a 1500-kg car, it is desired to bring the car to a stop from a speed of 8 km/h in a distance equal to 150 mm of spring deformation. Specify the required stiffness k for each of the two springs behind the bumper. The springs are undeformed at the start of impact.
  Ans. $k = 164.6$ kN/m

Problem 3/92

3/92 The 0.8-kg collar slides with negligible friction on the fixed rod in the vertical plane. If the collar starts from rest at A under the action of the constant 8-N horizontal force, calculate its velocity v as it hits the stop at B.   *Ans. $v = 4.73$ m/s*

Problem 3/93

3/93 The crawler wrecking crane is moving with a constant speed of 3 km/h when it is suddenly brought to a stop. Compute the maximum angle through which the cable of the wrecking ball swings.

Problem 3/94

3/94 The 30-kg crate slides down the curved path in the vertical plane. If the crate has a velocity of 1.2 m/s down the incline at A and a velocity of 8 m/s at B, compute the work U_f done on the crate by friction during the motion from A to B.
  Ans. $U_f = -827$ J

3/95 Floors which support heavy machinery must be designed to withstand the loads induced when a machine is accidentally dropped. If a machine of mass m is dropped by an overhead traveling crane from a negligible height above the floor, show that the maximum force transmitted to the floor is twice the weight of the machine. Assume that the floor deformation is proportional to the force applied to it.

3/96 The man and his bicycle have a combined mass of 95 kg. What power P is the man developing in riding up a 5-percent grade at a constant speed of 20 km/h?   *Ans. P = 259 W*

Problem 3/96

3/97 The position vector of a particle is given by $\mathbf{r} = 8t\mathbf{i} + 1.2t^2\mathbf{j} - 0.5(t^3 - 1)\mathbf{k}$ where t is the time in seconds from the start of the motion and where $\mathbf{r}$ is expressed in meters. For the condition when $t = 4$ s, determine the power P developed by the force $\mathbf{F} = 40\mathbf{i} - 20\mathbf{j} - 36\mathbf{k}$ N which acts on the particle.

3/98 A department store escalator handles a steady load of 30 people per minute in elevating them from the first to the second floor through a vertical rise of 7 m. The average person has a mass of 65 kg. If the motor which drives the unit delivers 3 kW, calculate the mechanical efficiency e of the system.   *Ans. e = 0.744*

Problem 3/98

3/99 The 4-kg ball and the attached light rod rotate in the vertical plane about the fixed axis at O. If the assembly is released from rest at $\theta = 0$ and moves under the action of the 60-N force, which is maintained normal to the rod, determine the velocity v of the ball as θ approaches 90°. Treat the ball as a particle.

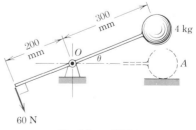

Problem 3/99

3/100 A car with a mass of 1500 kg starts from rest at the bottom of a 10-percent grade and acquires a speed of 50 km/h in a distance of 100 m with constant acceleration up the grade. What is the power P delivered to the drive wheels by the engine when the car reaches this speed?   *Ans.* $P = 40.4$ kW

3/101 The motor unit A is used to elevate the 300-kg cylinder at a constant rate of 2 m/s. If the power meter B registers an electrical input of 2.20 kW, calculate the combined electrical and mechanical efficiency e of the system.

Problem 3/101

3/102 The 6-kg cylindrical collar is released from rest in the position shown and drops onto the spring. Calculate the velocity v of the cylinder when the spring has been compressed 50 mm.   *Ans.* $v = 2.41$ m/s

Problem 3/102

3/103 The resistance R to penetration x of a 0.25-kg projectile fired with a velocity of 600 m/s into a certain block of fibrous material is shown in the graph. Represent this resistance by the dotted line and compute the velocity v of the projectile for the instant when $x = 25$ mm if the projectile is brought to rest after a total penetration of 75 mm.

Problem 3/103

3/104 The pressure p in a certain rifle barrel varies with the position of the bullet as shown in the graph. If the diameter of the bore is 7.5 mm and the mass of the bullet is 14 g, determine the muzzle velocity v of the bullet. Neglect the effect of friction in the barrel compared with the force of the gases on the bullet. Observe that one megapascal (MPa) equals 10^6 newtons per square meter (N/m²).

Ans. $v = 940$ m/s

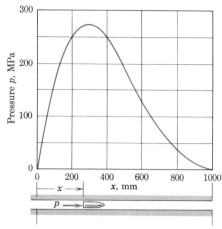

Problem 3/104

3/105 Each of the two systems is released from rest. Calculate the velocity v of each 25-kg cylinder after the 20-kg cylinder has dropped 2 m. The 10-kg cylinder of case (*a*) is replaced by a 10(9.81)-N force in case (*b*).

Problem 3/105

3/106 The ball is released from position A with a velocity of 3 m/s and swings in a vertical plane. At the bottom position the cord strikes the fixed bar at B, and the ball continues to swing in the dotted arc. Calculate the velocity v_C of the ball as it passes position C.

Ans. $v_C = 3.59$ m/s

Problem 3/106

3/107 Small metal blocks are discharged with a velocity of 0.45 m/s to a ramp by the upper conveyor shown. If the coefficient of kinetic friction between the blocks and the ramp is 0.30, calculate the angle θ which the ramp must make with the horizontal so that the blocks will transfer without slipping to the lower conveyor moving at the speed of 0.15 m/s.

Problem 3/107

Problem 3/108

Problem 3/109

Problem 3/110

3/108 A constant horizontal force $P = 700$ N is applied to the linkage as shown. With the 14-kg ball initially at rest on its support with $\theta = 60°$, calculate the velocity v of the ball as θ approaches zero where the ball reaches its highest position.

Ans. $v = 3.88$ m/s

3/109 The 0.5-kg collar slides with negligible friction along the fixed spiral rod which lies in the vertical plane. The rod has the shape of the spiral $r = 0.3\theta$ where r is in meters and θ is in radians. The collar is released from rest at A and slides to B under the action of a constant radial force $T = 10$ N. Calculate the velocity v of the slider as it reaches B.

Ans. $v = 5.30$ m/s

3/110 The carriage of mass m moves with negligible friction and has an initial horizontal velocity v_0, to the left in case (*a*) and to the right in case (*b*). If each carriage has the same force P applied to it as indicated, show that they reach the same position x with equal velocities v_1 and v_2 but at different times. If the carriages were sliding with friction instead of rolling, would your conclusion be the same?

3/111 Skid marks left by all four wheels of a car on level pavement are measured by a police officer to be 20 m in length. The driver of the car claims that she was not exceeding the speed limit of 80 km/h before she applied the brakes. Furthermore the driver released the brakes before coming to a stop and was observed to continue at a speed of 40 km/h after her skid. It is known from prevailing conditions that the coefficient of kinetic friction between the tires and the road was at least 0.80. Can it be shown that the driver was guilty of speeding?

3/112 A car is traveling at 60 km/h down a 10-percent grade when the brakes on all four wheels lock. If the kinetic coefficient of friction between the tires and the road is 0.70, find the distance s measured along the road which the car skids before coming to a stop.

Ans. $s = 23.7$ m

3/113 In a railroad classification yard a 68-Mg freight car moving at 0.5 m/s at A encounters a retarder section of track at B which exerts a retarding force of 32 kN on the car in the direction opposite to motion. Over what distance x should the retarder be activated in order to limit the speed of the car to 3 m/s at C?   *Ans.* $x = 53.2$ m

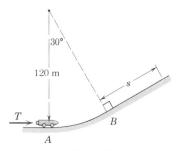

Problem 3/113

3/114 A small rocket-propelled test vehicle with a total mass of 100 kg starts from rest at A and moves with negligible friction along the track in the vertical plane as shown. If the propelling rocket exerts a constant thrust T of 1.5 kN from A to position B where it is shut off, determine the distance s which the vehicle rolls up the incline before stopping. The loss of mass due to the expulsion of gases by the rocket is small and may be neglected.

Problem 3/114

3/115 Determine the velocity v of the 60-kg sliding block after it has moved 1.4 m along the incline from rest under the action of the constant force $P = 600$ N. The light cables are wrapped securely around the integral pulleys of negligible mass. Pulley diameters are in the ratio of 2:1, friction in the bearing is negligible, and the coefficient of kinetic friction μ_k between the block and the incline is 0.40. Note that the center of the pulley moves up the incline the same distance that the point of application of P on the cable moves down the incline.   *Ans.* $v = 2.18$ m/s

Problem 3/115

3/116 Each of the sliders A and B has a mass of 2 kg and moves with negligible friction in its respective guide, with y being in the vertical direction. A 20-N horizontal force is applied to the midpoint of the connecting link of negligible mass, and the assembly is released from rest with $\theta = 0$. Calculate the velocity v_A with which A strikes the horizontal guide when $\theta = 90°$.

Problem 3/116

Problem 3/117

3/117 The 0.2-kg slider moves freely along the fixed curved rod from *A* to *B* in the vertical plane under the action of the constant 5-N tension in the cord. If the slider is released from rest at *A*, calculate its velocity *v* as it reaches *B*.   *Ans. v* = 4.48 m/s

Problem 3/118

3/118 The 7-kg collar *A* slides with negligible friction on the fixed vertical shaft. When the collar is released from rest at the bottom position shown, it moves up the shaft under the action of the constant force *F* = 200 N applied to the cable. Calculate the stiffness *k* which the spring must have if its maximum compression is to be limited to 75 mm. The position of the small pulley at *B* is fixed.

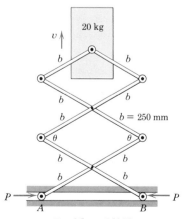

Problem 3/119

3/119 The vertical motion of the 20-kg block is controlled by the two forces *P* applied to the ends *A* and *B* of the linkage where *A* and *B* are constrained to move in the horizontal guide. If forces *P* = 1100 N are applied with the linkage initially at rest with $\theta = 60°$, determine the upward velocity *v* of the block as θ approaches 180°. Neglect friction and the mass of the links and note that *P* is greater than its equilibrium value of $(5W/2)$ ctn 30° = 850 N.   *Ans. v* = 5.95 m/s

3/120 The 10-kg block is released from rest on the horizontal surface at point B where the spring has been stretched a distance of 0.5 m from its neutral position A. The coefficient of kinetic friction between the block and the plane is 0.30. Calculate (*a*) the velocity v of the block as it passes point A and (*b*) the maximum distance x to the left of A which the block goes.

Problem 3/120

3/121 Calculate the horizontal velocity v with which the 20-kg carriage must strike the spring in order to compress it a maximum of 100 mm. The spring is known as a "hardening" spring, since its stiffness increases with deflection as shown in the accompanying graph.   *Ans. v = 2.38 m/s*

Problem 3/121

3/122 The car of mass m accelerates on a level road under the action of the driving force F from a speed v_1 to a higher speed v_2 in a distance s. If the engine develops a constant power output P, determine v_2. Treat the car as a particle under the action of the single horizontal force F.

Problem 3/122

3/123 The piston and ram of the steam hammer have a combined mass of 180 kg. The absolute pressure p acting on the 300-mm-diameter piston is 1.4 MPa for the first 100 mm of its downward stroke beginning at $x = 50$ mm. For the remaining 250 mm of the 350-mm stroke, the pressure p varies according to $pV =$ constant where V is the total volume of the cylinder above the piston. Determine the velocity v of impact of the ram with the work at the bottom of the stroke. Atmospheric pressure of 0.1014 MPa acts on the ram and the lower side of the piston.   *Ans. v = 15.83 m/s*

Problem 3/123

3/124 The third stage of a rocket fired vertically up over the north pole coasts to a maximum altitude of 500 km following burnout of its rocket motor. Calculate the downward velocity v of the rocket when it has fallen 100 km from its position of maximum altitude. (Use the mean value of 9.825 m/s² for g and 6371 km for the mean radius of the earth.) *Ans. $v = 1.309$ km/s*

3/125 The system is released from rest in the position shown. The 15-kg cylinder falls through the hole in the support, but the 15-kg collar (shown in section) is removed from the cylinder as it hits the support. Determine the distance s which the 50-kg block moves up the incline. The coefficient of kinetic friction between the block and the incline is 0.30, and the mass of the pulley is negligible. *Ans. $s = 1.67$ m*

Problem 3/125

3/126 The 10-kg cylinder is released from rest with $x = 1$ m where the spring is unstretched. Determine (a) the maximum velocity v of the cylinder and the corresponding value of x and (b) the maximum value of x during the motion. The stiffness of the spring is 450 N/m.
Ans. (a) $v = 1.462$ m/s, $x = 1.218$ m
(b) $x = 1.436$ m

Problem 3/126

Problem 3/127

3/127 The carriage of mass m is released from rest in its horizontal guide against the spring of stiffness k which has been compressed an amount x_0. Derive an expression for the power P developed by the spring in terms of the deflection x of the spring. Also determine the maximum power P_{max} and the corresponding value of x.

Ans. $$P = kx\sqrt{\frac{k}{m}(x_0{}^2 - x^2)}$$

$$P_{max} = \frac{k}{2}\sqrt{\frac{k}{m}}x_0{}^2 \text{ at } x = \frac{x_0}{\sqrt{2}}$$

3/7 POTENTIAL ENERGY

In the previous article on work and kinetic energy a particle or a combination of joined particles was isolated, and the work done by gravity forces, spring forces, and other externally applied forces acting on the particle or system was determined when evaluating U in the work-energy equation. In the present article we shall treat the work done by gravity forces and by spring forces by introducing the concept of *potential energy*. This concept will simplify the analysis of many problems.

(*a*) *Gravitational potential energy.*   We consider, first, the motion of a particle of mass m in close proximity to the earth's surface where the gravitational attraction (weight) mg is essentially constant, Fig. 3/6*a*. The *gravitational potential energy* V_g of the particle is defined as the work mgh done *against* the gravitational field to elevate the particle a distance h above some arbitrary reference plane where V_g is taken to be zero. Thus we write the potential energy as

$$\boxed{V_g = mgh} \tag{3/14}$$

This work is called potential energy because it may be converted into energy if the particle is allowed to do work on a supporting body while it returns to its lower original datum plane. In going from one level at $h = h_1$ to a higher level at $h = h_2$ the *change* in potential energy becomes

$$\Delta V_g = mg(h_2 - h_1) = mg\,\Delta h$$

The corresponding work done *by* the gravitational force on the particle is $-mg\,\Delta h$. Thus the work done by the gravitational force is the negative of the change in potential energy.

When large changes in altitude in the earth's field are encountered, Fig. 3/6*b*, the gravitational force $Gmm_e/r^2 = mgR^2/r^2$ is no longer constant. The work done *against* this force to change the radial position of the particle from r to r' is the change $V_g' - V_g$ in gravitational potential energy

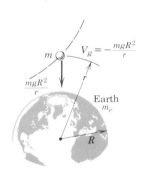

(*b*)

Figure 3/6

$$\int_r^{r'} mgR^2\,\frac{dr}{r^2} = mgR^2\left(\frac{1}{r} - \frac{1}{r'}\right) = V_g' - V_g$$

It is customary to take $V_g' = 0$ when $r' = \infty$, so that with this datum we have

$$\boxed{V_g = -\frac{mgR^2}{r}} \tag{3/15}$$

In going from r_1 to r_2 the corresponding change in potential energy is

$$\Delta V_g = mgR^2 \left(\frac{1}{r_1} - \frac{1}{r_2} \right)$$

which, again, is the *negative* of the work done *by* the gravitational force. We note that the potential energy of a given particle depends only on its position, h or r, and not on the particular path it followed in reaching that position.

(*b*) **Elastic potential energy.** The second example of potential energy is found in the deformation of an elastic body, such as a spring. The work which is done on the spring to deform it is stored in the spring and is called its *elastic potential energy* V_e. This energy is recoverable in the form of work done by the spring on the body attached to its movable end during the release of the spring's deformation. For the one-dimensional linear spring of stiffness k, which we discussed in Art. 3/6 and illustrated in Fig. 3/4, the force supported by the spring at any deformation x, tensile or compressive, from its undeformed position is $F = kx$. Thus we define the elastic potential energy of the spring as the work done on it to deform it an amount x, and we have

$$V_e = \frac{1}{2} \int_0^x kx \, dx = \frac{1}{2}kx^2 \qquad (3/16)$$

If the deformation, either tensile or compressive, of a spring increases from x_1 to x_2 during the motion interval, then the change in potential energy of the spring is its final value minus its initial value or

$$\Delta V_e = \frac{1}{2}k(x_2{}^2 - x_1{}^2)$$

which is positive. Conversely, if the deformation of a spring decreases from x_2 to x_1 during the motion interval, then the change in potential energy of the spring becomes negative. The magnitude of these changes is represented by the shaded trapezoidal area in the *F-x* diagram of Fig. 3/4.

Since the force exerted *on* the spring *by* the moving body is equal and opposite to the force F exerted *on* the body *by* the spring (Fig. 3/4),it follows that the work done on the spring is the negative of the work done on the body. Therefore we may replace the work U done by the spring on the body by $-\Delta V_e$, the negative of the potential energy change for the spring, provided the spring is now included within the system.

(*c*) **Work-energy equation.** With the elastic member in-cluded in the system, we now modify the work-energy equation to account for the potential energy terms. If U stands for the work of all external forces *other than* gravitational forces and spring forces, we may write Eq. 3/11 as $U + (-\Delta V_g) + (-\Delta V_e) = \Delta T$ or

$$U = \Delta T + \Delta V_g + \Delta V_e \qquad (3/17)$$

This alternative form of the work-energy equation is often far more convenient to use than Eq. 3/11, since the work of both gravity and spring forces is accounted for by focusing attention on the end-point positions of the particle and on the length of the elastic spring. The path followed between these end-point positions is of no consequence in the evaluation of ΔV_g and ΔV_e.

To help clarify the difference between the use of Eqs. 3/11 and 3/17, Fig. 3/7 shows schematically a particle of mass m constrained to move along a fixed path under the action of forces F_1 and F_2, the gravitational force $W = mg$, the spring force F, and the normal reaction N. In the b-part of the figure the particle is isolated with its free-body diagram, and the work done by each of the forces F_1, F_2, W, and the spring force $F = kx$ is evaluated for the interval of motion in question, say, from A to B, and equated to the change ΔT in kinetic energy using Eq. 3/11. The constraint reaction N, if normal to the path, will do no work. In the c-part of the figure for the alternative approach the spring is included as a part of the isolated system. The work done during the interval by F_1 and F_2 constitutes the U-term of Eq. 3/17 with the changes in elastic and gravitational potential energies included on the energy side of the equation. We note with the first approach that the work done by $F = kx$ could require a somewhat awkward integration to account for the changes in magnitude and direction of F as the particle moved from A to B. With the second approach, however, only the initial and final lengths of the spring would be required to evaluate ΔV_e.

We may rewrite the alternative work-energy relation, Eq. 3/17, for a particle-and-spring system as

$$U = \Delta(T + V_g + V_e) = \Delta E \qquad (3/17a)$$

where $E = T + V_g + V_e$ is the total mechanical energy of the particle and its attached linear spring. Equation 3/17a states that the net work done on the system by all forces other than gravitational forces and elastic forces equals the change in the total mechanical energy of the system. For problems where the only forces are gravitational, elastic, and nonworking constraint forces, the U-term will be zero, and the energy equation becomes merely

$$\Delta E = 0 \qquad \text{or} \qquad E = \text{constant} \qquad (3/18)$$

When E is constant, it is seen that transfers of energy between kinetic and potential may take place as long as the total mechanical energy $T + V_g + V_e$ does not change. Equation 3/18 expresses the *law of conservation of dynamical energy.*

(a)

(b)   (c)

Figure 3/7

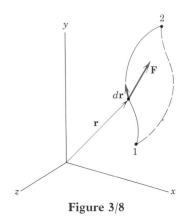

Figure 3/8

(d) Conservative force fields. We have observed that the work done against a gravitational or an elastic force depends only on the net change of position and not on the particular path followed in reaching the new position. Forces with this characteristic are associated with *conservative force fields* which possess an important mathematical property. Consider a force field where the force **F** is a function of the coordinates, Fig. 3/8. The work done by **F** during a displacement $d\mathbf{r}$ of its point of application is $dU = \mathbf{F} \cdot d\mathbf{r}$. The total work done along its path from 1 to 2 is

$$U = \int \mathbf{F} \cdot d\mathbf{r} = \int (F_x \, dx + F_y \, dy + F_z \, dz)$$

The integral $\int \mathbf{F} \cdot d\mathbf{r}$ is a line integral dependent, in general, upon the particular path followed between any two points 1 and 2 in space. If, however, $\mathbf{F} \cdot d\mathbf{r}$ is an *exact differential*[†] $- dV$ of some scalar function V of the coordinates, then

$$U = \int_{V_1}^{V_2} - dV = -(V_2 - V_1) \qquad (3/19)$$

which depends only on the end points of the motion and which is thus *independent* of the path followed. The minus sign before dV is arbitrary but is chosen to agree with the customary designation of the sign of potential energy change in the earth's gravity field. If V exists, the differential change in V becomes

$$dV = \frac{\partial V}{\partial x} \, dx + \frac{\partial V}{\partial y} \, dy + \frac{\partial V}{\partial z} \, dz$$

Comparison with $- dV = \mathbf{F} \cdot d\mathbf{r} = F_x \, dx + F_y \, dy + F_z \, dz$ gives us

$$F_x = -\frac{\partial V}{\partial x} \qquad F_y = -\frac{\partial V}{\partial y} \qquad F_z = -\frac{\partial V}{\partial z}$$

The force may also be written as the vector

$$\mathbf{F} = -\nabla V \qquad (3/20)$$

where the symbol ∇ stands for the vector operator "del" which is

$$\nabla = \mathbf{i} \, \frac{\partial}{\partial x} + \mathbf{j} \, \frac{\partial}{\partial y} + \mathbf{k} \, \frac{\partial}{\partial z}$$

The quantity V is known as the *potential function*, and the expression ∇V is known as the *gradient of the potential function*.

When force components are derivable from a potential as described, the force is said to be *conservative*, and the work done by **F** between any two points is independent of the path followed.

---

* Optional.
[†] Recall that a function $d\phi = P \, dx + Q \, dy + R \, dz$ is an exact differential in the coordinates x-y-z if

$$\frac{\partial P}{\partial y} = \frac{\partial Q}{\partial x} \qquad \frac{\partial P}{\partial z} = \frac{\partial R}{\partial x} \qquad \frac{\partial Q}{\partial z} = \frac{\partial R}{\partial y}$$

Sample Problem 3/16

The 10-kg slider A moves with negligible friction up the inclined guide. The attached spring has a stiffness of 60 N/m and is stretched 0.6 m in position A where the slider is released from rest. The 250-N force is constant and the pulley offers negligible resistance to the motion of the cord. Calculate the velocity v of the slider as it passes point C.

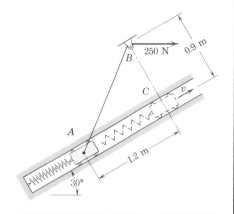

Solution. The slider and inextensible cord together with the attached spring will be analyzed as a system, which permits the use of Eq. 3/17. The only force acting on this system which does work is the 250-N tension applied to the cord. While the slider moves from A to C the point of application of the 250-N force moves a distance of $\overline{AB} - \overline{BC}$ or $1.5 - 0.9 = 0.6$ m. Thus

$$U = 250(0.6) = 150 \text{ J}$$

The change in kinetic energy of the slider is

$$\Delta T = \tfrac{1}{2}m(v^2 - v_0{}^2) = \tfrac{1}{2}(10)v^2$$

where the initial velocity v_0 is zero. The change in gravitational potential energy is

$$\Delta V_g = mg(\Delta h) = 10(9.81)(1.2 \sin 30°) = 58.9 \text{ J}$$

The change in elastic potential energy is

$$\Delta V_e = \tfrac{1}{2}k(x_2{}^2 - x_1{}^2) = \tfrac{1}{2}(60)([1.2 + 0.6]^2 - [0.6]^2) = 86.4 \text{ J}$$

Substitution into the alternative work-energy equation gives

$$[U = \Delta T + \Delta V_g + \Delta V_e] \qquad 150 = \tfrac{1}{2}(10)v^2 + 58.9 + 86.4$$

$$v = 0.974 \text{ m/s} \qquad\qquad Ans.$$

① The reactions of the guides on the slider are normal to the direction of motion and do no work.

② Since the center of mass of the slider has an upward component of displacement, ΔV_g is positive.

③ Be very careful not to make the mistake of using $\tfrac{1}{2}k(x_2 - x_1)^2$ for ΔV_e. We want the difference of the squares and not the square of the difference.

Sample Problem 3/17

The 3-kg slider is released from rest at point A and slides with negligible friction in a vertical plane along the circular rod. The attached spring has a stiffness of 350 N/m and has an unstretched length of 0.6 m. Determine the velocity of the slider as it passes position B.

Solution. The work done by the weight and the spring force on the slider will be treated as changes in the potential energies, and the reaction of the rod on the slider is normal to the motion and does no work. Hence $U = 0$. The changes in the potential and kinetic energies for the system of slider and spring are

$$\Delta V_e = \tfrac{1}{2}k(x_B{}^2 - x_A{}^2) = \tfrac{1}{2}(350)\{(0.6[\sqrt{2} - 1])^2 - (0.6)^2\} = -52.2 \text{ J}$$

$$\Delta V_g = W\Delta h = 3(9.81)(-0.6) = -17.66 \text{ J}$$

$$\Delta T = \frac{1}{2}m(v_B{}^2 - v_A{}^2) = \frac{1}{2}3(v_B{}^2 - 0) = 1.5v_B{}^2$$

$$[\Delta T + \Delta V_g + \Delta V_e = 0] \qquad 1.5v_B{}^2 - 17.66 - 52.2 = 0$$

$$v_B = 6.82 \text{ m/s} \qquad\qquad Ans.$$

① Note that if we evaluated the work done by the spring force acting on the slider by means of the integral $\int \mathbf{F} \cdot d\mathbf{r}$ it would necessitate a lengthy computation to account for the change in the magnitude of the force along with the change in the angle between the force and the tangent to the path. Note further that v_B depends only on the end conditions of the motion and does not require knowledge of the shape of the path.

PROBLEMS

Problem 3/128

Problem 3/129

Problem 3/130

3/128 The 4-kg slider is released from rest at A and slides with negligible friction down the circular rod in the vertical plane. Determine (a) the velocity v of the slider as it reaches the bottom at B and (b) the maximum deformation x of the spring.

Ans. (a) $v = 3.43$ m/s, (b) $x = 48.5$ mm

3/129 Point P on the 2-kg cylinder has an initial velocity $v_0 = 0.8$ m/s as it passes position A. Neglect the mass of the pulleys and cable and determine the distance y of point P below A when the 3-kg cylinder has acquired an upward velocity of 0.6 m/s. *Ans.* $y = 0.224$ m

3/130 The light rod is pivoted at O and carries the 2- and 4-kg particles. If the rod is released from rest at $\theta = 60°$ and swings in the vertical plane, calculate (a) the velocity v of the 2-kg particle just before it hits the spring in the dotted position and (b) the maximum compression x of the spring. Assume that x is small so that the position of the rod when the spring is compressed is essentially horizontal.

Ans. (a) $v = 1.162$ m/s, (b) $x = 12.07$ mm

3/131 When the mechanism is released from rest in the position where $\theta = 60°$, the 4-kg carriage drops and the 6-kg sphere rises. Determine the velocity v of the sphere when $\theta = 180°$. Neglect the mass of the links and treat the sphere as a particle.

Ans. $v = 0.990$ m/s

Problem 3/131

3/132 The 7-kg collar slides freely on the fixed vertical rod and is given an upward velocity $v_0 = 2.5$ m/s in the position shown. The collar compresses the upper spring and is then projected downward. Calculate the maximum resulting deformation x of the lower spring. *Ans. $x = 100.6$ mm*

Problem 3/132

3/133 The identical links of negligible mass are released simultaneously from rest at $\theta = 30°$ and rotate in the vertical plane. Determine the velocity v of each 1.2-kg sphere when θ reaches $90°$. The spring is unstretched when $\theta = 90°$.

Problem 3/133

3/134 The 10-kg collar slides on the smooth vertical rod and has a velocity $v_1 = 2$ m/s in position A where each spring is stretched 0.1 m. Calculate the velocity v_2 of the collar as it passes point B. *Ans. $v_2 = 2.26$ m/s*

3/135 The 1.5-kg slider C moves along the fixed rod under the action of the spring whose unstretched length is 0.3 m. If the velocity of the slider is 2 m/s at point A and 3 m/s at point B, calculate the work U_f done by friction between these two points. Also, determine the average friction force acting on the slider between A and B if the length of the path is 0.70 m. The x-y plane is horizontal.

Problem 3/135

Problem 3/134

Problem 3/136

Problem 3/137

Problem 3/138

Problem 3/139   **Problem 3/140**

3/136 The collar has a mass of 2 kg and is attached to the light spring which has a stiffness of 30 N/m and an unstretched length of 1.5 m. The collar is released from rest at A and slides up the smooth rod under the action of the constant 50-N force. Calculate the velocity v of the collar as it passes position B.   *Ans.* $v = 4.93$ m/s

3/137 The 1.5-kg ball is given an initial velocity $v_A = 2.5$ m/s in the vertical plane at position A where the two horizontal attached springs are unstretched. The ball follows the dotted path shown and crosses point B which is 125 mm directly below A. Calculate the velocity v_B of the ball at B. Each spring has a stiffness of 1800 N/m.

Ans. $v_B = 2.68$ m/s

3/138 The chain of length L is released from rest on a smooth horizontal surface with the overhang shown. Determine the velocity v of the last link to leave the surface.

$$Ans.\ v = \sqrt{\frac{g}{L}(L^2 - 4b^2)}$$

3/139 When the 5-kg plunger is released from rest in its vertical guide at $\theta = 0$, each spring of stiffness $k = 3.5$ kN/m is uncompressed. The links are free to slide through their pivoted collars and compress their springs. Calculate the velocity v of the plunger when the position $\theta = 30°$ is passed.

3/140 When $\theta = 90°$, the spring of stiffness $k = 2$ kN/m is uncompressed. If the balls are released from rest in their top position, calculate their velocity v as they strike the horizontal surface with $\theta = 0$. Treat each ball as a particle and neglect the mass of the right-angled links.   *Ans.* $v = 1.368$ m/s

3/141 The mechanism shown lies in the vertical plane and is released from rest in the position for which $\theta = 60°$. In this position the spring is unstretched. Calculate the velocity of the 5-kg sphere when $\theta = 90°$. The mass of the links is small and may be neglected.

Problem 3/141

3/142 The mechanism is released from rest with $\theta = 180°$ where the uncompressed spring of stiffness $k = 900$ N/m is just touching the underside of the 4-kg collar. Determine the angle θ corresponding to the maximum compression of the spring. Motion is in the vertical plane, and the mass of the links may be neglected.   *Ans.* $\theta = 43.8°$

3/143 The 3-kg sphere is carried by the parallelogram linkage where the spring is unstretched when $\theta = 90°$. If the mechanism is released from rest at $\theta = 90°$, calculate the velocity v of the sphere when the position $\theta = 135°$ is passed. The links are in the vertical plane, and their mass is small and may be neglected.

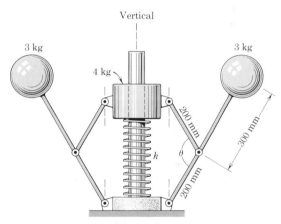

Problem 3/142

3/144 The cars of an amusement park ride have a speed $v_1 = 90$ km/h at the lowest part of the track. Determine their speed v_2 at the highest part of the track. Neglect energy loss due to friction. (*Caution:* Give careful thought to the change in potential energy of the system of cars.)

Ans. $v_2 = 35.1$ km/h

Problem 3/144

Problem 3/143

Problem 3/145

Problem 3/146

Problem 3/147

Problem 3/148

Problem 3/149

3/145 A particle of mass m is released from rest in the position shown and is accelerated to the left by the action of the elastic cord ABC which lies in the horizontal plane and which is unstretched at $x = 0$. The cord has a unit stiffness K, which is the ratio of its tension force to its unit elongation or strain (change in length divided by original length). Determine the expression for the velocity v reached by the particle at the instant x returns to zero.

3/146 A projectile is fired vertically upward from the north pole with a velocity v_0. Calculate the least value of v_0 which will allow the projectile to escape from the earth's gravitational pull assuming no atmospheric resistance. The earth's radius is 6371 km. Use the absolute value $g = 9.825$ m/s^2.
Ans. $v_0 = 11.19$ km/s

3/147 Upon its return voyage from a space mission, the spacecraft has a velocity of 24 000 km/h at point A, which is 7000 km from the center of the earth. **Determine the velocity of the spacecraft when it reaches point B, which is 6500 km from the center of the earth.** The trajectory between these two points is outside the effect of the earth's atmosphere.
Ans. $v_B = 26$ 300 km/h

3/148 A satellite is put into an elliptical orbit around the earth and has a velocity v_P at the perigee position P. Determine the expression for the velocity v_A at the apogee position A. The radii to A and P are, respectively, r_A and r_P. Note that the total energy remains constant.

3/149 A rocket launches an unpowered space capsule at point A with an absolute velocity $v_A = 13$ 000 km/h at an altitude of 40 km. After the capsule has traveled a distance of 400 km measured along its absolute space trajectory, its velocity at B is 12 400 km/h and its altitude is 80 km. Determine the average resistance P to motion in the rarified atmosphere. The mass of the capsule is 22 kg, and the earth's mean radius is 6371 km. Consider the center of the earth as fixed in space.
Ans. $P = 11.16$ N

3/150 The chain of length L is released from rest on the smooth incline with $x = 0$. Determine the velocity v of the links in terms of x.

Problem 3/150

3/151 The block of mass m is supported by the two fingers and is barely in contact with the uncompressed elastic spring of stiffness k and negligible mass. If the fingers are suddenly removed, determine the maximum velocity v reached by the block, the maximum force R transmitted to the floor through the spring, and the maximum deformation δ of the spring.

Ans. $v = g\sqrt{m/k}$, $R = 2mg$, $\delta = 2mg/k$

Problem 3/151

3/152 The shank of the 2-kg vertical plunger occupies the dotted position when resting in equilibrium against the spring of stiffness $k = 1.6$ kN/m. The upper end of the spring is welded to the plunger, and the lower end is welded to the base plate. If the plunger is lifted 40 mm above its equilibrium position and released from rest, calculate its velocity v as it strikes the button A. Friction is negligible.

Ans. $v = 1.119$ m/s

Problem 3/152

3/153 The rope shown in the *a*-part of the figure is released from rest in the smooth semicircular guide and acquires a velocity v when all of it is clear of the guide as seen in the *b*-part of the figure. Find v. (*Hint:* In finding the potential energy change imagine that the semicircular portion of the rope in the initial position becomes the bottom part of the rope of equal length in the final position.)

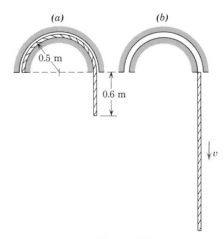

Problem 3/153

3/154 An instrument package of mass m is attached at A to two springs each of stiffness k and unstretched length b. The package is released from rest at this point and falls a short distance. The deflection y at any instant of time is very small compared with b, so that the stretch of the spring is very nearly given by $x = y \sin \theta$ where $\theta = \cos^{-1}(c/b)$. Determine the velocity $\dot{y}$ of the package as a function of y and find the maximum deflection $y_{\max}$ of the package.

Ans. $\dot{y} = \sqrt{2y\left(g - \dfrac{k}{m}\, y\, \dfrac{b^2 - c^2}{b^2}\right)},$

$$y_{\max} = \frac{mg}{k}\,\frac{b^2}{b^2 - c^2}$$

Problem 3/154

Problem 3/155

3/155 Calculate the maximum velocity of slider B if the system is released from rest with $x = y$. Motion is in the vertical plane. Assume friction is negligible.

Ans. $v_{B_{\max}} = 0.962$ m/s

Problem 3/156

▶**3/156** The chain starts from rest with a sufficient number of links hanging over the edge to barely initiate motion in overcoming friction between the remainder of the chain and the horizontal supporting surface. Determine the velocity v of the chain as the last link leaves the edge. The coefficient of kinetic friction is μ_k. Neglect any friction at the edge.

Ans. $v = \sqrt{\dfrac{gL}{1 + \mu_k}}$

SECTION C. IMPULSE AND MOMENTUM

3/8 IMPULSE AND MOMENTUM

   In the previous two articles we focused attention on the equations of work and energy which are obtained by integrating the equation of motion $\mathbf{F} = m\mathbf{a}$ with respect to the displacement of the particle. In consequence we found that the velocity changes could be expressed directly in terms of the work done or in terms of the overall changes in energy. In the present article we will direct our attention to the integration of the equation of motion with respect to time rather than displacement. This approach leads to the equations of impulse and momentum. We will find that these equations greatly facilitate the solution of many problems where the applied forces act during specified intervals of time.

   (*a*) *Linear impulse and linear momentum.*   Consider again the general curvilinear motion in space of a particle of mass m, Fig. 3/9, where the particle is located by its position vector $\mathbf{r}$ measured from a fixed origin O. The velocity of the particle is $\mathbf{v} = \dot{\mathbf{r}}$ and is tangent to its path (shown as a dashed line). The resultant $\Sigma\mathbf{F}$ of all forces on m is in the direction of its acceleration $\dot{\mathbf{v}}$. We may now write the basic equation of motion for the particle, Eq. 3/3, as

$$\Sigma\mathbf{F} = m\dot{\mathbf{v}} = \frac{d}{dt}(m\mathbf{v}) \qquad \text{or} \qquad \boxed{\Sigma\mathbf{F} = \dot{\mathbf{G}}} \qquad (3/21)$$

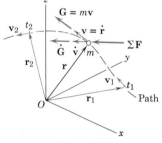

Figure 3/9

where the product of the mass and velocity is defined as the *linear momentum* $\mathbf{G} = m\mathbf{v}$ of the particle. Equation 3/21 states that *the resultant of all forces acting on a particle equals its time rate of change of linear momentum.* In SI the units of linear momentum $m\mathbf{v}$ are seen to be kg·m/s, which also equals N·s.

   Since Eq. 3/21 is a vector equation, we recognize that, in addition to the equality of the magnitudes of $\Sigma\mathbf{F}$ and $\dot{\mathbf{G}}$, the direction of the resultant force coincides with the direction of the *change* in linear momentum, which is the direction of the change in velocity. Equation 3/21 is one of the most useful and important relationships in dynamics, and it is valid as long as the mass m of the particle is not changing with time. The case where m changes with time is discussed in Art. 4/7 of Chapter 4. We now write the three scalar components of Eq. 3/21 as

$$\Sigma F_x = \dot{G}_x \qquad \Sigma F_y = \dot{G}_y \qquad \Sigma F_z = \dot{G}_z \qquad (3/22)$$

These equations may be applied independently of one another.
   All that we have done so far in this article is to rewrite Newton's second law in an alternative form in terms of momentum.

But we are now able to describe the effect of the resultant force $\Sigma\mathbf{F}$ on the linear momentum of the particle over a finite period of time simply by integrating Eq. 3/21 with respect to the time t. Multiplying the equation by dt gives $\Sigma\mathbf{F}\,dt = d\mathbf{G}$ which we integrate from time t_1 to time t_2 to obtain

$$\int_{t_1}^{t_2}\Sigma\mathbf{F}\,dt = \mathbf{G}_2 - \mathbf{G}_1 \qquad (3/23)$$

Here the linear momentum at time t_2 is $\mathbf{G}_2 = m\mathbf{v}_2$ and the linear momentum at time t_1 is $\mathbf{G}_1 = m\mathbf{v}_1$. The product of force and time is defined as *linear impulse* of the force, and Eq. 3/23 states that *the total linear impulse on m equals the corresponding change in linear momentum of m.*

Alternatively we may write Eq. 3/23 as

$$\mathbf{G}_1 + \int_{t_1}^{t_2}\Sigma\mathbf{F}\,dt = \mathbf{G}_2 \qquad (3/23a)$$

which merely says that the initial linear momentum of the body plus the linear impulse applied to it equals its final linear momentum.

The impulse integral is a vector which, in general, may involve changes in both magnitude and direction during the time interval. Under these conditions it will be necessary to express $\Sigma\mathbf{F}$ and $\mathbf{G}$ in component form and then combine the integrated components. The components of Eq. 3/23 become the scalar equations

$$\int_{t_1}^{t_2}\Sigma F_x\,dt = (mv_x)_2 - (mv_x)_1$$

$$\int_{t_1}^{t_2}\Sigma F_y\,dt = (mv_y)_2 - (mv_y)_1 \qquad (3/24)$$

$$\int_{t_1}^{t_2}\Sigma F_z\,dt = (mv_z)_2 - (mv_z)_1$$

These three scalar impulse-momentum equations are completely independent.

There are times when a force acting on a particle varies with the time in a manner determined by experimental measurements or by other approximate means. In this case a graphical or numerical integration must be accomplished. If, for example, a force F acting on a particle in a given direction varies with the time t as indicated in Fig. 3/10, then the impulse, $\int_{t_1}^{t_2} F\,dt$, of this force from t_1 to t_2 becomes the shaded area under the curve.

In evaluating the resultant impulse it is necessary to include the effect of *all* forces acting on m except those whose magnitudes are negligible. The student should be fully aware at this point that the only reliable method of accounting for the effects

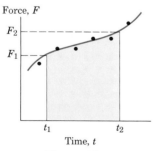

Figure 3/10

of *all* forces is to isolate the particle in question by drawing its *free-body diagram.*

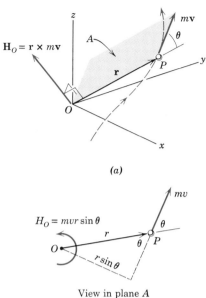

(a)

$\mathbf{H}_O = \mathbf{r} \times m\mathbf{v}$

$H_O = mvr \sin \theta$

View in plane *A*

(b)

Figure 3/11

   (*b*) *Angular impulse and angular momentum.*   In addition to the equations of linear impulse and linear momentum, there exists a parallel set of equations for angular impulse and angular momentum. First, we define the term angular momentum. In Fig. 3/11*a* is shown a particle *P* of mass *m* moving along a curve in space. The particle is located by its position vector **r** with respect to a convenient origin *O* of fixed coordinates *x-y-z*. The velocity of the particle is $\mathbf{v} = \dot{\mathbf{r}}$, and its linear momentum is $\mathbf{G} = m\mathbf{v}$. The *moment* of the *linear momentum vector m***v** about the origin *O* is defined as the *angular momentum* $\mathbf{H}_O$ of *P* about *O* and is given by the cross-product relation for the moment of a vector

$$\mathbf{H}_O = \mathbf{r} \times m\mathbf{v} \qquad (3/25)$$

The angular momentum, then, is a vector perpendicular to plane *A* defined by **r** and **v**. The sense of $\mathbf{H}_O$ is clearly defined by the right-hand rule for cross products.

   The scalar components of angular momentum may be obtained from the expansion

$$\mathbf{H}_O = \mathbf{r} \times m\mathbf{v} = m(v_z y - v_y z)\mathbf{i} + m(v_x z - v_z x)\mathbf{j} + m(v_y x - v_x y)\mathbf{k}$$

$$\mathbf{H}_O = m \begin{vmatrix} \mathbf{i} & \mathbf{j} & \mathbf{k} \\ x & y & z \\ v_x & v_y & v_z \end{vmatrix} \qquad (3/26)$$

so that

$$H_x = m(v_z y - v_y z) \qquad H_y = m(v_x z - v_z x) \qquad H_z = m(v_y x - v_x y)$$

Each of these expressions for angular momentum may be checked easily from Fig. 3/12, which shows the three linear-momentum components, by taking the moments of these components about the respective axes.

   To help visualize angular momentum we show in Fig. 3/11*b* a two-dimensional representation in plane *A* of the vectors shown in the *a*-part of the figure. The motion is viewed in plane *A* defined by **r** and **v**. The magnitude of the moment of *m***v** about *O* is simply the linear momentum *mv* times the moment arm $r \sin \theta$ or $mvr \sin \theta$, which is the magnitude of the cross product $\mathbf{H}_O = \mathbf{r} \times m\mathbf{v}$.

   Angular momentum is the moment of linear momentum and must not be confused with linear momentum. Angular momentum has the units $\text{kg} \cdot (\text{m/s}) \cdot \text{m} = \text{kg} \cdot \text{m}^2/\text{s} = \text{N} \cdot \text{m} \cdot \text{s}$.

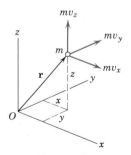

Figure 3/12

We are now ready to relate the moment of the forces acting on the particle P to its angular momentum. If $\Sigma\mathbf{F}$ represents the resultant of *all* forces acting on the particle P of Fig. 3/11, the moment $\mathbf{M}_O$ about the origin O is the vector cross product

$$\Sigma\mathbf{M}_O = \mathbf{r} \times \Sigma\mathbf{F} = \mathbf{r} \times m\dot{\mathbf{v}}$$

where Newton's second law $\mathbf{F} = m\dot{\mathbf{v}}$ has been substituted. We now differentiate Eq. 3/25 with time, using the rule for the differentiation of a cross product (see item 9, Art. C/7, Appendix C) and get

$$\dot{\mathbf{H}}_O = \dot{\mathbf{r}} \times m\mathbf{v} + \mathbf{r} \times m\dot{\mathbf{v}} = \mathbf{v} \times m\mathbf{v} + \mathbf{r} \times m\dot{\mathbf{v}}$$

The term $\mathbf{v} \times m\mathbf{v}$ is zero since the cross product of parallel vectors is identically zero. Substituting into the expression for $\Sigma\mathbf{M}_O$ gives

$$\boxed{\Sigma\mathbf{M}_O = \dot{\mathbf{H}}_O} \tag{3/27}$$

Equation 3/27 states that the *moment about the fixed point O of all forces acting on m equals the time rate of change of angular momentum of m about O*. This relation, particularly when extended to a system of particles, rigid or nonrigid, constitutes one of the most powerful tools of analysis in all of dynamics. Equation 3/27 is a vector equation with scalar components

$$\Sigma M_{O_x} = \dot{H}_{O_x} \qquad \Sigma M_{O_y} = \dot{H}_{O_y} \qquad \Sigma M_{O_z} = \dot{H}_{O_z} \tag{3/28}$$

Equation 3/27 gives the instantaneous relation between the moment and the time rate of change of angular momentum. To obtain the effect of the moment $\Sigma\mathbf{M}_O$ on the angular momentum of the particle over a finite period of time, Eq. 3/27 may be integrated from time t_1 to time t_2. Multiplying the equation by dt gives $\Sigma\mathbf{M}_O\,dt = d\mathbf{H}_O$ which we integrate to obtain

$$\boxed{\int_{t_1}^{t_2}\Sigma\mathbf{M}_O\,dt = \mathbf{H}_{O_2} - \mathbf{H}_{O_1}} \tag{3/29}$$

where $\mathbf{H}_{O_2} = \mathbf{r}_2 \times m\mathbf{v}_2$ and $\mathbf{H}_{O_1} = \mathbf{r}_1 \times m\mathbf{v}_1$. The product of moment and time is defined as *angular impulse*, and Eq. 3/29 states that the *total angular impulse on m about the fixed point O equals the corresponding change in angular momentum of m about O*.

Alternatively we may write Eq. 3/29 as

$$\mathbf{H}_{O_1} + \int_{t_1}^{t_2}\Sigma\mathbf{M}_O\,dt = \mathbf{H}_{O_2} \tag{3/29a}$$

which merely states that the initial angular momentum of the particle plus the angular impulse applied to it equals its final angular momentum. The units of angular impulse are clearly those of angular momentum, which are N·m·s or kg·m²/s.

As in the case of linear impulse and linear momentum, the equation of angular impulse and angular momentum is a vector equation where changes in direction as well as magnitude may occur during the interval of integration. Under these conditions it is necessary to express $\Sigma \mathbf{M}_O$ and $\mathbf{H}_O$ in component form and then combine the integrated components. Thus the x-component of Eq. 3/29 becomes

$$\int_{t_1}^{t_2} \Sigma M_{O_x}\, dt = (H_{O_x})_2 - (H_{O_x})_1$$
$$= m[(v_z y - v_y z)_2 - (v_z y - v_y z)_1]$$

where the subscripts 1 and 2 refer to the values of the respective quantities at times t_1 and t_2. Similar expressions exist for the y- and z-components of the angular momentum integral.

The foregoing angular-impulse and angular-momentum relations have been developed in their general three-dimensional sense. Most of the applications with which we shall be concerned, on the other hand, can be analyzed as plane-motion problems where moments are taken about a single axis normal to the plane of motion. In this event the angular momentum may change magnitude and sense, but the direction of the vector remains unaltered. Thus for a particle of mass m moving along a curved path in the x-y plane, Fig. 3/13, the angular momenta about O at points 1 and 2 have the magnitudes $H_{O_1} = |\mathbf{r}_1 \times m\mathbf{v}_1| = mv_1 d_1$ and $H_{O_2} = |\mathbf{r}_2 \times m\mathbf{v}_2| = mv_2 d_2$, respectively. In the illustration both H_{O_1} and H_{O_2} are represented in the counterclockwise sense according to the direction of the moment of the linear momentum. The scalar form of Eq. 3/29 applied to the motion between points 1 and 2 during the time interval t_1 to t_2 becomes

$$\Sigma M_O = \Sigma F r \sin \theta$$

Figure 3/13

$$\int_{t_1}^{t_2} \Sigma M_O\, dt = H_{O_2} - H_{O_1} \quad \text{or} \quad \int_{t_1}^{t_2} \Sigma F r \sin \theta\, dt = mv_2 d_2 - mv_1 d_1$$

This example should help clarify the relation between the scalar and vector forms of the angular impulse-momentum relations.

Equations 3/21 and 3/27 add no new basic information since they are merely alternative forms of Newton's second law. However, we will discover in subsequent chapters that the motion equations expressed in terms of time rate of change of momentum are applicable to the motion of rigid and nonrigid bodies and constitute a very general and powerful approach to many problems. The full generality of Eq. 3/27 is usually not required to describe the motion of a single particle or the plane motion of rigid bodies, but it does find important use in the analysis of the space motion of rigid bodies introduced in Chapter 7.

Sample Problem 3/18

A 0.2-kg particle moves in the vertical y-z plane (z up, y horizontal) under the action of its weight and a force $\mathbf{F}$ which varies with the time. The linear momentum of the particle in N·s is given by the expression $\mathbf{G} = \frac{3}{2}(t^2 + 3)\mathbf{j} - \frac{2}{3}(t^3 - 4)\mathbf{k}$ where t is the time in seconds. Determine $\mathbf{F}$ for the instant when $t = 2$ s.

Solution. The weight expressed as a vector is $-0.2(9.81)\mathbf{k}$ N. Thus the force-momentum equation becomes

① $[\Sigma\mathbf{F} = \dot{\mathbf{G}}]$ $\quad \mathbf{F} - 0.2(9.81)\mathbf{k} = \dfrac{d}{dt}\left[\frac{3}{2}(t^2 + 3)\mathbf{j} - \frac{2}{3}(t^3 - 4)\mathbf{k}\right]$

$$= 3t\mathbf{j} - 2t^2\mathbf{k}$$

For $t = 2$ s, $\quad \mathbf{F} = 0.2(9.81)\mathbf{k} + 3(2)\mathbf{j} - 2(2^2)\mathbf{k} = 6\mathbf{j} - 6.04\mathbf{k}$ N

Thus $\quad\quad F = \sqrt{6^2 + (6.04)^2} = 8.51$ N $\quad\quad$ *Ans.*

① Don't forget that $\Sigma\mathbf{F}$ includes *all* forces on the particle of which the weight is one.

Sample Problem 3/19

A particle with a mass of 0.5 kg has a velocity $u = 10$ m/s in the x-direction at time $t = 0$. Forces $\mathbf{F}_1$ and $\mathbf{F}_2$ act on the particle, and their magnitudes change with time according to the graphical schedule shown. Determine the velocity $\mathbf{v}$ of the particle at the end of the 3 seconds.

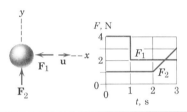

Solution. The impulse-momentum equation is applied in component form and gives for the x- and y-directions, respectively,

$\left[\int \Sigma F_x \, dt = m \, \Delta v_x\right] \quad -[4(1) + 2(3-1)] = 0.5(v_x - 10)$

① $\quad\quad\quad\quad\quad\quad\quad v_x = -6$ m/s

$\left[\int \Sigma F_y \, dt = m \, \Delta v_y\right] \quad [1(2) + 2(3-2)] = 0.5(v_y - 0)$

$\quad\quad\quad\quad\quad\quad\quad v_y = 8$ m/s

Thus

$\quad\quad \mathbf{v} = -6\mathbf{i} + 8\mathbf{j}$ m/s $\quad$ and $\quad v = \sqrt{6^2 + 8^2} = 10$ m/s

$\quad\quad \theta_x = \tan^{-1}\dfrac{8}{-6} = 126.9°$ $\quad\quad$ *Ans.*

② $\quad$ Although not called for, the path of the particle for the first 3 seconds is plotted in the figure. The velocity at $t = 3$ s is shown together with its components.

① The impulse in each direction is the corresponding area under the force-time graph. Note that F_1 is in the negative x-direction so its impulse is negative.

② It is important to note that the algebraic signs must be carefully respected in applying the momentum equations. Also we must recognize that impulse and momentum are vector quantities in contrast to work and energy, which are scalars.

Sample Problem 3/20

The loaded 150-kg skip is rolling down the incline at 4 m/s when a force P is applied to the cable as shown at time $t = 0$. The force P is increased uniformly with the time until it reaches 600 N at $t = 4$ s after which time it remains constant at this value. Calculate (a) the time t_1 at which the skip reverses its direction and (b) the velocity v of the skip at $t = 8$ s. Treat the skip as a particle.

Solution. The stated variation of P with the time is plotted, and the free-body diagram of the skip is drawn.

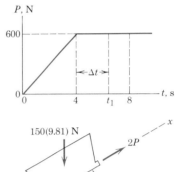

Part (a). The skip reverses direction when its velocity becomes zero. We will assume that this condition occurs at $t = 4 + \Delta t$ seconds. The impulse-momentum equation applied consistently in the positive x-direction gives

$$\left[\int \Sigma F_x \, dt = m \, \Delta v_x \right]$$

① $\frac{1}{2}(4)(2)(600) + 2(600)\,\Delta t - 150(9.81)\sin 30°(4 + \Delta t) = 150(0 - [-4])$

$$464\Delta t = 1143 \qquad \Delta t = 2.46 \text{ s}$$

$$t = 4 + 2.46 = 6.46 \text{ s} \qquad\qquad Ans.$$

Part (b). Applying the impulse-momentum equation to the entire interval gives

$$\left[\int \Sigma F_x \, dt = m \, \Delta v_x \right]$$

$\frac{1}{2}(4)(2)(600) + 4(2)(600) - 150(9.81)\sin 30°(8) = 150(v - [-4])$

$$150v = 714 \qquad v = 4.76 \text{ m/s} \qquad\qquad Ans.$$

The same result is obtained by analyzing the interval from t_1 to 8 s.

① The free-body diagram keeps us from making the error of using the impulse of P rather than 2P or of forgetting the impulse of the component of the weight. The first term in the equation is the triangular area of the P-t relation for the first 4 s, doubled for the force of 2P.

Sample Problem 3/21

The small 2-kg block slides on a smooth horizontal surface under the action of the force in the spring and a force F. The angular momentum of the block about O varies with time as shown in the graph. When $t = 6.5$ s, it is known that $r = 150$ mm and $\beta = 60°$. Determine F for this instant.

Solution. The only moment of the forces about O is due to F since the spring force passes through O. Thus $\Sigma M_O = Fr \sin \beta$. From the graph the time rate of change of H_O for $t = 6.5$ s is very nearly $(8 - 4)/(7 - 6)$ or $\dot{H}_O = 4$ kg·m²/s². The moment-angular momentum relation gives

① $[\Sigma M_O = \dot{H}_O] \qquad F(0.150)\sin 60° = 4 \qquad F = 30.8 \text{ N} \qquad Ans.$

① We do not need vector notation here since we have plane motion where the direction of the vector $\mathbf{H}_O$ does not change.

Problem 3/157

PROBLEMS

3/157 The jet fighter has a mass of 6450 kg and requires 10 seconds from rest to reach its takeoff speed of 250 km/h under the constant jet thrust $T = 48$ kN. Compute the time average R of the combined air and ground resistance during takeoff.
Ans. $R = 3.21$ kN

3/158 The two orbital maneuvering engines of the space shuttle develop 26 kN of thrust each. If the shuttle is traveling in orbit at a speed of 26 000 km/h, how long would it take to reach a speed of 26 200 km/h after the two engines are fired? The mass of the shuttle is 90 Mg. *Ans.* $t = 1$ min 36 s

Problem 3/159

3/159 A rifle fires a 60-g bullet with a horizontal muzzle velocity of 600 m/s. If it takes $3(10^{-3})$ s for the bullet to travel the length of the barrel, determine the average horizontal force P applied to the gun to hold it in place during the firing.

Problem 3/160

3/160 A 3-Mg experimental rocket sled is propelled by 6 rocket motors each with an impulse rating of 100 kN·s. The rockets are fired at $\frac{1}{4}$-s intervals, and the duration of each rocket firing is 1.5 s. If the velocity of the sled 3 s after starting from rest is 150 m/s, determine the time average R of the total aerodynamic and mechanical resistance to motion. Neglect the loss of mass due to exhausted fuel compared with the mass of the sled.
Ans. $R = 50$ kN

3/161 The lunar landing craft has a mass of 320 kg and is equipped with a retro-rocket that provides a constant upward thrust T capable of yielding a total impulse of 24 kN·s during a 10-s period. If the lunar craft was 180 m above the surface of the moon and was descending at the rate of 40 m/s, determine the time t required to reduce the descent rate to 2 m/s. The absolute acceleration due to gravity near the surface of the moon is 1.62 m/s².

Problem 3/161

3/162 The supertanker has a total displacement (mass) of $150(10^3)$ metric tons (one ton equals 1000 kg) and is lying still in the water when the tug commences a tow. If a constant tension of 200 kN is developed in the tow cable, compute the time required to bring the tanker to a speed of 1 knot from rest. At this low speed hull resistance to motion through the water is very small and may be neglected. (1 knot = 1.852 km/h)

Ans. $t = 6.84$ min

Problem 3/162

3/163 A ship with a displacement (mass) of 15 000 metric tons is cruising at a speed of 10 knots when the propeller-shaft thrust is increased to a value of 280 kN. If the resistance R to motion of the ship through the water varies with speed as shown, estimate an average value of R and approximate the time t required for the ship to increase its speed from 10 to 12 knots. (1 knot = 1.852 km/h)

Problem 3/163

3/164 The 9-kg block is moving to the right with a velocity of 0.6 m/s on a horizontal surface when a force P is applied to it at time $t = 0$. Calculate the velocity v of the block when $t = 0.4$ s. The kinetic coefficient of friction is $\mu_k = 0.3$.

Ans. $v = 1.823$ m/s

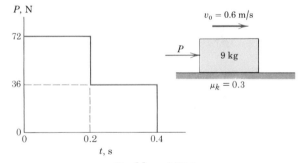

Problem 3/164

3/165 A ship with a displacement of 60,000 metric tons (1 ton = 1000 kg) is moving slowly at 5 knots in dense fog when danger is sighted and full speed in reverse is signaled to the engine room. The total reverse thrust of its two propellers equals 18 MN. Determine the time t required for the ship to acquire a speed of 3 knots astern. At these low hull speeds the hydrodynamic resistance to motion of the hull is small compared with the propeller thrust and may be neglected here. (1 knot = 1.852 km/h)

Problem 3/166

Problem 3/167

Problem 3/168

Problem 3/170

3/166 The space shuttle launches an 800-kg satellite by ejecting it from the cargo bay as shown. The ejection mechanism is activated and is in contact with the satellite for 4 seconds to give it a velocity of 0.3 m/s in the z-direction relative to the shuttle. The mass of the shuttle is 90 Mg. Determine the component of velocity v_f of the shuttle in the minus z-direction resulting from the ejection. Also find the time average F of the ejection force.

Ans. $v_f = 0.00264$ m/s, $F_{av} = 59.5$ N

3/167 The vertical motion of the 3-kg cylinder is controlled by the forces P applied to the end rollers of the extensible framework shown. If the upward velocity v of the cylinder is increased from 2 m/s to 4 m/s in 2 seconds, calculate the average force R_{av} under each of the two rollers during the 2-s interval. Analyze the entire system as a unit and neglect the small mass of the framework.

3/168 The resistance to motion of a certain racing toboggan is 2 percent of the normal force on its runners. Calculate the time t required for the toboggan to reach a speed of 100 km/h down the slope if it starts from rest.   *Ans.* $t = 7.73$ s

3/169 Determine the time required by a diesel-electric locomotive, which produces a constant drawbar pull of 270 kN, to increase the speed of a 1600-Mg freight train from 32 km/h to 48 km/h up a 1-percent grade. Train resistance is 50 N per megagram.

3/170 The inspection gondola for a cableway is being drawn up the sloping cable at the speed of 4 m/s. If the control cable at A suddenly breaks, calculate the time t after the break occurs for the gondola to reach a speed of 8 m/s down the inclined cable. Neglect friction and treat the gondola as a particle.

Ans. $t = 3.18$ s

3/171 Careful measurements made during the impact of the 200-g metal cylinder with the spring-loaded plate reveal a semielliptical relation between the contact force F and the time t of impact as shown. Determine the rebound velocity v of the cylinder if it strikes the plate with a velocity of 6 m/s.

3/172 A particle of mass m starts from rest and moves in a horizontal straight line under the action of a constant force P. If the resistance to motion is proportional to the square of the velocity and is $R = kv^2$, determine the total impulse I on the particle from the time it starts until it reaches its maximum velocity.

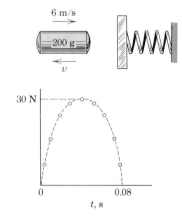

Problem 3/171

3/173 The third and fourth stages of a rocket are coasting in space with a velocity of 15 000 km/h when the fourth-stage engine ignites causing separation of the stages under a thrust T and its reaction. If the velocity v of the fourth stage is 10 m/s greater than the velocity v_1 of the third stage at the end of the $\frac{1}{2}$-s separation interval, calculate the average thrust T during this period. The masses of the third and fourth stages at separation are 40 kg and 60 kg, respectively. Assume that the entire blast of the fourth-stage engine impinges against the third stage with a force equal and opposite to T.

Ans. $T = 480$ N

Problem 3/173

3/174 The hydraulic braking system for the truck and trailer is set to produce equal braking forces for the two units. If the brakes are applied uniformly for 5 s to bring the rig to a stop from a speed of 30 km/h down the 10-percent grade, determine the force P in the coupling between the trailer and the truck. The mass of the truck is 10 Mg and that of the trailer is 7.5 Mg.

Problem 3/174

3/175 At a certain instant the linear momentum of a particle is given by $\mathbf{G} = -3\mathbf{i} - 2\mathbf{j} + 3\mathbf{k}$ kg·m/s and its position vector is $\mathbf{r} = 3\mathbf{i} + 4\mathbf{j} - 3\mathbf{k}$ m. Determine the magnitude H_O of the angular momentum of the particle about the origin of coordinates.

Ans. $H_O = 8.49$ kg·m^2/s

Problem 3/176

3/176 A 4-kg object, which is moving on a smooth horizontal surface with a velocity of 10 m/s in the $-x$-direction, is subjected to the force F_x which varies with the time as shown. Approximate the experimental data by the dotted line and determine the velocity of the object (*a*) at $t = 0.6$ s and (*b*) at $t = 0.9$ s.

Problem 3/177

3/177 The only force acting on an earth satellite traveling outside of the earth's atmosphere is the radial gravitational attraction. The moment of this force is zero about the earth's center taken as a fixed point. Prove that $r^2\dot{\theta}$ remains constant for the motion of the satellite.

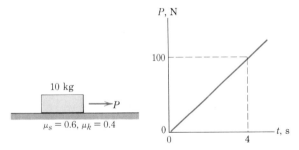

Problem 3/178

3/178 The force P, which is applied to the 10-kg block initially at rest, varies linearly with the time as indicated. If the coefficients of static and kinetic friction between the block and the horizontal surface are 0.6 and 0.4, respectively, determine the velocity of the block when $t = 4$ s.

Ans. $v = 6.61$ m/s

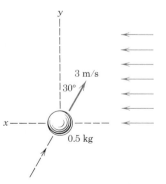

Problem 3/179

3/179 A 0.5-kg particle moving in the horizontal x-y plane with a velocity of 3 m/s in the direction shown encounters a steady flow of air in the x-direction. If the air stream exerts a force on the particle in the x-direction of 0.6 N, essentially constant, determine the time t required for the particle to cross the y-axis again.

3/180 A particle with a mass of 4 kg has a position vector in meters given by $\mathbf{r} = 3t^2\mathbf{i} - 2t\mathbf{j} - 3t\mathbf{k}$ where t is the time in seconds. For $t = 3$ s determine the magnitude of the angular momentum of the particle and the magnitude of the moment of all forces on the particle, both about the origin of coordinates. *Ans.* $H = 389$ N·m·s, $M = 260$ N·m

3/181 A 10-kg body is traveling in a horizontal straight line with a velocity of 3 m/s when a horizontal force P is applied to it at right angles to the initial direction of motion. If P varies according to the accompanying graph, remains constant in direction, and is the only force acting on the body in its plane of motion, find the magnitude of the velocity of the body when $t = 2$ s and the angle θ it makes with the direction of P.

Ans. $v = 3.91$ m/s, $\theta = 50.2°$

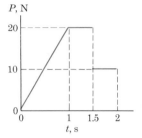

Problem 3/181

3/182 A 0.5-kg body oscillates along the horizontal x-axis under the sole action of an alternating force in the x-direction which decreases in amplitude with time t as shown and is given by $F_x = 4e^{-t}\cos 2\pi t$ where F_x is in newtons and t is in seconds. If the body has a velocity of 1.2 m/s in the negative x-direction at time $t = 0$, determine its velocity v_x at time $t = 2$ s.

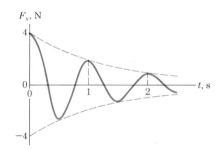

Problem 3/182

3/183 The ice-hockey puck with a mass of 0.20 kg has a velocity of 12 m/s before being struck by the hockey stick. After the impact the puck moves in the new direction shown with a velocity of 18 m/s. If the stick is in contact with the puck for 0.04 s, compute the magnitude of the average force $\mathbf{F}$ exerted by the stick on the puck during contact, and find the angle β made by $\mathbf{F}$ with the x-direction.

Ans. $F = 147.8$ N, $\beta = 12.02°$

Problem 3/183

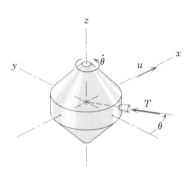

Problem 3/184

3/184 A spacecraft with a mass of 260 kg is moving with a velocity $u = 30\ 000$ km/h in the fixed x-direction remote from any attracting celestial body. The spacecraft is spin-stabilized and rotates about the z-axis at the constant rate $\dot{\theta} = \pi/10$ rad/s. During a quarter of a revolution from $\theta = 0$ to $\theta = \pi/2$, a jet is activated which produces a 600-N thrust of constant magnitude. Determine the y-component of the velocity of the spacecraft when $\theta = \pi/2$. Neglect the small change in mass due to the loss of exhaust gas through the control nozzle and treat the spacecraft as a particle.   *Ans.* $v_y = 7.35$ m/s

3/185 If the resistance R to the motion of a freight train of total mass m increases with velocity according to $R = R_0 + Kv$, where R_0 is the initial resistance to be overcome in starting the train and K is a constant, find the time t required for the train to reach a velocity v from rest on a level track under the action of a constant tractive force F.

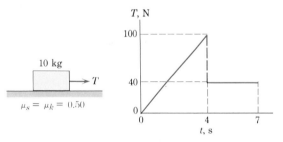

Problem 3/186

3/186 The 10-kg block is resting on the horizontal surface when the force T is applied to it for 7 seconds. The variation of T with time is shown. Calculate the maximum velocity reached by the block and the total time Δt during which the block is in motion. The coefficients of static and kinetic friction are both 0.50.   *Ans.* $v_{max} = 5.19$ m/s, $\Delta t = 5.54$ s

Problem 3/187

3/187 A charged particle of mass m starts from rest at time $t = 0$. It is subjected to a radial force F due to a rotating field of force whose intensity increases linearly with time, $F = Kt$, and whose direction with the x-axis is given by the angle $\theta = \omega t$ where $\omega = \dot{\theta} = $ constant. Determine the magnitude of the velocity $\mathbf{v}$ acquired by the particle during one half of a revolution of the force field starting from $t = 0$. Also specify the corresponding angle β made by $\mathbf{v}$ with the positive x-axis at this instant.

$$Ans.\ v = \frac{K}{m\omega^2}\sqrt{4 + \pi^2},\ \beta = 122.5°$$

3/9 CONSERVATION OF MOMENTUM

If the resultant force on a particle is zero during an interval of time, we see that Eq. 3/21 requires that its linear momentum **G** remain constant. Similarly if the resultant moment about a fixed point O of all forces acting on a particle is zero during an interval of time, Eq. 3/27 requires that its angular momentum $\mathbf{H}_O$ about that point remain constant. In the first case the linear momentum of the particle is said to be conserved, and in the second case the angular momentum of the particle is said to be conserved. Linear momentum may be conserved in one coordinate direction, such as x, but not necessarily in the y- or z-directions. Also angular momentum may be conserved about one axis but not about another axis.

A careful examination of the free-body diagram of the particle will disclose whether the total linear impulse on the particle in a particular direction is zero. If it is, the corresponding linear momentum is unchanged (conserved). Similarly, the free-body diagram will tell us whether the moment of the resultant force on the particle about a fixed point is zero, in which case the angular momentum about that point is unchanged (conserved).

Consider now the motion of two particles a and b which interact during an interval of time. If the interactive forces **F** and $-\mathbf{F}$ between them are the only unbalanced forces acting on the particles during the interval, it follows that the linear impulse on particle a is the negative of the linear impulse on particle b. Therefore, from Eq. 3/23 the change in linear momentum $\Delta\mathbf{G}_a$ of particle a is the negative of the change $\Delta\mathbf{G}_b$ in linear momentum of particle b. So we have $\Delta\mathbf{G}_a = -\Delta\mathbf{G}_b$ or $\Delta(\mathbf{G}_a + \mathbf{G}_b) = \mathbf{0}$. Thus the total linear momentum $\mathbf{G} = \mathbf{G}_a + \mathbf{G}_b$ for the system of the two particles remains constant during the interval, and we write

$$\boxed{\Delta\mathbf{G} = \mathbf{0}} \qquad (3/30)$$

Equation 3/30 expresses the *principle of conservation of linear momentum*.

For the interaction of the two particles it follows also that the moments of the equal and opposite interacting forces about any fixed point O not on their line of action are equal and opposite. Therefore, from Eq. 3/29 the corresponding changes in angular momentum of the two particles about the fixed point O are equal and opposite. So we may also write $\Delta(\mathbf{H}_a + \mathbf{H}_b) = \mathbf{0}$ or

$$\boxed{\Delta\mathbf{H}_O = \mathbf{0}} \qquad (3/30a)$$

which expresses the *principle of conservation of angular momentum*.

Sample Problem   3/22

The 50-g bullet traveling at 600 m/s strikes the 4-kg block centrally and is embedded within it. If the block is sliding on a smooth horizontal plane with a velocity of 12 m/s in the direction shown just before the impact, determine the velocity **v** of the block and bullet combined and its direction θ immediately after impact.

Solution.   Since the force of impact is internal to the system composed of the block and bullet and since there are no other external forces acting on the system in the plane of motion, it follows that the linear momentum of the system is conserved in both the x- and y-directions. Thus

$$[\Delta G_x = 0] \qquad 4(12 \sin 30°) + 0 = (4 + 0.050)v_x$$

$$v_x = 10.26 \text{ m/s}$$

$$[\Delta G_y = 0] \qquad 4(12 \sin 30°) + 0.050(600) = (4 + 0.050)v_y$$

$$v_y = 13.33 \text{ m/s}$$

The final velocity and its direction are given by

$$[v = \sqrt{v_x^2 + v_y^2}] \qquad v = \sqrt{(10.26)^2 + (13.33)^2} = 16.83 \text{ m/s} \qquad Ans.$$

$$\left[\tan \theta = \frac{v_y}{v_x} \right] \qquad \tan \theta = \frac{13.33}{10.26} = 1.30 \qquad \theta = 52.4° \qquad Ans.$$

① Writing the change in momentum equals zero, $\Delta G = 0$, is the same as saying that the initial and the final momenta are equal.

Sample Problem   3/23

A small mass particle is given an initial velocity $\mathbf{v}_0$ tangent to the horizontal rim of a smooth hemispherical bowl at a radius r_0 from the vertical centerline, as shown at point A. As the particle slides past point B, a distance h below A and a distance r from the vertical centerline, its velocity **v** makes an angle θ with the horizontal tangent to the bowl through B. Determine θ.

Solution.   The forces on the particle are its weight and the normal reaction exerted by the smooth surface of the bowl. Neither force exerts a moment about the axis O-O, so that angular momentum is conserved about that axis. Thus

① $[\Delta H_O = 0] \qquad mv_0 r_0 = mvr \cos \theta$

Also, energy is conserved so that

$$[\Delta T + \Delta V = 0] \qquad \tfrac{1}{2}m(v^2 - v_0^2) - mgh = 0 \qquad v = \sqrt{v_0^2 + 2gh}$$

Eliminating v and substituting $r^2 = r_0^2 - h^2$ give

$$v_0 r_0 = \sqrt{v_0^2 + 2gh} \sqrt{r_0^2 - h^2} \cos \theta$$

$$\theta = \cos^{-1} \frac{1}{\sqrt{1 + \dfrac{2gh}{v_0^2}} \sqrt{1 - \dfrac{h^2}{r_0^2}}} \qquad Ans.$$

① The angle θ is measured in the plane tangent to the hemispherical surface at B.

PROBLEMS

3/188 The 12-Mg truck drives onto the 350-Mg barge from the dock at 20 km/h and brakes to a stop on the deck. The barge is free to move in the water, which offers negligible resistance to motion at low speeds. Calculate the speed of the barge after the truck has come to rest on it.

Ans. v = 0.663 km/h

20 km/h

350 Mg

12 Mg

v

Problem 3/188

3/189 The 140-g projectile is fired with a velocity of 600 m/s and picks up three washers, each with a mass of 100 g. Find the common velocity v of the projectile and washers. Determine also the loss $|\Delta E|$ of energy during the interaction.

Ans. v = 190.9 m/s, $|\Delta E| = 17.18(10)^3$ J

600 m/s

Problem 3/189

3/190 The third and fourth stages of a rocket are coasting in space with a velocity of 18 000 km/h when a small explosive charge between the stages separates them. Immediately after separation the fourth stage has increased its velocity to $v_4 =$ 18 060 km/h. What is the corresponding velocity v_3 of the third stage? At separation the third and fourth stages have masses of 400 and 200 kg, respectively.

Ans. v_3 = 17 970 km/h

4th stage

3rd stage

v_4

v_3

Problem 3/190

3/191 The 450-kg ram of a pile driver falls 1.4 m from rest and strikes the top of a 240-kg pile embedded 0.9 m in the ground. Upon impact the ram is seen to move with the pile with no noticeable rebound. Determine the velocity v of the pile and ram immediately after impact. Can you justify using the principle of conservation of momentum even though the weights act during the impact?

1.4 m

0.9 m

Problem 3/191

3/192 Freight car A with a total mass of 80 Mg is moving along the horizontal track in a switching yard at 3 km/h. Freight car B with a total mass of 60 Mg and moving at 5 km/h overtakes car A and is coupled to it. Determine (*a*) the common velocity v of the two cars as they move together after being coupled and (*b*) the loss of energy $|\Delta E|$ due to the impact.

Ans. v = 3.86 km/h, $|\Delta E|$ = 5291 J loss

5 km/h

3 km/h

B

A

Problem 3/192

Problem 3/193

Problem 3/194

Problem 3/195

Problem 3/196

3/193 An emergency evacuation system at the launch tower for astronauts consists of a long slide-wire cable down which the escape cage travels to a safe distance from the tower. The cage, together with its two occupants, has a mass of 320 kg and approaches the netting horizontally at a speed of 28 m/s. The netting is held to the cable by a breakaway lashing and is attached to 20 m of heavy chain with a mass of 18 kg/m. The coefficient of kinetic friction between the chain and the ground is 0.70. Determine the initial velocity v of the chain when the cage has engaged the net, and find the time t to bring the cage to a stop after engagement. Assume all links of the chain remain in contact with the ground.

3/194 The central attractive force F on an earth satellite can exert no moment about the center O of the earth. For the particular elliptical orbit with major and minor axes as shown, a satellite will have a velocity of 33 880 km/h at the perigee altitude of 390 km. Determine the velocity of the satellite at point B and at apogee A. The radius of the earth is 6371 km.

Ans. $v_B = 19\ 540$ km/h
$v_A = 11\ 290$ km/h

3/195 The two spheres of equal mass m are able to slide along the horizontal rotating rod. If they are initially latched in position a distance r from the rotating axis with the assembly rotating freely with an angular velocity ω_0, determine the new angular velocity ω after the spheres are released and finally assume positions at the ends of the rod at a radial distance of $2r$. Also find the fraction n of the initial kinetic energy of the system which is lost. Neglect the small mass of the rod and shaft.

3/196 The 6-kg sphere and the 4-kg block (shown in section) are secured to the arm of negligible mass which rotates in the vertical plane about a horizontal axis at O. The 2-kg plug is released from rest at A and falls into the recess in the block when the arm has reached the horizontal position. An instant before engagement, the arm has an angular velocity $\omega_0 = 2$ rad/s. Determine the angular velocity ω of the arm immediately after the plug has wedged itself in the block.

Ans. $\omega = 0.1721$ rad/s CW

Problem 3/197

3/197 After release from rest at *B* the 0.8-kg cylindrical plug *A* slides down the smooth path and embeds itself in the 1.8-kg block *C*. Determine the velocity *v* of the block and embedded plug immediately after engagement and find the maximum deflection *x* of the spring. Neglect any friction under block *C*. What fraction *n* of the original energy of the system is lost?

Problem 3/198

3/198 An 8-Mg truck is resting on the deck of a barge which displaces 240 Mg and is at rest in still water. If the truck starts and drives toward the bow at a speed relative to the barge $v_{rel} = 6$ km/h, calculate the speed *v* of the barge. The resistance to motion of the barge through the water is negligible at low speeds. *Ans.* $v = 0.194$ km/h

Problem 3/199

3/199 Car *B* (1500 kg) traveling west at 48 km/h collides with car *A* (1600 kg) traveling north at 32 km/h as shown. If the two cars become entangled and move together as a unit after the crash, compute the magnitude *v* of their common velocity immediately after the impact and the angle θ made by the velocity vector with the north direction.

Problem 3/200

3/200 The ballistic pendulum is a simple device for determining ballistic velocities. The 60-g bullet is fired into the 30-kg box of sand suspended as a simple pendulum. If the pendulum is observed to swing through a maximum angle $\theta = 15°$ after the impact, compute the muzzle velocity of the bullet and the percentage loss of energy of the system during impact with the sand. *Ans.* $v = 710$ m/s, 99.8% loss of energy

Problem 3/201

3/201 The 500-ton tug is towing the 900-ton coal barge at a steady speed of 6 knots. For a short period of time the stern winch takes in the towing cable at the rate of 0.5 m/s. Calculate the reduced speed v of the tug during this interval. Assume the tow cable to be horizontal. (Recall 1 knot = 1.852 km/h and that 1 ton = 1000 kg.)

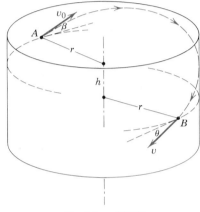

Problem 3/202

3/202 A particle is released on the smooth inside wall of a cylindrical tank at A with a velocity v_0 which makes an angle β with the horizontal tangent. When the particle reaches a point B a distance h below A, determine the expression for the angle θ made by its velocity with the horizontal tangent at B.

3/203 For Prob. 3/202 with given values of v_0, β, and r, find the maximum height b above point A reached by the particle and the magnitude v of its velocity at this point.

$$Ans.\ b = \frac{v_0^{\,2}}{2g} \sin^2 \beta, \ v = v_0 \cos \beta$$

Problem 3/204

3/204 The two mine cars of equal mass are connected by a rope which is initially slack. Car A is given a shove which imparts to it a velocity of 1.2 m/s with car B initially at rest. When the slack is taken up the rope suffers a tension impact which imparts a velocity to car B and reduces the velocity of car A. (*a*) If 40 percent of the kinetic energy of car A is lost during the rope impact, calculate the velocity v_B imparted to car B. (*b*) Following the initial impact car B overtakes car A and the two are coupled together. Calculate their final common velocity v_C.

$$Ans.\ v_B = 0.868 \text{ m/s}, \ v_C = 0.6 \text{ m/s}$$

Problem 3/205

3/205 The simple pendulum A of mass m_A and length l is suspended from the trolley B of mass m_B. If the system is released from rest at $\theta = 0$, determine the velocity v_B of the trolley when $\theta = 90°$. Friction is negligible.

$$Ans.\ v_B = \frac{m_A}{m_B} \sqrt{\frac{2gl}{1 + \dfrac{m_A}{m_B}}}$$

3/206 The small particle of mass m and its restraining cord are spinning with an angular velocity ω on the horizontal surface of a smooth disk, shown in section. As the force F is slightly relaxed, r increases and ω changes. Determine the rate of change of ω with respect to r and show that the work done by F during a movement dr equals the change in kinetic energy of the particle.

Problem 3/206

3/207 A torpedo boat with a displacement of 60 metric tons is moving at 18 knots when it fires a 140-kg torpedo horizontally with the launch tube at the 30° angle shown. If the torpedo has a velocity relative to the boat of 6 m/s as it leaves the tube, determine the momentary reduction Δv in forward speed of the boat. Recall that 1 knot = 1.852 km/h and 1 metric ton = 1000 kg. Solve the problem also by referring the motion to a coordinate system moving with the initial speed of the boat.

Ans. $\Delta v = 0.0121$ m/s

Problem 3/207

3/208 The assembly of two 5-kg spheres is rotating freely about the vertical axis at 40 rev/min with $\theta = 90°$. If the force F necessary to maintain the given position is increased to raise the base collar and reduce θ to 60°, determine the new angular velocity ω. Also determine the work U done by F in changing the configuration of the system. Assume that the mass of the arms and collars is negligible.

Ans. $\omega = 3.00$ rad/s, $U = 5.34$ J

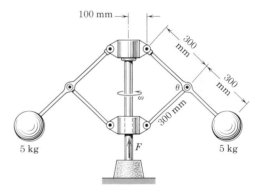

Problem 3/208

3/209 Two barges each with a displacement (mass) of 500 metric tons are loosely moored in calm water. A stunt driver starts his 1500-kg car from rest at A, drives along the deck, and leaves the end of the 15° ramp at a speed of 50 km/h relative to the barge and ramp. The driver successfully jumps the gap and brings his car to rest relative to barge 2 at B. Calculate the velocity v_2 imparted to barge 2 just after the car has come to rest on the barge. Neglect the resistance of the water to motion at the low velocities involved.

Ans. $v_2 = 40.0$ mm/s

Problem 3/209

SECTION D. SPECIAL APPLICATIONS

3/10 *INTRODUCTION*

The basic principles and methods of particle kinetics were developed and illustrated in the first three sections of this chapter. This treatment included the direct use of Newton's second law, the equations of work and energy, and the equations of impulse and momentum. Special attention was paid to the kind of problem for which each of the approaches was most appropriate.

There are several topics of specialized interest in particle kinetics which will be briefly treated in Section D. These topics involve further extension and application of the fundamental principles of dynamics, and their study will help to broaden one's background in mechanics.

3/11 *IMPACT*

The principles of impulse and momentum find important use in describing the behavior of colliding bodies. *Impact* refers to the collision between two bodies and is characterized by the generation of relatively large contact forces which act over a very short interval of time. Before we begin treatment of the subject, it should be stated that an impact is a very complex event involving material deformation and recovery and the generation of heat and sound. Small changes in the impact conditions may cause large changes in the impact process and thus in the conditions immediately following the impact. Hence we must be careful not to place excessive reliance upon the results of impact calculations.

(*a*) *Direct central impact.*   As an introduction to impact we consider the collinear motion of two spheres of masses m_1 and m_2, Fig. 3/14*a*, traveling with velocities v_1 and v_2. If v_1 is greater than v_2, collision occurs with the contact forces directed along

Figure 3/14

the line of centers. This condition is called *direct central impact.* Following initial contact a short period of increasing deformation takes place until the contact area between the spheres ceases to increase. At this instant both spheres, Fig. 3/14*b*, are moving with the same velocity v_0. During the remainder of contact a period of restoration occurs during which the contact area decreases to zero. The final condition is shown in the *c*-part of the figure where the spheres now have new velocities v_1' and v_2', where of necessity v_1' is less than v_2'. All velocities are arbitrarily assumed positive to the right, so that with this scalar notation a velocity to the left would carry a negative sign. If the impact is not overly severe and if the spheres are highly elastic, they will regain their original shape following the restoration. With a more severe impact and with less elastic bodies a permanent deformation may result.

Inasmuch as the contact forces are equal and opposite during impact, the linear momentum of the system remains unchanged, as discussed in Art. 3/9. Thus we apply the law of conservation of linear momentum and write

$$m_1 v_1 + m_2 v_2 = m_1 v_1' + m_2 v_2' \qquad (3/31)$$

It is assumed that any forces other than the large internal forces of contact which may act on the spheres during impact are relatively small and produce negligible impulses compared with the impulse associated with each internal impact force. In addition, we assume that no appreciable position change occurs during the short duration of the impact.

For given masses and initial conditions, the momentum equation contains two unknowns, v_1' and v_2'. Clearly an additional relationship is required before the final velocities can be found. This relationship must reflect the capacity of the contacting bodies to recover from the impact and can be expressed by the ratio e of the magnitude of the restoration impulse to the magnitude of the deformation impulse. This ratio is called the *coefficient of restitution.* If F_r and F_d represent the magnitudes of the contact forces during the restoration and deformation periods, respectively, as shown in Fig. 3/15, for particle 1 the definition of e together with the impulse-momentum equation give us

$$e = \frac{\int_{t_0}^{t} F_r \, dt}{\int_{0}^{t_0} F_d \, dt} = \frac{m_1(-v_1' - [-v_0])}{m_1(-v_0 - [-v_1])} = \frac{v_0 - v_1'}{v_1 - v_0}$$

Similarly for particle 2 we have

$$e = \frac{\int_{t_0}^{t} F_r \, dt}{\int_{0}^{t_0} F_d \, dt} = \frac{m_2(v_2' - v_0)}{m_2(v_0 - v_2)} = \frac{v_2' - v_0}{v_0 - v_2}$$

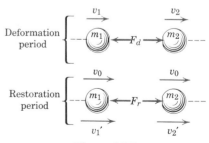

Figure 3/15

We are careful in these equations to express the change of momentum (and hence Δv) in the same direction as the impulse (and hence the force). The time for the deformation is taken as t_0 and the total time of contact is t. Eliminating v_0 between the two expressions for e gives us

$$e = \frac{v_2' - v_1'}{v_1 - v_2} = \frac{|\text{relative velocity of separation}|}{|\text{relative velocity of approach}|} \quad (3/32)$$

In addition to the initial conditions, if e is known for the impact condition at hand, then Eqs. 3/31 and 3/32 give us two equations in the two unknown final velocities v_1' and v_2'.

Impact phenomena are almost always accompanied by energy loss, which may be calculated by subtracting the kinetic energy of the system just after impact from that just before impact. Energy is lost through the generation of heat during the localized inelastic deformation of the material, through the generation and dissipation of elastic stress waves within the bodies, and through the generation of sound energy.

According to this classical theory of impact the value $e = 1$ means that the capacity of the two particles to recover equals their tendency to deform. This condition is one of *elastic impact* with no energy loss. The value $e = 0$, on the other hand, describes *inelastic* or *plastic impact* where the particles cling together after collision and the loss of energy is a maximum. All impact conditions lie somewhere between these two extremes. Also it should be noted that a coefficient of restitution must be associated with a *pair* of contacting bodies.

The coefficient of restitution is frequently considered a constant for given geometries and a given combination of contacting materials. Actually it depends upon the impact velocity and approaches unity as the impact velocity approaches zero as shown schematically in Fig. 3/16. A handbook value for e is generally unreliable.

Coefficient of
restitution, e

Perfectly elastic

Glass on glass

Steel on steel

Lead on lead

Perfectly plastic

Relative impact velocity

Figure 3/16

(*b*) *Oblique central impact.* We now extend the relationships developed for direct central impact to the case where the initial and final velocities are not parallel, Fig. 3/17. Here spherical particles of mass m_1 and m_2 have initial velocities $\mathbf{v}_1$ and $\mathbf{v}_2$ in the same plane and approach each other on a collision course, as shown in the *a*-part of the figure. The directions of the velocity vectors are measured from the direction tangent to the contacting surfaces, Fig. 3/17*b*. Thus the initial velocity components along the *t*- and *n*-axes are $(v_1)_n = -v_1 \sin \theta_1$, $(v_1)_t = v_1 \cos \theta_1$, $(v_2)_n = v_2 \sin \theta_2$, and $(v_2)_t = v_2 \cos \theta_2$. Note that $(v_1)_n$ is a negative quantity for the particular coordinate system and initial velocities

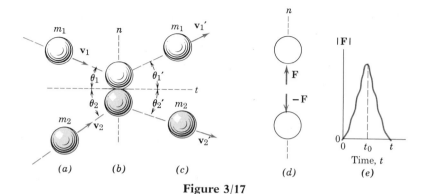

Figure 3/17

shown. The final rebound conditions are shown in the c-part of the figure. The impact forces are **F** and −**F** as seen in the d-part of the figure. They vary from zero to their peak value during the deformation portion of the impact and back again to zero during the restoration period as indicated in the e-part of the figure where t is the duration of the impact interval.

For given initial conditions of m_1, m_2, $(v_1)_n$, $(v_1)_t$, $(v_2)_n$, and $(v_2)_t$, there will be four unknowns, namely, $(v_1')_n$, $(v_1')_t$, $(v_2')_n$, and $(v_2')_t$. The four needed equations are:

(1) Momentum of the system is conserved in the n-direction giving

$$m_1(v_1)_n + m_2(v_2)_n = m_1(v_1')_n + m_2(v_2')_n$$

(2) and (3) The momentum for each particle is conserved in the t-direction since there is no impulse on either particle in the t-direction. Thus

$$m_1(v_1)_t = m_1(v_1')_t$$

$$m_2(v_2)_t = m_2(v_2')_t$$

(4) The coefficient of restitution, as in the case of direct central impact, is the positive ratio of the recovery impulse to the deformation impulse. Equation 3/32 applies, then, to the velocity components in the n-direction, and for the notation adopted with Fig. 3/17 we have

$$e = \frac{(v_2')_n - (v_1')_n}{(v_1)_n - (v_2)_n}$$

Once the four final velocity components are found, the angles θ_1' and θ_2' of Fig. 3/17 may be easily determined.

Sample Problem 3/24

The ram of a pile driver has a mass of 800 kg and is released from rest 2 m above the top of the 2400-kg pile. If the ram rebounds to a height of 0.1 m after impact with the pile, calculate (a) the velocity v_p' of the pile immediately after impact, (b) the coefficient of restitution e, and (c) the percentage loss of energy due to the impact.

2 m drop

0.1 m rebound

Solution. Conservation of energy during free fall gives the initial and final velocities of the ram from $v = \sqrt{2gh}$. Thus

$$v_r = \sqrt{2(9.81)(2)} = 6.26 \text{ m/s} \qquad v_r' = \sqrt{2(9.81)(0.1)} = 1.40 \text{ m/s}$$

(a) Conservation of momentum ($\Delta G_y = 0$) for the system of the ram and pile gives

$$\textcircled{1} \qquad 800(6.26) + 0 = 800(-1.40) + 2400v_p' \qquad v_p' = 2.55 \text{ m/s} \qquad Ans.$$

(b) The coefficient of restitution gives

$$e = \frac{|\text{rel. vel. separation}|}{|\text{rel. vel. approach}|} \qquad e = \frac{2.55 + 1.40}{6.26 + 0} = 0.63 \qquad Ans.$$

(c) The kinetic energy of the system just before impact is the same as the potential energy of the ram above the pile and is

$$T = V_g = mgh = 800(9.81)(2) = 15\ 700 \text{ J}$$

The kinetic energy T' just after impact is

$$T' = \tfrac{1}{2}(800)(1.40)^2 + \tfrac{1}{2}(2400)(2.55)^2 = 8620 \text{ J}$$

The percentage loss of energy is, therefore,

$$\frac{15\ 700 - 8620}{15\ 700}(100) = 45.1\% \qquad Ans.$$

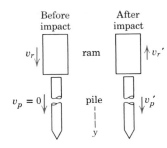

Before impact — After impact

v_r — ram — v_r'

$v_p = 0$ — pile — v_p'

y

$\textcircled{1}$ The impulses of the weights of the ram and pile are very small compared with the impulses of the impact forces and thus are neglected during the impact.

Sample Problem 3/25

A ball is projected onto the heavy plate with a velocity of 16 m/s at the 30° angle shown. If the effective coefficient of restitution is 0.5, compute the rebound velocity v' and its angle θ'.

16 m/s — n

v'

1

30° — θ'

2 — t

Solution. Let the ball be denoted body 1 and the plate body 2. The mass of the heavy plate may be considered infinite and its corresponding velocity zero after impact. The coefficient of restitution is applied to the velocity components normal to the plate in the direction of the impact force and gives

$$\textcircled{1} \qquad e = \frac{(v_2')_n - (v_1')_n}{(v_1)_n - (v_2)_n} \qquad 0.5 = \frac{0 - (v_1')_n}{-16 \sin 30° - 0} \qquad (v_1')_n = 4 \text{ m/s}$$

Momentum of the ball in the t-direction is unchanged since, with assumed smooth surfaces, there is no force acting on the ball in that direction. Thus

$$m(v_1)_t = m(v_1')_t \qquad (v_1')_t = (v_1)_t = 16 \cos 30° = 13.86 \text{ m/s}$$

The rebound velocity v' and its angle θ' are then

$$v' = \sqrt{(v_1')_n^2 + (v_1')_t^2} = \sqrt{4^2 + (13.86)^2} = 14.42 \text{ m/s} \qquad Ans.$$

$$\theta' = \tan^{-1}\left(\frac{(v_1')_n}{(v_1')_t}\right) = \tan^{-1}\left(\frac{4}{13.86}\right) = 16.1° \qquad Ans.$$

$W \ll F_{\text{impact}}$

F_{impact}

$\textcircled{1}$ We observe here that for infinite mass there is no way of applying the principle of conservation of momentum for the system in the n-direction. From the free-body diagram of the ball during impact we note that the impulse of the weight W is neglected since W is very small compared with the impact force.

Sample Problem 3/26

Spherical particle 1 has a velocity $v_1 = 6$ m/s in the direction shown and collides with spherical particle 2 of equal mass and diameter and initially at rest. If the coefficient of restitution for these conditions is $e = 0.6$, determine the resulting motion of each particle following impact. Also calculate the percentage loss of energy due to the impact.

Solution. The geometry at impact indicates that the normal n to the contacting surfaces makes an angle $\theta = 30°$ with the direction of $\mathbf{v}_1$, as indicated in the figure. Thus the initial velocity components are $(v_1)_n = v_1 \cos 30° = 6 \cos 30° = 5.196$ m/s, $(v_1)_t = v_1 \sin 30° = 6 \sin 30° = 3$ m/s, and $(v_2)_n = (v_2)_t = 0$.

Momentum conservation for the two-particle system in the n-direction gives

$$m_1(v_1)_n + m_2(v_2)_n = m_1(v_1')_n + m_2(v_2')_n$$

or, with $m_1 = m_2$, $\quad 5.196 + 0 = (v_1')_n + (v_2')_n \qquad (a)$

The coefficient of restitution relationship is

$$e = \frac{(v_2')_n - (v_1')_n}{(v_1)_n - (v_2)_n} \qquad 0.6 = \frac{(v_2')_n - (v_1')_n}{5.196 - 0} \qquad (b)$$

Simultaneous solution of Eqs. a and b yields

$$(v_1')_n = 1.039 \text{ m/s} \qquad (v_2')_n = 4.16 \text{ m/s}$$

Conservation of momentum for each particle holds in the t-direction because, with assumed smooth surfaces, there is no force in the t-direction. Thus for particles 1 and 2, we have

$$m_1(v_1)_t = m_1(v_1')_t \qquad (v_1')_t = (v_1)_t = 3 \text{ m/s}$$
$$m_2(v_2)_t = m_2(v_2')_t \qquad (v_2')_t = (v_2)_t = 0$$

The final speeds of the particles are

$$v_1' = \sqrt{(v_1')_n{}^2 + (v_1')_t{}^2} = \sqrt{(1.039)^2 + 3^2} = 3.17 \text{ m/s} \qquad \textit{Ans.}$$
$$v_2' = \sqrt{(v_2')_n{}^2 + (v_2')_t{}^2} = \sqrt{(4.16)^2 + 0^2} = 4.16 \text{ m/s} \qquad \textit{Ans.}$$

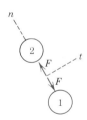

The angle θ' which $\mathbf{v}_1'$ makes with the t-direction is

$$\theta' = \tan^{-1}\left(\frac{(v_1')_n}{(v_1')_t}\right) = \tan^{-1}\left(\frac{1.039}{3}\right) = 19.1° \qquad \textit{Ans.}$$

The kinetic energies just before and just after impact, with $m = m_1 = m_2$, are

$$T = \tfrac{1}{2}m_1v_1{}^2 + \tfrac{1}{2}m_2v_2{}^2 = \tfrac{1}{2}m(6)^2 + 0 = 18m$$
$$T' = \tfrac{1}{2}m_1v_1'{}^2 + \tfrac{1}{2}m_2v_2'{}^2 = \tfrac{1}{2}m(3.17)^2 + \tfrac{1}{2}m(4.16)^2 = 13.68m$$

The percentage energy loss is then

$$\frac{\Delta E}{E}(100) = \frac{T - T'}{T}(100) = \frac{18m - 13.68m}{18m}(100) = 24\% \qquad \textit{Ans.}$$

① Be sure to set up n- and t-coordinates which are respectively normal to and tangent to the contacting surfaces. Calculation of the 30° angle is critical to all that follows.

② Note that, even though there are four equations in four unknowns for the standard problem of oblique central impact, only one pair of the equations is coupled.

③ We note that particle 2 has no initial or final velocity component in the t-direction. Hence its final velocity $\mathbf{v}_2'$ is restricted to the n-direction.

Problem 3/210

Problem 3/211

Problem 3/212

Problem 3/213

Problem 3/214

PROBLEMS

3/210 Tennis balls are usually rejected if they fail to rebound to waist level when dropped from shoulder level. If a ball just passes the test as indicated in the figure, determine the coefficient of restitution e and the percentage n of the original energy lost during the impact.

Ans. $e = 0.829$, $n = 31.2\%$

3/211 Compute the final velocities v_1' and v_2' of the two cylinders which slide on the smooth horizontal shaft. The coefficient of restitution is $e = 0.5$.

3/212 The 250-kg ram of a pile driver falls 1.2 m from rest and strikes the top of a 400-kg pile embedded 0.9 m in the ground. Immediately after impact the ram is seen to have no velocity. Determine the coefficient of restitution e and the velocity v' of the pile immediately after impact.

Ans. $e = 0.625$, $v' = 3.03$ m/s

3/213 A railroad car of mass m and initial speed v collides with and becomes coupled to a group of three identical cars. Compute the final speed v' of the group of four cars and the fractional loss n of energy if (*a*) there is no slack in the coupling mechanisms of the initially stationary group and (*b*) there is slack initially in all couplings. Assume that, when the slack is removed between two coupled cars, the effect is the same as an energy-dissipating collision with coefficient of restitution $e = 0$.

3/214 Three identical steel cylinders are free to slide horizontally on the fixed horizontal shaft. Cylinders 2 and 3 are at rest and are approached by cylinder 1 at a speed u. Express the final speed v of cylinder 3 in terms of u and the coefficient of restitution e.

Ans. $v = \dfrac{u}{4}(1 + e)^2$

3/215 The sphere of mass m_1 travels with an initial velocity v_1 directed as shown and strikes the sphere of mass m_2. For a given coefficient of restitution e, determine the mass ratio m_1/m_2 which results in m_1 being motionless after the impact.

Problem 3/215

3/216 Consider the two spheres of Prob. 3/215. For a given coefficient of restitution e, what should the mass ratio m_1/m_2 be so that the final velocity of mass m_2 is $1.5v_1$? Note any restrictions on your solution.

$$Ans. \frac{m_1}{m_2} = \frac{1.5}{e - 0.5}, \; e > 0.5$$

3/217 For the two spheres of Prob. 3/215, determine the maximum final velocity attainable by sphere 2. For what values of the coefficient of restitution e and the mass ratio m_1/m_2 does this maximum result occur?

3/218 A 1000-kg spacecraft is traveling in deep space with a speed of 2000 m/s when a 100-kg meteor moving at a right angle to its path strikes and becomes imbedded in the spacecraft. If the resulting path is as indicated in the figure, determine the speed v_m of the meteor just prior to impact.

Ans. $v_m = 7280$ m/s

Problem 3/218

3/219 Cylinder 1 of mass m moving with a velocity v_1 strikes cylinder 2 of mass $2m$ initially at rest. The impact force F varies with time as shown, where t_d is the duration of the deformation period and t_r is the duration of the restoration period. Determine the velocity v_2' of cylinder 2 immediately after impact in terms of the initial velocity v_1 of cylinder 1 for (a) $t_r = t_d$, (b) $t_r = 0.5t_d$, and (c) $t_r = 0$.

Ans. (a) $v_2' = 2v_1/3$, (b) $v_2' = v_1/2$, (c) $v_2' = v_1/3$

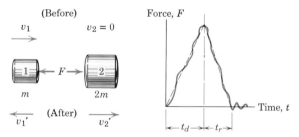

Problem 3/219

3/220 Two steel balls of the same diameter are connected by a rigid bar of negligible mass as shown and are dropped in the horizontal position from a height of 150 mm above the heavy steel and brass base plates. If the coefficient of restitution between the ball and the steel base is 0.6 and that between the other ball and the brass base is 0.4, determine the angular velocity ω of the bar immediately after impact. Assume the two impacts are simultaneous.

Problem 3/220

3/221 In the selection of the ram of a pile driver, similar to that shown with Prob. 3/212, it is desired that the ram lose all of its kinetic energy at each blow. Hence the velocity of the ram is zero immediately after impact. The mass of each pile to be driven is 300 kg, and experience has shown that a coefficient of restitution of 0.3 can be expected. What should be the mass m of the ram? Compute the velocity v of the pile immediately after impact if the ram is dropped from a height of 4 m onto the pile. Also compute the energy loss ΔE due to impact at each blow.

> *Ans.* $m = 90$ kg
> $v = 2.66$ m/s
> $\Delta E = 2470$ J

3/222 The steel ball strikes the heavy steel plate with a velocity $v_0 = 24$ m/s at an angle of 60° with the horizontal. If the coefficient of restitution is $e = 0.8$, compute the velocity v and its direction θ with which the ball rebounds from the plate.

> *Ans.* $v = 20.5$ m/s, $\theta = 54.2°$

Problem 3/222

3/223 The previous problem is modified in that the plate struck by the ball now has a mass equal to that of the ball and is supported as shown. Compute the final velocities of both masses immediately after impact if the plate is initially stationary and all other conditions are the same as stated in the previous problem.

Problem 3/223

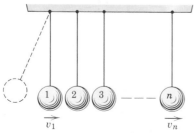

3/224 The figure shows n spheres of equal mass m suspended in a line by wires of equal length so that the spheres are almost touching each other. If sphere 1 is released from the dotted position and strikes sphere 2 with a velocity v_1, write an expression for the velocity v_n of the nth sphere immediately after being struck by the one adjacent to it. The common coefficient of restitution is e.

> *Ans.* $v_n = \left(\dfrac{1 + e}{2}\right)^{n-1} v_1$

Problem 3/224

3/225 Six identical spheres are arranged as shown in the figure. The two spheres at the left end are released from the dotted positions and strike sphere 3 with speed v_1. Assuming that the common coefficient of restitution is $e = 1$, explain why two spheres leave the right end of the row with speed v_1 instead of one sphere leaving the right end with speed $2v_1$.

Problem 3/225

3/226 The 2-kg piece of putty is dropped 2 m onto the 18-kg block initially at rest on the two springs, each with a stiffness $k = 1.2$ kN/m. Calculate the additional deflection δ of the springs due to the impact of the putty which adheres to the block upon contact.   *Ans.* $\delta = 65.9$ mm

Problem 3/226

3/227 Sphere A has a mass of 23 kg and a radius of 75 mm, while sphere B has a mass of 4 kg and a radius of 50 mm. If the spheres are traveling initially along the parallel paths with the speeds shown, determine the velocities of the spheres immediately after impact. The coefficient of restitution is 0.4 and friction is neglected.

Problem 3/227

3/228 During a pregame warmup period, two basketballs collide above the hoop when in the positions shown. Just before impact, ball 1 has a velocity $\mathbf{v}_1$ which makes a 30° angle with the horizontal. If the velocity $\mathbf{v}_2$ of ball 2 just before impact has the same magnitude as $\mathbf{v}_1$, determine the two possible values of the angle θ, measured from the horizontal, which will cause ball 1 to go directly through the center of the basket. The coefficient of restitution is $e = 0.8$.
          Ans. $\theta = 82.3°$ or $-22.3°$

Problem 3/228

Problem 3/229

Problem 3/230

Problem 3/231

Problem 3/233

3/229 The two identical steel balls moving with initial velocities v_1 and v_2 as shown collide in such a way that the line joining their centers is in the direction of v_2. From previous experiment the coefficient of restitution is known to be 0.60. Determine the velocity of each ball immediately after impact and find the percentage loss of kinetic energy of the system as a result of the impact.

3/230 To pass inspection, steel balls for ball bearings must clear the fixed bar A at the top of their rebound when dropped from rest through the vertical distance $H = 900$ mm onto the heavy inclined steel plate. If balls which have a coefficient of restitution of less than 0.7 with the rebound plate are to be rejected, determine the position of the bar by specifying h and s. Neglect any friction during impact.   *Ans. $h = 379$ mm, $s = 339$ mm*

3/231 In a game of pool, the eight ball is to be struck by the cue ball A so that the eight ball enters the right corner pocket B. Specify the distance x from the center of the left corner pocket C to the point where the cue ball strikes the cushion after impact with the eight ball. The equal-mass balls are 50 mm in diameter, and the coefficient of restitution is $e = 0.9$.   *Ans. $x = 410$ mm*

3/232 Show that the loss of energy due to direct central impact of two masses m_1 and m_2 having velocities v_1 and v_2 directed toward each other is given by

$$\Delta E = \frac{1 - e^2}{2} \frac{m_1 m_2}{m_1 + m_2} (v_1 + v_2)^2$$

where e is the coefficient of restitution for these particular impact conditions and the internal vibrational energy is neglected. (*Hint:* The energy loss depends on the relative impact velocity $v_1 + v_2$. Thus the center of mass of the system may be taken at rest to simplify the algebra so that $m_1 v_1 = m_2 v_2$.)

▶ **3/233** A child throws a ball from point A with a speed of 15 m/s. It strikes the wall at point B and then returns exactly to point A. Determine the necessary angle α if the coefficient of restitution in the wall impact is $e = 0.5$.   *Ans. $\alpha = 11.55°$ or $78.4°$*

3/12 *CENTRAL-FORCE MOTION*

When a particle moves under the influence of a force directed toward a fixed center of attraction, the motion is called *central-force motion*. The most common example of central-force motion is found with the orbits of planets and satellites. The laws which govern this motion were deduced from observation of the motions of the planets by J. Kepler (1571–1630). The dynamics of central-force motion is basic to the design of high-altitude rockets, earth satellites, and space vehicles of all types.

Consider a particle of mass m, Fig. 3/18, moving under the action of the central gravitational attraction

$$F = G\frac{mm_0}{r^2}$$

Figure 3/18

where m_0 is the mass of the attracting body assumed to be fixed, G is the universal gravitational constant, and r is the distance between the centers of the masses. The particle of mass m could represent the earth moving about the sun, the moon moving about the earth, or a satellite in its orbital motion about the earth above the atmosphere. The most convenient coordinate system to use is polar coordinates in the plane of motion since **F** will always be in the negative r-direction and there is no force in the θ-direction.

Equations 3/8 may be applied directly for the r- and θ-directions to give

$$-G\frac{mm_0}{r^2} = m(\ddot{r} - r\dot{\theta}^2)$$

$$0 = m(r\ddot{\theta} + 2\dot{r}\dot{\theta})$$

$$(3/33)$$

The second of the two equations when multiplied by r/m is seen to be the same as $d(r^2\dot{\theta})/dt = 0$, which is integrated to give

$$r^2\dot{\theta} = h, \qquad \text{a constant} \qquad (3/34)$$

The physical significance of Eq. 3/34 is made clear when we note that the angular momentum $\mathbf{r} \times m\mathbf{v}$ of m about m_0 has the magnitude $mr^2\dot{\theta}$. Thus Eq. 3/34 merely states that the angular momentum of m about m_0 remains constant, or is conserved. This statement is easily deduced from Eq. 3/27 where it is observed that the angular momentum $\mathbf{H}_O$ remains constant (is conserved) if there is no moment acting on the particle about a fixed point O.

We observe that during time dt the radius vector sweeps out an area, shaded in Fig. 3/18, equal to $dA = (\frac{1}{2}r)(r\,d\theta)$. Therefore the rate at which area is swept by the radius vector is $\dot{A} = \frac{1}{2}r^2\dot{\theta}$, which is constant according to Eq. 3/34. This conclusion is expressed in Kepler's *second law* of planetary motion which states that the areas swept through in equal times are equal.

The shape of the path followed by m may be obtained by solving the first of Eqs. 3/33 with the time t eliminated through combination with Eq. 3/34. To this end the mathematical substitution $r = 1/u$ is useful. Thus $\dot{r} = -(1/u^2)\dot{u}$, which from Eq. 3/34 becomes $\dot{r} = -h(\dot{u}/\dot{\theta})$ or $\dot{r} = -h(du/d\theta)$. The second time derivative is $\ddot{r} = -h(d^2u/d\theta^2)\dot{\theta}$, which by combining with Eq. 3/34, becomes $\ddot{r} = -h^2u^2(d^2u/d\theta^2)$. Substitution into the first of Eqs. 3/33 now gives

$$-Gm_0u^2 = -h^2u^2\frac{d^2u}{d\theta^2} - \frac{1}{u}h^2u^4$$

or

$$\frac{d^2u}{d\theta^2} + u = \frac{Gm_0}{h^2} \qquad (3/35)$$

which is a nonhomogeneous linear differential equation. The solution of this familiar second-order equation may be verified by direct substitution and is

$$u = \frac{1}{r} = C\cos(\theta + \delta) + \frac{Gm_0}{h^2}$$

where C and δ are the two integration constants. The phase angle δ may be eliminated by choosing the x-axis so that r is a minimum when $\theta = 0$. Thus

$$\frac{1}{r} = C\cos\theta + \frac{Gm_0}{h^2} \qquad (3/36)$$

The interpretation of Eq. 3/36 requires a knowledge of the equations for conic sections. We recall that a conic section is formed by the locus of a point which moves so that the ratio e of its distance from a point (focus) to a line (directrix) is constant. Thus from Fig. 3/18, $e = r/(d - r\cos\theta)$, which may be rewritten as

$$\frac{1}{r} = \frac{1}{d}\cos\theta + \frac{1}{ed} \qquad (3/37)$$

which is the same form as Eq. 3/36. Hence it is seen that the motion of m is along a conic section with $d = 1/C$ and $ed = h^2/(Gm_0)$ or

$$e = \frac{h^2C}{Gm_0} \qquad (3/38)$$

There are three cases to be investigated corresponding to $e < 1$ (ellipse), $e = 1$ (parabola), and $e > 1$ (hyperbola). The trajectory for each of these cases is shown in Fig. 3/19.

Case I: ellipse $(e < 1)$. From Eq. 3/37 it is seen that r is a minimum when $\theta = 0$ and is a maximum when $\theta = \pi$. Thus,

$$2a = r_{min} + r_{max} = \frac{ed}{1 + e} + \frac{ed}{1 - e} \quad \text{or} \quad a = \frac{ed}{1 - e^2}$$

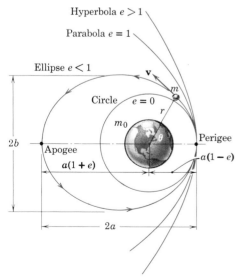

Hyperbola $e > 1$

Parabola $e = 1$

Ellipse $e < 1$

Circle $e = 0$

m_0

Apogee

$a(1 + e)$

$2b$

$2a$

Perigee

$a(1 - e)$

Figure 3/19

With the distance d expressed in terms of a, Eq. 3/37 and the maximum and minimum values of r may be written as

$$\frac{1}{r} = \frac{1 + e \cos \theta}{a(1 - e^2)}$$

$$r_{\min} = a(1 - e) \qquad r_{\max} = a(1 + e)$$

(3/39)

In addition, the relation $b = a\sqrt{1 - e^2}$, which comes from the geometry of the ellipse, gives the expression for the semiminor axis. It is seen that the ellipse becomes a circle with $r = a$ when $e = 0$. Equation 3/39 is an expression of Kepler's *first law* which says that the planets move in elliptical orbits around the sun as a focus.

The period τ for the elliptical orbit is the total area A of the ellipse divided by the constant rate $\dot{A}$ at which area is swept through. Thus

$$\tau = A/\dot{A} = \frac{\pi a b}{\frac{1}{2}r^2\dot\theta} \qquad \text{or} \qquad \tau = \frac{2\pi a b}{h}$$

by Eq. 3/34. Substitution of Eq. 3/38, the identity $d = 1/C$, the geometric relationships $a = ed/(1 - e^2)$ and $b = a\sqrt{1 - e^2}$ for the ellipse, and the equivalence $Gm_0 = gR^2$ yields upon simplification

$$\boxed{\tau = 2\pi \frac{a^{3/2}}{R\sqrt{g}}}$$

(3/40)

In this equation it is noted that R is the mean radius of the central attracting body and g is the absolute value of the acceleration due to gravity at the surface of the attracting body.

Equation 3/40 expresses Kepler's *third law* of planetary motion which states that the square of the period of motion is proportional to the cube of the semimajor axis of the orbit.

Case II: parabola $(e = 1)$.   Equations 3/37 and 3/38 become

$$\frac{1}{r} = \frac{1}{d}(1 + \cos \theta) \qquad \text{and} \qquad h^2 C = Gm_0$$

The radius vector and the dimension a become infinite as θ approaches π.

Case III: hyperbola $(e > 1)$.   From Eq. 3/37 it is seen that the radial distance r becomes infinite for the two values of the polar angle θ_1 and $-\theta_1$ defined by $\cos \theta_1 = -1/e$. Only branch I corresponding to $-\theta_1 < \theta < \theta_1$, Fig. 3/20, represents a physically possible motion. Branch II corresponds to angles in the remaining sector (with r negative). For this branch positive r's may be used if θ is replaced by $\theta - \pi$ and $-r$ by r. Thus Eq. 3/37

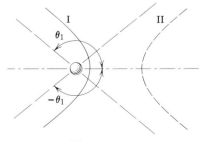

Figure 3/20

becomes

$$\frac{1}{-r} = \frac{1}{d}\cos(\theta - \pi) + \frac{1}{ed} \qquad \text{or} \qquad \frac{1}{r} = -\frac{1}{ed} + \frac{\cos\theta}{d}$$

But this expression contradicts the form of Eq. 3/36 where Gm_0/h^2 is necessarily positive. Hence branch II does not exist (except for repulsive forces).

Now consider the energies of particle m. The system is conservative, and the constant energy E of m is the sum of its kinetic energy T and potential energy V. The kinetic energy is $T = \frac{1}{2}mv^2 = \frac{1}{2}m(\dot{r}^2 + r^2\dot{\theta}^2)$ and the potential energy from Eq. 3/15 is $V = -mgR^2/r$. Recall that g is the absolute acceleration due to gravity measured at the surface of the attracting body, R is the radius of the attracting body, and $Gm_0 = gR^2$. Thus

$$E = \frac{1}{2}m(\dot{r}^2 + r^2\dot{\theta}^2) - \frac{mgR^2}{r}$$

This constant value of E can be determined from its value at $\theta = 0$ where $\dot{r} = 0$, $1/r = C + gR^2/h^2$ from Eq. 3/36, and $r\dot{\theta} = h/r$ from Eq. 3/34. Substitution into the expression for E and simplification yield

$$\frac{2E}{m} = h^2C^2 - \frac{g^2R^4}{h^2}$$

Now eliminate C by substitution of Eq. 3/38, which may be written as $h^2C = egR^2$, and obtain

$$e = +\sqrt{1 + \frac{2Eh^2}{mg^2R^4}} \qquad (3/41)$$

The plus value of the radical is mandatory since by definition e is positive. It is now seen that for the

elliptical orbit	$e < 1$,	E is negative
parabolic orbit	$e = 1$,	E is zero
hyperbolic orbit	$e > 1$,	E is positive

These conclusions, of course, depend on the arbitrary selection of the datum condition for zero potential energy ($V = 0$ when $r = \infty$).

The expression for the velocity v of m may be found from the energy equation, which is

$$\frac{1}{2}mv^2 - \frac{mgR^2}{r} = E$$

The total energy E is obtained from Eq. 3/41 by combining Eq. 3/38 and $1/C = d = a(1 - e^2)/e$ to give for the elliptical orbit

$$E = -\frac{gR^2m}{2a} \qquad (3/42)$$

Substitution into the energy equation yields

$$v^2 = 2gR^2 \left(\frac{1}{r} - \frac{1}{2a} \right)$$ (3/43)

from which the magnitude of the velocity may be computed for a particular orbit in terms of the radial distance r. Combination of the expressions for r_{min} and r_{max} corresponding to perigee and apogee, Eq. 3/39, with Eq. 3/43 gives for the respective velocities at these two positions for the elliptical orbit

$$v_P = R \sqrt{\frac{g}{a}} \sqrt{\frac{1+e}{1-e}} = R \sqrt{\frac{g}{a}} \sqrt{\frac{r_{max}}{r_{min}}}$$

$$v_A = R \sqrt{\frac{g}{a}} \sqrt{\frac{1-e}{1+e}} = R \sqrt{\frac{g}{a}} \sqrt{\frac{r_{min}}{r_{max}}}$$ (3/44)

Selected numerical data pertaining to the solar system are included in Appendix D and will be found useful in applying the foregoing relationships to problems in planetary motion.

The foregoing analysis is based upon three assumptions:

1. The two bodies possess spherical mass symmetry so that they may be treated as if their masses were concentrated at their centers, i.e., as if they were particles.

2. There are no forces present except the gravitational force which each mass exerts on the other.

3. Mass m_0 is fixed in space.

Assumption (1) is excellent for bodies which are distant from the central attracting body, which is the case for most heavenly bodies. A significant class of problems for which assumption (1) is poor is that of artificial satellites in the very near vicinity of oblate planets. As a comment on assumption (2), we note that aerodynamic drag on a low-altitude earth satellite is a force which usually cannot be ignored in the orbital analysis. For an artificial satellite in earth orbit, the error of assumption (3) is negligible because the ratio of the mass of the satellite to that of the earth is very small. On the other hand, for the earth–moon system, a small but significant error is introduced if assumption (3) is invoked—note that the lunar mass is about 1/81 times that of the earth.

Two-body problem.   We now account for the motion of both masses and allow the presence of forces in addition to those of mutual attraction by considering the "perturbed two-body" problem. Figure 3/21 depicts the major mass m_0, the minor mass

Figure 3/21

m, their respective position vectors $\mathbf{r}_1$ and $\mathbf{r}_2$ measured relative to an inertial frame, the gravitational forces $\mathbf{F}$ and $-\mathbf{F}$, and a non-two-body force $\mathbf{P}$ which is exerted on mass m. The force $\mathbf{P}$ might be due to aerodynamic drag, solar pressure, the presence of a third body, on-board thrusting activities, a nonspherical gravitational field, or a combination of these and other sources. Application of Newton's second law to each mass results in

$$G\frac{mm_0}{r^3}\mathbf{r} = m_0\ddot{\mathbf{r}}_1 \qquad \text{and} \qquad -G\frac{mm_0}{r^3}\mathbf{r} + \mathbf{P} = m\ddot{\mathbf{r}}_2$$

Dividing the first equation by m_0, the second equation by m, and subtracting the first equation from the second give

$$-G\frac{(m_0 + m)}{r^3}\mathbf{r} + \frac{\mathbf{P}}{m} = \ddot{\mathbf{r}}_2 - \ddot{\mathbf{r}}_1 = \ddot{\mathbf{r}}$$

or

$$\ddot{\mathbf{r}} + G\frac{(m_0 + m)}{r^3}\mathbf{r} = \frac{\mathbf{P}}{m} \tag{3/45}$$

Equation 3/45 is a second-order differential equation which, when solved, yields the relative position vector $\mathbf{r}$ as a function of time. Numerical techniques are usually required for the integration of the scalar differential equations which are equivalent to the vector Eq. 3/45, especially if $\mathbf{P}$ is nonzero. If $m_0 \gg m$ and $\mathbf{P} = \mathbf{0}$, we have the restricted two-body problem, the equation of motion of which is

$$\ddot{\mathbf{r}} + G\frac{m_0}{r^3}\mathbf{r} = \mathbf{0} \tag{3/45a}$$

With $\mathbf{r}$ and $\ddot{\mathbf{r}}$ expressed in polar coordinates, Eq. 3/45a becomes

$$(\ddot{r} - r\dot{\theta}^2)\mathbf{e}_r + (r\ddot{\theta} + 2\dot{r}\dot{\theta})\mathbf{e}_\theta + G\frac{m_0}{r^3}(r\mathbf{e}_r) = \mathbf{0}$$

When we equate coefficients of like unit vectors, we recover Eqs. 3/33.

Comparison of Eq. 3/45 (with $\mathbf{P} = \mathbf{0}$) and Eq. 3/45a allows us to relax the assumption that mass m_0 is fixed in space. If we replace m_0 with $(m_0 + m)$ in the expressions derived with the assumption of m_0 fixed, then we have expressions which account for the motion of m_0. For example, the corrected expression for the period of elliptical motion of m about m_0 is, from Eq. 3/40,

$$\tau = 2\pi\frac{a^{3/2}}{\sqrt{G(m_0 + m)}} \tag{3/45b}$$

where the equality $R^2g = Gm_0$ has been used.

Sample Problem   3/27

An artificial satellite is launched from point B on the equator by its carrier rocket and inserted into an elliptical orbit with a perigee altitude of 2000 km. If the apogee altitude is to be 4000 km, compute (a) the necessary perigee velocity v_P and the corresponding apogee velocity v_A, (b) the velocity at point C where the altitude of the satellite is 2500 km, and (c) the period τ for a complete orbit.

Solution. (a) The perigee and apogee velocities for specified altitudes are given by Eqs. 3/44 where

$$r_{max} = 6371 + 4000 = 10\ 371 \text{ km}$$
$$r_{min} = 6371 + 2000 = 8371 \text{ km}$$
$$a = (r_{min} + r_{max})/2 = 9371 \text{ km}$$

Thus

$$v_P = R\sqrt{\frac{g}{a}}\sqrt{\frac{r_{max}}{r_{min}}} = 6371(10^3)\sqrt{\frac{9.825}{9371(10^3)}}\sqrt{\frac{10\ 371}{8371}}$$
$$= 7261 \text{ m/s} \quad\text{or}\quad 26\ 140 \text{ km/h} \quad Ans.$$

$$v_A = R\sqrt{\frac{g}{a}}\sqrt{\frac{r_{min}}{r_{max}}} = 6371(10^3)\sqrt{\frac{9.825}{9371(10^3)}}\sqrt{\frac{8371}{10\ 371}}$$
$$= 5861 \text{ m/s} \quad\text{or}\quad 21\ 099 \text{ km/h} \quad Ans.$$

(b) For an altitude of 2500 km the radial distance from the earth's center is $r = 6371 + 2500 = 8871$ km. From Eq. 3/43 the velocity at point C becomes

$$v_C{}^2 = 2gR^2\left(\frac{1}{r} - \frac{1}{2a}\right) = 2(9.825)[(6371)(10^3)]^2\left(\frac{1}{8871} - \frac{1}{18\ 742}\right)\frac{1}{10^3}$$
$$= 47.353(10^6)\ (\text{m/s})^2$$
$$v_C = 6881 \text{ m/s} \quad\text{or}\quad 24\ 773 \text{ km/h} \quad Ans.$$

(c) The period of the orbit is given by Eq. 3/40, which becomes

$$\tau = 2\pi\frac{a^{3/2}}{R\sqrt{g}} = 2\pi\frac{[(9371)(10^3)]^{3/2}}{(6371)(10^3)\sqrt{9.825}} = 9026 \text{ s}$$
$$\text{or}\quad \tau = 2.507 \text{ h} \quad Ans.$$

① The mean radius of $12\ 742/2 = 6371$ km from Table D/2 in Appendix D is used. Also the absolute acceleration due to gravity $g = 9.825$ m/s² from Art. 1/5 will be used.

② We must be careful with units. It is often safer to work in base units, meters in this case, and convert later.

③ We should observe here that the time interval between successive overhead transits of the satellite as recorded by an observer on the equator is longer than the period calculated here since the observer will have moved in space due to the counterclockwise rotation of the earth as seen looking down on the north pole.

PROBLEMS

(Unless otherwise indicated the velocities mentioned in the problems which follow are measured from a nonrotating reference frame moving with the center of the attracting body. Also, aerodynamic drag is to be neglected unless stated otherwise. Use $g = 9.825$ m/s^2 for the absolute gravitational acceleration at the surface of the earth and $R = 6371$ km for the radius of the earth.)

3/234 Determine the speed v of the earth in its orbit about the sun. Assume a circular orbit of radius $150(10^6)$ kilometers. *Ans. $v = 29.75$ km/s*

3/235 Identify the location on the earth and the direction of flight trajectory which require the least expenditure of fuel for the launching of a satellite into an earth orbit.

3/236 A satellite is in a circular polar orbit of altitude 300 km. Determine the separation d at the equator between the ground tracks associated with two successive overhead passes of the satellite. *Ans. $d = 2520$ km*

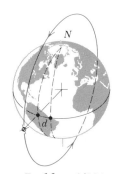

Problem 3/236

3/237 Determine the necessary velocity v_B in the direction indicated so that the spacecraft path will be tangent to the earth's surface at point C.

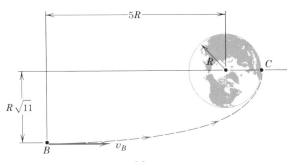

Problem 3/237

3/238 A satellite is in a circular earth orbit of radius $2R$, where R is the radius of the earth. What is the minimum velocity boost Δv necessary to reach point B, which is a distance of $6R$ from the earth's center? At what point in the original circular orbit should the velocity increment be added? *Ans. $\Delta v = 1257$ m/s*

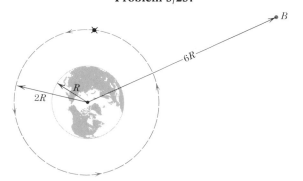

Problem 3/238

3/239 A rocket is orbiting the earth in a circular path of altitude H. If the jet engine is activated to give the rocket a sudden burst of speed, write the expression for the velocity v which must be attained in order to escape from the influence of the earth.

3/240 The Mars orbiter for the Viking mission was designed to make one complete trip around the planet in exactly the same time that it takes Mars to revolve once about its own axis. This time is 24 h, 37 min, 23 s. In this way it is possible for the orbiter to pass over the landing site of the lander capsule at the same time in each Martian day at the orbiter's minimum (periapsis) altitude. For the Viking I mission the periapsis altitude of the orbiter was 1508 km. Make use of the data in Table D/2 in Appendix D and compute the maximum (apoapsis) altitude h_a for the orbiter in its elliptical path.   *Ans.* $h_a = 32\ 600$ km

Problem 3/240

3/241 If the perigee altitude of an earth satellite is 240 km and the apogee altitude is 400 km, compute the eccentricity e of the orbit and the period τ of one complete orbit in space.

3/242 If an earth satellite has a velocity v whose direction makes an angle α with the radius r from the center of the earth to the satellite, derive an expression for the radius of curvature ρ of the path at this particular position. The radius of the earth is R.

$$Ans.\ \rho = \frac{r^2 v^2}{gR^2 \sin \alpha}$$

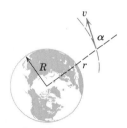

Problem 3/242

3/243 A satellite passes over the north pole at the perigee altitude of 500 km in an elliptical orbit of eccentricity $e = 0.7$. Calculate the absolute velocity v of the satellite as it crosses the equator.

3/244 In one of the orbits of the Apollo spacecraft about the moon its distance from the lunar surface varied from 100 to 300 km. Compute the maximum velocity of the spacecraft in this orbit.
          Ans. $v_P = 6024$ km/h

Problem 3/245

3/245 A "drag-free" satellite is one that carries a small mass inside a chamber as shown. If the satellite speed decreases due to drag, the mass speed will not, and so the mass moves relative to the chamber as indicated. Sensors detect this change in the position of the mass within the chamber, and the satellite thruster is periodically fired to recenter the mass. In this manner, compensation is made for drag. If the satellite is in a circular earth orbit of 200 km altitude and a total thruster burn time of 300 seconds occurs during 10 orbits, determine the drag force D acting on the 100-kg satellite. The thruster force T is 5 N.

Problem 3/246

3/246 Determine the angle β made by the velocity vector **v** with respect to the θ-direction for an earth satellite traveling in an elliptical orbit of eccentricity e. Express β in terms of the angle θ measured from perigee.

$$Ans. \ \tan \beta = \frac{e \sin \theta}{1 + e \cos \theta}$$

3/247 If the satellite of Sample Problem 3/27 has a mass of 500 kg, compute the net amount of energy E imparted to the satellite from earth sources to put it into its orbit.

Problem 3/248

3/248 Two satellites B and C are in the same circular orbit of altitude 800 kilometers. Satellite B is 2000 kilometers ahead of satellite C as indicated. Show that C can catch up to B by "putting on the brakes." Specifically, by what amount Δv should the circular orbit velocity of C be reduced so that it will rendezvous with B after one period in its new elliptical orbit? Check to see that C does not strike the earth in the elliptical orbit.

$$Ans. \ \Delta v = 115.5 \text{ m/s}$$

3/249 Determine the necessary amount Δv by which the circular orbit velocity of satellite C should be reduced if the catchup maneuver of Prob. 3/248 is to be accomplished with not one but two periods in a new elliptical orbit.

3/250 Derive an expression for the radius of curvature ρ of the path of an earth satellite at its nearest point of approach to the earth. The perigee altitude above the earth is H, the radius of the earth is R, and the eccentricity of the orbit is e.

$$Ans. \ \rho = (R + H)(1 + e)$$

3/251 A satellite is moving in a circular orbit at an altitude of 320 km above the surface of the earth. If a rocket motor on the satellite is activated to produce a velocity increase of 300 m/s in the direction of its motion during a very short interval of time, calculate the altitude H of the satellite at its new apogee position.

3/252 The 80-Mg space shuttle orbiter is in a circular orbit of altitude 320 km. The two orbital maneuvering system (OMS) engines, each of which has a thrust of 27 kN, are fired in retro-thrust for 150 seconds. Determine the angle β which locates the intersection of the shuttle trajectory with the earth's surface. Assume that the shuttle position B corresponds to the completion of the OMS burn and that no loss of altitude occurs during the burn.

$$Ans. \ \beta = 151.3°$$

Problem 3/252

3/253 A satellite is placed in a circular polar orbit a distance H above the earth. As the satellite goes over the north pole at A its retrojet is activated to produce a burst of negative thrust which reduces its velocity to a value that will ensure an equatorial landing. Derive the expression for the required reduction Δv_A of velocity at A. Note that A is the apogee of the elliptical path.

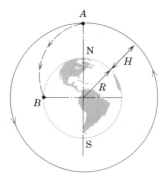

Problem 3/253

3/254 A projectile is launched from B with a speed of 2000 m/s at an angle α of 30° with the horizontal as shown. Determine the maximum altitude h_{max}.

$$Ans. \ h_{max} = 53.9 \text{ km}$$

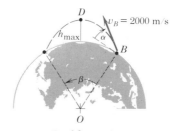

Problem 3/254

3/255 For the launch conditions at point B of Prob. 3/254, determine the angle β which locates the intersection of the trajectory with the surface of the earth.

3/256 Compute the magnitude of the necessary launch velocity at B if the projectile trajectory is to intersect the earth's surface such that the angle β equals 90°. The altitude at the highest point of the trajectory is $0.5R$.   *Ans.* $v_B = 7560$ m/s

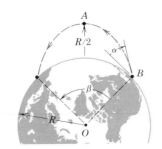

Problem 3/256

3/257 Compute the necessary launch angle α at point B for the trajectory prescribed in Prob. 3/256.

3/258 A satellite is in a circular earth orbit of 500 km altitude when it is suddenly given an outward radial component of velocity of 500 m/s when at point B. Determine the semimajor axis length a of the resulting elliptical orbit and find the orientation α of this axis relative to line l.
                    Ans. $a = 6900.7$ km, $\alpha = 2.6(10^{-7})°$

Problem 3/258

3/259 A spacecraft with a mass of 800 kg is traveling in a circular orbit 6000 km above the earth. It is desired to change the orbit to an elliptical one with a perigee altitude of 3000 km as shown. The transition is made by firing its retro-engine at A with a reverse thrust of 2000 N. Calculate the required time t for the engine to be activated.
                    Ans. $t = 162$ s

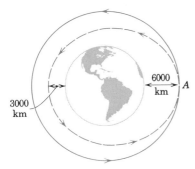

Problem 3/259

3/260 Compare the orbital period of the moon calculated with the assumption of a fixed earth with the period calculated without this assumption.

3/261 A space vehicle moving in a circular orbit of radius r_1 transfers to a larger circular orbit of radius r_2 by means of an elliptical path between A and B. (This transfer path is known as the Hohmann transfer ellipse.) The transfer is accomplished by a burst of speed Δv_A at A and a second burst of speed Δv_B at B. Write expressions for Δv_A and Δv_B in terms of the radii shown and the value g of the acceleration due to gravity at the earth's surface. If each Δv is positive, how can the velocity for path 2 be less than the velocity for path 1? Compute each Δv if $r_1 = (6371 + 500)$ km and $r_2 = (6371 + 1000)$ km.

$$Ans. \ \Delta v_A = R\sqrt{\frac{g}{r_1}}\left(\sqrt{\frac{2r_2}{r_1 + r_2}} - 1\right)$$
$$= 132.6 \ \text{m/s}$$
$$\Delta v_B = R\sqrt{\frac{g}{r_2}}\left(1 - \sqrt{\frac{2r_1}{r_1 + r_2}}\right)$$
$$= 130.3 \ \text{m/s}$$

Problem 3/261

3/262 A space station S is being assembled in its circular orbit 1200 km above the earth. The next payload P of parts along with its carrier rocket has a mass of 800 kg and is injected into a coasting orbit at P, which is 450 km above the earth. Determine the angle θ which specifies the relative position between S and P at injection in order that rendezvous will occur at A with parallel paths. If the carrier rocket can develop a thrust of 900 N, determine the time t during which the engines should be fired as P approaches S at A in order that the relative velocity between P and S will be zero at rendezvous.   *Ans. $\theta = 13.2°$, $t = 170$ s*

Problem 3/262

3/263 The satellite has a velocity at B of 3200 m/s in the direction indicated. Determine the angle β which locates the point C of impact with the earth.
   Ans. $\beta = 109.1°$

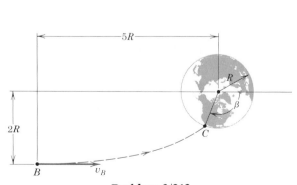

Problem 3/263

3/13 RELATIVE MOTION

Up to this point in our development of the kinetics of particle motion, we have applied Newton's second law and the equations of work-energy and impulse-momentum to problems where all measurements of motion were made with respect to a reference system which was considered fixed. The nearest we can come to a "fixed" reference system is the primary inertial system or astronomical frame of reference, which is an imaginary set of axes attached to the fixed stars. All other reference systems, then, are considered to have motion in space, including any reference system attached to the moving earth.

The accelerations of points attached to the earth as measured in the primary system are quite small, however, and we normally neglect them for most earth-surface measurements. For example, the acceleration of the center of the earth in its near-circular orbit around the sun considered fixed is 0.00593 m/s^2, and the acceleration of a point on the equator at sea level with respect to the center of the earth considered fixed is 0.0339 m/s^2. Clearly these accelerations are small compared with g and with most other significant accelerations in engineering work. Thus we make only a small error when we assume that our earth-attached reference axes are equivalent to a fixed reference system.

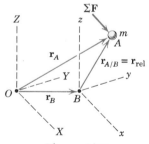

Figure 3/22

(a) Relative-motion equation. We now consider a particle A of mass m, Fig. 3/22, whose motion is observed from a set of axes x-y-z which have translatory motion with respect to a fixed reference frame X-Y-Z. Thus the x-y-z directions always remain parallel to the X-Y-Z directions. A discussion of motion relative to a rotating reference system is reserved for a later treatment in Arts. 5/7 and 7/7. The acceleration of the origin B of x-y-z is $\mathbf{a}_B$. The acceleration of A as observed from or relative to x-y-z is $\mathbf{a}_{rel} = \mathbf{a}_{A/B} = \ddot{\mathbf{r}}_{A/B}$, and by the relative-motion principle of Art. 2/8 the absolute acceleration of A is

$$\mathbf{a}_A = \mathbf{a}_B + \mathbf{a}_{rel}$$

Thus Newton's second law $\Sigma\mathbf{F} = m\mathbf{a}_A$ becomes

$$\Sigma\mathbf{F} = m(\mathbf{a}_B + \mathbf{a}_{rel}) \tag{3/46}$$

The force sum $\Sigma\mathbf{F}$ is disclosed, as always, by a complete free-body diagram which will appear the same to an observer in x-y-z or to one in X-Y-Z as long as only the real forces acting on the particle are represented. We can conclude immediately that Newton's second law does not hold with respect to an accelerating system since $\Sigma\mathbf{F} \neq m\mathbf{a}_{rel}$.

D'Alembert's principle. When a particle is observed from a fixed set of axes X-Y-Z, Fig. 3/23a, its absolute acceleration $\mathbf{a}$ is measured and the familiar relation $\Sigma\mathbf{F} = m\mathbf{a}$ is applied. When the particle is observed from a moving system x-y-z to which it is attached at the origin, Fig. 3/23b, the particle of necessity appears to be at rest or in equilibrium in x-y-z. Thus the observer who is accelerating with x-y-z concludes that a force $-m\mathbf{a}$ acts on the particle to balance $\Sigma\mathbf{F}$. This point of view which permits the treatment of a dynamics problem by the methods of statics was an outgrowth of the work of D'Alembert contained in his *Traité de Dynamique* published in 1743. This approach merely amounts to rewriting the equation of motion as $\Sigma\mathbf{F} - m\mathbf{a} = \mathbf{0}$, which assumes the form of a zero force summation if $-m\mathbf{a}$ is treated as a force. This fictitious force is known as the *inertia force*, and the artificial state of equilibrium created is known as *dynamic equilibrium*. The apparent transformation of a problem in dynamics to one in statics has become known as *D'Alembert's principle*.

Figure 3/23

Opinion differs concerning the original interpretation of D'Alembert's principle, but the principle in the form in which it is generally known is regarded in this book as being mainly of historical interest. It was evolved during a time when understanding and experience with dynamics were extremely limited and was a means of explaining dynamics in terms of the principles of statics which during earlier times were more fully understood. This excuse for using an artificial situation to describe a real one is open to question, as there is today a wealth of knowledge and experience with the phenomena of dynamics to support strongly the direct approach of thinking in terms of dynamics rather than in terms of statics. It is somewhat difficult to justify the long persistence in the acceptance of statics as a way of understanding dynamics particularly in view of the continued search for the understanding and description of physical phenomena in their most direct form.

Only one simple example of the method known as D'Alembert's principle will be cited. The conical pendulum of mass m, Fig. 3/24a, is swinging in a horizontal circle with its radial line r having an angular velocity ω. In the straightforward application of the equation of motion $\Sigma\mathbf{F} = m\mathbf{a}_n$ in the direction n of the acceleration, the free-body diagram in the b-part of the figure shows that $T\sin\theta = mr\omega^2$. When the equilibrium requirement in the y-direction, $T\cos\theta - mg = 0$, is introduced, the unknowns T and θ can be found. But if the reference axes are attached to the particle, the particle will appear to be in equilibrium relative to these axes. Accordingly, the inertia force $-m\mathbf{a}$ must be added, which amounts to visualizing the application of $mr\omega^2$ in the direction opposite to the acceleration, as shown in the c-part of the

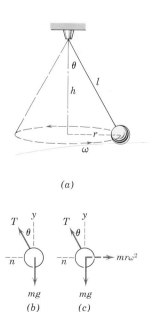

Figure 3/24

figure. With this pseudo free-body diagram a zero force summation in the n-direction gives $T \sin \theta - mr\omega^2 = 0$ which, of course, gives us the same result as before. We may conclude that no advantage results from this alternative formulation. The authors recommend against using it since it introduces no simplification and adds a nonexistent force to the diagram. In the case of a particle moving in a circular path this hypothetical inertia force is known as the *centrifugal force* since it is directed away from the center and is opposite to the direction of the acceleration. The student is urged to recognize that there is no actual centrifugal force acting on the particle. The only actual force which may properly be called centrifugal is the horizontal component of the tension T exerted *on* the cord *by* the particle.

(b) *Constant-velocity translating systems.* In discussing particle motion relative to moving reference systems, it is of importance for us to note the special case where the reference system has a constant translational velocity. If the x-y-z axes of Fig. 3/22 have a constant velocity, then $\mathbf{a}_B = \mathbf{0}$ and the acceleration of the particle is $\mathbf{a}_A = \mathbf{a}_{\text{rel}}$. Therefore we may write Eq. 3/46 as

$$\boxed{\Sigma \mathbf{F} = m\mathbf{a}_{\text{rel}}} \tag{3/47}$$

which tells us that Newton's second law holds for measurements made in a system moving with a constant velocity. Such a system is known as a Newtonian frame of reference. Observers in the moving system and in the fixed system will also agree on the designation of the resultant force acting on the particle from their identical free-body diagrams, provided they avoid the use of any so-called "inertia forces."

We will now examine the parallel question concerning the validity of the work-energy equation and the impulse-momentum equation relative to a constant velocity, nonrotating system. Again we take the x-y-z axes of Fig. 3/22 to be moving with a constant velocity $\mathbf{v}_B = \dot{\mathbf{r}}_B$ relative to the fixed axes X-Y-Z. The path of the particle A relative to x-y-z is governed by $\mathbf{r}_{\text{rel}}$ and is represented schematically in Fig. 3/25. The work done by $\Sigma \mathbf{F}$ relative to x-y-z is $dU_{\text{rel}} = \Sigma \mathbf{F} \cdot d\mathbf{r}_{\text{rel}}$. But $\Sigma \mathbf{F} = m\mathbf{a}_A = m\mathbf{a}_{\text{rel}}$ since $\mathbf{a}_B = \mathbf{0}$. Also $\mathbf{a}_{\text{rel}} \cdot d\mathbf{r}_{\text{rel}} = \mathbf{v}_{\text{rel}} \cdot d\mathbf{v}_{\text{rel}}$ for the same reason that $a_t \, ds = v \, dv$ in Art. 2/5 on curvilinear motion. Thus we have

$$dU_{\text{rel}} = m\mathbf{a}_{\text{rel}} \cdot d\mathbf{r}_{\text{rel}} = mv_{\text{rel}} \, dv_{\text{rel}} = d(\tfrac{1}{2}mv_{\text{rel}}^2)$$

We define the kinetic energy relative to x-y-z as $T_{\text{rel}} = \tfrac{1}{2}mv_{\text{rel}}^2$ so that we now have

$$\boxed{dU_{\text{rel}} = dT_{\text{rel}}} \quad \text{or} \quad \boxed{U_{\text{rel}} = \Delta T_{\text{rel}}} \tag{3/48}$$

Figure 3/25

which shows that the work-energy equation holds for measurements made relative to a constant-velocity system.

Relative to *x-y-z* the impulse on the particle during time dt is $\Sigma \mathbf{F}\, dt = m\mathbf{a}_A\, dt = m\mathbf{a}_{\text{rel}}\, dt$. But $m\mathbf{a}_{\text{rel}}\, dt = m\, d\mathbf{v}_{\text{rel}} = d(m\mathbf{v}_{\text{rel}})$ so

$$\Sigma \mathbf{F} \cdot dt = d(m\mathbf{v}_{\text{rel}})$$

We define the linear momentum of the particle relative to *x-y-z* as $\mathbf{G}_{\text{rel}} = m\mathbf{v}_{\text{rel}}$ which gives us $\Sigma \mathbf{F}\, dt = d\mathbf{G}_{\text{rel}}$. Dividing by dt and integrating give

$$\boxed{\Sigma \mathbf{F} = \dot{\mathbf{G}}_{\text{rel}}} \quad \text{and} \quad \boxed{\int \Sigma \mathbf{F}\, dt = \Delta \mathbf{G}_{\text{rel}}} \quad (3/49)$$

Thus the impulse-momentum equations for a fixed reference system also hold for measurements made relative to a constant-velocity system.

Finally, we define the relative angular momentum of the particle about a point in *x-y-z*, such as the origin *B*, as the moment of the relative linear momentum. Thus $\mathbf{H}_{B_{\text{rel}}} = \mathbf{r}_{\text{rel}} \times \mathbf{G}_{\text{rel}}$. The time derivative gives $\dot{\mathbf{H}}_{B_{\text{rel}}} = \dot{\mathbf{r}}_{\text{rel}} \times \mathbf{G}_{\text{rel}} + \mathbf{r}_{\text{rel}} \times \dot{\mathbf{G}}_{\text{rel}}$. The first term is nothing more than $\mathbf{v}_{\text{rel}} \times m\mathbf{v}_{\text{rel}} = \mathbf{0}$ and the second term is $\mathbf{r}_{\text{rel}} \times \Sigma \mathbf{F} = \Sigma \mathbf{M}_B$, the sum of the moments about *B* of all forces on *m*. Thus we have

$$\boxed{\Sigma \mathbf{M}_B = \dot{\mathbf{H}}_{B_{\text{rel}}}} \quad (3/50)$$

which shows that the moment-angular momentum relation holds with respect to a constant-velocity system.

Although the work-energy and impulse-momentum equations hold relative to a system translating with a constant velocity, the individual expressions for work, kinetic energy, and momentum differ between the fixed and the moving systems. Thus

$$(dU = \Sigma \mathbf{F} \cdot d\mathbf{r}_A) \neq (dU_{\text{rel}} = \Sigma \mathbf{F} \cdot d\mathbf{r}_{\text{rel}})$$

$$(T = \tfrac{1}{2}mv_A{}^2) \neq (T_{\text{rel}} = \tfrac{1}{2}mv_{\text{rel}}{}^2)$$

$$(\mathbf{G} = m\mathbf{v}_A) \neq (\mathbf{G}_{\text{rel}} = m\mathbf{v}_{\text{rel}})$$

Equations 3/47 through 3/50 are formal proof of the validity of the Newtonian equations of kinetics in any constant-velocity, nonrotating system. We might have surmised these conclusions from the fact that $\Sigma \mathbf{F} = m\mathbf{a}$ depends on acceleration and not velocity. We are also ready to conclude that there is no experiment which can be conducted in and relative to a constant-velocity system (Newtonian frame of reference) which discloses its absolute velocity. Any mechanical experiment will achieve the same results in any Newtonian system.

Sample Problem 3/28

A simple pendulum of mass m and length r is mounted on the flatcar which has a constant horizontal acceleration a_0 as shown. If the pendulum is released from rest relative to the flatcar at the position $\theta = 0$, determine the expression of the tension T in the supporting light rod for any value of θ. Also find T for $\theta = \pi/2$ and $\theta = \pi$.

Solution. We attach our moving x-y coordinate system to the translating car with origin at O for convenience. Relative to this system n- and t-coordinates are the natural ones to use since the motion is circular within x-y. The acceleration of m is given by the relative-acceleration equation

$$\mathbf{a} = \mathbf{a}_0 + \mathbf{a}_{\text{rel}}$$

where $\mathbf{a}_{\text{rel}}$ is the acceleration which would be measured by an observer riding with the car. He would measure an n-component equal to $r\dot{\theta}^2$ and a t-component equal to $r\ddot{\theta}$. The three components of the absolute acceleration of m are shown in the separate view.

First, we apply Newton's second law to the t-direction and get

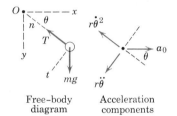

Free–body diagram   Acceleration components

① $[\Sigma F_t = ma_t]$ $mg \cos \theta = m(r\ddot{\theta} - a_0 \sin \theta)$

$$r\ddot{\theta} = g \cos \theta + a_0 \sin \theta$$

Integrating to obtain $\dot{\theta}$ as a function of θ yields

$[\dot{\theta}\,d\dot{\theta} = \ddot{\theta}\,d\theta]$ $\displaystyle\int_0^{\dot{\theta}} \dot{\theta}\,d\dot{\theta} = \int_0^{\theta} \frac{1}{r}(g \cos \theta + a_0 \sin \theta)\,d\theta$

$$\frac{\dot{\theta}^2}{2} = \frac{1}{r}[g \sin \theta + a_0(1 - \cos \theta)]$$

We now apply Newton's second law to the n-direction noting that the n-component of the absolute acceleration is $r\dot{\theta}^2 - a_0 \cos \theta$.

② $[\Sigma F_n = ma_n]$ $T - mg \sin \theta = m(r\dot{\theta}^2 - a_0 \cos \theta)$

$$= m[2g \sin \theta + 2a_0(1 - \cos \theta) - a_0 \cos \theta]$$

$$T = m[3g \sin \theta + a_0(2 - 3 \cos \theta)] \qquad Ans.$$

For $\theta = \pi/2$ and $\theta = \pi$ we have

$$T_{\pi/2} = m[3g(1) + a_0(2 - 0)] = m(3g + 2a_0) \qquad Ans.$$

$$T_{\pi} = m[3g(0) + a_0(2 - 3[-1])] = 5ma_0 \qquad Ans.$$

① We choose the t-direction first since the n-direction equation, which contains the unknown T, will involve $\dot{\theta}^2$, which, in turn, is obtained from an integration of $\ddot{\theta}$.

② Be sure to recognize that $\dot{\theta}\,d\dot{\theta} = \ddot{\theta}\,d\theta$ may be obtained from $v\,dv = a_t\,ds$ by dividing by r^2.

Sample Problem 3/29

The flatcar moves with a constant speed v_0 and carries a winch which produces a constant tension P in the cable attached to the small carriage. The carriage has a mass m and rolls freely on the horizontal surface starting from rest relative to the flatcar at $x = 0$ at which instant $X = x_0 = b$. Apply the work-energy equation to the carriage, first, as an observer moving with the frame of reference of the car and, second, as an observer on the ground. Show the compatibility of the two expressions.

Solution. To the observer on the flatcar the work done by P is

① $$U_{\text{rel}} = \int_0^x P\,dx = Px \qquad \text{for constant } P$$

The change in kinetic energy relative to the car is

$$\Delta T_{\text{rel}} = \tfrac{1}{2}m(\dot{x}^2 - 0)$$

The work-energy equation for the moving observer becomes

$$[U_{\text{rel}} = \Delta T_{\text{rel}}] \qquad Px = \tfrac{1}{2}m\dot{x}^2$$

To the observer on the ground the work done by P is

$$U = \int_b^X P\,dX = P(X - b)$$

The change in kinetic energy for the ground measurement is

② $$\Delta T = \tfrac{1}{2}m(\dot{X}^2 - v_0^2)$$

The work-energy equation for the fixed observer gives

$$[U = \Delta T] \qquad P(X - b) = \tfrac{1}{2}m(\dot{X}^2 - v_0^2)$$

To reconcile this equation with that for the moving observer we can make the following substitutions:

$$X = x_0 + x, \qquad \dot{X} = v_0 + \dot{x}, \qquad \ddot{X} = \ddot{x}$$

③ Thus $\quad P(X - b) = Px + P(x_0 - b) = Px + m\ddot{x}(x_0 - b)$

$$= Px + m\ddot{x}v_0 t = Px + mv_0\dot{x}$$

and $\quad \dot{X}^2 - v_0^2 = (v_0^2 + \dot{x}^2 + 2v_0\dot{x} - v_0^2) = \dot{x}^2 + 2v_0\dot{x}$

The work-energy equation for the fixed observer now becomes

$$Px + mv_0\dot{x} = \tfrac{1}{2}m\dot{x}^2 + mv_0\dot{x}$$

which is merely $Px = \tfrac{1}{2}m\dot{x}^2$ as concluded by the moving observer. We see, therefore, that the difference between the two work-energy expressions is

$$U - U_{\text{rel}} = T - T_{\text{rel}} = mv_0\dot{x}$$

① The only coordinate which the moving observer can measure is x.

② To the ground observer the initial velocity of the carriage is v_0 so its initial kinetic energy is $\tfrac{1}{2}mv_0^2$.

③ The symbol t stands for the time of motion from $x = 0$ to $x = x$. The displacement $x_0 - b$ of the carriage is its velocity v_0 times the time t or $x_0 - b = v_0 t$. Also, since the constant acceleration times the time equals the velocity change, $\ddot{x}t = \dot{x}$.

PROBLEMS

Problem 3/264

3/264 If the spring of constant k is compressed a distance δ as indicated, calculate the acceleration a_{rel} of the block of mass m_1 relative to the frame of mass m_2 upon release of the spring. The system is initially stationary.

$$Ans. \; a_{\text{rel}} = k\delta\left(\frac{1}{m_1} + \frac{1}{m_2}\right)$$

3/265 Apply the equation of impulse and momentum to the carriage of Sample Problem 3/29 for the first t seconds of the motion starting when $x = 0$. Write the equation both as an observer riding with the car and as a fixed observer on the ground. Reconcile the two relations.

Problem 3/266

3/266 The aircraft carrier is moving at a constant speed and launches a jet plane with a mass of 3 Mg in a distance of 75 m along the deck by means of a steam-driven catapult. If the plane leaves the deck with a velocity of 240 km/h relative to the carrier and if the jet thrust is constant at 22 kN during takeoff, compute the constant force P exerted by the catapult on the airplane during the 75-m travel of the launch carriage.   *Ans. P = 66.9 kN*

Problem 3/267

3/267 The slider A of mass m is released at B with no velocity relative to the moving box and moves down the slot with negligible friction. Use the principle of work-energy to determine the speed of the slider when it reaches position C if (a) the box moves with constant velocity $v = v_0$ and (b) the box has a velocity v_0 when the slider is released and moves with a constant acceleration a_0 to the right.

3/268 Two spacecraft A and B are traveling on the same straight-line path in deep space. Spacecraft B is 20 km directly ahead of A and has a forward velocity relative to A of 2 km/s. If it is desired to stabilize the separation distance at 1000 km, determine the magnitude F of the constant retro-thrust which must be applied to the 200-kg spacecraft B. Calculate the necessary time duration t of this thrusting activity.

Ans. F = 408 N, t = 980 s

3/269 The slider A has a mass of 2 kg and moves with negligible friction in the 30° slot in the vertical sliding plate. What horizontal acceleration a_0 should be given to the plate so that the absolute acceleration of the slider will be vertically down? What is the value of the corresponding force R exerted on the slider by the slot?

Ans. $a_0 = 16.99$ m/s², $R = 0$

Problem 3/269

3/270 If the aircraft carrier of Prob. 3/266 is moving with a constant speed u and its catapult launches a plane of mass m with a velocity v relative to the carrier in the direction of u, the kinetic energy of the plane with respect to the land is $\frac{1}{2}m(v + u)^2 = \frac{1}{2}mv^2 + \frac{1}{2}mu^2 + muv$ when launching is complete. If the plane is launched by the same catapult with the same accelerating force when the carrier is not moving, the kinetic energy of the plane is $\frac{1}{2}mv^2$. Explain the difference between the two kinetic-energy expressions.

Problem 3/271

3/271 The ball A of mass $m = 2$ kg and attached light rod of length $l = 0.6$ m are free to rotate in a horizontal plane about the vertical shaft attached to the vehicle. The vehicle, rod, and ball are initially at rest with $\theta = 0$ when the vehicle is given a constant acceleration $a_0 = 1.2$ m/s². Write an expression for the tension T in the rod as a function of θ and calculate T for the position $\theta = \pi/2$.

Ans. $T = 3ma_0 \sin \theta$, $T_{\pi/2} = 7.2$ N

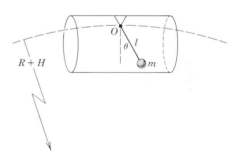

3/272 A simple pendulum of mass m and length l is mounted in a spacecraft which has curvilinear translation with its mass center in a circular earth orbit. If the pendulum pivot O is a distance H above the surface of the earth, describe what happens if the pendulum is displaced an amount θ_0 from the vertical and released from rest relative to the spacecraft.

Problem 3/272

3/273 A simple pendulum is placed on an elevator which accelerates upward as shown. If the pendulum is displaced an amount θ_0 and released from rest relative to the elevator, find the tension T_0 in the supporting light rod when $\theta = 0$. Evaluate your result for $\theta_0 = \pi/2$.

Ans. $T_0 = m(g + a_0)(3 - 2 \cos \theta_0)$

Problem 3/273

Problem 3/274

3/274 The coefficients of friction between the flat bed of the truck and crate are $\mu_s = 0.8$ and $\mu_k = 0.7$. The coefficient of kinetic friction between the truck tires and the road surface is 0.9. If the truck stops from an initial speed of 15 m/s with maximum braking (wheels skidding), determine where on the bed the crate finally comes to rest or the velocity v_{rel} relative to the truck with which the crate strikes the wall at the forward edge of the bed. *Ans.* $v_{rel} = 2.46$ m/s

Problem 3/275

3/275 The capsule of mass m is released from rest relative to the horizontal tube with the spring of stiffness k initially compressed a distance δ. The flatcar has a constant velocity v_0 to the right. Apply the work-energy equation to the motion of the capsule both as an observer riding with the car and as an observer fixed to the ground. Reconcile the two equations. (*Hint:* For the second case include the spring with the capsule as your system.)

Problem 3/276

3/276 The sliding block of mass m is attached to the frame by a spring of stiffness k and slides horizontally along the frame with negligible friction. The frame and block are both initially at rest with $x = x_0$, the unstretched length of the spring. If the frame is given a constant acceleration a_0, determine the maximum velocity of the block $(v_{rel})_{max}$ relative to the frame.

Ans. $(v_{rel})_{max} = a_0\sqrt{m/k}$

Problem 3/277

3/277 The small slider A moves with negligible friction down the inclined block which moves to the right with constant speed $v = v_0$. Use the principle of work-energy to determine the magnitude v_A of the absolute velocity of the slider as it passes point C if it is released at point B with no velocity relative to the block. Apply the equation both as an observer fixed to the block and as an observer fixed to the ground, and reconcile the two relations.

Ans. $v_A = [v_0{}^2 + 2gl \sin \theta$
$+ 2v_0 \cos \theta \sqrt{2lg \sin \theta}]^{1/2}$

▶ **3/278** Repeat Prob. 3/277, except let the inclined block have a small constant acceleration a_0 to the right. The velocity of the block is v_0 when the small slider is released with no velocity relative to the block.

▶ **3/279** When a particle is dropped from rest relative to the earth's surface at a latitude γ, the initial apparent acceleration is the relative acceleration due to gravity g_{rel}. The absolute acceleration due to gravity g is directed toward the center of the earth. Derive an expression for g_{rel} in terms of g, R, ω, and γ where R is the radius of the earth treated as a sphere and ω is the constant angular velocity of the earth about the polar axis considered as fixed. (Although axes x-y-z are attached to the earth and, hence, rotate, we may use Eq. 3/46 as long as the particle has no velocity relative to x-y-z.) (*Hint:* Use the first two terms of the binomial expansion for the approximation.)

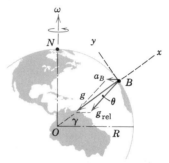

Problem 3/279

$$Ans.\ g_{rel} = g - R\omega^2 \cos^2 \gamma \left(1 - \frac{R\omega^2}{2g}\right) + \cdots$$
$$= 9.825 - 0.03382 \cos^2 \gamma \ \text{m/s}^2$$

▶ **3/280** A plumb bob is suspended from a fixed point on the surface of the earth at north latitude γ. Refer to the figure for Prob. 3/279 and determine the expression for the small angle θ made by the cord which supports the bob with the true vertical due to the angular velocity ω of the earth. The radius of the earth is R. Calculate the value of θ for $\gamma = 30°$. Let g stand for the absolute acceleration due to gravity. (Draw the free-body diagram of the bob and relate the vectors to the acceleration vectors shown with Prob. 3/279.)

$$Ans.\ \tan \theta = \frac{R\omega^2}{2g} \frac{\sin 2\gamma}{1 - (R\omega^2/g) \cos^2 \gamma}$$
$$\theta = 5'$$

3/14 PROBLEM FORMULATION AND REVIEW

In Chapter 3 we have developed the three basic methods of solution to problems in particle kinetics. This experience is central to the study of dynamics and lays the foundation for the subsequent study of rigid-body and nonrigid-body dynamics. First, we applied Newton's second law $\Sigma\mathbf{F} = m\mathbf{a}$ to determine the instantaneous relation between forces and the acceleration they produce. With the background of Chapter 2 for identifying the kind of motion and with the aid of our familiar free-body diagram to be certain that all forces are accounted for, we were able to solve a large variety of problems using *x-y*, *n-t*, and *r-θ* coordinates for plane-motion problems and *x-y-z*, *r-θ-z*, and *R-θ-φ* coordinates for space problems.

Second, we integrated the basic equation of motion $\Sigma\mathbf{F} = m\mathbf{a}$ with respect to displacement and derived the scalar equations for work and energy. These equations enabled us to relate the initial and final velocities to the work done during the interval by forces external to our defined system. We expanded this approach to include potential energy, both elastic and gravitational. With these tools we discovered that the energy approach is especially valuable for conservative systems, that is, systems wherein the loss of energy due to friction or other forms of dissipation is negligible.

Third, we rewrote Newton's second law in the form of force equals time rate of change of linear momentum and moment equals time rate of change of angular momentum. Then we integrated these relations with respect to time and derived the impulse and momentum equations. These equations were then applied to motion intervals where the forces were functions of the time. We also investigated the interactions between particles under conditions where the linear momentum is conserved and where the angular momentum is conserved.

In the final section of Chapter 3, we employed the three basic methods in specific application areas. In particular, we noted that the impulse-momentum method is convenient in developing the relations governing particle impacts. We observed that the direct application of Newton's second law allows the determination of the trajectory properties associated with a particle under central-force attraction. And lastly, we have seen that any of the three basic methods may be applied to particle motion relative to a translating frame of reference.

It is clear that the successful solution of problems in particle kinetics depends upon knowledge of the prerequisite particle kinematics. Analogously, the analytical developments associated with particle systems and rigid bodies, which are covered in the remainder of *Dynamics*, are heavily dependent upon the fundamental principles of particle kinetics as treated in Chapter 3.

REVIEW PROBLEMS

3/281 Upon landing, a small 1200-kg airplane has a touch-down speed of 100 km/h and requires 10 seconds to reduce its speed to 50 km/h along the runway. Compute the time average R of the total retarding force (aerodynamic plus mechanical braking) which acts on the airplane during this interval.   *Ans. $R = 1667$ N*

3/282 The small 2-kg carriage is moving freely along the horizontal with a speed of 4 m/s at time $t = 0$. A force applied to the carriage in the direction opposite to motion produces two impulse "peaks," one after the other, as shown by the graphical plot of the readings of the instrument which measured the force. Approximate the loading by the dotted lines and determine the velocity v of the carriage for $t = 1.5$ s.

Problem 3/282

3/283 The figure shows a centrifugal clutch consisting in part of a rotating spider A which carries four plungers B. As the spider is made to rotate about its center with a speed ω, the plungers move outward and bear against the interior surface of the rim of wheel C, causing it to rotate. The wheel and spider are independent except for frictional contact. If each plunger has a mass of 2 kg with center of mass at G, and if the coefficient of kinetic friction between the plungers and the wheel is 0.40, calculate the maximum moment M which can be transmitted to wheel C for a spider speed of 3000 rev/min.   *Ans. $M = 18.96$ kN·m*

Problem 3/283

3/284 A bead is released from rest at point A and slides down the wire bent into the form of a helix, as shown in the *a*-part of the figure. The experiment is repeated in the *b*-part of the figure where the wire has been unwrapped to form a straight line with a slope equal to that of the helix. What is the velocity reached by the bead as it passes point B or B' in each case in the absence of friction? In the presence of some friction does the velocity depend on the path, and if so, for which case would the velocity at the lower position be greater?

Problem 3/284

Problem 3/285

Problem 3/286

Problem 3/287

Problem 3/288

Problem 3/290

3/285 A spacecraft is designed for Martian impact on a course heading directly toward the center of the planet. If the craft is 10 000 km from the planet and is traveling at a relative speed of 20 000 km/h, compute the impact velocity v. The gravitational acceleration at the surface of Mars is 3.73 m/s^2, and the planet has a diameter of approximately 6800 km. Neglect any atmospheric resistance.

Ans. $v = 25\,400$ km/h

3/286 The cart of mass m is initially stationary with the spring undeformed and is acted upon by a horizontal force F which varies with time as shown. Determine the velocity of the cart at time $t = t_3$. Do not solve, but comment on the difficulty of solution with any one of the kinetics methods developed in Chapter 3.

3/287 The two freight cars of equal mass approach each other with velocities v_1 and v_2 as shown and are coupled together. Derive an expression for the fraction $\Delta T/T$ of the original kinetic energy which is lost during the coupling.

$$Ans. \; \frac{\Delta T}{T} = \frac{1}{2} + \frac{v_1 v_2}{v_1{}^2 + v_2{}^2}$$

3/288 For a perigee altitude of 300 km and an apogee altitude of 600 km an earth satellite must have a velocity at perigee $v_P = 28\,137$ km/h. Calculate the apogee velocity v_A.

3/289 The last two appearances of Comet Halley were in 1910 and 1986. The distance of its closest approach to the sun averages about one-half of the distance between the earth and the sun. Determine its maximum distance from the sun. Neglect the gravitational effects of the planets.

Ans. $5.37(10^9)$ km

3/290 The 60-g bullet is fired at the two blocks resting on a surface where the coefficient of kinetic friction is 0.50. The bullet passes through the 8-kg block and lodges in the 6-kg block. The blocks slide the distances shown. Compute the initial velocity v of the bullet.

3/291 The frame of mass $6m$ is initially at rest. A particle of mass m is attached to the end of the light rod which pivots freely at A. If the rod is released from rest in the horizontal position shown, determine the velocity v_{rel} of the particle with respect to the frame when the rod is vertical.

Ans. $v_{\text{rel}} = \sqrt{\frac{7}{3} gl}$ to the left

6 m

Problem 3/291

3/292 An automobile accident occurs as follows: The driver of a full-size car (vehicle A, 1800 kg) is traveling on a dry, level road and approaches a stationary compact car (vehicle B, 900 kg). Just 15 meters before collision, he applies the brakes, skidding all wheels. After impact, vehicle A skids an additional 15 m and vehicle B, whose driver had all brakes fully applied, skids 30 m. The final positions of the vehicles are shown in the figure. If the coefficient of kinetic friction is 0.9, was the driver of vehicle A exceeding the speed limit of 90 km/h before the initial application of his brakes?

Problem 3/292

3/293 The quarter-circular hollow tube of circular cross section starts from rest at time $t = 0$ and rotates about point O in a horizontal plane with a constant counterclockwise angular acceleration $\ddot{\theta} = 2$ rad/s². At what time t will the 0.5-kg particle P slip relative to the tube? The coefficient of static friction between the particle and the tube is $\mu_s = 0.8$. *Ans. $t = 2.02$ s*

Problem 3/293

3/294 The system is released from rest with $\theta = 0$. The cord to the 1.5-kg cylinder is securely wound around the light 50-mm-diameter pulley at O, to which are attached the light arms and their 2-kg spheres. The centers of the spheres are 250 mm and 375 mm from the axis at O. Determine the downward velocity of the 1.5-kg cylinder when $\theta = 30°$.

Problem 3/294

60 kg

0.3 m 0.3 m

0.3 m θ 0.3 m

P P

Problem 3/295

3/295 The vertical motion of the 60-kg cylinder is controlled by the two forces P applied to the bottom rollers of the symmetrical frame. Determine the constant force P which, if applied when the frame is at rest with $\theta = 120°$, will give the cylinder an upward velocity of 3 m/s when the position $\theta = 60°$ is passed. The links are very light so that their mass may be neglected. Also, for the instant when the upward acceleration of the cylinder is 20 m/s^2, find the corresponding force R under each of the supporting rollers.   *Ans.* $P = 2.70$ kN
$$R = 894 \text{ N}$$

(a)   (b)

0.6 m

A   C

Problem 3/296

3/296 A slider C has a speed of 3 m/s as it passes point A of the guide which lies in a horizontal plane. The coefficient of kinetic friction between the slider and the guide is $\mu_k = 0.6$. Compute the tangential deceleration a_t of the slider just after it passes point A if (a) the slider hole and guide cross section are both circular and (b) the slider hole and guide cross section are both square. In case (b), the sides of the square are vertical and horizontal. Assume a slight clearance between the slider and the guide.

B

A

C

1280 km   1280 km 320 km

Problem 3/297

3/297 Satellite A moving in the circular orbit and satellite B moving in the elliptical orbit collide and become entangled at point C. If the masses of the satellites are equal, determine the maximum altitude h_{max} of the resulting orbit.   *Ans.* $h_{max} = 1438$ km

B

0.48 m

A

Problem 3/298

3/298 The 3-Mg anvil A for a drop forge is mounted on a nest of four springs with a combined stiffness of 2.88 MN/m. The 500-kg hammer B falls 0.48 m from rest and strikes the anvil which is then observed to deflect 20 mm from its equilibrium position. Determine the height h of rebound of the hammer and the coefficient of restitution e which applies.

3/299 The object of the pinball-type game is to project the particle so that it enters the hole at E. When the spring is compressed and suddenly released, the particle is projected along the track, which is smooth except for the rough portion between points B and C, where the coefficient of kinetic friction is μ_k. The particle becomes a projectile at point D. Determine the correct spring compression δ so that the particle enters the hole at E. State any necessary conditions relating the lengths d and ρ.

$$\text{Ans. } \delta = \sqrt{\frac{mg}{k}\left[\frac{d^2}{2\rho} + 2\rho(1 + \mu_k)\right]}$$

$$d \geq 2\sqrt{2}\rho$$

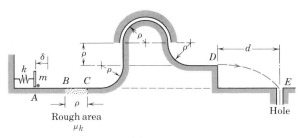

Problem 3/299

3/300 The small pickup truck is used to hoist the 40-kg bale of hay as shown. If the truck has reached a constant velocity $v = 5$ m/s when $x = 12$ m, compute the corresponding tension T in the rope.

Ans. $T = 424$ N

Problem 3/300

3/301 A bullet of mass m is fired with a velocity v from a rifle at rest relative to the ground. With the rifle attached to a vehicle moving with a speed u an identical bullet is fired in the direction of u. The kinetic energy of the bullet fired from the ground is $\frac{1}{2}mv^2$ and that of the bullet fired from the vehicle is $\frac{1}{2}m(u + v)^2 = \frac{1}{2}mu^2 + \frac{1}{2}mv^2 + muv$. Explain the significance of the muv term. Assume the velocity of the vehicle remains constant during the firing.

3/302 One of the functions of the space shuttle is to release communications satellites at low altitude. A booster rocket is fired at B, placing the satellite in an elliptical transfer orbit, the apogee of which is the altitude necessary for a geosynchronous orbit. (A geosynchronous orbit is an equatorial-plane circular orbit whose period is equal to the absolute rotational period of the earth. A satellite in such an orbit appears to remain stationary to an earth-fixed observer.) A second booster rocket is then fired at C, and the final circular orbit is achieved. On one of the early space-shuttle missions, a 700-kg satellite was released from the shuttle at B where $h_1 = 275$ km. The booster rocket was to fire for $t = 90$ seconds, forming a transfer orbit with $h_2 = 35\,900$ km. The rocket failed during its burn. Radar observations determined the apogee altitude of the transfer orbit to be only 1125 km. Determine the actual time t' that the rocket motor operated before failure. Assume negligible mass change during the booster rocket firing.

Ans. $t' = 8.47$ s

Problem 3/302

Problem 3/303

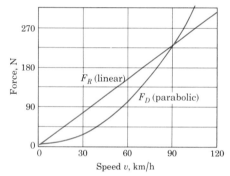

Force, N
Speed v, km/h

F_R (linear)
F_D (parabolic)

Problem 3/304

Problem 3/305

▶ **3/303** The small particle of mass m is given an initial high velocity in the horizontal plane and winds its cord around the fixed vertical shaft of radius a. All motion occurs essentially in the horizontal plane. If the angular velocity of the cord is ω_0 when the distance from the particle to the tangency point is r_0, determine the angular velocity ω of the cord and its tension T after it has turned through an angle θ. Does either of the principles of momentum conservation apply?

$$Ans. \ \omega = \frac{\omega_0}{1 - \dfrac{a}{r_0}\theta}, \ T = mr_0\omega_0\omega$$

▶ **3/304** Extensive wind-tunnel and coast-down studies of a 1000-kg automobile reveal the aerodynamic drag force F_D and the total nonaerodynamic rolling resistance force F_R to vary with speed as shown in the plot. Determine (a) the power P required for steady speeds of 45 km/h and 90 km/h and (b) the time t and the distance s required for the car to coast down to a speed of 5 km/h from an initial speed of 90 km/h. Assume a straight, level road and no wind.

$$Ans. \ P_{45} = 2.11 \text{ kW}, \ P_{90} = 11.25 \text{ kW}$$
$$t = 250 \text{ s}, \ s = 1775 \text{ m}$$

▶ **3/305** A flexible bicycle-type chain of length $\pi r/2$ and mass ρ per unit length is released from rest with $\theta = 0$ in the smooth circular channel and falls through the hole in the supporting surface. Determine (a) the common tangential acceleration a_t of all links as a function of θ, (b) the tension T in the chain for any position β from $\beta = \theta$ to $\beta = \pi/2$, and (c) the common speed v of the links as a function of θ.

Ans.

$$(a) \ a_t = \frac{2g}{\pi}(\cos\theta + \theta)$$

$$(b) \ T = \rho gr\left[\frac{2}{\pi}(\cos\theta + \theta)(\beta - \theta) + \cos\beta - \cos\theta\right]$$

$$(c) \ v = 2\sqrt{\frac{gr}{\pi}(\sin\theta + \theta^2/2)}$$

Computer-oriented problems

*3/306 An aerial ride of an amusement park is shown. Determine and plot the angle θ as a function of the rotational speed N for $0 \leq N \leq 12$ rev/min. Neglect the mass of the arms to which the gondolas are attached and treat each gondola as a particle.

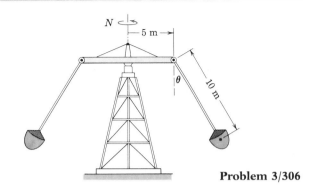

Problem 3/306

*3/307 The system of Prob. 3/86 is repeated here. A particle of mass m is introduced with zero velocity at $r = 0$ when $\theta = 0$. It slides outward through the smooth hollow tube, which is driven at the constant angular velocity ω_0 about a horizontal axis through point O. If $l = 1$ m and $\omega_0 = 0.5$ rad/s, determine the time t after release and the angular displacement θ for which the particle exits the tube. *Ans.* $t = 1.069$ s, $\theta = 30.6°$

Problem 3/307

*3/308 The mechanism of Prob. 3/141 is repeated here. The 5-kg sphere is released from rest at $\theta = 60°$ where the spring is unstretched. Plot the speed of the sphere as a function of θ and determine (a) the maximum speed of the sphere with the corresponding value of θ and (b) the maximum angle θ reached by the link. Friction and the mass of the links are small and may be neglected.

Ans. (a) $v_{max} = 1.515$ m/s, $\theta = 82.0°$
(b) $\theta_{max} = 108.6°$

Problem 3/308

*3/309 A 9-kg sphere A is held at the 60° angle shown and released. It strikes the 4.5-kg sphere B. The coefficient of restitution for this collision is $e = 0.75$. Sphere B is attached to the end of a light rod which pivots freely about point O. If the spring of constant $k = 1.5$ kN/m is initially unstretched, determine the maximum rotation angle θ of the light rod after impact. *Ans.* $\theta = 21.4°$

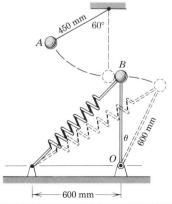

Problem 3/309

KINETICS OF SYSTEMS OF PARTICLES 4

4/1 *4/1 INTRODUCTION*

In the previous two chapters we have dealt with the principles of dynamics for the motion of a particle. Although attention was focused primarily on the kinetics of a single particle in Chapter 3, reference to the motion of two particles, considered together as a system, was made in the discussions of work-energy and impulse-momentum. Our next major step in the development of the subject of dynamics is to extend these principles, which we applied to a single particle, to describe the motion of a general system of particles. This extension provides a unity to the remaining sections of dynamics and enables us to treat the motion of rigid bodies and the motion of nonrigid systems as well. We recall that a rigid body is a solid system of particles wherein the distances between particles remain essentially unchanged. The overall motions found with machines, land and air vehicles, rockets and spacecraft, and many moving structures provide examples of rigid-body problems. A nonrigid body, on the other hand, may be a solid body where the object of investigation is the time dependence of the changes in shape due to elastic or nonelastic deformations. Or a nonrigid body may be a defined mass of liquid or gaseous particles which are flowing at a specified rate. Examples would be the air and fuel flowing through the turbine of an aircraft engine, the burned gases issuing from the nozzle of a rocket motor, or the water passing through a rotary pump.

Although the extension of the equations for single-particle motion to a general system of particles is accomplished without undue difficulty, it cannot be expected that the full generality and significance of these extended principles will be fully understood without considerable problem experience. For this reason we strongly recommend that the general results obtained in the following articles be reviewed frequently during the remainder of the study of dynamics. In this way the unity contained in these broader principles of dynamics will be brought into sharper focus and a more fundamental view of the subject will be acquired.

4/2 GENERALIZED NEWTON'S SECOND LAW

Newton's second law of motion will now be extended to cover a general mass system which we model by considering n mass particles bounded by a closed surface in space, Fig. 4/1. This bounding envelope, for example, might be the exterior surface of a given rigid body, the bounding surface of an arbitrary portion of the body, the exterior surface of a rocket containing both rigid and flowing particles, or a particular volume of fluid particles. In each case the system considered is the mass within the envelope, and that mass must be clearly defined and isolated.

Figure 4/1 shows a representative particle of mass m_i of the system isolated with forces $\mathbf{F}_1$, $\mathbf{F}_2$, $\mathbf{F}_3$, . . . acting on m_i from sources *external* to the envelope, and forces $\mathbf{f}_1$, $\mathbf{f}_2$, $\mathbf{f}_3$, . . . acting on m_i from sources *internal* to the envelope. The external forces are due to contact with external bodies or to external gravitational, electric, or magnetic forces. The internal forces are forces of reaction with other mass particles within the envelope. A representative particle of mass m_i is located by its position vector $\mathbf{r}_i$ measured from the nonaccelerating origin O of a Newtonian set of reference axes.* The center of mass G of the system of particles isolated is located by the position vector $\bar{\mathbf{r}}$ which, from the definition of the mass center as covered in statics, is given by

$$m\bar{\mathbf{r}} = \Sigma m_i \mathbf{r}_i$$

where the total system mass is $m = \Sigma m_i$. The summation sign Σ represents the summation $\Sigma_{i=1}^{n}$ over all n particles. Newton's second law, Eq. 3/3, when applied to m_i gives

$$\mathbf{F}_1 + \mathbf{F}_2 + \mathbf{F}_3 + \cdots + \mathbf{f}_1 + \mathbf{f}_2 + \mathbf{f}_3 + \cdots = m_i \ddot{\mathbf{r}}_i$$

where $\ddot{\mathbf{r}}_i$ is the acceleration of m_i. A similar equation may be written for each of the particles of the system. If these equations written for *all* particles of the system are added together, there results

$$\Sigma\mathbf{F} + \Sigma\mathbf{f} = \Sigma m_i \ddot{\mathbf{r}}_i$$

The term $\Sigma\mathbf{F}$ then becomes the vector sum of *all* forces acting on all particles of the isolated system from sources external to the system, and $\Sigma\mathbf{f}$ becomes the vector sum of all forces on all particles produced by the internal actions and reactions between particles. This last sum is identically zero since all internal forces

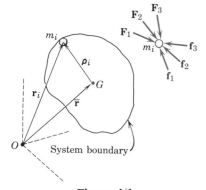

Figure 4/1

---

* It was shown in Art. 3/13 that any nonrotating and nonaccelerating set of axes constitutes a Newtonian reference system in which the principles of Newtonian mechanics are valid.

occur in pairs of equal and opposite actions and reactions. By differentiating the equation defining $\bar{\mathbf{r}}$ twice with time, we have $m\ddot{\bar{\mathbf{r}}} = \Sigma m_i \ddot{\mathbf{r}}_i$ where m has no time derivative as long as mass is not entering or leaving the system.* Substitution into the summation of the equations of motion gives

$$\boxed{\Sigma\mathbf{F} = m\ddot{\bar{\mathbf{r}}} \quad \text{or} \quad \Sigma\mathbf{F} = m\bar{\mathbf{a}}} \qquad (4/1)$$

where $\bar{\mathbf{a}}$ is the acceleration $\ddot{\bar{\mathbf{r}}}$ of the center of mass of the system.

Equation 4/1 is the generalized Newton's second law of motion for a mass system and is referred to as the *equation of motion* of m. The equation states that the resultant of the external forces on *any* system of mass equals the total mass of the system times the acceleration of the center of mass. This law expresses the so-called *principle of motion of the mass center*. It should be observed that $\bar{\mathbf{a}}$ is the acceleration of the mathematical point which represents instantaneously the position of the mass center for the given n particles. For a nonrigid body this acceleration need not represent the acceleration of any particular particle. It should be noted further that Eq. 4/1 holds for each instant of time and is therefore an instantaneous relationship. Equation 4/1 for the mass system had to be proved as it may not be inferred directly from Eq. 3/3 for the single particle. Equation 4/1 may be expressed in component form using x-y-z coordinates or whatever coordinate system is most convenient for the problem at hand. Thus

$$\Sigma F_x = m\bar{a}_x \qquad \Sigma F_y = m\bar{a}_y \qquad \Sigma F_z = m\bar{a}_z \qquad (4/1a)$$

Although Eq. 4/1, as a vector equation, requires that the acceleration vector $\bar{\mathbf{a}}$ must have the same direction as the resultant external force $\Sigma\mathbf{F}$, it does not follow that $\Sigma\mathbf{F}$ necessarily passes through G. In general, in fact, $\Sigma\mathbf{F}$ does not pass through G, as will be shown later.

4/3 WORK-ENERGY

In Art. 3/6 the work-energy relation for a single particle was developed. Applicability to a system of two joined particles was also noted. We now turn our attention to the general system of Fig. 4/1, where the work-energy relation for the representative particle of mass m_i is $U_i = \Delta T_i$. Here U_i is the work done on m_i during an interval of motion by all forces $\mathbf{F}_i = \mathbf{F}_1 + \mathbf{F}_2 + \mathbf{F}_3 + \cdots$ applied from sources external to the system and by all forces $\mathbf{f}_i = \mathbf{f}_1 + \mathbf{f}_2 + \mathbf{f}_3 + \cdots$ applied from sources internal to the

---

* If m is a function of time, a more complex situation develops; this situation is discussed in Art. 4/7 on variable mass.

system. The change in kinetic energy of m_i is $\Delta T_i = \Delta(\frac{1}{2}m_i v_i^2)$ where v_i is the magnitude of the particle velocity $\mathbf{v}_i = \dot{\mathbf{r}}_i$.

For the entire system the sum of the work-energy equations written for all particles is $\Sigma U_i = \Sigma\,\Delta T_i$ which may be represented by the same expression as Eq. 3/11 of Art. 3/6, namely,

$$\boxed{U = \Delta T}\tag{4/2}$$

where $U = \Sigma U_i$, the work done by all forces on all particles, and ΔT is the change in the total kinetic energy $T = \Sigma T_i$ of the system. As was pointed out in Art. 3/6, the work-energy equation may be expressed alternatively as

$$T_{\text{initial}} + U = T_{\text{final}}$$

For a rigid body or a system of rigid bodies joined by ideal frictionless connections no net work is done by the internal interacting forces or moments in the connections. We see that the work done by all pairs of internal forces $\mathbf{f}_i$ and $-\mathbf{f}_i$ at a typical connection, Fig. 4/2, in the system is zero since their points of application have identical displacement components in the direction of the forces. For this situation U becomes the work done on the system by the external forces only.

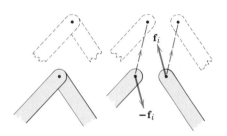

Figure 4/2

For a nonrigid mechanical system which has elastic members capable of storing energy, a part of the work done by the external forces goes into changing the internal elastic potential energy V_e. Also, if the works done by the gravity forces are *excluded* from the U-term and are accounted for by the changes in gravitational potential energy V_g, then we may equate the total work U done on the system during an interval of motion to the change ΔE in the mechanical energy of the system. Thus

$$U = \Delta E = \Delta(T + V_g + V_e)$$

or

$$\boxed{U = \Delta T + \Delta V_g + \Delta V_e}\tag{4/3}$$

which is the same as Eq. 3/17 in Art. 3/7. Alternatively we may express the work-energy relation as

$$(T + V_g + V_e)_{\text{initial}} + U = (T + V_g + V_e)_{\text{final}}$$

which, of course, is the same as Eq. 4/3.

We now examine the expression $T = \Sigma\frac{1}{2}m_i v_i^2$ for the kinetic energy of the mass system in more detail. By our principle of relative motion discussed in Art. 2/8, we may write the velocity of the representative particle as

$$\mathbf{v}_i = \bar{\mathbf{v}} + \dot{\boldsymbol{\rho}}_i$$

where $\bar{\mathbf{v}}$ is the velocity of the mass center G and $\dot{\boldsymbol{\rho}}_i$ is the velocity of m_i with respect to a translating reference frame moving with the mass center G. We recall the identity $v_i^2 = \mathbf{v}_i \cdot \mathbf{v}_i$ and write the kinetic energy of the system as

$$T = \Sigma \tfrac{1}{2} m_i \mathbf{v}_i \cdot \mathbf{v}_i = \Sigma \tfrac{1}{2} m_i (\bar{\mathbf{v}} + \dot{\boldsymbol{\rho}}_i) \cdot (\bar{\mathbf{v}} + \dot{\boldsymbol{\rho}}_i)$$
$$= \Sigma \tfrac{1}{2} m_i \bar{v}^2 + \Sigma \tfrac{1}{2} m_i |\dot{\boldsymbol{\rho}}_i|^2 + \Sigma m_i \bar{\mathbf{v}} \cdot \dot{\boldsymbol{\rho}}_i$$

Since $\boldsymbol{\rho}_i$ is measured from the mass center, $\Sigma m_i \boldsymbol{\rho}_i = \mathbf{0}$ and the third term is $\bar{\mathbf{v}} \cdot \Sigma m_i \dot{\boldsymbol{\rho}}_i = \bar{\mathbf{v}} \cdot \dfrac{d}{dt} \Sigma (m_i \boldsymbol{\rho}_i) = \mathbf{0}$. Also $\Sigma \tfrac{1}{2} m_i \bar{v}^2 = \tfrac{1}{2} \bar{v}^2 \Sigma m_i = \tfrac{1}{2} m \bar{v}^2$. Therefore the total kinetic energy becomes

$$\boxed{T = \tfrac{1}{2} m \bar{v}^2 + \Sigma \tfrac{1}{2} m_i |\dot{\boldsymbol{\rho}}_i|^2} \tag{4/4}$$

This equation expresses the fact that the total kinetic energy of a mass system equals the energy of mass-center translation of the system as a whole plus the energy due to motion of all particles relative to the mass center.

4/4 IMPULSE-MOMENTUM

(a) *Linear momentum.* From our definition in Art. 3/8 the linear momentum of the representative particle of the system depicted in Fig. 4/1 is $\mathbf{G}_i = m_i \mathbf{v}_i$ where the velocity of m_i is $\mathbf{v}_i = \dot{\mathbf{r}}_i$. The linear momentum of the system is defined as the vector sum of the linear momenta of all of its particles or $\mathbf{G} = \Sigma m_i \mathbf{v}_i$. By substituting the relative velocity relation $\mathbf{v}_i = \bar{\mathbf{v}} + \dot{\boldsymbol{\rho}}_i$ and noting again that $\Sigma m_i \boldsymbol{\rho}_i = m \bar{\boldsymbol{\rho}} = \mathbf{0}$ we get

$$\mathbf{G} = \Sigma m_i (\bar{\mathbf{v}} + \dot{\boldsymbol{\rho}}_i) = \Sigma m_i \bar{\mathbf{v}} + \frac{d}{dt} \Sigma m_i \boldsymbol{\rho}_i$$

$$= \bar{\mathbf{v}} \Sigma m_i + \frac{d}{dt} (\mathbf{0})$$

or

$$\boxed{\mathbf{G} = m \bar{\mathbf{v}}} \tag{4/5}$$

Thus the linear momentum of any mass system of constant mass is the product of the mass and the velocity of its center of mass.

The time derivative of $\mathbf{G}$ is $m \dot{\bar{\mathbf{v}}} = m \bar{\mathbf{a}}$, which by Eq. 4/1 is the resultant external force acting on the system. Thus we have

$$\boxed{\Sigma \mathbf{F} = \dot{\mathbf{G}}} \tag{4/6}$$

which has the same form as Eq. 3/21 for a single particle. Equation 4/6 states that the resultant of the external forces on any mass system equals the time rate of change of the linear momentum of the system and is an alternative form of the generalized second law of motion, Eq. 4/1. As was noted at the end of the last article, $\Sigma\mathbf{F}$, in general, does not pass through G. In Eq. 4/6 the total mass was assumed constant during differentiation with time, so that the equation does not apply to systems whose mass changes with time.

(*b*) *Angular momentum.*   The angular momentum of our general mass system will now be determined about the fixed point O, about the mass center G, and about an arbitrary point P, shown in Fig. 4/3, which may have an acceleration $\mathbf{a}_P = \ddot{\mathbf{r}}_P$.

(*About O*).   The angular momentum of the mass system about the point O, fixed in the Newtonian reference system, is defined as the vector sum of the moments of the linear momenta about O of all particles of the system and is

$$\mathbf{H}_O = \Sigma(\mathbf{r}_i \times m_i\mathbf{v}_i)$$

The time derivative of the vector product is $\dot{\mathbf{H}}_O = \Sigma(\dot{\mathbf{r}}_i \times m_i\mathbf{v}_i) + \Sigma(\mathbf{r}_i \times m_i\dot{\mathbf{v}}_i)$. The first summation vanishes since the cross product of two equal vectors $\dot{\mathbf{r}}_i$ and $\mathbf{v}_i$ is zero. The second summation is $\Sigma(\mathbf{r}_i \times m_i\mathbf{a}_i) = \Sigma(\mathbf{r}_i \times \mathbf{F}_i)$ which is the vector sum of the moments about O of all forces acting on all particles of the system. This moment sum $\Sigma\mathbf{M}_O$ represents only the moments of forces external to the system, since the internal forces cancel one another and their moments add up to zero. Thus the moment sum is

$$\boxed{\Sigma\mathbf{M}_O = \dot{\mathbf{H}}_O} \qquad (4/7)$$

which has the same form as Eq. 3/27 for a single particle. Equation 4/7 states that the resultant vector moment about any fixed point of all external forces on any system of mass equals the time rate of change of angular momentum of the system about the fixed point. As in the linear-momentum case, Eq. 4/7 does not apply if the total mass of the system is changing with time.

(*About G*).   The angular momentum about the mass center G is the sum of the moments of the linear momenta about G of all particles of the system and is

$$\mathbf{H}_G = \Sigma\boldsymbol{\rho}_i \times m_i\dot{\mathbf{r}}_i \qquad (4/8)$$

Alternatively we may express $\mathbf{H}_G$ in terms of the moments about G of linear momenta relative to P or relative to G where the linear momentum of each particle is its mass times its velocity $\dot{\boldsymbol{\rho}}_i'$ relative to P or its velocity $\dot{\boldsymbol{\rho}}_i$ relative to G. With the notation "rel" designating the relative quantity, we may write

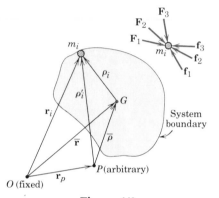

Figure 4/3

(Relative to P) $(\mathbf{H}_G)_{\mathrm{rel}} = \Sigma \boldsymbol{\rho}_i \times m_i \dot{\boldsymbol{\rho}}_i{}' = \Sigma \boldsymbol{\rho}_i \times m_i (\dot{\mathbf{r}}_i - \dot{\mathbf{r}}_P)$

$\qquad\qquad\qquad = \mathbf{H}_G + \dot{\mathbf{r}}_P \times \Sigma m_i \boldsymbol{\rho}_i = \mathbf{H}_G$

(Relative to G) $(\mathbf{H}_G)_{\mathrm{rel}} = \Sigma \boldsymbol{\rho}_i \times m_i \dot{\boldsymbol{\rho}}_i = \Sigma \boldsymbol{\rho}_i \times m_i (\dot{\mathbf{r}}_i - \dot{\bar{\mathbf{r}}})$

$\qquad\qquad\qquad = \mathbf{H}_G + \dot{\bar{\mathbf{r}}} \times \Sigma m_i \boldsymbol{\rho}_i = \mathbf{H}_G$

where $\Sigma m_i \boldsymbol{\rho}_i = \mathbf{0}$ by definition of the mass center. These equalities hold only for angular momentum about the mass center.

Differentiating Eq. 4/8 with respect to time gives

$$\dot{\mathbf{H}}_G = \Sigma \boldsymbol{\rho}_i \times m_i \ddot{\bar{\mathbf{r}}}_i + \Sigma \dot{\boldsymbol{\rho}}_i \times m_i (\dot{\bar{\mathbf{r}}} + \dot{\boldsymbol{\rho}}_i)$$

where $\bar{\mathbf{r}} + \boldsymbol{\rho}_i$ has been substituted for $\mathbf{r}_i$. The first summation is the sum of the moments $\Sigma \mathbf{M}_G$ of all external forces about G. (The moments of internal forces cancel one another as previously noted.) The first term of the second summation becomes

$-\dot{\bar{\mathbf{r}}} \times \Sigma m_i \dot{\boldsymbol{\rho}}_i = -\dot{\bar{\mathbf{r}}} \times \dfrac{d}{dt} \Sigma m_i \boldsymbol{\rho}_i = \mathbf{0}$ since $\Sigma m_i \boldsymbol{\rho}_i = \mathbf{0}$. The second

term of the second summation is also zero since the cross product of two identical vectors is zero. We are left with

$$\boxed{\Sigma \mathbf{M}_G = \dot{\mathbf{H}}_G} \qquad\qquad (4/9)$$

where we may use either the absolute or the relative angular momentum.

Equations 4/7 and 4/9 are among the most powerful of the governing relations in dynamics and form the basis for much of the analysis in three-dimensional problems.

(*About P*).   It is frequently convenient to write the expression for the angular momentum about some arbitrary point P, Fig. 4/3, in terms of the angular momentum about the mass center G. The absolute angular momentum about P is

$$\mathbf{H}_P = \Sigma \boldsymbol{\rho}_i{}' \times m_i \dot{\mathbf{r}}_i$$

With the substitution of $\bar{\boldsymbol{\rho}} + \boldsymbol{\rho}_i$ for $\boldsymbol{\rho}_i{}'$, the first term is $\bar{\boldsymbol{\rho}} \times \Sigma m_i \mathbf{v}_i = \bar{\boldsymbol{\rho}} \times m\bar{\mathbf{v}}$, and the second term is $\Sigma \boldsymbol{\rho}_i \times m_i \mathbf{v}_i = \mathbf{H}_G$. Thus

$$\boxed{\mathbf{H}_P = \mathbf{H}_G + \bar{\boldsymbol{\rho}} \times m\bar{\mathbf{v}}} \qquad\qquad (4/10)$$

Similarly, the angular momentum relative to P becomes $(\mathbf{H}_P)_{\mathrm{rel}} = \Sigma \boldsymbol{\rho}_i{}' \times m_i \dot{\boldsymbol{\rho}}_i{}'$. Substituting $\boldsymbol{\rho}_i{}' = \bar{\boldsymbol{\rho}} + \boldsymbol{\rho}_i$ and expanding gives

$$\boxed{(\mathbf{H}_P)_{\mathrm{rel}} = (\mathbf{H}_G)_{\mathrm{rel}} + \bar{\boldsymbol{\rho}} \times m\bar{\mathbf{v}}_{\mathrm{rel}}} \qquad\qquad (4/11)$$

where we have observed that $\dot{\bar{\boldsymbol{\rho}}} = \bar{\mathbf{v}}_{\mathrm{rel}}$, $\Sigma m_i \dot{\boldsymbol{\rho}}_i = \dfrac{d}{dt} \Sigma m_i \boldsymbol{\rho}_i = \dfrac{d}{dt} (\mathbf{0})$,

and $\Sigma \boldsymbol{\rho}_i \times m_i \dot{\boldsymbol{\rho}}_i = (\mathbf{H}_G)_{\mathrm{rel}}$. All quantities in Eq. 4/11 are measured relative to a nonrotating system attached to P.

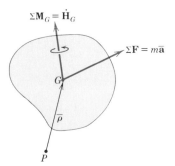

Figure 4/4

We now make use of the principle of moments developed in our study of statics. Figure 4/4 represents the resultants of the external forces acting on the system expressed in terms of the resultant force $\Sigma\mathbf{F}$ shown through G and the corresponding couple $\Sigma\mathbf{M}_G$. We see that the sum of the moments about P of all forces external to the system must equal the moment of their resultants about P. Therefore we may write

$$\Sigma\mathbf{M}_P = \Sigma\mathbf{M}_G + \bar{\rho} \times \Sigma\mathbf{F}$$

which, by Eqs. 4/9 and 4/1, becomes

$$\Sigma\mathbf{M}_P = \dot{\mathbf{H}}_G + \bar{\rho} \times m\bar{\mathbf{a}} \qquad (4/12)$$

Equation 4/12 enables us to write the moment equation about any convenient moment center P and is easily visualized with the aid of Fig. 4/4.

The moment equation about P may also be expressed in terms of the angular momentum about P. With $\mathbf{H}_G = (\mathbf{H}_G)_{\mathrm{rel}}$ and Eq. 4/11 we have $\dot{\mathbf{H}}_G = (\dot{\mathbf{H}}_G)_{\mathrm{rel}} = (\dot{\mathbf{H}}_P)_{\mathrm{rel}} - \dot{\bar{\rho}} \times m\bar{\mathbf{v}}_{\mathrm{rel}} - \bar{\rho} \times m\dot{\bar{\mathbf{v}}}_{\mathrm{rel}}$. Noting that $\dot{\bar{\rho}} \times m\bar{\mathbf{v}}_{\mathrm{rel}} = \dot{\bar{\rho}} \times m\dot{\bar{\rho}} = 0$ and substituting $\dot{\mathbf{H}}_G$ into Eq. 4/12, we have

$$\Sigma\mathbf{M}_P = (\dot{\mathbf{H}}_P)_{\mathrm{rel}} - \bar{\rho} \times m\dot{\bar{\mathbf{v}}}_{\mathrm{rel}} + \bar{\rho} \times m\bar{\mathbf{a}} = (\dot{\mathbf{H}}_P)_{\mathrm{rel}} + \bar{\rho} \times m(\bar{\mathbf{a}} - \dot{\bar{\mathbf{v}}}_{\mathrm{rel}})$$

But the relative acceleration relation between P and G gives $\bar{\mathbf{a}} = \mathbf{a}_P + \ddot{\bar{\rho}} = \mathbf{a}_P + \dot{\bar{\mathbf{v}}}_{\mathrm{rel}}$. With this substitution the moment equation becomes

$$\Sigma\mathbf{M}_P = (\dot{\mathbf{H}}_P)_{\mathrm{rel}} + \bar{\rho} \times m\mathbf{a}_P \qquad (4/13)$$

We see that Eq. 4/13 reduces to the simpler form under any one of the following conditions:

$$\Sigma\mathbf{M}_P = (\dot{\mathbf{H}}_P)_{\mathrm{rel}} \qquad \text{if} \begin{cases} \mathbf{a}_P = \mathbf{0} \text{ (becomes equivalent to Eq. 4/7)} \\ \bar{\rho} = \mathbf{0} \text{ (becomes equivalent to Eq. 4/9)} \\ \bar{\rho} \text{ and } \mathbf{a}_P \text{ are parallel } (\mathbf{a}_P \text{ directed toward or} \\ \qquad\qquad\qquad\qquad\qquad\qquad \text{away from } G) \end{cases}$$

4/5 CONSERVATION OF ENERGY AND MOMENTUM

Conditions frequently occur when there is no net change in the total mechanical energy of a system during an interval of motion. Other conditions exist when there is no net change in the momentum of a system. These conditions are treated separately as follows.

(a) *Conservation of energy.* A mass system is said to be *conservative* if it does not lose energy by virtue of internal friction forces which do negative work or by virtue of nonelastic members which dissipate energy upon cycling. If no work is done on a conservative system during an interval of motion by external forces (other than gravity or other potential forces), then no part of the energy of the system is lost. In this case $E_{\text{initial}} = E_{\text{final}}$ or $\Delta E = 0$. Therefore we may write Eq. 4/3 as

$$\Delta T + \Delta V_e + \Delta V_g = 0 \qquad (4/14)$$

which expresses the *law of conservation of dynamical energy.* This law holds only in the ideal case where internal kinetic friction is sufficiently small to be neglected.

(*b*) *Conservation of momentum.* If, for a certain interval of time, the resultant external force $\Sigma\mathbf{F}$ acting on a conservative or nonconservative mass system is zero, Eq. 4/6 requires that $\dot{\mathbf{G}} = \mathbf{0}$, so that during this interval

$$\Delta\mathbf{G} = \mathbf{0} \qquad (4/15)$$

which expresses the *principle of conservation of linear momentum.* Thus in the absence of an external impulse the linear momentum of a system remains unchanged.

Similarly, if the resultant moment about a fixed point O or about the mass center G of all external forces on any mass system is zero, Eq. 4/7 or 4/9 requires, respectively, that

$$\Delta\mathbf{H}_O = \mathbf{0} \quad \text{or} \quad \Delta\mathbf{H}_G = \mathbf{0} \qquad (4/16)$$

These relations express the *principle of conservation of angular momentum* for a general mass system in the absence of an angular impulse. Thus, if there is no angular impulse about a fixed point (or about the mass center), the angular momentum of the system about the fixed point (or about the mass center) remains unchanged. Either equation may hold without the other.

It was proved in Art. 3/12 that the basic laws of Newtonian mechanics hold for measurements made relative to a set of axes which translate with a constant velocity. Thus Eqs. 4/1 through 4/16 are valid provided all quantities are expressed relative to the translating axes.

Equations 4/1 through 4/16 are among the most important of the basic derived laws of mechanics. In this chapter these laws have been derived for the most general system of constant mass in order that the generality of the laws will be established. These laws find repeated use when applied to specific mass systems such as rigid and nonrigid solids and certain fluid systems which are discussed in the articles that follow. The reader is urged to study these laws carefully and to compare them with their more restricted forms encountered earlier in Chapter 3.

Sample Problem 4/1

Each of the three balls has a mass m and is welded to the rigid equiangular frame of negligible mass. The assembly rests on a smooth horizontal surface. If a force $\mathbf{F}$ is suddenly applied to the one bar as shown, determine (a) the acceleration of point O and (b) the angular acceleration $\ddot{\theta}$ of the frame.

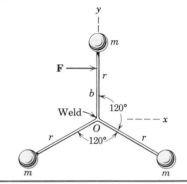

Solution. (a) Point O is the mass center of the system of the three balls, so that its acceleration is given by Eq. 4/1.

① $[\Sigma \mathbf{F} = m\bar{\mathbf{a}}]$ $\qquad F\mathbf{i} = 3m\bar{\mathbf{a}}$ $\qquad \bar{\mathbf{a}} = \mathbf{a}_O = \dfrac{F}{3m}\mathbf{i}$ $\qquad$ *Ans.*

(b) We determine $\ddot{\theta}$ from the moment principle, Eq. 4/9. To find $\mathbf{H}_G$ we note that the velocity of each ball relative to the mass center O as measured in the nonrotating axes x-y is $r\dot{\theta}$ where $\dot{\theta}$ is the common angular velocity of the spokes. The angular momentum of the system about O is the sum of the moments of the relative linear momenta as shown by Eq. 4/8 so is expressed by

$$H_O = H_G = 3(mr\dot{\theta})r = 3mr^2\dot{\theta}$$

② Equation 4/9 now gives

$[\Sigma \mathbf{M}_G = \dot{\mathbf{H}}_G]$

$$Fb = \frac{d}{dt}(3mr^2\dot{\theta}) = 3mr^2\ddot{\theta} \qquad \text{so} \qquad \ddot{\theta} = \frac{Fb}{3mr^2} \qquad Ans.$$

① We note that the result depends only on the magnitude and direction of $\mathbf{F}$ and not on b, which locates the line of action of $\mathbf{F}$.

② Although $\dot{\theta}$ is initially zero we need the expression for $H_O = H_G$ in order to get $\dot{H}_G$. We observe also that $\ddot{\theta}$ is independent of the motion of O.

Sample Problem 4/2

Consider the same conditions as for Sample Problem 4/1 except that the spokes are freely hinged at O and so do not constitute a rigid system. Explain the difference between the two problems.

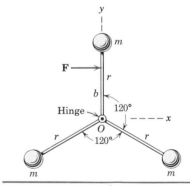

Solution. The generalized Newton's second law holds for any mass system, so that the acceleration $\bar{\mathbf{a}}$ of the mass center G is the same as with Sample Problem 4/1, namely,

$$\bar{\mathbf{a}} = \frac{F}{3m}\mathbf{i} \qquad Ans.$$

Although G coincides with O at the instant represented, the motion of the hinge O is not the same as the motion of G since O will not remain the center of mass as the angles between the spokes change.

① Both ΣM_G and $\dot{H}_G$ have the same values for the two problems at the instant represented. However, the angular motions of the spokes in this problem are all different and are not easily determined.

① This present system could be dismembered and the motion equations written for each of the parts with the unknowns eliminated one by one. Or a more sophisticated method using the equations of Lagrange could be employed. (See the senior author's *Dynamics, 2nd Edition SI Version* 1975, for a discussion of this approach.)

Sample Problem 4/3

A shell with a mass of 20 kg is fired from point O with a velocity $u = 300$ m/s in the vertical x-z plane at the inclination shown. When it reaches the top of its trajectory at P it explodes into three fragments A, B, and C. Immediately after the explosion fragment A is observed to rise vertically a distance of 500 m above P, and fragment B is seen to have a horizontal velocity $\mathbf{v}_B$ and eventually land at point Q. When recovered the masses of the fragments A, B, and C are found to be 5, 9, and 6 kg, respectively. Calculate the velocity which fragment C has immediately after the explosion. Neglect atmospheric resistance.

Solution. From our knowledge of projectile motion (Sample Problem 2/6) the time required for the shell to reach P and its vertical rise are

$$t = u_z/g = 300(4/5)/9.81 = 24.5 \text{ s}$$

$$h = \frac{u_z^2}{2g} = \frac{[(300)(4/5)]^2}{2(9.81)} = 2936 \text{ m}$$

The velocity of A has the magnitude

$$v_A = \sqrt{2gh_A} = \sqrt{2(9.81)(500)} = 99.0 \text{ m/s}$$

With no z-component of velocity initially fragment B requires 24.5 s to return to the ground. Thus its horizontal velocity, which remains constant, is

$$v_B = s/t = 4000/24.5 = 163.5 \text{ m/s}$$

Since the force of the explosion is internal to the system of the shell and its three fragments, the linear momentum of the system remains unchanged during the explosion. Thus

① $[\Delta \mathbf{G} = 0]$ $m\mathbf{v} = m_A\mathbf{v}_A + m_B\mathbf{v}_B + m_C\mathbf{v}_C$

$$20(300)(\tfrac{3}{5})\mathbf{i} = 5(99.0\mathbf{k}) + 9(163.5)(\mathbf{i} \cos 45° + \mathbf{j} \sin 45°) + 6\mathbf{v}_C$$

$$6\mathbf{v}_C = 2560\mathbf{i} - 1040\mathbf{j} - 495\mathbf{k}$$

② $$\mathbf{v}_C = 427\mathbf{i} - 173\mathbf{j} - 82.5\mathbf{k} \text{ m/s}$$

$$v_C = \sqrt{(427)^2 + (173)^2 + (82.5)^2} = 468 \text{ m/s} \qquad \textit{Ans.}$$

① The velocity $\mathbf{v}$ of the shell at the top of its trajectory is, of course, the constant horizontal component of its initial velocity $\mathbf{u}$, which becomes $u(3/5)$.

② We note that the mass center of the three fragments while still in flight continues to follow the same trajectory that the shell would have followed if it had not exploded.

Sample Problem 4/4

The 16-kg carriage A moves horizontally in its guide with a speed of 1.2 m/s and carries two assemblies of balls and light rods which rotate about a shaft at O in the carriage. Each of the four balls has a mass of 1.6 kg. The assembly on the front face rotates counterclockwise at a speed of 80 rev/min, and the assembly on the back side rotates clockwise at a speed of 100 rev/min. For the entire system calculate (a) the kinetic energy T, (b) the magnitude G of the linear momentum, and (c) the magnitude H_O of the angular momentum about point O.

Solution. (a) **Kinetic energy.** The velocity of the balls with respect to O is

$$[|\dot{\boldsymbol{\rho}}_i| = v_{\text{rel}} = r\dot{\theta}] \quad (v_{\text{rel}})_{1,2} = 0.450\,\frac{80(2\pi)}{60} = 3.77 \text{ m/s}$$

$$(v_{\text{rel}})_{3,4} = 0.300\,\frac{100(2\pi)}{60} = 3.14 \text{ m/s}$$

The kinetic energy of the system is given by Eq. 4/4. The translational part is

$$\tfrac{1}{2}m\bar{v}^2 = \tfrac{1}{2}[16 + 4(1.6)](1.2)^2 = 16.13 \text{ J}$$

The rotational part of the kinetic energy depends on the squares of the relative velocities and is

$$\Sigma\tfrac{1}{2}m_i|\dot{\boldsymbol{\rho}}_i|^2 = 2\left[\frac{1}{2}\,1.6(3.77)^2\right]_{(1,2)} + 2\left[\frac{1}{2}\,1.6(3.14)^2\right]_{(3,4)}$$

$$= 22.74 + 15.79 = 38.53 \text{ J}$$

The total kinetic energy is

$$T = \tfrac{1}{2}m\bar{v}^2 + \Sigma\tfrac{1}{2}m_i|\dot{\boldsymbol{\rho}}_i|^2 = 16.13 + 38.53 = 54.66 \text{ J} \qquad Ans.$$

(b) **Linear momentum.** The linear momentum of the system by Eq. 4/5 is the total mass times v_O, the velocity of the center of mass. Thus

$$[\mathbf{G} = m\bar{\mathbf{v}}] \qquad G = [16 + 4(1.6)](1.2) = 26.88 \text{ kg·m/s}$$

(c) **Angular momentum about O.** The angular momentum about O is due to the moments of the linear momenta of the balls. Taking counterclockwise as positive, we have

$$H_O = \Sigma|\mathbf{r}_i \times m_i\mathbf{v}_i|$$

$$H_O = [2(1.6)(0.450)(3.77)]_{(1,2)} - [2(1.6)(0.300)(3.14)]_{(3,4)}$$

$$= 5.43 - 3.02 = 2.41 \text{ kg·m}^2/\text{s} \qquad Ans.$$

① Note that the mass m is the total mass, carriage plus the four balls, and that $\bar{v}$ is the velocity of the mass center O which is the carriage velocity.

② Note that the direction of rotation, clockwise or counterclockwise, makes no difference in the calculation of kinetic energy, which depends on the square of the velocity.

③ There is a temptation to overlook the contribution of the balls since their linear momenta relative to O in each pair are in opposite directions and cancel. However, each ball also has a velocity component $\bar{v}$ and hence a momentum component $m_i\bar{v}$.

④ Contrary to the case of kinetic energy where the direction of rotation was immaterial, angular momentum is a vector quantity and the direction of rotation must be accounted for.

PROBLEMS

4/1 The two 2-kg balls are initially at rest on the horizontal surface when a vertical force $F = 60$ N is applied to the junction of the attached wires as shown. Compute the vertical component a_y of the initial acceleration of each ball by considering the system as a whole.   *Ans.* $a_y = 5.19$ m/s^2

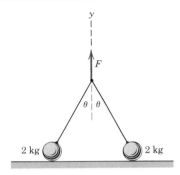

Problem 4/1

4/2 Three monkeys A, B, and C with masses of 10, 15, and 8 kg, respectively, are climbing up and down the rope suspended from D. At the instant represented, A is descending the rope with an acceleration of 2 m/s^2, and C is pulling himself up with an acceleration of 1.5 m/s^2. Monkey B is climbing up with a constant speed of 0.8 m/s. Treat the rope and monkeys as a complete system and calculate the tension T in the rope at D.
Ans. $T = 316$ N

Problem 4/2

4/3 Calculate the acceleration of the center of mass of the system of the four 10-kg cylinders. Neglect friction and the mass of the pulleys and cables.
Ans. $\bar{a} = 15.19$ m/s^2

Problem 4/3

Problem 4/4

4/4 The two spheres, each of mass m, are connected by the spring and hinged bars of negligible mass. The spheres are free to slide in the smooth guides up the incline θ. Determine the acceleration a_C of the center C of the spring.

Problem 4/5

4/5 The three identical steel balls are welded to the two connecting rods of negligible mass to form a rigid unit. The assembly is released from rest in the position shown and slides in the vertical plane. In the absence of friction determine the common velocity v of the balls when they have reached the horizontal surface. *Ans.* $v = \sqrt{bg\sqrt{2}}$

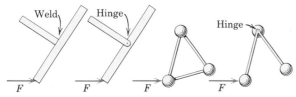

Problem 4/6

4/6 The four systems slide on a smooth horizontal surface and have the same mass m. The configurations of mass in the two pairs are identical. What can be said about the acceleration of the mass center for each system? Explain any difference in the accelerations of the members.

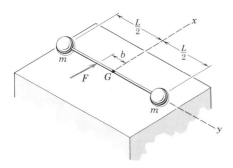

Problem 4/7

4/7 Two steel balls, each of mass m, are welded to a light rod of length L and negligible mass and are initially at rest on a smooth horizontal surface. A horizontal force of magnitude F is suddenly applied to the rod as shown. Determine (*a*) the instantaneous acceleration $\bar{a}$ of the mass center G and (*b*) the corresponding rate $\ddot{\theta}$ at which the angular velocity of the assembly about G is changing with time. *Ans.* (*a*) $\bar{a} = \dfrac{F}{2m}$, (*b*) $\ddot{\theta} = \dfrac{2Fb}{mL^2}$

Problem 4/8

4/8 The four 3-kg balls are rigidly mounted to the rotating frame and shaft, which are initially rotating freely about the vertical z-axis at the angular rate of 20 rad/s clockwise when viewed from above. If a constant torque $M = 30$ N·m is applied to the shaft, calculate the time t to reverse the direction of rotation and reach an angular velocity of $\dot{\theta} = 20$ rad/s in the same sense as M.

4/9 The three freight cars are rolling along the horizontal track with the velocities shown. After the impacts occur the three cars become coupled together and move with a common velocity v. The loaded cars A, B, and C have masses of 65 Mg, 50 Mg, and 75 Mg, respectively. Determine v and calculate the percentage loss of energy of the system due to coupling.

Ans. $v = 0.355$ km/h, 95.0% loss

Problem 4/9

4/10 The 300-kg and 400-kg mine cars are rolling in opposite directions along the horizontal track with the respective speeds of 0.6 m/s and 0.3 m/s. Upon impact the cars become coupled together. Just prior to impact a 100-kg boulder leaves the delivery chute with a speed of 1.2 m/s in the direction shown and lands in the 300-kg car. Calculate the velocity of the system after the boulder has come to rest relative to the car. Would the final velocity be the same if the cars were coupled before the boulder dropped?

Problem 4/10

4/11 The man of mass m_1 and the woman of mass m_2 are standing on opposite ends of the platform of mass m_0 which moves with negligible friction and is initially at rest with $s = 0$. The man and woman begin to approach each other. Derive an expression for the displacement s of the platform when the two meet in terms of the displacement x_1 of the man relative to the platform.

$$Ans. \ s = \frac{(m_1 + m_2)x_1 - m_2 l}{m_0 + m_1 + m_2}$$

Problem 4/11

4/12 Determine the expression for the speed $\dot{s}$ of the platform of Prob. 4/11 if the man runs at a speed $\dot{x}_1$ and the woman walks at the speed $\dot{x}_2$ both relative to the platform.

4/13 Prove that the linear momentum **G** of a system of mass m with respect to fixed axes is given by

$$\mathbf{G} = \mathbf{G}_{rel} + m\mathbf{v}_0$$

where $\mathbf{G}_{rel}$ is the linear momentum of the system relative to a set of translating axes moving with a velocity $\mathbf{v}_0$.

4/14 For any system of mass show that the angular momenta about a fixed point O and about the mass center G are related by

$$\mathbf{H}_O = \mathbf{H}_G + \bar{\mathbf{r}} \times \mathbf{G}$$

where $\bar{\mathbf{r}}$ is the position vector from O to G and $\mathbf{G}$ is the linear momentum of the system.

4/15 For any system of mass show that the angular momenta about a fixed point O and about an arbitrary point P are related by

$$\mathbf{H}_O = \mathbf{H}_P + \mathbf{r}_P \times \mathbf{G}$$

where $\mathbf{r}_P$ is the position vector from O to P and $\mathbf{G}$ is the linear momentum of the system.

4/16 Show that the sum of the moments about an arbitrary point P of the external forces acting on a general system of mass is given by

$$\Sigma\mathbf{M}_P = \dot{\mathbf{H}}_P + \mathbf{v}_P \times \mathbf{G}$$

where $\mathbf{H}_P$ is the angular momentum about P using absolute velocities, $\mathbf{v}_P$ is the velocity of P, and $\mathbf{G}$ is the linear momentum of the system. Identify the three conditions under which the equation reduces to $\Sigma\mathbf{M}_P = \dot{\mathbf{H}}_P$.

175 mm

75 mm

F

150 mm

2 kg

250 mm

1 kg

Problem 4/17

4/17 The two balls are attached to the light rigid rod, which is suspended by a cord from the support above it. If the balls and rod, initially at rest, are struck by the force $F = 60$ N, calculate the corresponding acceleration $\bar{a}$ of the mass center and the rate $\ddot{\theta}$ at which the angular velocity of the bar is changing.

Ans. $\bar{a} = 20$ m/s^2, $\ddot{\theta} = 336$ rad/s^2

4/18 The small car which has a mass of 20 kg rolls freely on the horizontal track and carries the 5-kg sphere mounted on the light rotating rod with $r = 0.4$ m. A geared motor drive maintains a constant angular speed $\dot{\theta} = 4$ rad/s of the rod. If the car has a velocity $v = 0.6$ m/s when $\theta = 0$, calculate v when $\theta = 60°$. Neglect the mass of the wheels and any friction.

$\dot{\theta}$

r

θ

v

Problem 4/18

4/19 Determine the maximum value of the velocity v of the car in Prob. 4/18 and the corresponding angle θ.   *Ans.* $v = 0.920$ m/s, $\theta = 90°$

4/20 The cars of a roller-coaster ride have a speed of 30 km/h as they pass over the top of the circular track. Neglect any friction and calculate their speed v when they reach the horizontal bottom position. At the top position the radius of the circular path of their mass centers is 18 m, and all six cars have the same mass.

Ans. $v = 72.7$ km/h

Problem 4/20

4/21 The angular momentum of a system of 6 particles about a fixed point O at time $t = 4$ s is $\mathbf{H}_4 = 3.65\mathbf{i} + 4.27\mathbf{j} - 5.36\mathbf{k}$ kg·m²/s. At time $t = 4.1$ s the angular momentum is $\mathbf{H}_{4.1} = 3.67\mathbf{i} + 4.30\mathbf{j} - 5.20\mathbf{k}$ kg·m²/s. Determine the average value of the resultant moment about point O of all forces acting on all particles during the 0.1-s interval.

4/22 At a certain instant particles A, B, and C with masses of 2, 3, and 4 kg, respectively, have position vectors $\mathbf{r}_A = 2\mathbf{i} + \mathbf{j} - 2\mathbf{k}$ m, $\mathbf{r}_B = 2\mathbf{i} + 2\mathbf{j} - \mathbf{k}$ m, and $\mathbf{r}_C = \mathbf{i} - 2\mathbf{j} - 2\mathbf{k}$ m. The velocities of the particles at this same instant are $\mathbf{v}_A = \dot{\mathbf{r}}_A = 3\mathbf{i} + 2\mathbf{j} - 2\mathbf{k}$ m/s, $\mathbf{v}_B = \dot{\mathbf{r}}_B = 2\mathbf{i} + 3\mathbf{j} + \mathbf{k}$ m/s, and $\mathbf{v}_C = \dot{\mathbf{r}}_C = -2\mathbf{i} - 2\mathbf{j} + \mathbf{k}$ m/s. Calculate the angular momentum of the system of three particles about the origin O.   *Ans.* $\mathbf{H}_O = -5\mathbf{i} - 4\mathbf{j} - 16\mathbf{k}$ kg·m²/s

4/23 Each of the bars A and B has a mass of 10 kg and slides in its horizontal guideway with negligible friction. Motion is controlled by the lever of negligible mass connected to the bars as shown. Calculate the acceleration of point C on the lever when the 200-N force is applied as indicated. To verify your result analyze the kinetics of each member separately and determine a_C by kinematic considerations from the calculated accelerations of the two bars.

Problem 4/23

Problem 4/24

4/24 The 25-Mg flatcar supports a 7.5 Mg vehicle on a 5° ramp built on the flatcar. If the vehicle is released from rest with the flatcar also at rest, determine the velocity v of the flatcar when the vehicle has rolled $s = 12$ m down the ramp just before hitting the stop at B. Neglect all friction and treat the vehicle and the flatcar as particles.

Ans. $v = 1.186$ m/s

Problem 4/25

▶4/25 A horizontal bar of mass m_1 and small diameter is suspended by two wires of length l from a carriage of mass m_2, which is free to roll along the horizontal rails. If the bar and carriage are released from rest with the wires making an angle θ with the vertical, determine the velocity $v_{b/c}$ of the bar relative to the carriage and the velocity v_c of the carriage at the instant when $\theta = 0$. Neglect all friction and treat the carriage and the bar as particles in the vertical plane of motion.

Ans. $v_{b/c} = \sqrt{(1 + m_1/m_2)2gl(1 - \cos \theta)}$

$$v_c = \sqrt{\frac{2gl(1 - \cos \theta)}{(m_2/m_1)(1 + m_2/m_1)}}$$

Problem 4/26

▶4/26 A test firing of two projectiles each with a mass of 10 kg takes place from the 1-Mg vehicle which is moving with an initial velocity $v_0 = 1.2$ m/s in the direction opposite to the firing. The muzzle velocity of each projectile (relative to the barrel) is $v_r = 1200$ m/s. Calculate the velocity v' of the vehicle after the projectiles have been fired (*a*) simultaneously or (*b*) in sequence. Neglect the friction and mass of the wheels.

Ans. (*a*) $v' = 24.7$ m/s, (*b*) $v' = 24.8$ m/s

4/6 STEADY MASS FLOW

The momentum relation developed in Art. 4/4 for a general system of mass provides us with a direct means of analyzing the action of mass flow where a change of momentum occurs. The dynamics of mass flow is of great importance in the description of fluid machinery of all types including turbines, pumps, nozzles, air-breathing jet engines, and rockets. The treatment of mass flow in this article is not intended to take the place of a study of fluid mechanics but merely to present the basic principles and equations of momentum which find important usage in fluid mechanics and in the general flow of mass whether the form be liquid, gaseous, or granular.

One of the most important cases of mass flow occurs during steady-flow conditions where the rate at which mass enters a given volume equals the rate at which mass leaves the same volume. The volume in question may be enclosed by a rigid container, fixed or moving, such as the nozzle of a jet aircraft or rocket, the space between blades in a gas turbine, the volume within the casing of a centrifugal pump, or the volume within the bend of a pipe through which a fluid is flowing at a steady rate. The design of such fluid machines is dependent on the analysis of the forces and moments which are developed through the corresponding momentum changes of the flowing mass.

Consider a rigid container, shown in section in Fig. 4/5a, into which mass flows in a steady stream at the rate m' through the entrance section of area A_1. Mass leaves the container through the exit section of area A_2 at the same rate, so that there is no accumulation or depletion of the total mass within the container during the period of observation. The velocity of the entering stream is $\mathbf{v}_1$ normal to A_1 and that of the leaving stream is $\mathbf{v}_2$ normal to A_2. If ρ_1 and ρ_2 are the respective densities of the two streams, the continuity of flow requires that

$$\rho_1 A_1 v_1 = \rho_2 A_2 v_2 = m' \tag{4/17}$$

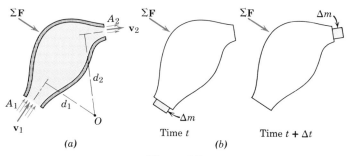

Time t Time $t + \Delta t$

(a)   (b)

Figure 4/5

The forces which act are described by isolating either the mass of fluid within the container or the entire container and the fluid within it. We would use the first approach if the forces between the container and the fluid were to be described, and we would adopt the second approach when the forces external to the container are desired. The latter situation is our primary interest in which case the *system isolated* consists of the fixed structure of the container and the fluid within it at a particular instant of time. This isolation is described by a free-body diagram of the mass within a closed space volume defined by the exterior surface of the container and the entrance and exit surfaces. We must account for all forces applied *externally* to this system, and in Fig. 4/5a the vector sum of this external force system is denoted by $\Sigma\mathbf{F}$. Included in $\Sigma\mathbf{F}$ are the forces exerted on the container at points of its attachment to other structures including attachments at A_1 and A_2, if present; the forces acting on the fluid within the container at A_1 and A_2 due to any static pressure which may exist in the fluid at these positions; and the weight of the fluid and structure if appreciable. The resultant $\Sigma\mathbf{F}$ of all of these external forces must equal $\dot{\mathbf{G}}$, the time rate of change of the linear momentum of the isolated system, by Eq. 4/6 which was developed in Art. 4/4 for any system of constant mass, rigid or nonrigid.

The expression for $\dot{\mathbf{G}}$ may be obtained by an incremental analysis. Figure 4/5b illustrates the system at time t when the system mass is that of the container, the mass within it, and an increment Δm about to enter during time Δt. At time $t + \Delta t$ the same total mass is that of the container, the mass within it, and an equal increment Δm which leaves the container in time Δt. The linear momentum of the container and mass within it between the two sections A_1 and A_2 remains unchanged during Δt so that the change in momentum of the system in time Δt is

$$\Delta\mathbf{G} = (\Delta m)\mathbf{v}_2 - (\Delta m)\mathbf{v}_1 = \Delta m(\mathbf{v}_2 - \mathbf{v}_1)$$

Division by Δt and passage to the limit yield $\dot{\mathbf{G}} = m'\,\Delta\mathbf{v}$ where

$$m' = \lim_{\Delta t \to 0}\left(\frac{\Delta m}{\Delta t}\right) = \frac{dm}{dt}$$

Thus by Eq. 4/6

$$\boxed{\Sigma\mathbf{F} = m'\,\Delta\mathbf{v}} \qquad\qquad (4/18)$$

Equation 4/18 establishes the relation between the resultant force on a steady-flow system and the corresponding mass flow rate and vector velocity increment.*

---

* We must be careful not to interpret dm/dt as the time derivative of the mass of the isolated system. That derivative is zero since the system mass is constant for a steady-flow process. To help avoid confusion, the symbol m' rather than dm/dt is used to represent the steady mass flow rate.

Alternatively we may note that the time rate of change of linear momentum is the vector difference between the rate at which linear momentum leaves the system and the rate at which linear momentum enters the system. Thus we may write $\dot{\mathbf{G}} = m'\mathbf{v}_2 - m'\mathbf{v}_1 = m'\,\Delta\mathbf{v}$, which agrees with the foregoing result.

We can now see one of the powerful applications of our general force-momentum equation which we derived for any mass system. Our system here includes particles which are rigid (the structural container for the mass stream) and particles which are in motion (the flow of mass). By defining the boundary of the system, the mass within which is constant for steady flow conditions, we are able to utilize the generality of Eq. 4/6. However, we must be very careful to account for *all* external forces acting on the system, and they become clear if our free-body diagram is correct.

A similar formulation is obtained for the case of angular momentum in steady-flow systems. The resultant moment of all external forces about some fixed point O on or off the system, Fig. 4/5a, equals the time rate of change of angular momentum of the system about O. This fact was established in Eq. 4/7 which, for the case of steady flow in a single plane, becomes

$$\Sigma M_O = m'(v_2 d_2 - v_1 d_1) \qquad (4/19)$$

When the velocities of the incoming and outgoing flows are not in the same plane, the equation may be written in vector form as

$$\boxed{\Sigma \mathbf{M}_O = m'(\mathbf{d}_2 \times \mathbf{v}_2 - \mathbf{d}_1 \times \mathbf{v}_1)} \qquad (4/19a)$$

where $\mathbf{d}_1$ and $\mathbf{d}_2$ are the position vectors to the centers of A_1 and A_2 from the fixed reference O. In both relations the mass center G may be used alternatively as a moment center by virtue of Eq. 4/9.

Equations 4/18 and 4/19 are very simple relations which find important use in describing relatively complex fluid actions. It should be noted that these equations relate *external* forces to the resultant changes in momentum and are independent of the flow path and momentum changes *internal* to the system.

The foregoing analysis may be applied to systems moving with constant velocity by noting that the basic relations $\Sigma\mathbf{F} = \dot{\mathbf{G}}$ and $\Sigma\mathbf{M}_O = \dot{\mathbf{H}}_O$ or $\Sigma\mathbf{M}_G = \dot{\mathbf{H}}_G$ apply to systems moving with constant velocity as discussed in Arts. 3/12 and 4/4. The only restriction is that the mass within the system remain constant with respect to time.

Three examples of the analysis of steady mass flow are given in the following sample problems which illustrate the application of the principles embodied in Eqs. 4/18 and 4/19.

Sample Problem   4/5

The smooth vane shown diverts the open stream of fluid of cross-sectional area A, mass density ρ, and velocity v. (*a*) Determine the force components R and F required to hold the vane in a fixed position. (*b*) Find the forces when the vane is given a constant velocity u less than v and in the direction of v.

Fixed vane

Solution.   Part(a).   The free-body diagram of the vane together with the fluid portion undergoing the momentum change is shown. The momentum equation may be applied to the isolated system for the change in motion in both the x- and y-directions. With the vane stationary the magnitude of the exit velocity v' equals that of the entering velocity v with fluid friction neglected. The changes in the velocity components are, then,

Moving vane

① $$\Delta v_x = v' \cos \theta - v = -v(1 - \cos \theta)$$

and

$$\Delta v_y = v' \sin \theta - 0 = v \sin \theta$$

The mass rate of flow is $m' = \rho A v$, and substitution into Eq. 4/18 gives

$[\Sigma F_x = m' \, \Delta v_x]$ $-F = \rho A v[-v(1 - \cos \theta)]$

$\qquad\qquad\qquad F = \rho A v^2 (1 - \cos \theta)$ *Ans.*

$[\Sigma F_y = m' \, \Delta v_y]$ $R = \rho A v[v \sin \theta]$

$\qquad\qquad\qquad R = \rho A v^2 \sin \theta$ *Ans.*

Part (b).   In the case of the moving vane the final velocity v' of the fluid upon exit is the vector sum of the velocity u of the vane plus the velocity of the fluid relative to the vane $v - u$. This combination is shown in the velocity diagram to the right of the figure for the exit conditions. The x-component of v' is the sum of the components of its two parts, so $v_x' = (v - u) \cos \theta + u$. The change in x-velocity of the stream is

$$\Delta v_x = (v - u) \cos \theta + (u - v) = -(v - u)(1 - \cos \theta)$$

The y-component of v' is $(v - u) \sin \theta$, so that the change in the y-velocity of the stream is $\Delta v_y = (v - u) \sin \theta$.

The mass rate of flow m' is the mass undergoing momentum change per unit of time. This rate is the mass flowing over the vane per unit time and *not* the rate of issuance from the nozzle. Thus

$$m' = \rho A(v - u)$$

The impulse-momentum principle of Eq. 4/18 applied in the positive coordinate direction gives

② $[\Sigma F_x = m' \, \Delta v_x]$ $-F = \rho A(v - u)[-(v - u)(1 - \cos \theta)]$

$\qquad\qquad\qquad F = \rho A(v - u)^2 (1 - \cos \theta)$ *Ans.*

$[\Sigma F_y = m' \, \Delta v_y]$ $R = \rho A(v - u)^2 \sin \theta$ *Ans.*

① Be careful with algebraic signs when using Eq. 4/18. The change in v_x is the final value minus the initial value measured in the positive x-direction. Also we must be careful to write $-F$ for ΣF_x.

② Observe that for given values of u and v the angle for maximum force F is $\theta = 180°$.

Sample Problem 4/6

For the moving vane of Sample Problem 4/5 determine the optimum speed u of the vane for the generation of maximum power by the action of the fluid on the vane.

Solution. The force R shown with the figure for Sample Problem 4/5 is normal to the velocity of the vane so does no work. The work done by the force F shown is negative, but the power developed by the force (reaction to F) exerted by the fluid on the moving vane is

$$[P = Fu] \qquad P = \rho A (v - u)^2 u (1 - \cos \theta)$$

The velocity of the vane for maximum power for the one blade in the stream is specified by

$$\left[\frac{dP}{du} = 0 \right] \qquad \rho A (1 - \cos \theta)(v^2 - 4uv + 3u^2) = 0$$

①
$$(v - 3u)(v - u) = 0 \qquad u = \frac{v}{3} \qquad \textit{Ans.}$$

The second solution $u = v$ gives a minimum condition of zero power. An angle $\theta = 180°$ completely reverses the flow and clearly produces both maximum force and maximum power for any value of u.

① The result here applies to a single vane only. In the case of multiple vanes, such as the blades on a turbine disk, the rate at which fluid issues from the nozzles is the same rate at which fluid is undergoing momentum change. Thus $m' = \rho A v$ rather than $\rho A (v - u)$. With this change the optimum value of u turns out to be $u = v/2$.

Sample Problem 4/7

The offset nozzle has a discharge area A at B and an inlet area A_0 at C. A liquid enters the nozzle at a static gage pressure p through the fixed pipe and issues from the nozzle with a velocity v in the direction shown. If the constant density of the liquid is ρ, write expressions for the tension T, shear Q, and bending moment M in the pipe at C.

Solution. The free-body diagram of the nozzle and the fluid within it shows the tension T, shear Q, and bending moment M acting on the flange of the nozzle where it attaches to the fixed pipe. The force pA_0 on the fluid within the nozzle due to the static pressure is an additional external force.

Continuity of flow with constant density requires that

$$Av = A_0 v_0$$

where v_0 is the velocity of the fluid at entrance to the nozzle. The momentum principle of Eq. 4/18 applied to the system in the two coordinate directions gives

$$[\Sigma F_x = m' \, \Delta v_x] \qquad pA_0 - T = \rho A v (v \cos \theta - v_0)$$

①
$$T = pA_0 + \rho A v^2 \left(\frac{A}{A_0} - \cos \theta \right) \qquad \textit{Ans.}$$

$$[\Sigma F_y = m' \, \Delta v_y] \qquad -Q = \rho A v (-v \sin \theta - 0)$$

$$Q = \rho A v^2 \sin \theta \qquad \textit{Ans.}$$

The moment principle of Eq. 4/19 applied in the clockwise sense gives

②
$$[\Sigma M_O = m'(v_2 d_2 - v_1 d_1)] \qquad M = \rho A v (va \cos \theta + vb \sin \theta - 0)$$

$$M = \rho A v^2 (a \cos \theta + b \sin \theta) \qquad \textit{Ans.}$$

① Again, be careful to observe the correct algebraic signs of the terms on both sides of Eqs. 4/18 and 4/19.

② The forces and moment acting on the pipe are equal and opposite to those shown acting on the nozzle.

Sample Problem 4/8

An air-breathing jet aircraft of total mass m flying with a constant speed v consumes air at the mass rate m_a' and exhausts burned gas at the mass rate m_g' with a velocity u relative to the aircraft. Fuel is consumed at the constant rate m_f'. The total aerodynamic forces acting on the aircraft are the lift L, normal to the direction of flight, and the drag D opposite to the direction of flight. Any force due to the static pressure across the inlet and exhaust surfaces is assumed to be included in D. Write the equation for the motion of the aircraft and identify the thrust T.

Solution. The free-body diagram of the aircraft together with the air, fuel, and exhaust gas within it is given and shows only the weight, lift, and drag forces as defined. We attach axes x-y to the aircraft and apply our momentum equation relative to the moving system.

The fuel will be treated as a steady stream entering the aircraft with no velocity relative to the system and leaving with a relative velocity u in the exhaust stream. We now apply Eq. 4/18 relative to the reference axes and treat the air and fuel flows separately. For the air flow the change in velocity in the x-direction relative to the moving system is

$$\Delta v_a = -u - (-v) = -(u - v)$$

and for the fuel flow the x-change in velocity relative to x-y is

$$\Delta v_f = -u - (0) = -u$$

Thus we have

$$[\Sigma F_x = m' \, \Delta v_x] \qquad -mg \sin \theta - D = -m_a'(u - v) - m_f'u$$
$$= -m_g'u + m_a'v$$

where the substitution $m_g' = m_a' + m_f'$ has been made. Changing signs gives

$$m_g'u - m_a'v = mg \sin \theta + D$$

which is simply an equation of equilibrium of the aircraft.

If we modify the boundaries of our system to expose the interior surfaces upon which the air and gas act, we will have the simulated model shown where the air exerts a force $m_a'v$ on the interior of the turbine and the exhaust gas reacts against the interior surfaces with the force $m_g'u$.

The commonly used model is shown in the final diagram where the net effect of air and exhaust momentum changes is replaced by a simulated thrust

$$T = m_g'u - m_a'v \qquad \qquad Ans.$$

applied to the aircraft from a presumed external source.

Inasmuch as m_f' is generally only 2% or less of m_a', we can use the approximation $m_g' \cong m_a'$ and express the thrust as

$$T \cong m_a'(u - v) \qquad \qquad Ans.$$

We have analyzed the case of constant velocity. Although our Newtonian principles do not generally hold relative to accelerating axes, it can be shown that we may use the $F = ma$ equation for the simulated model and write $T - mg \sin \theta - D = m\dot{v}$ with virtually no error.

① Note that the boundary of the system cuts across the air stream at the entrance to the air scoop and across the exhaust stream at the nozzle.

② We are permitted to use moving axes which translate with constant velocity. See Arts. 3/13 and 4/2.

③ Riding with the aircraft we observe the air entering our system with a velocity $-v$ measured in the plus x-direction and leaving the system with an x-velocity of $-u$. The final minus the initial values give the expression cited, namely, $-u - (-v) = -(u - v)$.

④ We now see that the "thrust" is, in reality, not a force external to the entire airplane shown in the first figure but can be modeled as an external force.

PROBLEMS

4/27 The Saturn V rocket has a total mass of approximately $2.7(10^3)$ Mg. At launch fuel is burned at the rate of 13 Mg/s with an exhaust velocity of 2400 m/s. Calculate the initial acceleration of the rocket. Assume atmospheric pressure across the exit plane of the exhaust nozzle.

Ans. $a = 1.746$ m/s^2

4/28 An experimental aircraft designed for vertical take-off and landing (VTOL) has a mass of 7 Mg. Movable vanes deflect the exhaust stream downward at takeoff. Each of the craft's two engines sucks in air at the rate of 60 kg/s at a density of 1.217 kg/m^3 and discharges the exhaust at a velocity of 600 m/s. Each engine uses fuel at the rate of 1 kg/s. Determine the initial vertical acceleration a.   *Ans.* $a = 0.647$ m/s^2

Problem 4/28

4/29 When the rocket reaches the position in its trajectory shown, it has a mass of 3 Mg and is beyond the effect of the earth's atmosphere. Gravitational acceleration is 9.60 m/s^2. Fuel is being consumed at the rate of 130 kg/s, and the exhaust velocity relative to the nozzle is 600 m/s. Compute the n- and t-components of acceleration of the rocket.

Ans. $a_n = 8.31$ m/s^2, $a_t = 21.2$ m/s^2

Problem 4/29

4/30 The figure shows a 90° bend in a closed-loop wind tunnel of 3.6-m diameter. Fixed curved vanes smooth out the change in direction of the flow. What forces F_x and F_y are applied to the bend by its supports to change the direction of the flow for operation of the tunnel at an air speed of 250 km/h and with a density of 1.206 kg/m^3 at atmospheric pressure? Neglect the small difference in pressure between sections A and B.

Problem 4/30

Problem 4/31

Problem 4/32

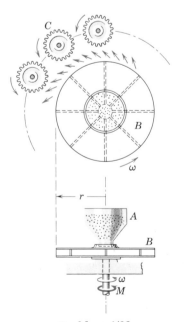

Problem 4/33

4/31 A rotary snow plow mounted on a large truck eats its way through a snow drift on a level road at a constant speed of 20 km/h. The plow discharges 60 Mg of snow per minute from its 45° chute with a velocity of 12 m/s relative to the plow. Calculate the tractive force *P* on the tires in the direction of motion necessary to move the plow and find the corresponding lateral force *R* between the tires and the road. *Ans. P* = 5.56 kN, *R* = 8.49 kN

4/32 Fresh water issues from the nozzle with a velocity of 30 m/s at the rate of 0.05 m³/s and is split into two equal streams by the fixed vane and deflected through 60° as shown. Calculate the force *F* required to hold the vane in place. The density of water is 1000 kg/m³.

4/33 Shot peening is used to increase the fatigue strength of gear teeth and other machine parts. The device shown consists of a reservoir *A* which feeds the shot pellets into the center of a rotating impeller *B* that flings them at high speed and produces a spray of shot. The spray of high-velocity pellets impinges against the rotating gears *C* that are being treated. If *m* kg of shot are consumed per second and if the angular velocity is ω, determine the torque *M* required to drive the impeller.

4/34 A 1500-kg rocket sled is propelled along its horizontal track by four rocket motors each providing a forward thrust of 8 kN. Frictional resistance to motion along the track is 10 percent of the total weight. The exhaust velocity through each nozzle is 480 m/s. Calculate the initial acceleration of the sled and the rate m' at which each motor uses fuel in kg/s.   *Ans. $a = 20.4$ m/s^2, $m' = 16.67$ kg/s*

4/35 A jet-engine noise suppressor consists of a movable duct which is secured directly behind the jet exhaust by cable A and deflects the blast directly upward. During ground test the engine sucks in air at the rate of 43 kg/s and burns fuel at the rate of 0.8 kg/s. The exhaust velocity is 720 m/s. Determine the tension T in the cable.

Problem 4/35

4/36 The fire tug discharges a stream of salt water (density 1030 kg/m^3) with a nozzle velocity of 40 m/s at the rate of 0.080 m^3/s. Calculate the propeller thrust T which must be developed by the tug to maintain a fixed position while pumping.
       Ans. $T = 2.85$ kN

Problem 4/36

4/37 The jet aircraft has a mass of 4.6 Mg and has a drag (air resistance) of 32 kN at a speed of 1000 km/h at a particular altitude. The aircraft consumes air at the rate of 106 kg/s through its intake scoop and uses fuel at the rate of 0.98 kg/s. If the exhaust has a rearward velocity of 680 m/s relative to the exhaust nozzle, determine the maximum angle of elevation α at which the jet can fly with a constant speed of 1000 km/h at the particular altitude in question.

Problem 4/37

4/38 The figure shows the top view of an experimental rocket sled which is traveling at a speed of 300 m/s when its forward scoop enters a water channel to act as a brake. The water is diverted at right angles relative to the sled. If the frontal flow area of the scoop is 10^{-2} m^2, calculate the initial braking force. The density of water is 1000 kg/m^3.   *Ans. $F = 900$ kN*

Problem 4/38

Problem 4/39

4/39 The experimental ground-effect machine has a total mass of 2.2 Mg. It hovers close to the ground by pumping air at atmospheric pressure through the circular intake duct at *B* and discharging it horizontally under the periphery of the skirt *C*. For an intake velocity *v* of 45 m/s calculate the average air pressure *p* under the 6-m-diameter machine at ground level. The density of the air is 1.206 kg/m³.   *Ans.* $p = 1.035$ kPa

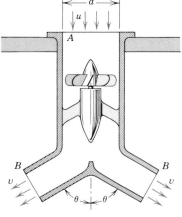

Problem 4/40

4/40 The ducted fan unit of mass *m* is supported in the vertical position on its flange at *A*. The unit draws in air with a density ρ and a velocity *u* through section *A* and discharges it through section *B* with a velocity *v*. Both inlet and outlet pressures are atmospheric. Write an expression for the force *R* applied to the flange of the fan unit by the supporting slab.

4/41 Find the force *F* required to move the vane of Prob. 4/32 to the left with a velocity of 30 m/s against the stream. What power *P* is required?
                Ans. $F = 3$ kN, $P = 90$ kW

Problem 4/42

4/42 The pipe bend shown has a cross-sectional area *A* and is supported in its plane by the tension *T* applied to its flanges by the adjacent connecting pipes (not shown). If the velocity of the liquid is *v*, its density ρ, and its static pressure *p*, determine *T* and show that it is independent of the angle θ.

4/43 In a test of the operation of a "cherry picker" fire truck the equipment is free to roll with its brakes released. For the position shown the truck is observed to deflect the spring of stiffness $k = 15$ kN/m a distance of 150 mm because of the action of the horizontal stream of water issuing from the nozzle when the pump is activated. If the exit diameter of the nozzle is 30 mm, calculate the velocity v of the stream as it leaves the nozzle. Also determine the added moment M which the joint at A must resist when the pump is in operation with the nozzle in the position shown.

Ans. $v = 56.4$ m/s, $M = 29.8$ kN·m

Problem 4/43

4/44 A marine terminal for unloading bulk wheat from a ship is equipped with a vertical pipe with a nozzle at A which sucks wheat up the pipe and transfers it to the storage building. Calculate the x- and y-components of the force $\mathbf{R}$ required to change the momentum of the flowing mass in rounding the bend. Identify all forces applied externally to the bend and mass within it. Air flows through the 350-mm-diameter pipe at the rate of 16 Mg per hour under a vacuum of 230 mm of mercury ($p = -30.7$ kPa) and carries with it 135 Mg of wheat per hour at a speed of 40 m/s.

Problem 4/44

4/45 The axial-flow fan C pumps air through the duct of circular cross section and exhausts it with a velocity v at B. The air densities at A and B are ρ_A and ρ_B, respectively, and the corresponding pressures are p_A and p_B. The fixed deflecting blades at D restore axial flow to the air after it passes through the propeller blades C. Write an expression for the resultant horizontal force R exerted on the fan unit by the flange and bolts at A.

Ans. $R = \dfrac{\pi d^2}{4}\left[\rho_B\left(1 - \dfrac{\rho_B}{\rho_A}\right)v^2 + (p_B - p_A)\right]$

Problem 4/45

4/46 For a jet acting on multiple vanes where each vane which enters the jet is followed immediately by another, as in a turbine or water wheel (see figures for Probs. 4/59 and 4/60), determine the maximum power *P* which can be developed for a given blade angle and the corresponding optimum peripheral speed *u* of the vanes in terms of the jet velocity *v* for maximum power. Modify Sample Problem 4/6 by assuming an infinite number of vanes so that the rate at which fluid leaves the nozzle equals the rate at which fluid passes over the vanes.

Ans. $P = \frac{1}{4}\rho A v^3 (1 - \cos\theta)$, $u = v/2$

Problem 4/47

4/47 Air enters the pipe at *A* at the rate of 6 kg/s under a pressure of 1400 kPa gage and leaves the whistle at atmospheric pressure through the opening at *B*. The entering velocity of the air at *A* is 45 m/s, and the exhaust velocity at *B* is 360 m/s. Calculate the tension *T*, shear *V*, and bending moment *M* in the pipe at *A*. The net flow area at *A* is 7500 mm².

Ans. $V = 1.871$ kN

$T = 9.69$ kN

$M = 1.122$ kN·m

Problem 4/48

4/48 In the static test of a jet engine and exhaust nozzle assembly, air is sucked into the engine at the rate of 30 kg/s and fuel is burned at the rate of 0.6 kg/s. The flow area, static pressure, and axial-flow velocity for the three sections shown are as follows:

	Sec. A	Sec. B	Sec. C
Flow area, m²	0.15	0.16	0.06
Static pressure, kPa	−14	140	14
Axial-flow velocity, m/s	120	315	600

Determine the tension *T* in the diagonal member of the supporting test stand and calculate the force *F* exerted on the nozzle flange at *B* by the bolts and gasket to hold the nozzle to the engine housing.

Problem 4/49

4/49 The open pipe bend *AB* is welded to the flange which is supported by the pin at *O* and the link at *C*. A stream of fresh water (density 1000 kg/m³) is directed into the pipe at *A* with a velocity of 24 m/s and emerges at *B* with essentially the same speed. If the flow rate is 0.20 m³/s, calculate the force *C* in the link. *Ans.* $C = 3.35$ kN

4/50 Air is pumped through the stationary duct A with a velocity of 15 m/s and exhausted through an experimental nozzle section BC. The average static pressure across section B is 1050 kPa gage, and air density at this pressure and at the temperature prevailing is 13.5 kg/m³. The average static pressure across the exit section C is measured to be 14 kPa gage, and the corresponding density of air is 1.217 kg/m³. Calculate the force T exerted on the nozzle flange at B by the bolts and the gasket to hold the nozzle in place.

Problem 4/50

4/51 The curved vane moves to the right with a velocity of 10 m/s in the direction of the stream of fresh water, which issues from the 25-mm-diameter nozzle with a velocity of 24 m/s and is diverted by the vane. Calculate the forces F_x and F_y on the vane required to support the motion. Friction along the vane is negligible.

Ans. $F_x = 179.5$ N, $F_y = 48.1$ N

Problem 4/51

4/52 The helicopter shown has a mass m and hovers in position by imparting downward momentum to a column of air defined by the slip-stream boundary shown. Find the downward velocity v given to the air by the rotor at a section in the stream below the rotor where the pressure is atmospheric and the stream radius is r. Also find the power P required of the engine. Neglect the rotational energy of the air, any temperature rise due to air friction, and any change in air density ρ.

Problem 4/52

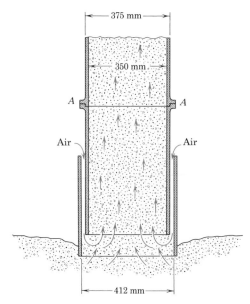

375 mm

350 mm

A   A

Air   Air

412 mm

Problem 4/53

4/53 An axial section of the suction nozzle *A* for the bulk wheat unloader described in Prob. 4/44 is shown here. The outer pipe is secured to the inner pipe by several longitudinal webs which do not restrict the flow of air. A vacuum of 230 mm of mercury ($p = -30.7$ kPa gage) is maintained in the inner pipe, and the pressure across the bottom of the outer pipe is atmospheric ($p = 0$). Air at 1.206 kg/m^3 is drawn in through the space between the pipes at a rate of 16 Mg per hour at atmospheric pressure and draws with it 135 Mg of wheat per hour up the pipe at a velocity of 40 m/s. If the nozzle unit below section *A-A* has a mass of 30 kg, calculate the compression *C* in the connection at *A-A*.   *Ans. C = 265 N*

250 km/h

Thrust
reverser

20°

Problem 4/54

4/54 Reverse thrust to decelerate a jet aircraft during landing is obtained by swinging the exhaust deflectors into place as shown, thus causing a partial reversal of the exhaust stream. For a jet engine which consumes 40 kg of air per second at a ground speed of 250 km/h and which uses 0.6 kg of fuel per second, determine the reverse thrust as a fraction *n* of the forward thrust with deflectors removed. The exhaust velocity relative to the nozzle is 600 m/s. Assume air enters the engine with a relative velocity equal to the ground speed of the aircraft.

4/55 A ball of mass *m* is supported in a vertical jet of water with density ρ. If the stream of water issuing from the nozzle has a diameter *d* and velocity *u*, determine the height *h* above the nozzle at which the ball is supported. Assume that the jet remains intact and that there is no energy loss in the jet.

$$Ans.\ h = \frac{u^2}{2g} - 8g\left(\frac{m}{\pi\rho u d^2}\right)^2$$

h

Problem 4/55

4/56 The fire hydrant is tested under a high standpipe pressure of 800 kPa. The total flow rate is 0.280 m³/s. The flow is divided equally between the two outlets each of which has a cross-sectional area of 3800 mm². The inlet cross-sectional area at the base is $6.80(10^4)$ mm². Neglect the weight of the hydrant and water within it and compute the tension T, the shear V, and the bending moment M in the base of the standpipe at B. The density of water is 1000 kg/m³. The static pressure of the water as it enters the base at B is 800 kPa.
Ans. $T = 53.0$ kN, $V = 0.691$ kN, $M = 2.48$ kN·m

Problem 4/56

4/57 The sprinkler is made to rotate at the constant angular velocity ω and distributes water at the volume rate Q. Each of the four nozzles has an exit area A. Write an expression for the torque M on the shaft of the sprinkler necessary to maintain the given motion. For a given pressure and, hence, flow rate Q, at what speed ω_0 will the sprinkler operate with no applied torque? Let ρ be the density of the water.

$$Ans. \; M = \rho Q\left(\frac{Qr}{4A} - [r^2 + b^2]\omega\right)$$

$$\omega_0 = \frac{Qr}{4A(r^2 + b^2)}$$

Problem 4/57

4/58 The centrifugal pump handles 20 m³ of fresh water per minute with inlet and outlet velocities of 18 m/s. The impeller is turned clockwise through the shaft at O by a motor which delivers 40 kW at a pump speed of 900 rev/min. With the pump filled but not turning, the vertical reactions at C and D are each 250 N. Calculate the forces exerted by the foundation on the pump at C and D while the pump is running. The tensions in the connecting pipes at A and B are exactly balanced by the respective forces due to the static pressure in the water. (*Suggestion:* Isolate the entire pump and water within it between sections A and B and apply the momentum principle to the entire system.)
Ans. $C = 4340$ N up, $D = 3840$ N down

Problem 4/58

Problem 4/59

▶ **4/59** In the figure is shown a detail of the stationary nozzle diaphragm *A* and the rotating blades *B* of a gas turbine. The products of combustion pass through the fixed diaphragm blades at the 27° angle and impinge on the moving rotor blades. The angles shown are selected so that the velocity of the gas relative to the moving blade at entrance is at the 20° angle for minimum turbulence, corresponding to a mean blade velocity of 315 m/s at a radius of 375 mm. If gas flows past the blades at the rate of 15 kg/s, determine the theoretical power output *P* of the turbine. Neglect fluid and mechanical friction with the resulting heat-energy loss, and assume that all the gases are deflected along the surfaces of the blades with a velocity of constant magnitude relative to the blade.

Ans. P = 1.197 MW

Problem 4/60

▶ **4/60** In the figure is shown an impulse-turbine wheel for a hydroelectric power plant which is to operate with a static head of water of 300 m at each of its 6 nozzles and is to rotate at the speed of 270 rev/min. Each wheel and generator unit is to develop an output power of 22 000 kW. The efficiency of the generator may be taken to be 0.90, and an efficiency of 0.85 for the conversion of the kinetic energy of the water jets to energy delivered by the turbine may be expected. The mean peripheral speed of such a wheel for greatest efficiency will be about 0.47 times the jet velocity. If each of the buckets is to have the shape shown, determine the necessary jet diameter *d* and wheel diameter *D*. Assume that the water acts on the bucket which is at the tangent point of each jet stream.

Ans. d = 165.3 mm, D = 2.55 m

4/7 VARIABLE MASS

In Art. 4/4 we extended the equations for the motion of a particle to include a system of particles. This extension led to the very general expressions $\Sigma\mathbf{F} = \dot{\mathbf{G}}$, $\Sigma\mathbf{M}_O = \dot{\mathbf{H}}_O$, and $\Sigma\mathbf{M}_G = \dot{\mathbf{H}}_G$, which are Eqs. 4/6, 4/7, and 4/9, respectively. In the derivation of these equations the summations were taken over a fixed number of particles, so that the mass of the system to be analyzed was constant. In Art. 4/6 these momentum principles were extended in Eqs. 4/18 and 4/19 to describe the action of forces on a system defined by a geometric volume through which passes a steady flow of mass. The amount of mass within the control volume was, therefore, constant with respect to time and we were, thus, permitted to use Eqs. 4/6, 4/7, and 4/9. When the mass within the boundary of a system under consideration is not constant, the foregoing relationships are no longer valid.*

We will now develop the equation for the linear motion of a system whose mass varies with time. Consider, first, a body which gains mass by virtue of its overtaking and swallowing a stream of matter, Fig. 4/6*a*. The mass of the body and its velocity at any instant are m and v, respectively. The stream of matter is assumed to be moving in the same direction as m with a constant velocity v_0 less than v. By virtue of Eq. 4/18 the force exerted by m on the particles of the stream to accelerate them from a velocity v_0 to a greater velocity v is $R = m'(v - v_0) = \dot{m}u$ where the time rate of increase of m is $m' = \dot{m}$ and where u is the relative velocity with which the particles approach m. In addition to R, all other forces acting on m in the direction of its motion are designated

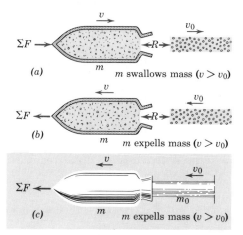

Figure 4/6

---

* In relativistic mechanics the mass is found to be a function of velocity, and its time derivative has a meaning different from that in Newtonian mechanics.

by ΣF. The equation of motion of m from Newton's second law is, therefore, $\Sigma F - R = m\dot{v}$ or

$$\boxed{\Sigma F = m\dot{v} + \dot{m}u} \qquad (4/20)$$

Similarly, if the body loses mass by expelling it at a velocity v_0 less than v, Fig. 4/6b, the force R required to decelerate the particles from a velocity v to a lesser velocity v_0 is $R = m'(-v_0 - [-v]) = m'(v - v_0)$. But $m' = -\dot{m}$ since m is decreasing. Also the relative velocity with which the particles leave m is $u = v - v_0$. Thus the force R becomes $R = -\dot{m}u$. If ΣF denotes the resultant of all other forces acting on m in the direction of its motion, Newton's second law requires $\Sigma F + R = m\dot{v}$ or

$$\Sigma F = m\dot{v} + \dot{m}u$$

which is the same relationship as in the case where m is gaining mass. We may use Eq. 4/20, therefore, as the equation of motion of m whether it is gaining or losing mass.

A frequent error in the use of the force-momentum equation is to express the partial force sum ΣF as

$$\Sigma F = \frac{d}{dt}(mv) = m\dot{v} + \dot{m}v$$

From this expansion we see that the direct differentiation of the linear momentum gives the correct force ΣF *only* when the body picks up mass initially at rest or when it expels mass at a zero absolute velocity. In both instances $v_0 = 0$ and $u = v$.

We may also obtain Eq. 4/20 by a direct differentiation of the momentum from the basic relation $\Sigma F = \dot{G}$ provided a proper system of constant total mass is chosen. To illustrate this approach, we take the case where m is losing mass and use Fig. 4/6c which shows the system of m and an arbitrary portion m_0 of the stream of ejected mass. The mass of this system is $m + m_0$ and is constant. The ejected stream of mass is assumed to move undisturbed once separated from m, and the only force external to the entire system is ΣF which is applied directly to m as before. The reaction $R = -\dot{m}u$ is internal to the system and is not disclosed as an external force on the system. With constant total mass the momentum principle $\Sigma F = \dot{G}$ is applicable and we have

$$\Sigma F = \frac{d}{dt}(mv + m_0v_0) = m\dot{v} + \dot{m}v + \dot{m}_0v_0 + m_0\dot{v}_0$$

Clearly $\dot{m}_0 = -\dot{m}$, and the velocity of the ejected mass with respect to m is $u = v - v_0$. Also $\dot{v}_0 = 0$ since m_0 moves undisturbed with no acceleration once free of m. Thus the relation becomes

$$\Sigma F = m\dot{v} + \dot{m}u$$

which is identical with the result of the previous formulation, Eq. 4/20.

The case of m losing mass is clearly descriptive of rocket propulsion. Figure 4/7a shows a vertically ascending rocket, the system for which is the mass within the volume defined by the exterior surface of the rocket and the exit plane across the nozzle. External to this system the free-body diagram discloses the instantaneous values of gravitational attraction mg, aerodynamic resistance R, and the force pA due to the average static pressure p across the nozzle exit plane of area A. Also the rate of mass flow is $m' = -\dot{m}$. Thus we may write the equation of motion of the rocket $\Sigma F = m\dot{v} + \dot{m}u$ as $pA - mg - R = m\dot{v} + \dot{m}u$ or

$$m'u + pA - mg - R = m\dot{v} \qquad (4/21)$$

Equation 4/21 is of the form "$\Sigma F = ma$" where the first term in "ΣF" is the thrust $T = m'u$. Thus the rocket may be simulated as a body to which an external thrust T is applied, Fig. 4/7b, and the problem may then be analyzed like any other "$F = ma$" problem except that m is a function of time.

It may be observed that, during the initial stages of motion when the magnitude of the velocity v of the rocket is less than the relative exhaust velocity u, the absolute velocity v_0 of the exhaust gases will be directed rearward. On the other hand, when the rocket reaches a velocity v whose magnitude is greater than u, the absolute velocity v_0 of the exhaust gases will be directed forward. For a given mass rate of flow the rocket thrust T depends only on the relative exhaust velocity u and not on the magnitude or on the direction of the absolute velocity v_0 of the exhaust gases.

In the foregoing treatment of bodies whose mass changes with time we have assumed that all elements of the mass m of the body were moving with the same velocity v at any instant of time and that the particles of mass which were added to or expelled from the body underwent an abrupt transition of velocity upon entering or leaving the body. Thus this velocity change has been modeled as a mathematical discontinuity. In reality this change in velocity cannot be discontinuous even though the transition may be rapid. In the case of a rocket, for example, the velocity change occurs continuously in the space between the combustion zone and the exit plane of the exhaust nozzle. A more general analysis* of variable-mass dynamics removes this restriction of discontinuous velocity change and introduces a slight correction to Eq. 4/20.

Actual system
(a)

Simulated system
(b)

Figure 4/7

---

* For a development of the equations which describe the general motion of a time-dependent system of mass see Art. 53 of the senior author's *Dynamics, 2nd Edition SI Version* 1975, John Wiley & Sons.

Sample Problem   4/9

The end of a chain of length L and mass ρ per unit length which is piled on a platform is lifted vertically with a constant velocity v by a variable force P. Find P as a function of the height x of the end above the platform. Also find the energy lost during the lifting of the chain.

Solution I (variable-mass approach).   Equation 4/20 will be used and applied to the moving part of the chain of length x which is gaining mass. The force summation ΣF includes all forces acting on the moving part except the force exerted by the particles which are being attached. From the diagram we have

$$\Sigma F_x = P - \rho g x$$

The velocity is constant so that $\dot{v} = 0$. The rate of increase of mass is $\dot{m} = \rho v$, and the relative velocity with which the attaching particles approach the moving part is $u = v - 0 = v$. Thus Eq. 4/20 becomes

$$[\Sigma F = m\dot{v} + \dot{m}u] \qquad P - \rho g x = 0 + \rho v(v) \qquad P = \rho(gx + v^2) \qquad Ans.$$

Thus we see that the force P consists of the two parts, $\rho g x$, which is the weight of the moving part of the chain, and ρv^2, which is the added force required to change the momentum of the links on the platform from a condition at rest to a velocity v.

Solution II (constant-mass approach).   The principle of impulse and momentum for a system of particles expressed by Eq. 4/6 will be applied to the entire chain considered as the system of constant mass. The free-body diagram of the system shows the unknown force P, the total weight of all links $\rho g L$, and the force $\rho g(L - x)$ exerted by the platform on those links which are at rest upon it. The momentum of the system at any position is $G_x = \rho x v$ and the momentum equation gives

$$\left[\Sigma F_x = \frac{dG_x}{dt}\right] \qquad P + \rho g(L - x) - \rho g L = \frac{d}{dt}(\rho x v) \qquad P = \rho(gx + v^2)$$

$$Ans.$$

Again the force P is seen to be equal to the weight of the portion of the chain which is off the platform plus the added term which accounts for the time rate of increase of momentum of the chain.

Energy loss.   Each link on the platform acquires its velocity abruptly through an impact with the link above it which lifts it off the platform. The succession of impacts gives rise to an energy loss ΔE. The work-energy equation may be written $U = \Delta T + \Delta V_g + \Delta E$.

$$U = \int P\,dx = \int_0^L (\rho g x + \rho v^2)\,dx = \tfrac{1}{2}\rho g L^2 + \rho v^2 L$$

$$\Delta T = \tfrac{1}{2}\rho L v^2 \qquad \Delta V_g = \rho g L \frac{L}{2} = \tfrac{1}{2}\rho g L^2$$

Substituting into the work-energy equation gives

$$\tfrac{1}{2}\rho g L^2 + \rho v^2 L = \tfrac{1}{2}\rho L v^2 + \tfrac{1}{2}\rho g L^2 + \Delta E, \qquad \Delta E = \tfrac{1}{2}\rho L v^2 \qquad Ans.$$

① The model of Fig. 4/6a shows the mass being added to the leading end of the moving part. With the chain the mass is added to the trailing end, but the effect is the same.

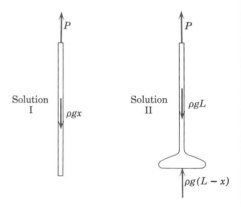

② We must be very careful not to use $\Sigma F = \dot{G}$ for a system whose mass is changing. Thus we have taken the total chain as the system since its mass is constant.

Sample Problem 4/10

Replace the open-link chain of Sample Problem 4/9 by a flexible but inextensible rope or bicycle-type chain of length L and mass ρ per unit length. Determine the force P required to elevate the end of the rope with a constant velocity v and determine the corresponding reaction R between the coil and the platform.

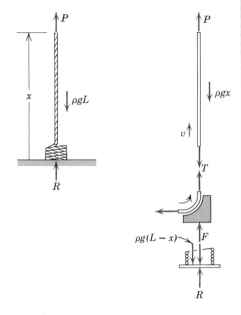

Solution. We shall neglect any horizontal motion of the rope as it uncoils. Since elements of the rope acquire their upward velocity in a smooth and continuous manner, there is no energy loss as was the case with the open-link chain. Therefore we apply the work-energy ① equation to the entire rope for an infinitesimal movement and get

② $[dU = dT + dV_g]$ $\quad P\,dx = d(\tfrac{1}{2}\rho x v^2) + d\left(\rho g x \dfrac{x}{2}\right)$

$$= \tfrac{1}{2}\rho v^2\,dx + \rho g x\,dx$$

$$P = \tfrac{1}{2}\rho v^2 + \rho g x \qquad\qquad Ans.$$

Next we apply the force-momentum equation to the entire system and get

③ $\left[\Sigma F = \dfrac{dG}{dt}\right]$ $\quad P + R - \rho g L = \dfrac{d}{dt}\,(\rho x v)$

$$P + R = \rho g L + \rho v^2$$

Eliminating P between the two equations gives

$$R = \rho g L + \rho v^2 - (\tfrac{1}{2}\rho v^2 + \rho g x)$$

$$R = \rho g(L - x) + \tfrac{1}{2}\rho v^2 \qquad\qquad Ans.$$

To understand more fully the mechanism of the ideal rope and the difference between it and the open-link chain we have drawn free-body diagrams of the vertical portion, the transition section with a vane for supporting its change of momentum, and the coil and supporting plate. Equilibrium of the vertical section ($\dot v = 0$) requires

$[\Sigma F = 0]$ $\quad T + \rho g x - P = 0$ $\quad T = P - \rho g x = \tfrac{1}{2}\rho v^2$

The force-momentum equation for the transition section gives

④ $[\Sigma F = m'\,\Delta v]$ $\quad F + T = m'(v - 0)$ $\quad F = -\tfrac{1}{2}\rho v^2 + \rho v(v - 0)$

$$F = \tfrac{1}{2}\rho v^2$$

Equilibrium of the base plate gives

$[\Sigma F = 0]$ $\quad R - F - \rho g(L - x) = 0$ $\quad R = \rho g(L - x) + \tfrac{1}{2}\rho v^2$

which checks the previous calculation. Lacking a central diverting vane, the force F would come from the coil of rope with R remaining the same.

① Since P is a variable we must evaluate the work during an infinitesimal displacement rather than a finite displacement.

② The center of mass of the length x of rope is a distance $x/2$ above the base, so the potential energy relative to the base is the weight $\rho g x$ times $x/2$. Note that v is constant so is not differentiated.

③ Note that $\dot x = v$.

④ We see that the force required to hoist the rope is less than that required to hoist the open-link chain of Sample Problem 4/9 by the amount $\tfrac{1}{2}\rho v^2$. Also, the reaction under the rope exceeds that under the chain by the same amount $\tfrac{1}{2}\rho v^2$. A part of the force required to change the momentum of the rope comes from the tension T in the rope and a part from the platform. The total upward force $P + R$ in the two cases is the same.

Sample Problem 4/11

A rocket of initial total mass m_0 is fired vertically up from the north pole and accelerates until the fuel, which burns at a constant rate, is exhausted. The relative nozzle velocity of the exhaust gas has a constant value u, and the nozzle exhausts at atmospheric pressure throughout the flight. If the residual mass of the rocket structure and machinery is m_b when burnout occurs, determine the expression for the maximum velocity reached by the rocket. Neglect atmospheric resistance and the variation of gravity with altitude.

Solution I ($F = ma$ solution). We adopt the approach illustrated with Fig. 4/7b and treat the thrust as an external force on the rocket. With the neglect of the back pressure p across the nozzle and the atmospheric resistance R, Eq. 4/21 or Newton's second law gives

$$T - mg = m\dot{v}$$

But the thrust is $T = m'u = -\dot{m}u$ so that the equation of motion becomes

$$-\dot{m}u - mg = m\dot{v}$$

Multiplication by dt, division by m, and rearrangement give

$$dv = -u\frac{dm}{m} - g\,dt$$

which is now in a form that can be integrated. The velocity v corresponding to the time t is given by the integration

$$\int_0^v dv = -u\int_{m_0}^m \frac{dm}{m} - g\int_0^t dt$$

or

$$v = u\ln\frac{m_0}{m} - gt$$

Since the fuel is burned at the constant rate $m' = -\dot{m}$, the mass at any time t is $m = m_0 + \dot{m}t$. If we let m_b stand for the mass of the rocket when burnout occurs, then the time at burnout becomes $t_b = (m_b - m_0)/\dot{m} = (m_0 - m_b)/(-\dot{m})$. This time gives the condition for maximum velocity which now becomes

$$v_{\max} = u\ln\frac{m_0}{m_b} + \frac{g}{\dot{m}}(m_0 - m_b) \qquad \text{Ans.}$$

The quantity $\dot{m}$ is a negative number since the mass decreases with time.

Solution II (variable-mass solution). If we use Eq. 4/20 then $\Sigma F = -mg$ and the equation becomes

$$[\Sigma F = m\dot{v} + \dot{m}u] \qquad -mg = m\dot{v} + \dot{m}u$$

But $\dot{m}u = -m'u = -T$ so that the equation of motion becomes

$$T - mg = m\dot{v}$$

which is the same as formulated with *Solution I*.

① The neglect of atmospheric resistance is not a bad assumption for a first approximation inasmuch as the velocity of the ascending rocket is smallest in the dense part of the atmosphere and greatest in the rarefied region. Also for an altitude of 320 km the acceleration due to gravity is 91% of the value at the surface of the earth.

② Vertical launch from the north pole is taken only to eliminate any complication due to the earth's rotation in figuring the absolute trajectory of the rocket.

PROBLEMS

4/61 The space shuttle, together with its central fuel tank and two booster rockets, has a total mass of $2.04(10^3)$ Mg at lift-off. Each of the two booster rockets produces a thrust of $11.8(10^3)$ kN, and each of the three main engines of the shuttle produces a thrust of 2000 kN. The specific impulse (ratio of exhaust velocity to gravitational acceleration) for each of the three main engines of the shuttle is 455 s. Calculate the initial vertical acceleration a of the assembly with all five engines operating, and find the rate m' at which fuel is being consumed by each of the shuttle's three engines.   *Ans. $a = 4.70$ m/s^2, $m' = 448$ kg/s*

Problem 4/61

4/62 A tank truck for washing down streets has a total mass of 10 Mg when its tank is full. With the spray turned on, 40 kg of water per second issue from the nozzle with a velocity of 20 m/s relative to the truck at the 30° angle shown. If the truck is to accelerate at the rate of 0.6 m/s^2 when starting on a level road, determine the required tractive force P between the tires and the road when (*a*) the spray is turned on and (*b*) when the spray is turned off.   *Ans. (a) $P = 5307$ N, (b) $P = 6000$ N*

Problem 4/62

4/63 A tank, which has a mass of 50 kg when empty, is propelled to the left by a force P and scoops up fresh water from a stream flowing in the opposite direction with a velocity of 1.5 m/s. The entrance area of the scoop is 2000 mm^2, and water enters the scoop at a rate equal to the velocity of the scoop relative to the stream. Determine the force P at a certain instant for which 80 kg of water have been ingested and the velocity and acceleration of the tank are 2 m/s and 0.4 m/s^2, respectively. Neglect the small impact pressure at the scoop necessary to elevate the water in the tank.   *Ans. $P = 76.5$ N*

Problem 4/63

Problem 4/64

4/64 A coil of heavy flexible cable with a total length of 100 m and a mass of 1.2 kg/m is to be laid along a straight horizontal line. The end is secured to a post at *A*, and the cable peels off the coil and emerges through the horizontal opening in the cart as shown. The cart and drum together have a mass of 40 kg. If the cart is moving to the right with a velocity of 2 m/s when 30 m of cable remain in the drum and the tension in the rope at the post is 2.4 N, determine the force *P* required to give the cart and drum an acceleration of 0.3 m/s². Neglect all friction.

Problem 4/65

4/65 The rocket has a mass of 8.5 Mg when launched. Fuel is burned at the rate of 200 kg/s with a nozzle exhaust velocity of 760 m/s. Calculate the initial acceleration *a* of the rocket along the ramp and find the angle θ with the horizontal made by the acceleration of the mass center of the rocket just after it clears the ramp.

Ans. $a = 9.39$ m/s², $\theta = 32.4°$

4/66 The mass *m* of a raindrop increases as it picks up moisture during its vertical descent through still air. If the air resistance to motion of the drop is *R* and its downward velocity is *v*, write the equation of motion for the drop and show that the relation $\Sigma F = d(mv)/dt$ is obeyed as a special case of the variable-mass equation.

4/67 A small rocket of initial mass m_0 is fired vertically upward near the surface of the earth (*g* constant). If air resistance is neglected, determine the manner in which the mass *m* of the rocket must vary as a function of the time *t* after launching in order that the rocket may have a constant vertical acceleration *a* with a constant relative velocity *u* of the escaping gases with respect to the nozzle.

4/68 The rate m' in kilograms per second at which fuel is burned in a certain rocket mounted vertically in a test stand is programmed so that the momentum thrust balances the decreasing weight of the rocket for the duration of the test. If the initial mass of the rocket and fuel is 6000 kg and the relative exhaust velocity is 800 m/s regardless of fuel rate, determine the manner in which m' must vary with the time t in seconds after firing. Also find the mass m_f of fuel used in the first 10 s.

Ans. $m' = 73.6e^{-0.0123t}$ kg/s, $m_f = 692$ kg

4/69 The upper end of the open-link chain of length L and mass ρ per unit length is lowered at a constant speed v by the force P. Determine the reading R of the platform scale in terms of x.

Ans. $R = \rho gx + \rho v^2$

Problem 4/69

4/70 The upper end of the open-link chain of length L and mass ρ per unit length is released from rest with the lower end just touching the platform of the scale. Determine the expression for the force F read on the scale as a function of the distance x through which the upper end has fallen. (*Comment:* The chain acquires a free-fall velocity of $\sqrt{2gx}$ since the links on the scale exert no force on those above, which are still falling freely. Work the problem in two ways, first, by evaluating the time rate of change of momentum for the entire chain and, second, by considering the force F to be composed of the weight of the links at rest on the scale plus the force necessary to divert an equivalent stream of fluid.) *Ans.* $F = 3\rho gx$

Problem 4/70

4/71 The end of a pile of loose-link chain of mass ρ per unit length is being pulled horizontally along the surface by a constant force P. If the coefficient of kinetic friction between the chain and the surface is μ_k, determine the acceleration a of the chain in terms of x and $\dot{x}$.

Problem 4/71

Problem 4/72

4/72 An open-link chain of length $L = 8$ m with a mass of 48 kg is resting on a smooth horizontal surface when end A is doubled back on itself by a force P applied to end A. (*a*) Calculate the required value of P to give A a constant velocity of 1.5 m/s. (*b*) Calculate the acceleration a of end A if $P = 20$ N and if $v = 1.5$ m/s when $x = 4$ m.
$$\textit{Ans. (a) } P = 6.75 \text{ N, } (b)\ a = 1.104 \text{ m/s}^2$$

Problem 4/73

4/73 An ideal rope or bicycle-type chain of length L and mass ρ per unit length is resting on a smooth horizontal surface when end A is doubled back on itself by a force P applied to end A. End B of the rope is secured to a fixed support. Determine the force P required to give A a constant velocity v. (*Hint:* The action of the loop can be modeled by inserting a circular disk of negligible mass as shown in the separate sketch and then taking the disk radius as zero. It is easily shown that the tensions in the rope at C, D, and B are all equal to P under the ideal conditions imposed and with constant velocity.) $\textit{Ans. } P = \rho v^2/4$

4/74 If Eq. 4/20 is applied to the moving part of the ideal rope of Sample Problem 4/10 using $\Sigma F = P - \rho g x$, $m = \rho x$, $\dot{m} = \rho v$, $\dot{v} = 0$, and $u = v$, explain why the result for P is invalid.

Problem 4/75

4/75 A coal car with an empty mass of 25 Mg is moving freely with a speed of 1.2 m/s under a hopper which opens and releases coal into the moving car at the constant rate of 4 Mg per second. Determine the distance x moved by the car during the time that 32 Mg of coal are deposited in the car. Neglect any frictional resistance to rolling along the horizontal track. $\textit{Ans. } x = 6.18$ m

4/76 The left end of the open-link chain of length L and mass ρ per unit length is released from rest at $x = 0$. Determine the expression for the tension T in the chain at its support at A in terms of x. Also determine the energy loss ΔE during the entire motion from $x = 0$ to $x = 2L$.

Problem 4/76

4/77 The open-link chain of total length L and of mass ρ per unit length is released from rest at $x = 0$ at the same instant that the platform starts from rest at $y = 0$ and moves vertically up with a constant acceleration a. Determine the expression for the total force R exerted on the platform by the chain t seconds after the motion starts.

$$Ans. \ R = \tfrac{3}{2}\rho(a + g)^2 t^2$$

Problem 4/77

4/78 An open-link chain of mass ρ per unit length is released from rest on the incline with $x = 0$. The coefficient of kinetic friction between the chain and the incline is μ_k. The chain piles up in a bucket at the base of the incline. Determine as a function of x the added force R on the bucket required to hold it in place against the impact of the chain. Note that the last link in motion at the bottom of the incline has lost contact with the link which immediately preceded it.

$$Ans. \ R = \rho gx(3 \sin \theta - 2\mu_k \cos \theta) \text{ with } \mu_k < \tan \theta$$

Problem 4/78

4/79 An open-link chain of length $\pi r/2$ and mass ρ per unit length is released from rest at $\theta = 0$ and slides down the smooth circular channel. The chain piles up at the bottom where the last moving link at $\beta = \pi/2$ has lost contact with the one before it, which has come to rest. Determine the expression for the tangential acceleration a_t common to all links as a function of θ. (*Hint:* Write the force equation of motion for a general element and integrate the tension from zero at $\beta = \theta$ to zero at $\beta = \pi/2$.) Also find the loss ΔE of energy during the motion from $\theta = 0$ to $\theta = \pi/2$.

$$Ans. \ a_t = g \frac{\cos \theta}{\pi/2 - \theta}, \ \Delta E = \rho gr^2$$

Last link in motion at $\beta = \pi/2$

Problem 4/79

Problem 4/80

4/80 In the figure is shown a system used to arrest the motion of an airplane landing on a field of restricted length. The plane of mass m rolling freely with a velocity v_0 engages a hook which pulls the ends of two heavy chains each of length L and mass ρ per unit length in the manner shown. A conservative calculation of the effectiveness of the device neglects the retardation of chain friction on the ground and any other resistance to the motion of the airplane. With these assumptions compute the velocity v of the airplane at the instant that the last link of each chain is put in motion. Also determine the relation between displacement x and the time t after contact with the chain. Assume each link of the chain acquires its velocity v suddenly upon contact with the moving links.

$$Ans.\ v = \frac{v_0}{1 + 2\rho L/m},\quad x = \frac{m}{\rho}\left[\sqrt{1 + \frac{2v_0 t\rho}{m}} - 1\right]$$

Problem 4/81

▶4/81 A metal canister of mass m_0 contains a coil of flexible cord of mass ρ per unit length. The end of the cord is secured to a support at the top of an elevator shaft, and the canister is released from rest. A thumb screw on the neck of the canister adjusts a spring clamp which applies a constant friction force $F/2$ on each side of the cord as it passes through the neck. Determine an expression for the acceleration a of the canister and the force P at the support in terms of x and the velocity v. The total length of the cord is L.

$$Ans.\ a = g + \frac{\rho v^2/2 - F}{m_0 + \rho(L - x)},\quad P = \rho gx + \rho v^2/2 + F$$

Problem 4/82

▶4/82 A rope or hinged-link bicycle-type chain of length L and mass ρ per unit length is released from rest with $x = 0$. Determine the expression for the total force R exerted on the fixed platform by the chain as a function of x. Note that the hinged-link chain is a conservative system during all but the last increment of motion. Compare the result with that of Prob. 4/77 if the upward motion of the platform in that problem is taken to be zero.

$$Ans.\ R = \rho gx \frac{4L - 3x}{2(L - x)}$$

▶**4/83** One end of the pile of chain falls through a hole in its support and pulls the remaining links after it in a steady flow. If the links which are initially at rest acquire the velocity of the chain suddenly and without frictional resistance or interference from the support or from adjacent links, find the velocity v of the chain as a function of x if $v = 0$ when $x = 0$. Also find the acceleration a of the falling chain and the energy ΔE lost from the system as the last link leaves the platform. (*Hint:* Apply Eq. 4/20 and treat the product xv as the variable when solving the differential equation. Also note at the appropriate step that $dx = v\,dt$.) The total length of the chain is L, and its mass per unit length is ρ.

$$Ans.\ v = \sqrt{\frac{2gx}{3}},\ a = \frac{g}{3},\ \Delta E = \frac{\rho g L^2}{6}$$

Problem 4/83

▶**4/84** Replace the pile of chain in Prob. 4/83 by a coil of rope of mass ρ per unit length and total length L as shown, and determine the velocity of the falling section in terms of x if it starts from rest at $x = 0$. Show that the acceleration is constant at $g/2$. The rope is considered to be perfectly flexible in bending but inextensible and constitutes a conservative system (no energy loss). Rope elements acquire their velocity in a continuous manner from zero to v in a small transition section of the rope at the top of the coil. For comparison with the chain of Prob. 4/83 this transition section may be considered to have negligible length without violating the requirement that there be no energy loss in the present problem. Also determine the force R exerted by the platform on the coil in terms of x and explain why R becomes zero when $x = 2L/3$. Neglect the dimensions of the coil compared with x.

$$Ans.\ v = \sqrt{gx},\ R = \rho g\left(L - \frac{3x}{2}\right)$$

Problem 4/84

PART II

DYNAMICS OF RIGID BODIES

PLANE KINEMATICS OF RIGID BODIES

5

5/1 INTRODUCTION

In Chapter 2 on particle kinematics we developed the relationships governing the displacement, velocity, and acceleration of points as they moved along straight or curved paths. In rigid-body kinematics we use these same relationships but must also account for the rotational motion of the body as well. Thus, rigid-body kinematics involves both linear and angular quantities.

The description of the motion of rigid bodies is useful in two important ways. First, it is frequently necessary to generate, transmit, or control certain desired motions by the use of cams, gears, and linkages of various types. Here an analysis of the displacement, velocity, and acceleration of the motion is necessary to determine the design geometry of the mechanical parts. Furthermore, as a result of the motion generated, forces are frequently developed which must be accounted for in the design of the parts. Second, it is often necessary to determine the motion of a rigid body resulting from the forces applied to it. Calculation of the motion of a rocket under the influence of its jet thrust and gravitational attraction is an example of such a problem. In both types of problems we need to have command of the principles of rigid-body kinematics. In this chapter we will cover the kinematics of motion which may be analyzed as taking place in a single plane. In Chapter 7 we will present a brief introduction to the kinematics of motion which requires three spatial coordinates.

A rigid body was defined in the previous chapter as a system of particles for which the distances between the particles remain unchanged. Thus if each particle of such a body is located by a position vector from reference axes attached to and rotating with the body, there would be no change in any position vector as measured from these axes. This formulation is, of course, an ideal one since all solid materials change shape to some extent when forces are applied to them. Nevertheless, if the movements associated with the changes in shape are very small compared with the overall movements of the body as a whole, then the ideal concept of rigidity is quite acceptable. As mentioned in Chapter 1 the displacements due to the flutter of an aircraft wing, for instance, are of no consequence in the description of the flight path of aircraft as a whole, for which the rigid-body assumption is clearly in order. On the other hand, if the problem is one of describing, as a function of time, the internal wing stress due to wing flutter, then the relative motions of portions of the wing become of prime importance and cannot be neglected. In this case the wing may not be considered a rigid body. In the present chapter and in the two that follow essentially all of the material is based upon the assumption of rigidity.

A rigid body executes plane motion when all parts of the body move in parallel planes. For convenience we generally consider the *plane of motion* to be the plane which contains the mass center, and we treat the body as a thin slab with motion confined to the plane of the slab. This idealization fits a very large category of rigid-body motions encountered in engineering.

The plane motion of a rigid body may be divided into several categories as represented in Fig. 5/1.

Translation is defined as any motion in which every line in the body remains parallel to its original position at all times. In translation there is *no rotation of any line in the body*. In *rectilinear translation*, part (*a*) of Fig. 5/1, all points in the body move in parallel straight lines. In *curvilinear translation*, part (*b*), all points move on congruent curves. We note that in each of the two cases of translation the motion of the body is completely specified by the motion of any point in the body, since all points have the same motion. Thus our earlier study of the motion of a point (particle) in Chapter 2 enables us to describe completely the translation of a rigid body.

Rotation about a fixed axis, part (*c*) of Fig. 5/1, is the angular motion about the axis. It follows that all particles move in circular paths about the axis of rotation, and all lines in the body (including those that do not pass through the axis) rotate through the same angle in the same time. Again, our discussion in Chapter 2 on the circular motion of a point enables us to describe the motion of a rotating rigid body, which is treated in the next article.

General plane motion of a rigid body, part (*d*) of Fig. 5/1, is a combination of translation and rotation. We will utilize the principles of relative motion covered in Art. 2/8 in the description of general plane motion.

Note that in each of the examples cited the actual motion of the body is projected onto the single plane of motion as represented in each figure.

Determination of the plane motion of rigid bodies is accomplished by either a direct calculation of the absolute displacements and their time derivatives from the geometry involved or by utiliz-

Type of Rigid–Body Plane Motion   Example

(a) Rectilinear translation	Rocket test sled
(b) Curvilinear translation	Parallel–link swinging plate
(c) Fixed–axis rotation	Compound pendulum
(d) General plane motion	Connecting rods in a reciprocating engine

Figure 5/1

ing the principles of relative motion. Each method is important and useful and will be taken up in turn in the articles which follow.

5/2 ROTATION

The rotation of a rigid body is described by its angular motion. Figure 5/2 shows a rigid body which is rotating as it undergoes plane motion in the plane of the figure. The angular positions of any two lines 1 and 2 attached to the body are specified by θ_1 and θ_2 measured from any fixed reference direction which is convenient. Since the angle β is invariant, the relation $\theta_2 = \theta_1 + \beta$ upon differentiation with respect to time gives $\dot{\theta}_2 = \dot{\theta}_1$ and $\ddot{\theta}_2 = \ddot{\theta}_1$ or, during a finite interval, $\Delta\theta_2 = \Delta\theta_1$. Thus *all lines in a rigid body in its plane of motion have the same angular displacement, the same angular velocity, and the same angular acceleration.* It should be noted that the angular motion of a line depends only on its angular displacement with respect to any arbitrary fixed reference and on the time derivatives of the displacement. Angular motion does not require the presence of a fixed axis, normal to the plane of motion, about which the line and the body rotate.

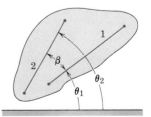

Figure 5/2

(*a*) *Angular motion relations.*   The angular velocity ω and angular acceleration α of a rigid body in plane rotation are, respectively, the first and second time derivatives of the angular position coordinate θ of any line in the body in the plane of motion. These definitions give

$$\omega = \frac{d\theta}{dt} = \dot{\theta}$$

$$\alpha = \frac{d\omega}{dt} = \dot{\omega} \quad \text{or} \quad \alpha = \frac{d^2\theta}{dt^2} = \ddot{\theta} \qquad (5/1)$$

$$\omega \, d\omega = \alpha \, d\theta \quad \text{or} \quad \dot{\theta} \, d\dot{\theta} = \ddot{\theta} \, d\theta$$

The third relation is obtained by eliminating dt from the first two. In each of these relations the positive direction for ω and α, clockwise or counterclockwise, is the same as that chosen for θ. Equations 5/1 should be recognized as analogous to the defining equations for the rectilinear motion of a particle expressed by Eqs. 2/1, 2/2, and 2/3. In fact all relations which were described for rectilinear motion in Art. 2/2 apply to the case of rotation in a plane if the linear quantities s, v, and a are replaced by their respective equivalent angular quantities θ, ω, and α.

For rotation with *constant* angular acceleration the integrals of Eqs. 5/1 become

$$\omega = \omega_0 + \alpha t$$

$$\omega^2 = \omega_0{}^2 + 2\alpha(\theta - \theta_0)$$

$$\theta = \theta_0 + \omega_0 t + \tfrac{1}{2}\alpha t^2$$

Here θ_0 and ω_0 are the values of the angular position coordinate and angular velocity, respectively, at $t = 0$ and t is the time of the interval of motion considered. The student should be able to carry out these integrations easily, as they are completely analogous to the corresponding equations for rectilinear motion with constant acceleration covered in Art. 2/2.

The graphical relationships described for s, v, a, and t in Figs. 2/3 and 2/4 may be used for θ, ω, and α merely by the substitution of corresponding symbols. The student should sketch these graphical relations for plane rotation. The mathematical procedures for obtaining velocity and displacement from acceleration described for rectilinear motion may be applied to rotation by merely replacing the linear quantities by their corresponding angular quantities.

(*b*) *Rotation about a fixed axis.*   When a body rotates about a fixed axis, all points other than those on the axis move in concentric circles about the fixed axis. Thus for the rigid body in Fig. 5/3 rotating about a fixed axis normal to the plane of the figure through O, any point such as A moves in a circle of radius r. From our previous discussion in Art. 2/5 we are already familiar with the relationships between the linear motion of A and the angular motion of the line normal to its path, which is also the angular motion of the rigid body. With the notation $\omega = \dot{\theta}$ and $\alpha = \dot{\omega} = \ddot{\theta}$ for the angular velocity and angular acceleration, respectively, of the body we have Eqs. 2/11 rewritten as

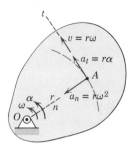

Figure 5/3

$$
\begin{aligned}
v &= r\omega \\
a_n &= r\omega^2 = v^2/r = v\omega \\
a_t &= r\alpha
\end{aligned}
\tag{5/2}
$$

These quantities may be expressed alternatively using the cross-product relationship of vector notation. The vector formulation is especially important in three-dimensional analysis of motion. The angular velocity of the rotating body may be expressed by the vector $\boldsymbol{\omega}$ normal to the plane of rotation and having a sense governed by the right-hand rule, as shown in Fig. 5/4a. From the definition of the vector cross product we see that the vector $\mathbf{v}$ is obtained by crossing $\boldsymbol{\omega}$ into $\mathbf{r}$. This cross product gives the correct magnitude and direction for $\mathbf{v}$ and we write

$$\mathbf{v} = \dot{\mathbf{r}} = \boldsymbol{\omega} \times \mathbf{r}$$

The order of the vectors to be crossed must be retained. The reverse order gives $\mathbf{r} \times \boldsymbol{\omega} = -\mathbf{v}$.

The acceleration of A is obtained by differentiating the cross-product expression for $\mathbf{v}$ which gives

$$
\begin{aligned}
\mathbf{a} = \dot{\mathbf{v}} &= \boldsymbol{\omega} \times \dot{\mathbf{r}} + \dot{\boldsymbol{\omega}} \times \mathbf{r} \\
&= \boldsymbol{\omega} \times (\boldsymbol{\omega} \times \mathbf{r}) + \dot{\boldsymbol{\omega}} \times \mathbf{r} \\
&= \boldsymbol{\omega} \times \mathbf{v} + \boldsymbol{\alpha} \times \mathbf{r}
\end{aligned}
$$

Here $\boldsymbol{\alpha} = \dot{\boldsymbol{\omega}}$ stands for the angular acceleration of the body. Thus the vector equivalents to Eqs. 5/2 are

$$
\begin{aligned}
\mathbf{v} &= \boldsymbol{\omega} \times \mathbf{r} \\
\mathbf{a}_n &= \boldsymbol{\omega} \times (\boldsymbol{\omega} \times \mathbf{r}) \\
\mathbf{a}_t &= \boldsymbol{\alpha} \times \mathbf{r}
\end{aligned}
\tag{5/3}
$$

and are shown in Fig. 5/4b.

For three-dimensional motion of a rigid body the angular velocity vector $\boldsymbol{\omega}$ may change direction as well as magnitude, and in this case the angular acceleration, which is the time derivative of angular velocity $\boldsymbol{\alpha} = \dot{\boldsymbol{\omega}}$, will no longer be in the same direction as $\boldsymbol{\omega}$.

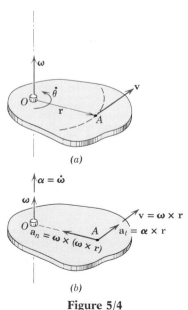

(a)

(b)

Figure 5/4

Sample Problem   5/1

A flywheel rotating freely at 1800 rev/min clockwise is subjected to a variable counterclockwise torque which is first applied at time $t = 0$. The torque produces a counterclockwise angular acceleration $\alpha = 4t$ rad/s^2 where t is the time in seconds during which the torque is applied. Determine (a) the time required for the flywheel to reduce its clockwise angular speed to 900 rev/min, (b) the time required for the flywheel to reverse its direction of rotation, and (c) the total number of revolutions, clockwise plus counterclockwise, turned by the flywheel during the first 14 seconds of torque application.

Solution.   The counterclockwise direction will be taken arbitrarily as positive.

(a) Since α is a known function of the time, we may integrate it to obtain angular velocity. With the initial angular velocity of ① $-1800(2\pi)/60 = -60\pi$ rad/s we have

$$[d\omega = \alpha \, dt] \qquad \int_{-60\pi}^{\omega} d\omega = \int_0^t 4t \, dt \qquad \omega = -60\pi + 2t^2$$

Substituting the clockwise angular speed of 900 rev/min or $\omega = -900(2\pi)/60 = -30\pi$ rad/s gives

$$-30\pi = -60\pi + 2t^2 \qquad t^2 = 15\pi \qquad t = 6.86 \text{ s} \qquad Ans.$$

(b) The flywheel changes direction when its angular velocity is momentarily zero. Thus

$$0 = -60\pi + 2t^2 \qquad t^2 = 30\pi \qquad t = 9.71 \text{ s} \qquad Ans.$$

(c) The total number of revolutions through which the flywheel turns during 14 seconds is the number of clockwise turns N_1 during the first 9.71 seconds plus the number of counterclockwise turns N_2 during the remainder of the interval. Integrating the expression for ω in terms of t gives us the angular displacement in radians. Thus for the first interval

$$[d\theta = \omega \, dt] \qquad \int_0^{\theta_1} d\theta = \int_0^{9.71} (-60\pi + 2t^2) \, dt$$

② $$\theta_1 = [-60\pi t + \tfrac{2}{3}t^3]_0^{9.71} = -1220 \text{ rad}$$

or $N_1 = 1220/2\pi = 194.2$ revolutions clockwise.

For the second interval

$$\int_0^{\theta_2} d\theta = \int_{9.71}^{14} (-60\pi + 2t^2) \, dt$$

③ $$\theta_2 = [-60\pi t + \tfrac{2}{3}t^3]_{9.71}^{14} = 410 \text{ rad}$$

or $N_2 = 410/2\pi = 65.3$ revolutions counterclockwise. Thus the total number of revolutions turned during the 14 seconds is

$$N = N_1 + N_2 = 194.2 + 65.3 = 259.5 \text{ rev} \qquad Ans.$$

We have plotted ω versus t and we see that θ_1 is represented by the negative area and θ_2 by the positive area. If we had integrated over the entire interval in one step, we would have obtained $|\theta_2| - |\theta_1|$.

① We must be very careful to be consistent with our algebraic signs. The lower limit is the negative (clockwise) value of the initial angular velocity. Also we must convert revolutions to radians since α is in radian units.

② Again note that the minus sign signifies clockwise in this problem.

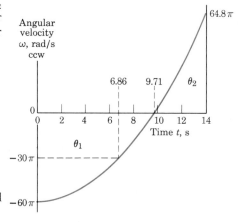

③ We could have converted the original expression for α into the units of rev/s^2, in which case our integrals would have come out directly in revolutions.

Sample Problem 5/2

The pinion A of the hoist motor drives gear B which is attached to the hoisting drum. The load L is lifted from its rest position and acquires an upward velocity of 2 m/s in a vertical rise of 0.8 m with constant acceleration. As the load passes this position compute (a) the acceleration of point C on the cable in contact with the drum and (b) the angular velocity and angular acceleration of the pinion A.

Solution. (a) If the cable does not slip on the drum, the vertical velocity and acceleration of the load L are, of necessity, the same as the tangential velocity v and tangential acceleration a_t of point C. For the rectilinear motion of L with constant acceleration the n- and t-components of the acceleration of C become

$[v^2 = 2as]$ $a = a_t = v^2/2s = 2^2/[2(0.8)] = 2.5 \text{ m/s}^2$

① $[a_n = v^2/r]$ $a_n = 2^2/(0.400) = 10 \text{ m/s}^2$

$[a = \sqrt{a_n{}^2 + a_t{}^2}]$ $a_C = \sqrt{(10)^2 + (2.5)^2} = 10.31 \text{ m/s}^2$ *Ans.*

(b) The angular motion of gear A is determined from the angular motion of gear B by the velocity v_1 and tangential acceleration a_1 of their common point of contact. First the angular motion of gear B is determined from the motion of point C on the attached drum. Thus

$[v = r\omega]$ $\omega_B = v/r = 2/(0.400) = 5 \text{ rad/s}$

$[a_t = r\alpha]$ $\alpha_B = a_t/r = 2.5/(0.400) = 6.25 \text{ rad/s}^2$

Then from $v_1 = r_A\omega_A = r_B\omega_B$ and $a_1 = r_A\alpha_A = r_B\alpha_B$ we have

$$\omega_A = \frac{r_B}{r_A}\omega_B = \frac{0.300}{0.100}5 = 15 \text{ rad/s CW}$$ *Ans.*

$$\alpha_A = \frac{r_B}{r_A}\alpha_B = \frac{0.300}{0.100}6.25 = 18.75 \text{ rad/s}^2 \text{ CW}$$ *Ans.*

① Recognize that a point on the cable changes the direction of its velocity after it contacts the drum and acquires a normal component of acceleration.

Sample Problem 5/3

The right-angled bar rotates clockwise with an angular velocity which is decreasing at the rate of 4 rad/s². Write the vector expressions for the velocity and acceleration of point A when $\omega = 2$ rad/s.

Solution. Using the right-hand rule gives

$$\omega = -2\mathbf{k} \text{ rad/s} \quad \text{and} \quad \alpha = +4\mathbf{k} \text{ rad/s}^2$$

The velocity and acceleration of A become

$[\mathbf{v} = \omega \times \mathbf{r}]$ $\mathbf{v} = -2\mathbf{k} \times (0.4\mathbf{i} + 0.3\mathbf{j}) = 0.6\mathbf{i} - 0.8\mathbf{j} \text{ m/s}$ *Ans.*

$[\mathbf{a}_n = \omega \times (\omega \times \mathbf{r})]$ $\mathbf{a}_n = -2\mathbf{k} \times (0.6\mathbf{i} - 0.8\mathbf{j}) = -1.6\mathbf{i} - 1.2\mathbf{j} \text{ m/s}^2$

$[\mathbf{a}_t = \alpha \times \mathbf{r}]$ $\mathbf{a}_t = 4\mathbf{k} \times (0.4\mathbf{i} + 0.3\mathbf{j}) = -1.2\mathbf{i} + 1.6\mathbf{j} \text{ m/s}^2$

$[\mathbf{a} = \mathbf{a}_n + \mathbf{a}_t]$ $\mathbf{a} = -2.8\mathbf{i} + 0.4\mathbf{j} \text{ m/s}^2$ *Ans.*

The magnitudes of $\mathbf{v}$ and $\mathbf{a}$ are

$$v = \sqrt{0.6^2 + 0.8^2} = 1.0 \text{ m/s} \quad \text{and} \quad a = \sqrt{2.8^2 + 0.4^2} = 2.83 \text{ m/s}^2$$

PROBLEMS

Problem 5/2

Problem 5/4

Problem 5/5

5/1 A torque applied to a flywheel causes it to accelerate uniformly from a speed of 300 rev/min to a speed of 900 rev/min in 6 seconds. Determine the number of revolutions N through which the wheel turns during this interval. (*Suggestion:* Use revolutions and minutes for units in your calculations.)

Ans. $N = 60$ rev

5/2 Magnetic tape is fed over and around the light pulleys mounted in a computer frame. If the speed v of the tape is constant and if the ratio of the magnitudes of the acceleration of points A and B is 2/3, determine the radius r of the larger pulley.

5/3 The counterclockwise angular position of a radial line on a rotating wheel is given by $\theta = 3t^3 - 2t^2 + 4$ where θ is in radians and t is in seconds. Determine the angular displacement $\Delta\theta$ of the wheel during the time $(t > 0)$ in which the angular acceleration changes from 32 to 68 rad/s², both counterclockwise. *Ans.* $\Delta\theta = 144$ rad

5/4 Experimental data for a rotating control element reveal the plotted relation between angular velocity and the angular coordinate θ as shown. Approximate the angular acceleration α of the element when $\theta = 6$ rad.

5/5 The rectangular plate rotates clockwise about its fixed bearing at O. If edge BC has a constant angular velocity of 6 rad/s, determine the vector expressions for the velocity and acceleration of point A using the coordinates given.

Ans. $\mathbf{v}_A = 1.68\mathbf{i} - 1.8\mathbf{j}$ m/s
$\mathbf{a}_A = -10.8\mathbf{i} - 10.08\mathbf{j}$ m/s²

5/6 The angular velocity of a gear is controlled according to $\omega = 12 - 3t^2$ where ω, in radians per second, is positive in the clockwise sense and where t is the time in seconds. Find the net angular displacement $\Delta\theta$ from the time $t = 0$ to $t = 3$ s. Also find the total number of revolutions N through which the gear turns during the 3 seconds.

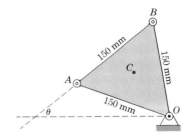

Problem 5/7

5/7 The plate *OAB* forms an equilateral triangle which rotates counterclockwise with increasing speed about point *O*. If the normal and tangential components of acceleration of the centroid *C* at a certain instant are 80 m/s^2 and 30 m/s^2, respectively, determine the values of $\dot{\theta}$ and $\ddot{\theta}$ at this same instant. The angle θ is the angle between line *AB* and the fixed horizontal axis.

Ans. $\dot{\theta} = 30.4$ rad/s, $\ddot{\theta} = 346$ rad/s^2

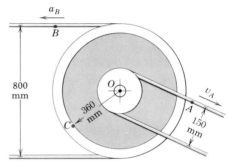

Problem 5/8

5/8 The two V-belt pulleys form an integral unit and rotate about the fixed axis at *O*. At a certain instant, point *A* on the belt of the smaller pulley has a velocity $v_A = 1.5$ m/s, and point *B* on the belt of the larger pulley has an acceleration $a_B = 45$ m/s^2 as shown. For this instant determine the magnitude of the acceleration $\mathbf{a}_C$ of point *C* and sketch the vector in your solution.

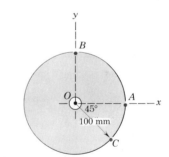

Problem 5/9

5/9 The circular disk rotates about its center *O*. For the instant represented, the velocity of *A* is $\mathbf{v}_A = 200\mathbf{j}$ mm/s and the tangential acceleration of *B* is $(\mathbf{a}_B)_t = 150\mathbf{i}$ mm/s^2. Write the vector expressions for the angular velocity $\boldsymbol{\omega}$ and angular acceleration $\boldsymbol{\alpha}$ of the disk. Use these results to write the vector expression for the acceleration of point *C*.

Ans. $\boldsymbol{\omega} = 2\mathbf{k}$ rad/s, $\boldsymbol{\alpha} = -1.5\mathbf{k}$ rad/s^2
$\mathbf{a}_C = 25\sqrt{2}(-11\mathbf{i} + 5\mathbf{j})$ mm/s^2

5/10 The belt-driven pulley and attached disk are rotating with increasing angular velocity. At a certain instant the speed v of the belt is 1.5 m/s, and the total acceleration of point *A* is 75 m/s^2. For this instant determine (*a*) the angular acceleration α of the pulley and disk, (*b*) the total acceleration of point *B*, and (*c*) the acceleration of point *C* on the belt.

Problem 5/10

5/11 A radial line on a gear, which rotates with constant angular acceleration, has an angular coordinate $\theta = 4$ rad when $t = 0$. After 10 seconds its angular coordinate is zero. Also, when $t = 4$ s, the gear reverses the direction of its motion. Calculate the angular velocity of the gear when $t = 10$ s.

Ans. $\omega = -2.4$ rad/s

Problem 5/12

5/12 The circular disk rotates about its z-axis with an angular velocity in the direction shown. At a certain instant the magnitude of the velocity of point A is 3 m/s and is decreasing at the rate of 7.2 m/s². Write the vector expressions for the angular acceleration α of the disk and the total acceleration of point B at this instant.

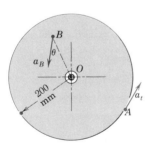

Problem 5/13

5/13 The circular disk rotates about its center O. At a certain instant the tangential component of acceleration of a point A on the rim is 6 m/s², and the tangent of the angle θ made by the acceleration of any point B in the disk with its radial line BO to the center is 0.75. Determine the velocity of point A. From the given information is it possible to tell whether the angular velocity is clockwise or counterclockwise?

Ans. $v_A = 1.265$ m/s, no

Problem 5/14

5/14 A variable torque is applied to a rotating wheel at time $t = 0$ and causes the clockwise angular acceleration to increase linearly with the clockwise angular displacement θ of the wheel during the next 30 revolutions. When the wheel has turned the additional 30 revolutions, its angular velocity is 90 rad/s. Determine its angular velocity ω_0 at the start of the interval at $t = 0$.

5/15 The motion of a rotating body is controlled so that the rate of change of its angular velocity ω with respect to its angular coordinate θ is a constant k. If the angular velocity of the body is ω_0 at the start of the interval where $\theta = 0$ and $t = 0$, derive the expressions for θ, ω, and the angular acceleration α as functions of the time t.

Ans. $\theta = \dfrac{\omega_0}{k}\left(e^{kt} - 1\right)$, $\omega = \omega_0 e^{kt}$, $\alpha = \omega_0 k e^{kt}$

5/16 The circular disk rotates with a constant angular velocity $\omega = 40$ rad/s about its axis, which is inclined in the y-z plane at the angle $\theta = \tan^{-1}\frac{3}{4}$. Determine the vector expressions for the velocity and acceleration of point P, whose position vector at the instant shown is $\mathbf{r} = 150\mathbf{i} + 160\mathbf{j} - 120\mathbf{k}$ mm. (Check the magnitudes of your results from the scalar values $v = r\omega$ and $a_n = r\omega^2$.)

Ans. $\mathbf{v} = 0.4(-20\mathbf{i} + 12\mathbf{j} - 9\mathbf{k})$ m/s
$\mathbf{a} = 16(-15\mathbf{i} - 16\mathbf{j} + 12\mathbf{k})$ m/s^2

Problem 5/16

5/17 A V-belt speed-reduction drive is shown where pulley A drives the two integral pulleys B which in turn drive pulley C. If A starts from rest at time $t = 0$ and is given a constant angular acceleration α_1, derive expressions for the angular velocity of C and the magnitude of the acceleration of a point P on the belt, both at time t.

Ans. $\omega_C = \left(\dfrac{r_1}{r_2}\right)^2 \alpha_1 t$

$a_P = \dfrac{r_1^2}{r_2}\alpha_1\sqrt{1 + \left(\dfrac{r_1}{r_2}\right)^4 \alpha_1^2 t^4}$

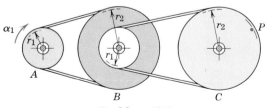

Problem 5/17

5/3 ABSOLUTE MOTION

We now develop the approach of absolute-motion analysis to describe the plane kinematics of rigid bodies. In this approach we make use of the geometric relations which define the configuration of the body involved and then proceed to take the time derivatives of the defining geometric relations to obtain velocities and accelerations. In Art. 2/9 of Chapter 2 on particle kinematics we introduced the application of absolute-motion analysis for the constrained motion of connected particles. For the pulley configurations treated, the relevant velocities and accelerations were determined by successive differentiation of the lengths of the connecting cables. In this earlier treatment the geometric relations were quite simple, and no angular quantities had to be considered. Now that we will be dealing with rigid-body motion, however, we find that our defining geometric relations include both linear and angular measurements and, therefore, the time derivatives of these quantities will involve both linear and angular velocities and linear and angular accelerations.

In absolute-motion analysis it is essential that we be totally consistent with the mathematics of the description. For example, if the angular position of a moving line in the plane of motion is specified by its counterclockwise angle θ measured from some convenient fixed reference axis, then the positive sense for both angular velocity $\dot{\theta}$ and angular acceleration $\ddot{\theta}$ will also be counterclockwise. A negative sign for either quantity will, of course, indicate a clockwise angular motion. The defining relations for linear motion, Eqs. 2/1, 2/2, and 2/3, and the relations involving angular motion, Eqs. 5/1 and 5/2 or 5/3, will find repeated use in the motion analysis and should be mastered thoroughly.

The absolute-motion approach to rigid-body kinematics is quite straightforward provided the configuration lends itself to a geometric description which is not overly complex. If the geometric configuration is awkward or complex, analysis by the principles of relative motion may be preferable. Relative-motion analysis is treated in this chapter beginning with Art. 5/4. The choice between absolute- or relative-motion analysis is best indicated after experience has been gained with both approaches.

The three sample problems which follow illustrate the application of absolute-motion analysis to three commonly encountered situations, and they should be studied thoroughly. The kinematics of a rolling wheel, treated in Sample Problem 5/4, is especially basic and will find repeated use in the problem work because the rolling wheel in various forms is such a common element in mechanical systems.

Sample Problem 5/4

A wheel of radius r rolls on a flat surface without slipping. Determine the angular motion of the wheel in terms of the linear motion of its center O. Also determine the acceleration of a point on the rim of the wheel as the point comes into contact with the surface upon which the wheel rolls.

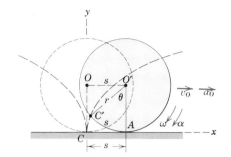

Solution. The figure shows the wheel rolling to the right from the dotted to the full position without slipping. The linear displacement of the center O is s, which is also the arc length $C'A$ along the rim upon which the wheel rolls. The radial line CO rotates to the new position $C'O'$ through the angle θ, where θ is measured from the vertical direction. If the wheel does not slip, the arc $C'A$ must equal the distance s. Thus the displacement relationship and its two time derivatives give

$$s = r\theta$$
$$v_O = r\omega \qquad \qquad \text{Ans.}$$
$$a_O = r\alpha$$

where $v_O = \dot{s}$, $a_O = \dot{v}_O = \ddot{s}$, $\omega = \dot{\theta}$, and $\alpha = \dot{\omega} = \ddot{\theta}$. The angle θ, of course, must be in radians. The acceleration a_O will be directed in the sense opposite to that of v_O if the wheel is slowing down. In this event the angular acceleration α will have the opposite sense to ω.

The origin of fixed coordinates is taken arbitrarily but conveniently at the point of contact between C on the rim of the wheel and the ground. When point C has moved along its cycloidal path to C', its new coordinates and their time derivatives become

$$x = s - r \sin \theta = r(\theta - \sin \theta) \qquad y = r - r \cos \theta = r(1 - \cos \theta)$$
$$\dot{x} = r\dot{\theta}(1 - \cos \theta) = v_O(1 - \cos \theta) \qquad \dot{y} = r\dot{\theta} \sin \theta = v_O \sin \theta$$
$$\ddot{x} = \dot{v}_O(1 - \cos \theta) + v_O \dot{\theta} \sin \theta \qquad \ddot{y} = \dot{v}_O \sin \theta + v_O \dot{\theta} \cos \theta$$
$$\quad = a_O(1 - \cos \theta) + r\omega^2 \sin \theta \qquad \quad = a_O \sin \theta + r\omega^2 \cos \theta$$

For the desired instant of contact, $\theta = 0$ and

$$\ddot{x} = 0 \qquad \text{and} \qquad \ddot{y} = r\omega^2 \qquad \qquad \text{Ans.}$$

Thus the acceleration of the point C on the rim at the instant of contact with the ground depends only on r and ω and is directed toward the center of the wheel. If desired, the velocity and acceleration of C at any position θ may be obtained by writing the expressions $\mathbf{v} = \dot{x}\mathbf{i} + \dot{y}\mathbf{j}$ and $\mathbf{a} = \ddot{x}\mathbf{i} + \ddot{y}\mathbf{j}$.

Application of the kinematic relationships for a wheel which rolls without slipping should be recognized for various configurations of rolling wheels such as those illustrated on the right. If a wheel slips as it rolls, the foregoing relations are no longer valid.

① These three relations are not entirely unfamiliar at this point, and their application to the rolling wheel should be mastered thoroughly.

② Clearly when $\theta = 0$, the point of contact has zero velocity so that $\dot{x} = \dot{y} = 0$. The acceleration of the contact point on the wheel will also be obtained by the principles of relative motion in Art. 5/6.

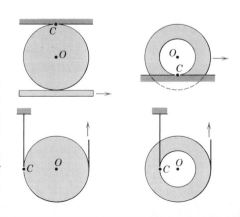

Sample Problem 5/5

The load L is being hoisted by the pulley and cable arrangement shown. Each cable is wrapped securely around its respective pulley so it does not slip. The two pulleys to which L is attached are fastened together to form a single rigid body. Calculate the velocity and acceleration of the load L and the corresponding angular velocity ω and angular acceleration α of the double pulley under the following conditions:

Case (a) Pulley 1: $\omega_1 = \dot{\omega}_1 = 0$ (pulley at rest)

Pulley 2: $\omega_2 = 2$ rad/s, $\alpha_2 = \dot{\omega}_2 = -3$ rad/s^2

Case (b) Pulley 1: $\omega_1 = 1$ rad/s, $\alpha_1 = \dot{\omega}_1 = 4$ rad/s^2

Pulley 2: $\omega_2 = 2$ rad/s, $\alpha_2 = \dot{\omega}_2 = -2$ rad/s^2

Solution. The tangential displacement, velocity, and acceleration of a point on the rim of pulley 1 or 2 equal the corresponding vertical motions of point A or B since the cables are assumed to be inextensible.

Case (a). With A momentarily at rest, line AB rotates to AB' through the angle $d\theta$ during time dt. From the diagram we see that the displacements and their time derivatives give

$$ds_B = \overline{AB}\,d\theta \qquad v_B = \overline{AB}\omega \qquad (a_B)_t = \overline{AB}\alpha$$
$$ds_O = \overline{AO}\,d\theta \qquad v_O = \overline{AO}\omega \qquad a_O = \overline{AO}\alpha$$

With $v_D = r_2\omega_2 = 0.1(2) = 0.2$ m/s, $a_D = r_2\alpha_2 = 0.1(-3) = -0.3$ m/s^2 we have for the angular motion of the double pulley

$$\omega = v_B/\overline{AB} = v_D/\overline{AB} = 0.2/0.3 = 0.667 \text{ rad/s (CCW)} \qquad Ans.$$
$$\alpha = (a_B)_t/\overline{AB} = a_D/\overline{AB} = -0.3/0.3 = -1 \text{ rad/s}^2 \text{ (CW)} \qquad Ans.$$

The corresponding motion of O and the load L is

$$v_O = \overline{AO}\omega = 0.1(0.667) = 0.0667 \text{ m/s} \qquad Ans.$$
$$a_O = \overline{AO}\alpha = 0.1(-1) = -0.1 \text{ m/s}^2 \qquad Ans.$$

Case (b). With point C, and hence point A, in motion line AB moves to $A'B'$ during time dt. From the diagram for this case we see that the displacements and their time derivatives give

$$ds_B - ds_A = \overline{AB}\,d\theta \qquad v_B - v_A = \overline{AB}\omega \qquad (a_B)_t - (a_A)_t = \overline{AB}\alpha$$
$$ds_O - ds_A = \overline{AO}\,d\theta \qquad v_O - v_A = \overline{AO}\omega \qquad a_O - (a_A)_t = \overline{AO}\alpha$$

With $v_C = r_1\omega_1 = 0.1(1) = 0.1$ m/s, $v_D = r_2\omega_2 = 0.1(2) = 0.2$ m/s

$a_C = r_1\alpha_1 = 0.1(4) = 0.4$ m/s^2, $a_D = r_2\alpha_2 = 0.1(-2) = -0.2$ m/s^2

we have for the angular motion of the double pulley

$$\omega = \frac{v_B - v_A}{\overline{AB}} = \frac{v_D - v_C}{\overline{AB}} = \frac{0.2 - 0.1}{0.3} = 0.333 \text{ rad/s (CCW)} \qquad Ans.$$

$$\alpha = \frac{(a_B)_t - (a_A)_t}{\overline{AB}} = \frac{a_D - a_C}{\overline{AB}} = \frac{-0.2 - 0.4}{0.3} = -2 \text{ rad/s}^2 \text{ (CW)} \qquad Ans.$$

The corresponding motion of O and the load L is

$$v_O = v_A + \overline{AO}\omega = v_C + \overline{AO}\omega = 0.1 + 0.1(0.333) = 0.1333 \text{ m/s} \qquad Ans.$$
$$a_O = (a_A)_t + \overline{AO}\alpha = a_C + \overline{AO}\alpha = 0.4 + 0.1(-2) = 0.2 \text{ m/s}^2 \qquad Ans.$$

① Recognize that the inner pulley is a wheel rolling along the fixed line of the left-hand cable. Thus the expressions of Sample Problem 5/4 hold.

Case (a)

② Since B moves along a curved path, in addition to its tangential component of acceleration $(a_B)_t$, it will also have a normal component of acceleration toward O which does not affect the angular acceleration of the pulley.

③ The diagrams show these quantities and the simplicity of their linear relationships. The visual picture of the motion of O and B as AB rotates through the angle $d\theta$ should clarify the analysis.

Case (b)

④ Again, as in case (a), the differential rotation of line AB as seen from the figure establishes the relation between the angular velocity of the pulley and the linear velocities of points A, O, and B. The negative sign for $(a_B)_t = a_D$ produces the acceleration diagram shown but does not destroy the linearity of the relationships.

Sample Problem   5/6

Motion of the equilateral triangular plate ABC in its plane is controlled by the hydraulic cylinder D. If the piston rod in the cylinder is moving upward at the constant rate of 0.3 m/s during an interval of its motion, calculate for the instant when $\theta = 30°$ the velocity and acceleration of the center of the roller B in the horizontal guide and the angular velocity and angular acceleration of edge CB.

Solution.   With the x-y coordinates chosen as shown the given motion of A is $v_A = \dot{y} = 0.3$ m/s and $a_A = \ddot{y} = 0$. The accompanying motion of B is given by x and its time derivatives which may be obtained from $x^2 + y^2 = b^2$. Differentiating gives

$$x\dot{x} + y\dot{y} = 0 \qquad \dot{x} = -\frac{y}{x}\dot{y}$$

①

$$x\ddot{x} + \dot{x}^2 + y\ddot{y} + \dot{y}^2 = 0 \qquad \ddot{x} = -\frac{\dot{x}^2 + \dot{y}^2}{x} - \frac{y}{x}\ddot{y}$$

With $y = b\sin\theta$, $x = b\cos\theta$, and $\ddot{y} = 0$ the expressions become

$$v_B = \dot{x} = -v_A \tan\theta$$

$$a_B = \ddot{x} = -\frac{v_A^{\,2}}{b}\sec^3\theta$$

Substituting the numerical values $v_A = 0.3$ m/s and $\theta = 30°$ gives

$$v_B = -0.3\left(\frac{1}{\sqrt{3}}\right) = -0.1732 \text{ m/s} \qquad\qquad Ans.$$

$$a_B = -\frac{(0.3)^2(2/\sqrt{3})^3}{0.2} = -0.693 \text{ m/s}^2 \qquad\qquad Ans.$$

The negative signs indicate that the velocity and the acceleration of B are both to the right since x and its derivatives are positive to the left.

The angular motion of CB is the same as that of every line on the plate including AB. Differentiating $y = b\sin\theta$ gives

$$\dot{y} = b\dot{\theta}\cos\theta \qquad \omega = \dot{\theta} = \frac{v_A}{b}\sec\theta$$

The angular acceleration is

$$\alpha = \dot{\omega} = \frac{v_A}{b}\dot{\theta}\sec\theta\tan\theta = \frac{v_A^{\,2}}{b^2}\sec^2\theta\tan\theta$$

Substitution of the numerical values gives

$$\omega = \frac{0.3}{0.2}\frac{2}{\sqrt{3}} = 1.732 \text{ rad/s} \qquad\qquad Ans.$$

$$\alpha = \frac{(0.3)^2}{(0.2)^2}\left(\frac{2}{\sqrt{3}}\right)^2\frac{1}{\sqrt{3}} = 1.732 \text{ rad/s}^2 \qquad\qquad Ans.$$

Both ω and α are counterclockwise since their signs are positive in the sense of positive measurement of θ.

---

① Observe that it is simpler to differentiate a product than a quotient. Thus differentiate $x\dot{x} + y\dot{y} = 0$ rather than $\dot{x} = -y\dot{y}/x$.

PROBLEMS

Problem 5/18

5/18 Show that the velocity of a point P on a vertical diameter of the wheel which rolls without slipping equals its distance from the constant point C times the angular velocity ω of the wheel. Also express the velocity of P in terms of the velocity of the center O.

Problem 5/19

5/19 The telephone-cable reel rolls without slipping on the horizontal surface. If point A on the cable has a velocity $v_A = 0.4$ m/s to the right, compute the velocity of the center O and the angular velocity ω of the reel. (Be careful not to make the mistake of assuming that the reel rolls to the left.)
Ans. $v_O = 0.6$ m/s, $\omega = 0.667$ rad/s CW

Problem 5/20

5/20 The elements of the electric motor drive for a diesel-electric locomotive are shown. The quill gear is integral with the 1000-mm-diameter wheel and is driven by the pinions of the two electric motors mounted to the wheel carriage. (Gear teeth are not shown.) For a locomotive speed of 100 km/h, what is the rotational speed N of the motor pinions?

Problem 5/21

5/21 The concentric wheels at A are riveted together and roll without slipping. Specify the angular velocity ω_B of drum B if drum C has an angular velocity ω_C. *Ans. $\omega_B = \frac{5}{4}\omega_C$*

5/22 The cables at A and B are wrapped securely around the rims and the hub of the integral pulley as shown. If the cables at A and B are given upward velocities of 0.3 m/s and 0.4 m/s, respectively, calculate the velocity of the center O and the angular velocity of the pulley.

Problem 5/22

5/23 The spool rolls on its hub up the inner cable A as the equalizer plate B pulls the outer cables down. The three cables are wrapped securely around their respective peripheries and do not slip. If, at the instant represented, B has moved down a distance of 1600 mm from rest with a constant acceleration of 0.2 m/s², determine the velocity of point C and the acceleration of the center O for this particular instant.

Ans. $a_O = 0.05$ m/s² up
$v_C = 1.2$ m/s up

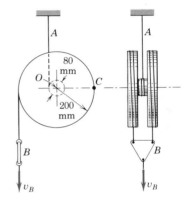

Problem 5/23

5/24 The double pulleys are fastened together to form a rigid unit. Compute the velocity of the load L and the corresponding angular velocity of the double pulley if $v_A = 8$ m/s and $v_B = 4$ m/s.

Ans. $v_L = 6.5$ m/s, $\omega = 15$ rad/s CW

Problem 5/24

5/25 Point A is given a constant acceleration a to the right starting from rest with x essentially zero. Determine the angular velocity ω of link AB in terms of x and a.

Problem 5/25

Problem 5/26

5/26 Calculate the angular velocity ω of the slender bar AB as a function of the distance x and the constant angular velocity ω_0 of the drum.

$$Ans. \ \omega = \frac{rh\omega_0}{x^2 + h^2}$$

Problem 5/27

5/27 The circular cam is mounted eccentrically about its fixed bearing at O and turns counterclockwise at the constant angular velocity ω. The cam causes the fork A and attached control rod to oscillate in the horizontal x-direction. Write the expressions for the velocity v_x and acceleration a_x of the control rod in terms of the angle θ measured from the vertical. The contact surfaces of the fork are vertical.

Problem 5/28

5/28 The crank OA starts from rest at $\theta = 0$ and turns clockwise with a constant angular acceleration α. Determine the acceleration a_x of the shaft B for the position $\theta = 90°$. *Ans.* $a_x = -\pi r \alpha$

Problem 5/29

5/29 The slotted arm OA rotates with a constant angular velocity $\omega = \dot{\theta}$ during a limited interval of its motion and moves the pivoted slider block along the horizontal slot. Write the expressions for the velocity v_B and acceleration a_B of the pin B in the slider block in terms of θ.

5/30 Rotation of the lever OA is controlled by the motion of the contacting circular disk whose center is given a horizontal velocity v. Determine the expression for the angular velocity ω of the lever OA in terms of x.

$$Ans.\ \omega = \frac{v}{x\sqrt{(x/r)^2 - 1}}$$

Problem 5/30

5/31 The slotted arm pivots about O and maintains the relation between the motions of sliders A and B and their control rods. Each small pivoted block is pinned to its respective slider and is constrained to slide in its rotating slot. Show that the displacement x is proportional to the reciprocal of y. Then establish the relation between the velocities v_A and v_B. Also, if v_A is constant for a short interval of motion, determine the acceleration of B.

Problem 5/31

5/32 Motion of the sliders B and C in the horizontal guide is controlled by the vertical motion of the slider A. If A is given an upward velocity v_A, determine as a function of θ the magnitude v of the equal and opposite velocities which B and C have as they move toward one another.

$$Ans.\ v = \tfrac{1}{2} v_A \tan \theta$$

Problem 5/32

5/33 The elements of a wheel-and-disk mechanical integrator are shown in the figure. The integrator wheel A turns about its fixed shaft and is driven by friction from disk B with no slipping occurring tangent to its rim. The distance y is a variable and can be controlled at will. Show that the angular displacement of the integrator wheel is given by $z = (1/b) \int y \, dx$, where x is the angular displacement of the disk B.

Problem 5/33

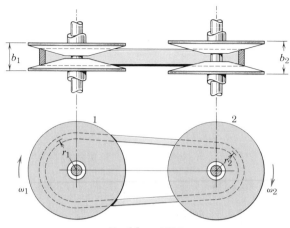

Problem 5/34

5/34 A variable-speed belt drive consists of the two pulleys each of which is constructed of two cones which turn as a unit but are capable of being drawn together or separated so as to change the effective radius of the pulley. If the angular velocity ω_1 of pulley 1 is constant, determine the expression for the angular acceleration $\alpha_2 = \dot{\omega}_2$ of pulley 2 in terms of the rates of change $\dot{r}_1$ and $\dot{r}_2$ of the effective radii.

$$Ans. \ \alpha_2 = \frac{\dot{r}_1 r_2 - r_1 \dot{r}_2}{r_2{}^2} \cdot \omega_1$$

Problem 5/35

5/35 Film passes through the guide rollers shown and is being wound onto the reel, which is turned at a constant angular velocity ω. Determine the acceleration $a = \dot{v}$ of the film as it enters the rollers. The thickness of the film is t, and s is sufficiently large so that the change in the angle made by the film with the horizontal is negligible.

5/36 If the film on the reel shown with Prob. 5/35 is being unwound by reversing the direction of motion, determine the angular acceleration α of the reel if the film is pulled through the guide rollers with a constant velocity v.

$$Ans. \ \alpha = \frac{tv^2}{2\pi r^3}$$

Problem 5/37

5/37 Link OB revolves clockwise with a constant angular velocity ω. Determine the acceleration of slider A for the position $\theta = 0$.

5/38 Link *OA* has an angular velocity $\omega_{OA} = 8$ rad/s as it passes the position shown. Determine the corresponding angular velocity ω_{CB} of the slotted link *CB*. Solve by considering the relation between the infinitesimal displacements involved.

Ans. $\omega_{CB} = 6.30$ rad/s

Problem 5/38

5/39 Link *OA* revolves counterclockwise with an angular velocity of 3 rad/s. Link *AB* slides through the pivoted collar at *C*. Determine the angular velocity ω of *AB* when $\theta = 40°$.

5/40 For the conditions of Sample Problem 5/6 determine the vector expression for the velocity of the mass center *G* of the plate. For a homogeneous plate *G* is located at the centroid.

Ans. $\bar{\mathbf{v}} = 0.1\mathbf{j}$ m/s

Problem 5/39

5/41 The Geneva wheel is a mechanism for producing intermittent rotation. Pin *P* in the integral unit of wheel *A* and locking plate *B* engages the radial slots in wheel *C* thus turning wheel *C* one-fourth of a revolution for each revolution of the pin. At the engagement position shown, $\theta = 45°$. For a constant clockwise angular velocity $\omega_1 = 2$ rad/s of wheel *A*, determine the corresponding counterclockwise angular velocity ω_2 of wheel *C* for $\theta = 20°$. (Note that the motion during engagement is governed by the geometry of triangle O_1O_2P with changing θ.)

Ans. $\omega_2 = 1.923$ rad/s

Problem 5/41

5/42 The wheel rolls to the right without slipping, and its center *O* has a constant velocity v_O. Determine the expressions for the magnitudes of the velocity **v** and acceleration **a** of point *A* as functions of the angle θ measured from the vertical. Use the moving coordinates *x-y* which translate with the center *O*.

Problem 5/42

Problem 5/43

5/43 Arm *OA* revolves about its bearing at *O* with the counterclockwise angular velocity $\omega_{OA} = 4$ rad/s. Determine the angular velocity ω_C of gear *C* if (*a*) gear *B* is fixed and does not rotate and (*b*) gear *B* rotates independently with a clockwise angular velocity $\omega_B = 2$ rad/s.

Ans. (a) $\omega_C = 8$ rad/s, (b) $\omega_C = 10$ rad/s

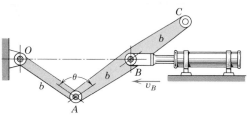

Problem 5/44

5/44 The piston rod of the hydraulic cylinder gives point *B* a velocity v_B as shown. Determine the magnitude v_C of the velocity of end *C* of link *ABC* in terms of θ.

Problem 5/45

5/45 The collar *A* slides on the pivoted arm O_1B and is pinned to the crank O_2A. For an interval of its motion crank O_2A has a constant clockwise angular velocity of 2 rad/s. Compute the angular velocity ω of arm O_1B when $\theta = 45°$.

Ans. $\omega = 0.381$ rad/s

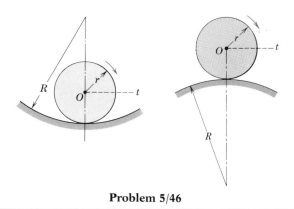

Problem 5/46

5/46 Show that the expressions $v = r\omega$ and $a_t = r\alpha$ hold for the motion of the center *O* of the wheel which rolls on the concave or convex circular arc, where ω and α are the absolute angular velocity and acceleration, respectively, of the wheel. (*Hint:* Follow the example of Sample Problem 5/4 and allow the wheel to roll a small distance. Be very careful to identify the correct *absolute* angle through which the wheel turns in each case in determining its angular velocity and angular acceleration.)

▶5/47 The punch is operated by a simple harmonic oscillation of the pivoted sector given by $\theta = \theta_0 \sin 2\pi t$ where the amplitude is $\theta_0 = \pi/12$ rad (15°) and the time for one complete oscillation is 1 second. Determine the acceleration of the punch when (*a*) $\theta = 0$ and (*b*) $\theta = \pi/12$.

Ans. (*a*) $a = 0.909$ m/s² up
(*b*) $a = 0.918$ m/s² down

Problem 5/47

▶5/48 One of the most common mechanisms is the slider-crank. Express the angular velocity ω_{AB} and angular acceleration α_{AB} of the connecting rod AB in terms of the crank angle θ for a given constant crank speed ω_0. Take ω_{AB} and α_{AB} to be positive counterclockwise.

Problem 5/48

Ans. $\omega_{AB} = \dfrac{r\omega_0}{l} \dfrac{\cos \theta}{\sqrt{1 - \dfrac{r^2}{l^2} \sin^2 \theta}}$

$$\alpha_{AB} = \frac{r\omega_0^2}{l} \sin \theta \frac{\dfrac{r^2}{l^2} - 1}{\left(1 - \dfrac{r^2}{l^2} \sin^2 \theta\right)^{3/2}}$$

5/4 RELATIVE VELOCITY

The second approach to rigid-body kinematics is to use the principles of relative motion. In Art. 2/8 we developed these principles for measurements relative to translating axes and applied the relative velocity equation

$$\mathbf{v}_A = \mathbf{v}_B + \mathbf{v}_{A/B} \qquad [2/20]$$

to the motions of two particles A and B.

We now choose two points on the *same* rigid body for our two particles. The consequence of this choice is that the motion of one point as seen by an observer translating with the other point must be circular since the radial distance to the observed point from the reference point does not change. This observation is the *key* to the successful understanding of a large majority of problems in the plane motion of rigid bodies. This concept is illustrated in Fig. 5/5a, which shows a rigid body moving in the plane of the figure from position AB to $A'B'$ during time Δt. This movement may be visualized as occurring in two parts. First, the body translates to the parallel position $A''B'$ with the displacement $\Delta \mathbf{r}_B$. Second, the body rotates about B' through the angle $\Delta \theta$. From the nonrotating reference axes x'-y' attached to the reference point B', it is seen that this remaining motion of the body is one of simple rotation about B', giving rise to the displacement $\Delta \mathbf{r}_{A/B}$ of A with respect to B. To the nonrotating observer attached to B the body appears to undergo fixed-axis rotation about B with A executing circular motion as emphasized in Fig. 5/5b. Therefore the relationships developed for circular motion in Arts. 2/5 and 5/2 and cited as Eqs. 2/11 and 5/2 (or 5/3) describe the relative portion of A's motion.

Point B was arbitrarily chosen as the reference point for attachment of our nonrotating reference axes x-y. Point A could have been used just as well, in which case we observe B to have

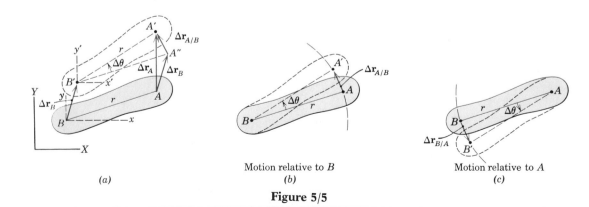

Motion relative to B
(b)

Motion relative to A
(c)

(a)

Figure 5/5

circular motion about *A* considered fixed as shown in Fig. 5/5*c*.
We see that the sense of the rotation, counterclockwise in this
example, is the same whether we choose *A* or *B* as the reference,
and we see that $\Delta\mathbf{r}_{B/A} = -\Delta\mathbf{r}_{A/B}$.

With *B* as the reference point the total displacement of *A* is
seen from Fig. 5/5*a* to be

$$\Delta\mathbf{r}_A = \Delta\mathbf{r}_B + \Delta\mathbf{r}_{A/B}$$

where $\Delta\mathbf{r}_{A/B}$ has the magnitude $r\,\Delta\theta$ as $\Delta\theta$ approaches zero. We
note that the *relative linear motion* $\Delta\mathbf{r}_{A/B}$ is accompanied by the
absolute angular motion $\Delta\theta$, as seen from the translating axes
x'-y'. Dividing the expression for $\Delta\mathbf{r}_A$ by the corresponding time
interval Δt and passing to the limit give the relative-velocity
equation

$$\mathbf{v}_A = \mathbf{v}_B + \mathbf{v}_{A/B} \qquad (5/4)$$

This expression is the same as Eq. 2/20 with the one restriction that the distance *r* between *A* and *B* remains constant. The
magnitude of the relative velocity is thus seen to be $v_{A/B} = \lim_{\Delta t \to 0}(|\Delta\mathbf{r}_{A/B}|/\Delta t) = \lim_{\Delta t \to 0}(r\,\Delta\theta/\Delta t)$ which, with $\omega = \dot\theta$, becomes

$$v_{A/B} = r\omega \qquad (5/5)$$

With **r** representing the vector $\mathbf{r}_{A/B}$, from the first of Eqs. 5/3 we
may write the relative velocity as the vector

$$\mathbf{v}_{A/B} = \omega \times \mathbf{r} \qquad (5/6)$$

where ω is the angular-velocity vector normal to the plane of the
motion in the sense determined by the right-hand rule. A critical
observation seen from Figs. 5/5*b* and 5/5*c* is that the relative linear
velocity is always perpendicular to the line joining the two points
in question.

The application of Eq. 5/4 is clarified by visualizing the separate translation and rotation components of the equation. These
components are emphasized in Fig. 5/6 which shows a rigid body
in plane motion. With *B* chosen as the reference point, the velocity of *A* is the vector sum of the translational portion $\mathbf{v}_B$ plus the
rotational portion $\mathbf{v}_{A/B} = \omega \times \mathbf{r}$ which has the magnitude $v_{A/B} = r\omega$, where $|\omega| = \dot\theta$, the *absolute* angular velocity of *AB*. The fact
that the *relative linear velocity* is *always perpendicular* to the
line joining the two points in question is an important key to the
solution of many problems. The student should draw the equivalent diagram where point *A* is used as the reference point rather
than *B*.

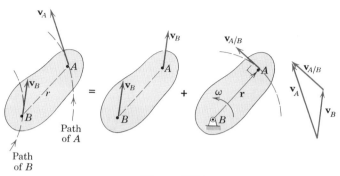

Figure 5/6

A second use of Eq. 5/4 for relative-velocity problems in plane motion may be made for constrained sliding contact between two links in a mechanism. In this case we choose points *A* and *B* as coincident points, one on each link, for the instant under consideration. In contrast to the previous example, in this case the two points are on different bodies so are not a fixed distance apart. This second use of the relative-velocity equation is illustrated in Sample Problem 5/10.

Solution of the relative-velocity equation may be carried out by scalar or vector algebra, or a graphical analysis may be employed. In any event a sketch of the vector polygon which represents the vector equation should be made to reveal the physical relationships involved. From this sketch, scalar component equations may be written by projecting the vectors along convenient directions. Usually a simultaneous solution may be avoided by a careful choice of the projections. Alternatively, each term in the relative-motion equation may be written in terms of its **i**- and **j**-components from which two scalar equations result when the equality is applied, separately, to the coefficients of the **i**- and **j**-terms. Many problems lend themselves to a graphical solution, particularly when the given geometry results in an awkward mathematical expression. In this case the known vectors are first constructed in their correct positions using a convenient scale. Next, the unknown vectors, which complete the polygon and satisfy the vector equation, are measured directly from the drawing. The choice of method to be used depends upon the particular problem at hand, the accuracy required, and upon individual preference and experience. All three of the approaches are illustrated in the sample problems which follow.

Regardless of which method of solution we employ, we note that the single vector equation in two dimensions is equivalent to two scalar equations, so that two scalar unknowns can be determined. The unknowns, for instance, might be the magnitude of one vector and the direction of another.

Sample Problem 5/7

The wheel of radius $r = 300$ mm rolls to the right without slipping and has a velocity $v_O = 3$ m/s of its center O. Calculate the velocity of point A on the wheel for the instant represented.

Solution I (scalar-geometric). The center O is chosen as the reference point for the relative-velocity equation since its motion is given. We therefore write

$$\mathbf{v}_A = \mathbf{v}_O + \mathbf{v}_{A/O}$$

where the relative-velocity term is observed from the translating axes x-y attached to O. The angular velocity of AO is the same as that of the wheel which, from Sample Problem 5/4, is $\omega = v_O/r = 3/0.3 = 10$ rad/s. Thus from Eq. 5/5 we have

$$[v_{A/O} = r_0\dot{\theta}] \qquad v_{A/O} = 0.2(10) = 2 \text{ m/s}$$

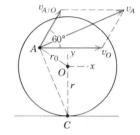

(1) which is normal to AO as shown. The vector sum $\mathbf{v}_A$ is shown on the diagram and may be calculated from the law of cosines. Thus

(2) $\quad v_A{}^2 = 3^2 + 2^2 + 2(3)(2)\cos 60° = 19 \text{ (m/s)}^2 \qquad v_A = 4.36 \text{ m/s} \quad Ans.$

The contact point C momentarily has zero velocity and can be used alternatively as the reference point, in which case the relative-velocity equation becomes $\mathbf{v}_A = \mathbf{v}_C + \mathbf{v}_{A/C} = \mathbf{v}_{A/C}$ where

$$v_{A/C} = \overline{AC}\omega = \frac{\overline{AC}}{\overline{OC}}v_O = \frac{0.436}{0.300}(3) = 4.36 \text{ m/s} \qquad v_A = v_{A/C} = 4.36 \text{ m/s}$$

The distance $\overline{AC} = 436$ mm is calculated separately. We see that $\mathbf{v}_A$
(3) is normal to AC since A is momentarily rotating about point C.

Solution II (vector). We will now use Eq. 5/6 and write

$$\mathbf{v}_A = \mathbf{v}_O + \mathbf{v}_{A/O} = \mathbf{v}_O + \boldsymbol{\omega} \times \mathbf{r}_0$$

(4) where $\quad \boldsymbol{\omega} = -10\mathbf{k}$ rad/s

$$\mathbf{r}_0 = 0.2(-\mathbf{i}\cos 30° + \mathbf{j}\sin 30°) = -0.173\mathbf{i} + 0.1\mathbf{j} \text{ m}$$
$$\mathbf{v}_O = 3\mathbf{i} \text{ m/s}$$

We now solve the vector equation

$$\mathbf{v}_A = 3\mathbf{i} + \begin{vmatrix} \mathbf{i} & \mathbf{j} & \mathbf{k} \\ 0 & 0 & -10 \\ -0.173 & 0.1 & 0 \end{vmatrix} = 3\mathbf{i} + 1.73\mathbf{j} + 1.0\mathbf{i}$$

$$= 4\mathbf{i} + 1.73\mathbf{j} \text{ m/s} \qquad\qquad Ans.$$

The magnitude $v_A = \sqrt{4^2 + (1.73)^2} = \sqrt{19} = 4.36$ m/s and direction agree with the previous solution.

(1) Be sure to visualize $v_{A/O}$ as the velocity which A appears to have in its circular motion relative to O.

(2) The vectors may also be laid off to scale graphically and the magnitude and direction of v_A measured directly from the diagram.

(3) The velocity of any point on the wheel is easily determined by using the contact point C as the reference point. The student should construct the velocity vectors for a number of points on the wheel for practice.

(4) The vector $\boldsymbol{\omega}$ is directed into the paper by the right-hand rule whereas the positive z-direction is out from the paper, hence the minus sign.

Sample Problem 5/8

Crank *CB* oscillates about *C* through a limited arc causing crank *OA* to oscillate about *O*. When the linkage passes the position shown with *CB* horizontal and *OA* vertical, the angular velocity of *CB* is 2 rad/s counterclockwise. For this instant determine the angular velocities of *OA* and *AB*.

Solution I (*vector*). The relative-velocity equation $\mathbf{v}_A = \mathbf{v}_B + \mathbf{v}_{A/B}$ is rewritten as

$$\boldsymbol{\omega}_{OA} \times \mathbf{r}_A = \boldsymbol{\omega}_{CB} \times \mathbf{r}_B + \boldsymbol{\omega}_{AB} \times \mathbf{r}_{A/B}$$

where $\boldsymbol{\omega}_{OA} = \omega_{OA}\mathbf{k}$, $\boldsymbol{\omega}_{CB} = 2\mathbf{k}$ rad/s, $\boldsymbol{\omega}_{AB} = \omega_{AB}\mathbf{k}$

$$\mathbf{r}_A = 100\mathbf{j} \text{ mm} \qquad \mathbf{r}_B = -75\mathbf{i} \text{ mm} \qquad \mathbf{r}_{A/B} = -175\mathbf{i} + 50\mathbf{j} \text{ mm}$$

Substitution gives

$$\omega_{OA}\mathbf{k} \times 100\mathbf{j} = 2\mathbf{k} \times (-75\mathbf{i}) + \omega_{AB}\mathbf{k} \times (-175\mathbf{i} + 50\mathbf{j})$$
$$-100\omega_{OA}\mathbf{i} = -150\mathbf{j} - 175\omega_{AB}\mathbf{j} - 50\omega_{AB}\mathbf{i}$$

Matching coefficients of the respective **i**- and **j**-terms gives

$$-100\omega_{OA} + 50\omega_{AB} = 0 \qquad 25(6 + 7\omega_{AB}) = 0$$

the solutions of which are

$$\omega_{AB} = -6/7 \text{ rad/s} \qquad \text{and} \qquad \omega_{OA} = -3/7 \text{ rad/s} \qquad Ans.$$

Solution II (*scalar-geometric*). Solution by the scalar geometry of the vector triangle is particularly simple here since $\mathbf{v}_A$ and $\mathbf{v}_B$ are at right angles for this special position of the linkages. First we compute v_B, which is

$$[v = r\omega] \qquad v_B = 0.075(2) = 0.150 \text{ m/s}$$

and represent it in its correct direction as shown. The vector $\mathbf{v}_{A/B}$ must be perpendicular to *AB*, and the angle θ between $\mathbf{v}_{A/B}$ and $\mathbf{v}_B$ is also the angle made by *AB* with the horizontal direction. This angle is given by

$$\tan\theta = \frac{100 - 50}{250 - 75} = \frac{2}{7}$$

The horizontal vector $\mathbf{v}_A$ completes the triangle for which we have

$$v_{A/B} = v_B/\cos\theta = 0.150/\cos\theta$$
$$v_A = v_B \tan\theta = 0.150(2/7) = 0.30/7 \text{ m/s}$$

The angular velocities become

$$[\omega = v/r] \qquad \omega_{AB} = v_{A/B}/\overline{AB} = \frac{0.150}{\cos\theta}\frac{\cos\theta}{0.250 - 0.075}$$
$$= 6/7 \text{ rad/s CW} \qquad Ans.$$

$$[\omega = v/r] \qquad \omega_{OA} = v_A/\overline{OA} = \frac{0.30}{7}\frac{1}{0.100} = 3/7 \text{ rad/s CW} \qquad Ans.$$

① We are using here the first of Eqs. 5/3 and Eq. 5/6.

② The minus signs in the answers indicate that the vectors $\boldsymbol{\omega}_{AB}$ and $\boldsymbol{\omega}_{OA}$ are in the negative **k**-direction. Hence, the angular velocities are clockwise.

③ Always make certain that the sequence of vectors in the vector polygon agrees with the equality of vectors specified by the vector equation.

Sample Problem 5/9

The common configuration of a reciprocating engine is that of the slider-crank mechanism shown. If the crank OB has a clockwise rotational speed of 1500 rev/min, determine for the position where $\theta = 60°$ the velocity of the piston A, the velocity of point G on the connecting rod, and the angular velocity of the connecting rod.

Solution. The velocity of the crank pin B as a point on AB is easily found, so that B will be used as the reference point for determining the velocity of A. The relative-velocity equation may now be written

$$\mathbf{v}_A = \mathbf{v}_B + \mathbf{v}_{A/B}$$

The crank-pin velocity is

① $[v = r\omega]$ $\qquad v_B = 0.125\dfrac{1500(2\pi)}{60} = 19.63 \text{ m/s}$

① Remember always to convert ω to radians per unit time when using $v = r\omega$.

and is normal to OB. The direction of $\mathbf{v}_A$ is, of course, along the horizontal cylinder axis. The direction of $\mathbf{v}_{A/B}$ must be perpendicular to the line AB as explained in the present article and as indicated on the diagram where the reference point B is shown as fixed. We obtain this direction by computing angle β from the law of sines which gives

$$\frac{125}{\sin \beta} = \frac{350}{\sin 60°} \qquad \beta = \sin^{-1} 0.3093 = 18.0°$$

We now complete the sketch of the velocity triangle where the angle between $\mathbf{v}_{A/B}$ and $\mathbf{v}_A$ is $90° - 18.0° = 72.0°$ and the third angle is $180° - 30° - 72.0° = 78.0°$. Vectors $\mathbf{v}_A$ and $\mathbf{v}_{A/B}$ are shown with their proper sense such that the head-to-tail sum of $\mathbf{v}_B$ and $\mathbf{v}_{A/B}$ equals $\mathbf{v}_A$. The magnitudes of the unknowns are now calculated from the trigonometry of the vector triangle or are scaled from the diagram if a graphical solution is used. Solving for $\mathbf{v}_A$ and $\mathbf{v}_{A/B}$ by the law of sines gives

② $\qquad \dfrac{v_A}{\sin 78.0°} = \dfrac{19.63}{\sin 72.0°} \qquad v_A = 20.2 \text{ m/s} \qquad Ans.$

② A graphical solution to this problem is the quickest to achieve although its accuracy is limited. Solution by vector algebra can, of course, be used but would involve somewhat more labor in this problem.

$$\frac{v_{A/B}}{\sin 30°} = \frac{19.63}{\sin 72.0°} \qquad v_{A/B} = 10.32 \text{ m/s}$$

The angular velocity of AB is counterclockwise as revealed by the sense of $\mathbf{v}_{A/B}$ and is

$[\omega = v/r]$ $\qquad \omega_{AB} = \dfrac{v_{A/B}}{AB} = \dfrac{10.32}{0.350} = 29.5 \text{ rad/s} \qquad Ans.$

We now determine the velocity of G by writing

$$\mathbf{v}_G = \mathbf{v}_B + \mathbf{v}_{G/B}$$

where $\quad v_{G/B} = \overline{GB}\omega_{AB} = \dfrac{\overline{GB}}{\overline{AB}} v_{A/B} = \dfrac{100}{350}(10.32) = 2.95 \text{ m/s}$

As seen from the diagram $\mathbf{v}_{G/B}$ has the same direction as $\mathbf{v}_{A/B}$. The vector sum is shown on the last diagram. We can calculate v_G with some geometric labor or simply measure its magnitude and direction from the velocity diagram drawn to scale. For simplicity we adopt the latter procedure here and obtain

$$v_G = 19.24 \text{ m/s} \qquad Ans.$$

As seen, the diagram may be superposed directly on the first velocity diagram.

Sample Problem 5/10

The power screw turns at a speed which gives the threaded collar C a velocity of 0.25 m/s vertically down. Determine the angular velocity of the slotted arm when $\theta = 30°$.

$\theta = 30°$

—450 mm—

Solution. The angular velocity of the arm can be found if the velocity of a point on the arm is known. We choose a point A on the arm coincident with the pin B of the collar for this purpose. If we use B as our reference point and write $\mathbf{v}_A = \mathbf{v}_B + \mathbf{v}_{A/B}$, we see from the diagram which shows the arm and points A and B an instant before and an instant after coincidence that $\mathbf{v}_{A/B}$ has a direction along the slot away from O.

The magnitudes of $\mathbf{v}_A$ and $\mathbf{v}_{A/B}$ are the only unknowns in the vector equation, so that it may now be solved. We draw the known vector $\mathbf{v}_B$ and then obtain the intersection P of the known directions of $\mathbf{v}_{A/B}$ and $\mathbf{v}_A$. The solution gives

$$v_A = v_B \cos \theta = 0.25 \cos 30° = 0.217 \text{ m/s}$$

$[\omega = v/r]$ $\qquad \omega = v_A/\overline{OA} = \dfrac{0.217}{0.450/\cos 30°}$

$$= 0.417 \text{ rad/s CCW} \qquad Ans.$$

We note the difference between this problem of constrained sliding contact between two links and the three preceding sample problems of relative velocity where no sliding contact occurred and where the points A and B were located on the same rigid body in each case.

① Physically, of course, this point does not exist, but we can imagine such a point in the middle of the slot and attached to the arm.

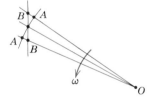

② Always identify the knowns and unknowns before attempting the solution of a vector equation.

PROBLEMS

5/49 If v_A and v_B are the instantaneous velocities of the respective ends of the rigid link, show that the angular velocity of the link is given by $(v_A \sin \theta + v_B \sin \beta)/L$ and that $v_A \cos \theta = v_B \cos \beta$.

Problem 5/49

5/50 At a particular instant end A of the bar AB has a velocity of 4 m/s in the direction shown. What is the minimum possible velocity which end B can have? What would be the corresponding angular velocity of the bar?

   Ans. $(v_B)_{min} = 2$ m/s, $\omega_{AB} = 4.33$ rad/s CCW

Problem 5/50

5/51 The magnitude of the absolute velocity of point A on the automobile tire is 10 m/s when A is in the position shown. What is the corresponding velocity v_O of the car and the angular velocity ω of the wheel? (The wheel rolls without slipping.)

   Ans. $v_O = 7.07$ m/s, $\omega = 23.6$ rad/s CW

Problem 5/51

5/52 The center O of the circular disk, which rolls without slipping, has a velocity of 0.3 m/s to the right. For the position shown, write the vector expressions for the velocity of A relative to B and for the velocity of P.

Problem 5/52

5/53 Determine the angular velocity ω of the telescoping link AB at the instant represented. The angular velocity of each of the driving links is shown.

   Ans. $\omega = 0.24$ rad/s CCW

Problem 5/53

Problem 5/54

5/54 The rider of the bicycle shown pumps steadily to maintain a constant speed of 16 km/h against a slight head wind. Calculate the maximum and minimum magnitudes of the absolute velocity of the pedal *A*.

5/55 Each of the sliding bars *A* and *B* engages its respective rim of the two riveted wheels without slipping. Determine the magnitude of the velocity of point *P* for the position shown.   *Ans.* $v_P = 0.900$ m/s

5/56 Determine the velocity of the piston and the angular velocity of the connecting rod *AB* in Sample Problem 5/9 for the position $\theta = 240°$.
Ans. $v_A = 13.81$ m/s, $\omega_{AB} = 29.5$ rad/s CW

Problem 5/55

5/57 The elements of a simplified clam-shell bucket for a dredge are shown. The cable which opens and closes the bucket passes through the block at *O*. With *O* as a fixed point, determine the angular velocity ω of the bucket jaws when $\theta = 45°$ as they are closing. The upward velocity of the control cable is 0.5 m/s as it passes through the block.
Ans. $\omega = 0.722$ rad/s

5/58 The pin in the rotating arm *OA* engages the slotted link and causes it to rotate. Show that the angular velocity of *CB* is one-half that of *OA* regardless of the angle θ.

Problem 5/57

Problem 5/58

5/59 The uniform square plate moves in the *x-y* plane and has a clockwise angular velocity. At the instant represented, point *A* has a velocity of 2 m/s to the right, and the velocity of *C* relative to a nonrotating observer at *B* has the magnitude of 1.2 m/s. Determine the vector expressions for the angular velocity of the plate and the velocity of its center *G*.

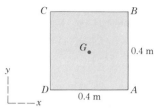

Problem 5/59

5/60 For the position shown, the two links are at right angles, and the hydraulic cylinder imparts a velocity $v_A = 0.5$ m/s to pin *A*. For this instant, determine the angular velocities of *CB* and *AB*.

Ans. $\omega_{CB} = 3.2$ rad/s CCW
$\omega_{AB} = 2.4$ rad/s CCW

5/61 A small geared motor of high torque mounted in the sliding block rotates the arm *BA* at a constant rate $\dot{\theta} = 0.3$ rad/s. Determine the angular velocity of link *OA* when the angle $\theta = 60°$ is passed.

Problem 5/60

5/62 The elements of a switching device are shown. If the vertical control rod has a downward velocity *v* of 0.9 m/s when $\theta = 60°$ and if roller *A* is in continuous contact with the horizontal surface, determine the magnitude of the velocity of *C* for this instant. *Ans.* $v_C = 1.873$ m/s

5/63 End *A* of the link has a downward velocity v_A of 2 m/s during an interval of its motion. For the position where $\theta = 30°$ determine the angular velocity ω of *AB* and the velocity v_G of the midpoint *G* of the link. Solve the relative-velocity equations, first, using the geometry of the vector polygon and, second, using vector algebra.

Problem 5/61

Problem 5/63

Problem 5/62

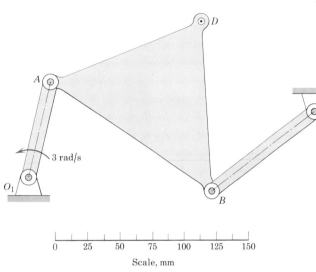

Scale, mm

Problem 5/64

5/64 Motion of the triangular plate ABD is controlled by links O_1A and O_2B. Determine the angular velocity ω of the plate and the magnitude of the velocity of point D. The drawing is represented to scale so that lengths and angles may be measured from the drawing to within available accuracy.

Ans. $\omega \cong 1.03$ rad/s CCW, $v_D \cong 315$ mm/s

5/65 The rotation of the gear is controlled by the horizontal motion of end A of the rack AB. If the piston rod has a constant velocity $\dot{x} = 300$ mm/s during a short interval of motion, determine the angular velocity ω_0 of the gear and the angular velocity ω_{AB} of AB at the instant when $x = 800$ mm.

5/66 The mechanism of Prob. 5/39 is repeated here. Determine by relative motion principles the angular velocity ω of AB when $\theta = 40°$. Note that the direction of the velocity of a point on the rod AB coincident with C is in the direction of the rod as it slides through the collar.

Ans. $\omega = 0.825$ rad/s CW

5/67 The Geneva mechanism of Prob. 5/41 is shown again here. By relative motion principles determine the angular velocity ω_2 of wheel C for $\theta = 20°$. Wheel A has a constant clockwise angular velocity $\omega_1 = 2$ rad/s.

Problem 5/65

Problem 5/66

Problem 5/67

5/68 The elements of the mechanism for deployment of a spacecraft magnetometer boom are shown. Determine the angular velocity of the boom when the driving link *OB* crosses the *y*-axis with an angular velocity $\omega_{OB} = 0.5$ rad/s if at this instant $\tan \theta = 4/3$.   *Ans.* $\omega_{CA} = 0.429\mathbf{k}$ rad/s

Problem 5/68

5/69 In the four-bar linkage shown control link *OA* has a counterclockwise angular velocity $\omega_0 = 10$ rad/s during a short interval of motion. When link *CB* passes the vertical position shown, point *A* has co-ordinates $x = -60$ mm and $y = 80$ mm. By means of vector algebra determine the angular velocity of *AB* and *BC*.
          Ans. $\omega_{AB} = 2.5\mathbf{k}$ rad/s, $\omega_{BC} = 5.83\mathbf{k}$ rad/s

5/70 Motion of the threaded collars *A* and *B* is controlled by the rotation of their respective lead screws. If *A* has a velocity to the right of 75 mm/s and *B* has a velocity to the left of 50 mm/s when *x* = 150 mm, determine the angular velocity ω of *ACD* at this instant.   *Ans.* $\omega = 0.295$ rad/s CCW

Problem 5/69

5/71 At the instant represented the slotted arm is revolving counterclockwise about *O* with an angular velocity of 2 rad/s. Also for this instant, $\theta = 30°$ and $\beta = 45°$. The pin *A* is attached to arm *BC*. Use a graphical solution to determine the velocity of point *B* for this position.

Problem 5/70

Problem 5/71

Problem 5/72

5/72 Pin *P* on the end of the horizontal rod slides freely in the slotted gear. The gear engages the moving rack *A* and the fixed rack *B* (teeth not shown) so rolls without slipping. If *A* has a velocity of 120 mm/s to the left for the instant shown, determine the velocity v_P of the rod for this position.

Ans. $v_P = 60$ mm/s

5/73 A mechanism for pushing small boxes from an assembly line onto a conveyor belt is shown with arm *OD* and crank *CB* in their vertical positions. The crank revolves clockwise at a constant rate of 1 revolution every 2 seconds. For the position shown, determine the speed at which the box is being shoved horizontally onto the conveyor belt.

Ans. $v_E = 0.514$ m/s

Problem 5/73

▶**5/74** At the instant represented, $a = 150$ mm and $b = 125$ mm, and the distance $a + b$ between *A* and *B* is decreasing at the rate of 0.2 m/s. Determine the common velocity v of points *B* and *D* for this instant.

Ans. $v = 0.0536$ m/s

▶**5/75** The wheel rolls without slipping. For the instant portrayed, when *O* is directly under point *C*, link *OA* has a velocity $v = 1.5$ m/s to the right and $\theta = 30°$. Determine the angular velocity ω of the slotted link.

Ans. $\omega = 18.22$ rad/s CCW

Problem 5/74

Problem 5/75

5/5 INSTANTANEOUS CENTER OF ZERO VELOCITY

In the previous article we determined the velocity of a point on a rigid body in plane motion by adding the relative velocity due to rotation about a convenient reference point to the velocity of the reference point. In the present article we will solve the problem by choosing a unique reference point which momentarily has zero velocity. Insofar as velocities are concerned the body may be considered to be in pure rotation about an axis, normal to the plane of motion, passing through this point. This axis is called the *instantaneous axis* of zero velocity, and the intersection of this axis with the plane of motion is known as the *instantaneous center* of zero velocity. This approach provides us with a valuable means for visualizing and analyzing velocities in plane motion.

The existence of the instantaneous center is easily shown. For the body in Fig. 5/7*a* let us assume that the directions of the absolute velocities of any two points *A* and *B* on the body are known and are not parallel. If there is a point about which *A* has absolute circular motion at the instant considered, this point must lie on the normal to $\mathbf{v}_A$ through *A*. Similar reasoning applies to *B*, and the intersection *C* of these two perpendiculars fulfills the requirement for an absolute center of rotation *at the instant considered.* Point *C* is the instantaneous center of zero velocity and may lie on or off the body. If it lies off the body, it may be visualized as lying on the body extended. The instantaneous center is not a fixed point in the body nor a fixed point in the plane.

If the magnitude of the velocity of one of the points, say, v_A, is also known, the angular velocity ω of the body and the linear velocity of every point in the body are easily obtained. Thus the angular velocity of the body, Fig. 5/7a, is

$$\omega = \frac{v_A}{r_A}$$

which, of course, is also the angular velocity of *every* line in the body. Therefore the velocity of *B* is $v_B = r_B\omega = (r_B/r_A)v_A$. Once

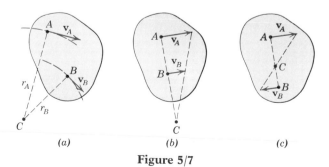

(a)   (b)   (c)

Figure 5/7

the instantaneous center is located, the direction of the instantaneous velocity of every point in the body is readily found since it must be perpendicular to the radial line joining the point in question with *C*.

If the velocities of two points in a body having plane motion are parallel, Fig. 5/7*b* or 5/7*c*, the line joining the points is perpendicular to the direction of the velocities, and the instantaneous center *C* is located by direct proportion as shown. We can readily see from Fig. 5/7*b* that as the parallel velocities approach equality in magnitude, the instantaneous center *C* moves farther away from the body and approaches infinity in the limit as the body gives up its angular velocity and translates only.

As the body changes its position, the instantaneous center *C* also changes its position both in space and on the body. The locus of the instantaneous centers in space is known as the *space centrode*, and the locus of the positions of the instantaneous centers on the body is known as the *body centrode*. At the instant considered the two curves are tangent at the position of point *C*. It may be shown that the body centrode curve rolls on the space centrode curve during the motion of the body, as indicated schematically in Fig. 5/8.

Whereas the instantaneous center of zero velocity as a point on the body is momentarily at rest, its acceleration generally is *not* zero. Thus this point may *not* be used as an instantaneous center of zero acceleration in a manner analogous to its use for finding velocity. An instantaneous center of zero acceleration does exist for bodies in general plane motion, but its location and use represent a specialized topic in mechanism kinematics and will not be treated here.

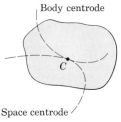

Body centrode

C

Space centrode

Figure 5/8

Sample Problem 5/11

The wheel of Sample Problem 5/7, shown again here, rolls to the right without slipping with its center O having a velocity $v_O = 3$ m/s. Locate the instantaneous center of zero velocity and use it to find the velocity of point A for the position indicated.

Solution. The point on the rim of the wheel in contact with the ground has no velocity if the wheel is not slipping; it is, therefore, the instantaneous center C of zero velocity. The angular velocity of the wheel becomes

$[\omega = v/r]$ $\qquad \omega = v_O/\overline{OC} = 3/0.300 = 10$ rad/s

The distance from A to C is

① $\qquad \overline{AC} = \sqrt{(0.300)^2 + (0.200)^2 - 2(0.300)(0.200)\cos 120°} = 0.436$ m

The velocity of A becomes

② $[v = r\omega]$ $\qquad v_A = \overline{AC}\omega = 0.436(10) = 4.36$ m/s $\qquad$ *Ans.*

The direction of $\mathbf{v}_A$ is perpendicular to AC as shown.

① Be sure to recognize that the cosine of $120°$ is itself negative.

② From the results of this problem you should be able to visualize and sketch the velocities of all points on the wheel.

Sample Problem 5/12

Arm OB of the linkage has a clockwise angular velocity of 10 rad/s in the position shown where $\theta = 45°$. Determine the velocity of A, the velocity of D, and the angular velocity of link AB for the instant shown.

Solution. The directions of the velocities of A and B are tangent to their circular paths about the fixed centers O' and O as shown. The intersection of the two perpendiculars to the velocities from A and B

① locates the instantaneous center C for the link AB. The distances $\overline{AC}$, $\overline{BC}$, and $\overline{DC}$ shown on the diagram are computed or scaled from the drawing. The angular velocity of BC, considered as a line on the body extended, is equal to the angular velocity of AC, DC, and AB and is

$[\omega = v/r]$ $\qquad \omega_{BC} = \dfrac{v_B}{\overline{BC}} = \dfrac{\overline{OB}\omega_{OB}}{\overline{BC}} = \dfrac{150\sqrt{2}(10)}{350\sqrt{2}}$

$\qquad\qquad = 4.29$ rad/s CCW $\qquad$ *Ans.*

Thus the velocities of A and D are

$[v = r\omega]$ $\qquad v_A = 0.350(4.29) = 1.500$ m/s $\qquad$ *Ans.*
$\qquad\qquad v_D = 0.381(4.29) = 1.632$ m/s $\qquad$ *Ans.*

in the directions shown.

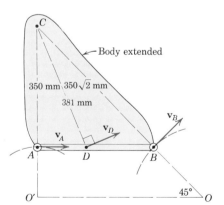

① For the instant depicted we should visualize link AB and its body extended to be rotating as a single unit about point C.

Problem 5/76

Problem 5/78

Problem 5/79

PROBLEMS

5/76 The bar *AB* has a counterclockwise angular velocity of 6 rad/s. Construct the velocity vectors for points *A* and *G* of the bar and specify their magnitudes if the instantaneous center of zero velocity for the bar is (*a*) at C_1, (*b*) at C_2, and (*c*) at C_3.
Ans. (*a*) $v_A = 150$ mm/s, $v_G = 150$ mm/s
(*b*) $v_A = 1050$ mm/s, $v_G = 750$ mm/s
(*c*) $v_A = 750$ mm/s, $v_G = 541$ mm/s

5/77 Solve Prob. 5/51 by the method of Art. 5/5.

5/78 For the instant represented point *A* of the equilateral plate has a velocity $v_A = 0.72$ m/s, and *BD* is observed to have a clockwise angular velocity. If the instantaneous center of zero velocity is 150 mm from *A*, determine the magnitude of the velocity of point *D* for this instant.
Ans. $v_D = 1.018$ m/s

5/79 The blade of a rotary power mower turns counterclockwise at the angular speed of 1800 rev/min. If the body centrode is a circle of radius 0.75 mm, compute the velocity v_O of the mower.

5/80 The rear wheels of a 0.5-Mg pickup truck have a diameter of 650 mm and are slipping on an icy road as torque is applied to them through the differential. If the body centrode for each wheel is a circle of diameter 50 mm and if each wheel is turning at the rate of 400 rev/min, determine the velocity of the truck. Also determine the velocity v_r with which the rubber slips on the icy surface.
Ans. $v = 3.77$ km/h, $v_r = 12.57$ m/s

5/81 Solve Prob. 5/63 by the method of Art. 5/5.
Ans. $\omega = 11.55$ rad/s CW, $v_G = 1.155$ m/s

5/82 Solve Prob. 5/57 by the method of Art. 5/5.

5/83 Solve Prob. 5/68 by the method of Art. 5/5.

5/84 Solve Prob. 5/60 by the method of Art. 5/5.

5/85 Solve Prob. 5/55 by the method of Art. 5/5.

5/86 The scale drawing of the linkage of Prob. 5/64 is shown again here. Locate the instantaneous center of zero velocity directly on the drawing and use the scale shown to determine the velocity of point D to a close approximation for the instant depicted. Also determine the angular velocity ω of the plate. After working the problem, compare your results to those cited with Prob. 5/64.

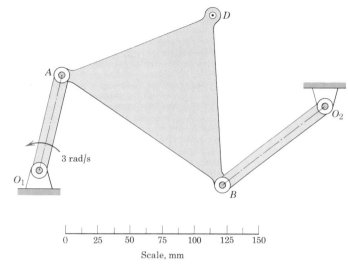

Problem 5/86

5/87 The flexible band F is attached at E to the rotating sector and leads over the guide pulley. Determine the angular velocities of AD and BD for the position shown if the band has a velocity of 4 m/s.

5/88 Solve Sample Problem 5/9 by using the method of Art. 5/5.

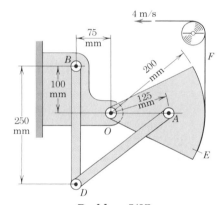

Problem 5/87

5/89 The gear D (teeth not shown) rotates counterclockwise about O with a constant angular velocity of 3 rad/s. The 90° sector AOB is mounted on an independent shaft at O, and each of the small gears at A and B meshes with gear D. If the sector has a clockwise angular velocity of 4 rad/s at the instant represented, determine the corresponding angular velocity ω of each of the small gears.

Ans. $\omega = 32$ rad/s CW

Problem 5/89

Problem 5/90

Problem 5/91

Problem 5/92

Problem 5/93

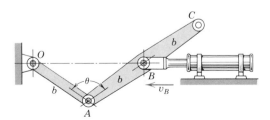

Problem 5/94

5/90 A device which tests the resistance to wear of two materials A and B is shown. If the link EO has a velocity of 1.2 m/s to the right when $\theta = 45°$, determine the rubbing velocity v_A.

5/91 The hydraulic cylinder produces a limited horizontal motion of point A. If $v_A = 4$ m/s when $\theta = 45°$, determine the magnitude of the velocity of D and the angular velocity ω of ABD for this position.
Ans. $v_D = 4.50$ m/s, $\omega = 7.47$ rad/s

5/92 Horizontal oscillation of the spring-loaded plunger E is controlled by varying the air pressure in the horizontal pneumatic cylinder F. If the plunger has a velocity of 2 m/s to the right when $\theta = 30°$, determine the downward velocity v_D of roller D in the vertical guide and find the angular velocity ω of ABD for this position.
Ans. $v_D = 2.31$ m/s, $\omega = 13.33$ rad/s

5/93 The mechanism of Prob. 5/30 is repeated here. Determine the expression for the angular velocity ω of lever AO in terms of x if the shaft and the center of the wheel have a velocity v to the right. Also determine the angular velocity ω_w of the wheel if it does not slip on the lever.

5/94 The mechanism of Prob. 5/44 is shown again here. By the method of this article determine the expression for the magnitude of the velocity of point C in terms of the velocity v_B of the piston rod and the angle θ.

$$\text{Ans. } v_C = \frac{v_B}{2}\sqrt{\sec^2\frac{\theta}{2} + 8}$$

5/95 Construct the space centrode and the body centrode for the link *AB* within the limits of its constrained motion. Show that the motion may be described by rolling the body centrode on the space centrode.

Problem 5/95

5/96 The three gears 1, 2, and 3 of equal radii are mounted on the rotating arm as shown. (Gear teeth are omitted from the drawing.) Arm *OA* rotates clockwise about *O* at the angular rate of 4 rad/s, while gear 1 rotates independently at the counterclockwise rate of 8 rad/s. Determine the angular velocity of gear 3.   *Ans.* $\omega_3 = 8$ rad/s CCW

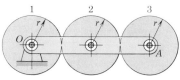

Problem 5/96

▶**5/97** The shaft at *O* drives the arm *OA* at a clockwise speed of 90 rev/min about the fixed bearing at *O*. Use the method of the instantaneous center of zero velocity to determine the rotational speed of gear *B* (gear teeth not shown) if (*a*) ring gear *D* is fixed and (*b*) ring gear *D* rotates counterclockwise about *O* with a speed of 80 rev/min.

$$\text{*Ans.* } (a) \ \omega_B = 360 \text{ rev/min}$$
$$(b) \ \omega_B = 600 \text{ rev/min}$$

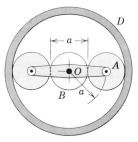

Problem 5/97

▶**5/98** The large roller bearing rolls to the left on its outer race with a velocity of its center *O* of 0.9 m/s. At the same time the central shaft and inner race rotate counterclockwise with an angular speed of 240 rev/min. Determine the angular velocity ω of each of the rollers.   *Ans.* $\omega = 10.73$ rad/s CW

Problem 5/98

5/6 *RELATIVE ACCELERATION*

To obtain the relative-acceleration relationship the equation $\mathbf{v}_A = \mathbf{v}_B + \mathbf{v}_{A/B}$ for relative velocities using nonrotating reference axes for two points A and B in plane motion may be differentiated with respect to time to get $\dot{\mathbf{v}}_A = \dot{\mathbf{v}}_B + \dot{\mathbf{v}}_{A/B}$ or

$$\mathbf{a}_A = \mathbf{a}_B + \mathbf{a}_{A/B} \tag{5/7}$$

In words Eq. 5/7 states that the acceleration of point A equals the acceleration of point B plus (vectorially) the acceleration which A appears to have to a nonrotating observer moving with B. If points A and B are located on the same rigid body in the plane of motion, the distance r between them remains constant so that the observer moving with B perceives A to have circular motion about B, as we saw in Art. 5/4 with the relative-velocity relationship. With the relative motion being circular it follows that the relative-acceleration term will have both a normal component directed from A toward B due to the change of direction of $\mathbf{v}_{A/B}$ and a tangential component perpendicular to AB due to the change in magnitude of $\mathbf{v}_{A/B}$. These acceleration components for circular motion, cited in Eqs. 5/2, were covered earlier in Art. 2/5 and should be thoroughly familiar by now. Thus we may write

$$\mathbf{a}_A = \mathbf{a}_B + (\mathbf{a}_{A/B})_n + (\mathbf{a}_{A/B})_t \tag{5/8}$$

where the magnitudes of the relative-acceleration components are

$$\begin{aligned} (a_{A/B})_n &= v_{A/B}{}^2/r = r\omega^2 \\ (a_{A/B})_t &= \dot{v}_{A/B} = r\alpha \end{aligned} \tag{5/9}$$

or in vector notation the acceleration components become

$$\begin{aligned} (\mathbf{a}_{A/B})_n &= \omega \times (\omega \times \mathbf{r}) \\ (\mathbf{a}_{A/B})_t &= \alpha \times \mathbf{r} \end{aligned} \tag{5/9a}$$

In these relationships ω is the angular velocity and α is the angular acceleration of the body. The vector locating A from B is $\mathbf{r}$. It is important to observe that the *relative* acceleration terms depend on the respective *absolute* angular velocity and *absolute* angular acceleration.

The meaning of Eqs. 5/8 and 5/9 is illustrated in Fig. 5/9, which shows a rigid body in plane motion with points A and B moving along separate curved paths with absolute accelerations

$\mathbf{a}_A$ and $\mathbf{a}_B$. Contrary to the case with velocities the accelerations $\mathbf{a}_A$ and $\mathbf{a}_B$ are, in general, not tangent to the paths described by A and B when these paths are curvilinear. The figure shows the acceleration of A to be composed of two parts, the acceleration of B and the acceleration of A with respect to B. A sketch showing the reference point as fixed is useful in disclosing the correct sense of each of the two components of the relative-acceleration term.

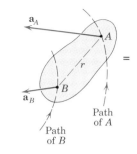

Alternatively we may write the acceleration equation in the reverse order, which puts the nonrotating reference axes on A rather than B. This order gives

$$\mathbf{a}_B = \mathbf{a}_A + \mathbf{a}_{B/A}$$

Here $\mathbf{a}_{B/A}$ and its n- and t-components are the negatives of $\mathbf{a}_{A/B}$ and its n- and t-components. The student should make a sketch corresponding to Fig. 5/9 for this inverse sequence of terms.

As in the case of the relative-velocity equation we can handle the solution to Eq. 5/8 in three different ways, namely, by scalar algebra and geometry, by vector algebra, or by graphical construction. The student will be well advised to become familiar with all three techniques. In any event a sketch of the vector polygon representing the vector equation should be made with close attention paid to the head-to-tail combination of vectors which agrees with the equation. Known vectors should be added first, and the unknown vectors will become the closing legs of the vector polygon. It is vital that we visualize the vectors in their geometrical sense, as only then can the full significance of the acceleration equation be perceived. Before attempting a solution it is well to identify the knowns and unknowns and to keep in mind that a solution to a vector equation in two dimensions can be carried out when the unknowns have been reduced to two scalar quantities. These quantities may be the magnitude or direction of any of the terms of the equation. We observe that when both points move on curved paths, there will, in general, be six scalar quantities to account for in Eq. 5/8.

Figure 5/9

Inasmuch as the normal acceleration components depend on velocities, it is generally necessary to solve for the velocities before the acceleration calculations can be made. The reference point in the relative-acceleration equation is chosen as some point on the body in question whose acceleration is either known or can be easily found. Care must be exercised *not* to use the instantaneous center of zero velocity as the reference point unless its acceleration is known and accounted for. An instantaneous center of zero acceleration exists for a rigid body in general plane motion but will not be discussed here since its use is somewhat specialized.

Sample Problem 5/13

The wheel of radius r rolls to the left without slipping and, at the instant considered, the center O has a velocity $\mathbf{v}_O$ and an acceleration $\mathbf{a}_O$ to the left. Determine the acceleration of points A and C on the wheel for the instant considered.

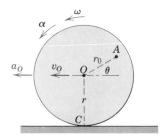

Solution. From our previous analysis of Sample Problem 5/4 we know that the angular velocity and angular acceleration of the wheel are

$$\omega = v_O/r \quad \text{and} \quad \alpha = a_O/r$$

The acceleration of A is written in terms of the given acceleration of O. Thus

$$\mathbf{a}_A = \mathbf{a}_O + \mathbf{a}_{A/O} = \mathbf{a}_O + (\mathbf{a}_{A/O})_n + (\mathbf{a}_{A/O})_t$$

The relative-acceleration terms are viewed as though O were fixed, and for this relative circular motion they have the magnitudes

$$(a_{A/O})_n = r_0\omega^2 = r_0\left(\frac{v_O}{r}\right)^2$$

$$(a_{A/O})_t = r_0\alpha = r_0\left(\frac{a_O}{r}\right)$$

① and the directions shown.

Adding the vectors head-to-tail gives $\mathbf{a}_A$ as shown. In a numerical problem we may obtain the combination algebraically or graphically. The algebraic expression for the magnitude of $\mathbf{a}_A$ is found from the square root of the sum of the squares of its components. If we use n- and t-directions we have

② $$a_A = \sqrt{(a_A)_n{}^2 + (a_A)_t{}^2} = \sqrt{[a_O \cos\theta + (a_{A/O})_n]^2 + [a_O \sin\theta + (a_{A/O})_t]^2}$$

$$= \sqrt{(r\alpha \cos\theta + r_0\omega^2)^2 + (r\alpha \sin\theta + r_0\alpha)^2} \qquad \textit{Ans.}$$

The direction of $\mathbf{a}_A$ can be computed if desired.

The acceleration of the instantaneous center C of zero velocity considered as a point on the wheel is obtained from the expression

$$\mathbf{a}_C = \mathbf{a}_O + \mathbf{a}_{C/O}$$

where the components of the relative-acceleration term are $(a_{C/O})_n = r\omega^2$ directed from C to O and $(a_{C/O})_t = r\alpha$ directed to the right on account of the counterclockwise angular acceleration of line CO about O. The terms are added together in the lower diagram, and it is seen that

③ $$a_C = r\omega^2 \qquad \textit{Ans.}$$

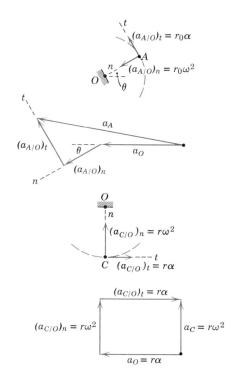

① The counterclockwise angular acceleration α of OA determines the positive direction of $(a_{A/O})_t$. The normal component $(a_{A/O})_n$ is, of course, directed toward the reference center O.

② If the wheel were rolling to the right with the same velocity v_O but still had an acceleration a_O to the left, note that the solution for a_A would be unchanged.

③ We note that the acceleration of the instantaneous center of zero velocity is independent of α and is directed toward the center of the wheel. This conclusion is a useful result to remember.

Sample Problem 5/14

The linkage of Sample Problem 5/8 is repeated here. Crank CB has a constant counterclockwise angular velocity of 2 rad/s in the position shown during a short interval of its motion. Determine the angular acceleration of links AB and OA for this position. Solve by using vector algebra.

Solution. We first solve for the velocities which were obtained in Sample Problem 5/8. They are

$$\omega_{AB} = -6/7 \text{ rad/s} \qquad \text{and} \qquad \omega_{OA} = -3/7 \text{ rad/s}$$

where the counterclockwise direction ($+\mathbf{k}$-direction) is taken as positive. The acceleration equation is

$$\mathbf{a}_A = \mathbf{a}_B + (\mathbf{a}_{A/B})_n + (\mathbf{a}_{A/B})_t$$

where, from Eqs. 5/3 and 5/9a, we may write

① $\qquad \mathbf{a}_A = \boldsymbol{\alpha}_{OA} \times \mathbf{r}_A + \boldsymbol{\omega}_{OA} \times (\boldsymbol{\omega}_{OA} \times \mathbf{r}_A)$

$\qquad\qquad = \alpha_{OA}\mathbf{k} \times 100\mathbf{j} + (-\tfrac{3}{7}\mathbf{k}) \times (-\tfrac{3}{7}\mathbf{k} \times 100\mathbf{j})$

$\qquad\qquad = -100\alpha_{OA}\mathbf{i} - 100(\tfrac{3}{7})^2\mathbf{j} \text{ mm/s}^2$

$\qquad \mathbf{a}_B = \boldsymbol{\alpha}_{CB} \times \mathbf{r}_B + \boldsymbol{\omega}_{CB} \times (\boldsymbol{\omega}_{CB} \times \mathbf{r}_B)$

$\qquad\qquad = 0 + 2\mathbf{k} \times (2\mathbf{k} \times [-75\mathbf{i}])$

$\qquad\qquad = 300\mathbf{i} \text{ mm/s}^2$

② $\qquad (\mathbf{a}_{A/B})_n = \boldsymbol{\omega}_{AB} \times (\boldsymbol{\omega}_{AB} \times \mathbf{r}_{A/B})$

$\qquad\qquad = -\tfrac{6}{7}\mathbf{k} \times [(-\tfrac{6}{7}\mathbf{k}) \times (-175\mathbf{i} + 50\mathbf{j})]$

$\qquad\qquad = (\tfrac{6}{7})^2(175\mathbf{i} - 50\mathbf{j}) \text{ mm/s}^2$

$\qquad (\mathbf{a}_{A/B})_t = \boldsymbol{\alpha}_{AB} \times \mathbf{r}_{A/B}$

$\qquad\qquad = \alpha_{AB}\mathbf{k} \times (-175\mathbf{i} + 50\mathbf{j})$

$\qquad\qquad = -50\alpha_{AB}\mathbf{i} - 175\alpha_{AB}\mathbf{j} \text{ mm/s}^2$

We now substitute these results into the relative-acceleration equation and equate separately the coefficients of the $\mathbf{i}$-terms and the coefficients of the $\mathbf{j}$-terms to give

$$-100\alpha_{OA} = 428.6 - 50\alpha_{AB}$$
$$-18.37 = -36.73 - 175\alpha_{AB}$$

The solutions are

$$\alpha_{AB} = -0.105 \text{ rad/s}^2 \qquad \text{and} \quad \alpha_{OA} = -4.34 \text{ rad/s}^2 \qquad \textit{Ans.}$$

Since the unit vector $\mathbf{k}$ points out from the paper in the positive z-direction, we see that the angular accelerations of AB and OA are both clockwise (negative).

It is recommended that the student sketch each of the acceleration vectors in its proper geometric relationship according to the relative-acceleration equation to help clarify the meaning of the solution.

① Remember to preserve the order of the factors in the cross products.

② In expressing the term $\mathbf{a}_{A/B}$ be certain that $\mathbf{r}_{A/B}$ is written as the vector from B to A not the reverse.

Sample Problem 5/15

The slider-crank mechanism of Sample Problem 5/9 is repeated here. The crank OB has a constant clockwise angular speed of 1500 rev/min. For the instant when the crank angle θ is 60° determine the acceleration of the piston A and the angular acceleration of the connecting rod AB.

Solution.   The acceleration of A may be expressed in terms of the acceleration of the crank pin B. Thus

$$\mathbf{a}_A = \mathbf{a}_B + (\mathbf{a}_{A/B})_n + (\mathbf{a}_{A/B})_t$$

Point B moves in a circle of 125-mm radius with a constant speed so that it has only a normal component of acceleration directed from B to O.

$$[a_n = r\omega^2] \qquad a_B = 0.125\left(\frac{1500[2\pi]}{60}\right)^2 = 3084 \text{ m/s}^2$$

The relative-acceleration terms are visualized with A rotating in a circle relative to B, which is considered fixed, as shown. From Sample Problem 5/9 the angular velocity of AB for these same conditions is $\omega_{AB} = 29.5$ rad/s so that

$$[a_n = r\omega^2] \qquad (a_{A/B})_n = 0.350(29.5)^2 = 305 \text{ m/s}^2$$

directed from A to B. The tangential component $(\mathbf{a}_{A/B})_t$ is known in direction only since its magnitude depends on the unknown angular acceleration of AB. We also know the direction of $\mathbf{a}_A$ since the piston is confined to move along the horizontal axis of the cylinder. There are now only two remaining unknowns left in the equation, namely, the magnitudes of $\mathbf{a}_A$ and $(\mathbf{a}_{A/B})_t$, so the solution may be carried out.

If we adopt an algebraic solution using the geometry of the acceleration polygon, we first compute the angle between AB and the horizontal. With the law of sines this angle becomes 18.0°. Equating separately the horizontal components and the vertical components of the terms in the acceleration equation, as seen from the acceleration polygon, gives

$$a_A = 3084 \cos 60° + 305 \cos 18.0° - (a_{A/B})_t \sin 18.0°$$
$$0 = 3084 \sin 60° - 305 \sin 18.0° - (a_{A/B})_t \cos 18.0°$$

The solution to these equations gives the magnitudes

$$(a_{A/B})_t = 2710 \text{ m/s}^2 \qquad \text{and} \qquad a_A = 994 \text{ m/s}^2 \qquad Ans.$$

With the sense of $(\mathbf{a}_{A/B})_t$ also determined from the diagram, the angular acceleration of AB is seen from the figure representing rotation relative to B to be

$$[\alpha = a_t/r] \qquad \alpha_{AB} = 2710/(0.350) = 7740 \text{ rad/s}^2 \text{ clockwise} \qquad Ans.$$

If we adopt a graphical solution, we begin with the known vectors $\mathbf{a}_B$ and $(\mathbf{a}_{A/B})_n$ and add them head-to-tail using a convenient scale. Next we construct the direction of $(\mathbf{a}_{A/B})_t$ through the head of the last vector. The solution of the equation is obtained by the intersection P of this last line with a horizontal line through the starting point representing the known direction of the vector sum $\mathbf{a}_A$. Scaling the magnitudes from the diagram gives values which agree with the calculated results

$$a_A = 994 \text{ m/s}^2 \qquad \text{and} \qquad (a_{A/B})_t = 2710 \text{ m/s}^2 \qquad Ans.$$

① If the crank OB had an angular acceleration, $\mathbf{a}_B$ would also have a tangential component of acceleration.

② Alternatively the relation $a_n = v^2/r$ may be used for calculating $(a_{A/B})_n$ provided the relative velocity $v_{A/B}$ is used for v. The equivalence is easily seen when it is recalled that $v_{A/B} = r\omega$.

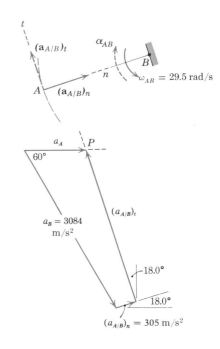

③ Except where extreme accuracy is required do not hesitate to use a graphical solution, as it is quick and reveals the physical relationships among the vectors. The known vectors, of course, may be added in any order as long as the governing equation is satisfied.

PROBLEMS

5/99 The 450-mm-radius fan blade rotates counterclockwise at the constant rate $\dot{\theta} = 2$ rad/s. The
block to which it is pivoted has an acceleration
$a_O = 2.4$ m/s^2. Determine the magnitudes of the
acceleration of points A and B on the blade tips
when $\theta = 30°$.
Ans. $a_A = 1.232$ m/s^2, $a_B = 4.06$ m/s^2

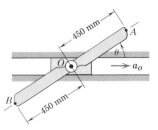

Problem 5/99

5/100 For the rotating blade of Prob. 5/99 with an acceleration $a_O = 2.4$ m/s^2 of its bearing O, determine
the constant angular velocity $\dot{\theta}$ for which the acceleration of A will be zero as the blade crosses the
position $\theta = 0$. What is the acceleration of B at this
same instant?

5/101 If the blade of Prob. 5/99 starts from rest with
$\theta = 0$ and in $\frac{1}{2}$ second acquires a velocity of its
center O of 1.5 m/s and an angular velocity $\dot{\theta} =$
4 rad/s, both with constant acceleration, determine the magnitude of the acceleration of A at
$t = \frac{1}{2}$ s. *Ans.* $a_A = 5.68$ m/s^2

5/102 Determine the angular acceleration α_{AB} of AB for
the position shown if link OB has a constant an-
gular velocity ω.

Problem 5/102

5/103 A container for waste materials is dumped by the
hydraulic-actuated linkage shown. If the piston
rod starts from rest in the position indicated and
has an acceleration of 0.5 m/s^2 in the direction
shown, compute the initial angular acceleration of
the container. *Ans.* $\alpha_{AB} = 0.177$ rad/s^2 CW

Problem 5/103

Problem 5/104

Problem 5/105

Problem 5/107

Problem 5/108

5/104 The sliding collar moves up and down the shaft, causing an oscillation of crank *OB*. If the velocity of *A* is not changing as it passes the null position where *AB* is horizontal and *OB* is vertical, determine the angular acceleration of *OB* in that position.

5/105 The punch of Prob. 5/47 is shown again here where the sector is given an oscillation specified by $\theta = \theta_0 \sin 2\pi t$. Calculate the acceleration of the punch when $\theta = 0$ if $\theta_0 = \pi/12$ rad.

Ans. $a_B = 0.909$ m/s² (up)

5/106 A car with tires of 600-mm diameter accelerates at a constant rate from rest to a velocity of 60 km/h in a distance of 40 m. Determine the magnitude of the acceleration of a point *A* on the top of the wheel as the car reaches a speed of 10 km/h.

5/107 The 9-m steel beam is being hoisted from its horizontal position by the two cables attached at *A* and *B*. If the initial angular accelerations of the hoisting drums are $\alpha_1 = 0.5$ rad/s² and $\alpha_2 = 0.2$ rad/s² in the directions shown, determine the corresponding angular acceleration α of the beam, the acceleration of *C*, and the distance *b* from *B* to a point *P* on the beam centerline that has no acceleration.

Ans. $\alpha = 0.05$ rad/s² CW, $a_C = 0.05$ m/s² down
$b = 2$ m right of *B*

5/108 The sheave with a diameter of 750 mm is elevated by the cable as shown. If the block *B* has a constant upward acceleration of 2 m/s², determine the acceleration of a point *A* which is on the top of the sheave at the time when *B* reaches an upward velocity of 1.5 m/s.

Ans. $\mathbf{a}_A = -1\mathbf{i} - 0.5\mathbf{j}$ m/s²

5/109 The circular disk rolls to the left without slipping. If $\mathbf{a}_{A/B} = -2.7\mathbf{j}$ m/s², determine the velocity and acceleration of the center O of the disk.

Problem 5/109

5/110 The figure for Prob. 5/55 is shown again here. If, in addition to the information shown, the bar A has an acceleration of 2 m/s² to the right and there is no acceleration of bar B, calculate the magnitude of the acceleration of P for the instant depicted. No slipping occurs between the two integral wheels and the bars.

Ans. $a_P = 3.62$ m/s²

Problem 5/110

5/111 The center O of the spool starts from rest and acquires a velocity of 1.2 m/s up the incline with constant acceleration in a distance s of 2.4 m under the action of a steady pull P applied to point A of the cable. The cable is wrapped securely around its hub, and the wheel rolls without slipping. For the 2.4-m position shown, calculate the acceleration of point A on the cable and point B on the spool.

Problem 5/111

5/112 The bicycle of Prob. 5/54 is repeated here. If the cyclist starts from rest and acquires a velocity of 2 m/s in a distance of 3 m with constant acceleration, compute the magnitude of the acceleration of pedal A for this condition if it is in the top position.

Ans. $a_A = 1.28$ m/s²

5/113 Calculate the angular acceleration of the plate in the position shown where control link AO has a constant angular velocity $\omega_{OA} = 4$ rad/s and $\theta = 60°$ for both links.

Problem 5/113

Problem 5/112

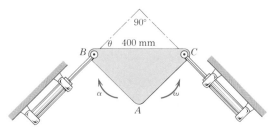

Problem 5/114

5/114 At the instant represented $\theta = 45°$, and the triangular plate ABC has a counterclockwise angular velocity of 20 rad/s and a clockwise angular acceleration of 100 rad/s². Determine the magnitudes of the corresponding velocity $\mathbf{v}$ and acceleration $\mathbf{a}$ of the piston rod of the hydraulic cylinder attached to C. *Ans.* $v = 5.66$ m/s, $a = 84.9$ m/s²

5/115 Determine the acceleration of the piston of Sample Problem 5/15 for (a) $\theta = 0°$, (b) $\theta = 90°$, and (c) $\theta = 180°$. Take the positive x-direction to the right.
Ans. (a) $\mathbf{a}_A = 4186\mathbf{i}$ m/s²
(b) $\mathbf{a}_A = -1179\mathbf{i}$ m/s²
(c) $\mathbf{a}_A = -1983\mathbf{i}$ m/s²

5/116 If the wheel in each case rolls on the circular surface without slipping, determine the acceleration of the point C on the wheel momentarily in contact with the circular surface. The wheel has an angular velocity ω and an angular acceleration α.

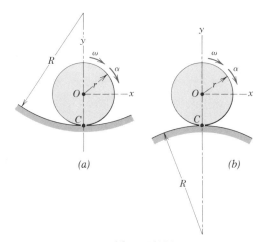

(a)   (b)

Problem 5/116

5/117 The linkage of Prob. 5/60 is shown again here. If $v_A = 0.5$ m/s and is constant when the two links become perpendicular to one another, determine the angular acceleration of CB for this position.
Ans. $\alpha_{CB} = 5.76$ rad/s² CW

5/118 The constrained link of Prob. 5/63 is repeated here. If end A has a constant downward velocity v_A of 2 m/s as the bar passes the position for which $\theta = 30°$, determine the acceleration of the mass center G in the middle of the link. Compare solution by vector algebra with solution by vector geometry.

Problem 5/117

Problem 5/118

5/119 The drive wheel for the diesel-electric locomotive of Prob. 5/20 is shown again here. When the locomotive has traveled 9 m from rest with constant acceleration, each of its driving pinions has a rotational speed of 160 rev/min. For this instant determine the vector expression for the acceleration of point *A*. No slipping occurs between the driving wheel and the track.

Ans. $\mathbf{a}_A = -34.1\mathbf{i} - 0.975\mathbf{j}$ m/s²

5/120 The simplified clam-shell bucket of Prob. 5/57 is shown again here. With the block at *O* considered fixed and with the constant velocity of the control cable at *C* equal to 0.5 m/s, determine the angular acceleration α of the right-hand bucket jaw when θ = 45° as the bucket jaws are closing.

Ans. α = 0.099 rad/s² CW

5/121 The hydraulic cylinder imparts motion to point *B* which causes link *OA* to rotate. For the instant shown where *OA* is vertical and *AB* is horizontal, the velocity v_B of pin *B* is 4 m/s and is increasing at the rate of 20 m/s². For this position determine the angular acceleration of *OA*.

5/122 The elements of a power hacksaw are shown in the figure. The saw blade is mounted in a frame which slides along the horizontal guide. If the motor turns the flywheel at a constant counterclockwise speed of 60 rev/min, determine the acceleration of the blade for the position where θ = 90°, and find the corresponding angular acceleration of the link *AB*.

Ans. $a_A = 4.89$ m/s², $\alpha_{AB} = 0.467$ rad/s² CCW

Problem 5/119

Problem 5/120

Problem 5/122

Problem 5/121

Problem 5/123

5/123 Elements of the switching device of Prob. 5/62 are shown again here. If the velocity v of the control rod is 0.9 m/s and is slowing down at the rate of 6 m/s² when $\theta = 60°$, determine the magnitude of the acceleration of C.

5/124 The deployment mechanism for the spacecraft magnetometer boom of Prob. 5/68 is shown again here. The driving link OB has a constant clockwise angular velocity ω_{OB} of 0.5 rad/s as it crosses the vertical position. Determine the angular acceleration $\boldsymbol{\alpha}_{CA}$ of the boom for the position shown where $\tan \theta = 4/3$.

> *Ans.* $\boldsymbol{\alpha}_{CA} = -0.0758\mathbf{k}$ rad/s²

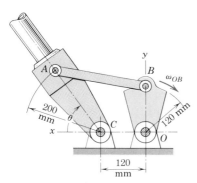

Problem 5/124

5/125 The linkage of Prob. 5/69 is shown again here. If OA has a constant counterclockwise angular velocity $\omega_0 = 10$ rad/s, calculate the angular acceleration of link AB for the position where the coordinates of A are $x = -60$ mm and $y = 80$ mm. Link BC is vertical for this position. Solve by vector algebra. (Use the results of Prob. 5/69 for the angular velocities of AB and BC.)

> *Ans.* $\boldsymbol{\alpha}_{AB} = 10.42\mathbf{k}$ rad/s², $\boldsymbol{\alpha}_{BC} = -19.21\mathbf{k}$ rad/s²

Problem 5/125

▶ **5/126** The assembly-line mechanism of Prob. 5/73 is re-peated here. For the configuration shown, crank *CB* has a constant clockwise angular velocity of π rad/s. Determine the acceleration of *E*. (From previous solution for the velocities in this position we have $\omega_{AB} = 0.286$ rad/s CW and $\omega_{OA} = 0.857$ rad/s CW.)　　　　*Ans.* $a_E = 0.285$ m/s²

Problem 5/126

▶ **5/127** An oil pumping rig is shown in the figure. The flexible pump rod *D* is fastened to the sector at *E* and is always vertical as it enters the fitting below *D*. The link *AB* causes the beam *BCE* to oscillate as the weighted crank *OA* revolves. If *OA* has a constant clockwise speed of 1 rev every 3 s, deter-mine the acceleration of the pump rod *D* when the beam and the crank *OA* are both in the horizontal position shown.　　　　*Ans.* $a_D = 0.568$ m/s² down

Problem 5/127

▶ **5/128** An intermittent-drive mechanism for perforated tape *F* consists of the link *DAB* driven by the crank *OB*. The trace of the motion of the finger at *D* is shown by the dotted line. Determine the accelera-tion of *D* at the instant shown when both *OB* and *CA* are horizontal if *OB* has a constant clockwise rotational velocity of 120 rev/min.　　　*Ans.* $a_D = 1997$ mm/s²

Problem 5/128

5/7 MOTION RELATIVE TO ROTATING AXES

In our discussion of the relative motion of particles in Art. 2/8 and in our use of the relative-motion equations for the plane motion of rigid bodies in this present chapter, we have made all of our relative-velocity and relative-acceleration measurements from *nonrotating* reference axes. The solution of many problems in kinematics where motion is generated within or observed from a system which itself is rotating is greatly facilitated by the use of rotating reference axes. An example of such a motion would be the movement of a fluid particle along the curved vane of a centrifugal pump where the path relative to the vanes of the impeller becomes an important design consideration.

We begin the description of motion using rotating axes by considering the plane motion of two particles *A* and *B* in the fixed *X-Y* plane, Fig. 5/10*a*. For the time being we will consider *A* and *B* to be moving independently of one another for the sake of generality. The motion of *A* will now be observed from a moving reference frame *x-y* which has its origin attached to *B* and which rotates with an angular velocity $\omega = \dot\theta$. We may write this angular velocity as the vector $\boldsymbol{\omega} = \omega\mathbf{k} = \dot\theta\mathbf{k}$ where the vector is normal to the plane of motion and where its positive sense is in the positive *z*-direction (out from the paper) as established by the right-hand rule. The absolute position vector of *A* is given by

$$\mathbf{r}_A = \mathbf{r}_B + \mathbf{r} = \mathbf{r}_B + (x\mathbf{i} + y\mathbf{j}) \tag{5/10}$$

(a)

(b)

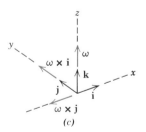

(c)

Figure 5/10

where **i** and **j** are unit vectors attached to the *x-y* frame and $\mathbf{r} = x\mathbf{i} + y\mathbf{j}$ stands for $\mathbf{r}_{A/B}$, the position vector of *A* with respect to *B*. The velocity and acceleration equations require successive differentiation of the position-vector equation with respect to time. In contrast to the case of translating axes treated in Art. 2/8, the unit vectors **i** and **j** are now rotating with the *x-y* axes and, therefore, have time derivatives which must be evaluated. These derivatives may be seen from Fig. 5/10*b* which shows the infinitesimal change in each unit vector during time *dt* as the reference axes rotate through an angle $d\theta = \omega\, dt$. The differential change in **i** is *d***i**, and it has the direction of **j** and a magnitude equal to the angle $d\theta$ times the length of the vector **i**, which is unity. Thus $d\mathbf{i} = d\theta\, \mathbf{j}$. Similarly the unit vector **j** has an infinitesimal change *d***j** which points in the negative *x*-direction, so that $d\mathbf{j} = -d\theta\, \mathbf{i}$. Dividing by *dt* and replacing $d\mathbf{i}/dt$ by $\dot{\mathbf{i}}$, $d\mathbf{j}/dt$ by $\dot{\mathbf{j}}$, and $d\theta/dt$ by $\dot\theta = \omega$ results in

$$\dot{\mathbf{i}} = \omega\mathbf{j} \qquad \text{and} \qquad \dot{\mathbf{j}} = -\omega\mathbf{i}$$

When the cross product is introduced, we see from Fig. 5/10*c* that $\boldsymbol\omega \times \mathbf{i} = \omega\mathbf{j}$ and $\boldsymbol\omega \times \mathbf{j} = -\omega\mathbf{i}$. Thus the time derivatives of the unit vectors may be written as

$$\dot{\mathbf{i}} = \omega \times \mathbf{i} \quad \text{and} \quad \dot{\mathbf{j}} = \omega \times \mathbf{j} \tag{5/11}$$

(*a*) *Relative velocity.* We will now use the expressions of Eqs. 5/11 when we take the time derivative of the position-vector equation for A and B to obtain the relative-velocity relation. Differentiation of Eq. 5/10 gives

$$\dot{\mathbf{r}}_A = \dot{\mathbf{r}}_B + \frac{d}{dt}(x\mathbf{i} + y\mathbf{j})$$

$$= \dot{\mathbf{r}}_B + (x\dot{\mathbf{i}} + y\dot{\mathbf{j}}) + (\dot{x}\mathbf{i} + \dot{y}\mathbf{j})$$

But $x\dot{\mathbf{i}} + y\dot{\mathbf{j}} = \omega \times x\mathbf{i} + \omega \times y\mathbf{j} = \omega \times (x\mathbf{i} + y\mathbf{j}) = \omega \times \mathbf{r}$. Also, since the observer in x-y measures velocity components $\dot{x}$ and $\dot{y}$, we see that $\dot{x}\mathbf{i} + \dot{y}\mathbf{j} = \mathbf{v}_{\text{rel}}$, which is the velocity relative to the x-y frame of reference. Thus the relative-velocity equation becomes

$$\mathbf{v}_A = \mathbf{v}_B + \omega \times \mathbf{r} + \mathbf{v}_{\text{rel}} \tag{5/12}$$

Comparison of Eq. 5/12 with Eq. 2/20 for nonrotating reference axes shows that $\mathbf{v}_{A/B} = \omega \times \mathbf{r} + \mathbf{v}_{\text{rel}}$ from which we conclude that the term $\omega \times \mathbf{r}$ is the difference between the relative velocity as measured from nonrotating and rotating axes.

To illustrate further the meaning of the last two terms in Eq. 5/12, the motion of particle A relative to the rotating x-y plane is shown in Fig. 5/11 as taking place in a curved slot in a plate which represents the rotating x-y reference system. The velocity of A as measured relative to the plate, $\mathbf{v}_{\text{rel}}$, would be tangent to the path fixed in the x-y plate and would have a magnitude $\dot{s}$ where s is measured along the path. This relative velocity may also be viewed as the velocity $\mathbf{v}_{A/P}$ relative to a point P attached to the plate and coincident with A at the instant under consideration. The term $\omega \times \mathbf{r}$ has a magnitude $r\dot{\theta}$ and a direction normal to $\mathbf{r}$ and is the velocity relative to B of point P as seen from nonrotating axes attached to B.

The following comparison will help to establish the equivalence and clarify the difference between the relative-velocity equations written for rotating and nonrotating reference axes.

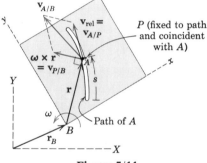

Figure 5/11

$$\mathbf{v}_A = \mathbf{v}_B + \omega \times \mathbf{r} + \mathbf{v}_{\text{rel}}$$

$$\mathbf{v}_A = \underbrace{\mathbf{v}_B + \mathbf{v}_{P/B}}_{} + \mathbf{v}_{A/P} \tag{5/12a}$$

$$\mathbf{v}_A = \underbrace{\mathbf{v}_P \qquad + \mathbf{v}_{A/P}}_{}$$

$$\mathbf{v}_A = \mathbf{v}_B \quad + \quad \mathbf{v}_{A/B}$$

In the second equation the term $\mathbf{v}_{P/B}$ is measured from a nonrotating position—otherwise it would be zero. The term $\mathbf{v}_{A/P}$ is the same as $\mathbf{v}_{\text{rel}}$ and is the velocity of A as measured in the x-y frame. In the third equation $\mathbf{v}_P$ is the absolute velocity of P and represents the effect of the moving coordinate system, both translational and rotational. The fourth equation is the same as that developed for nonrotating axes, Eq. 2/20, and it is seen that $\mathbf{v}_{A/B} = \mathbf{v}_{P/B} + \mathbf{v}_{A/P} = \boldsymbol{\omega} \times \mathbf{r} + \mathbf{v}_{\text{rel}}$.

(*b*) *Transformation of a time derivative.* Equation 5/12 represents a transformation of the time derivative of the position vector between rotating and nonrotating axes. We may easily generalize this result to apply to the time derivative of any vector quantity $\mathbf{V} = V_x \mathbf{i} + V_y \mathbf{j}$. Accordingly, the total time derivative as taken in X-Y is

$$\left(\frac{d\mathbf{V}}{dt}\right)_{XY} = (\dot{V}_x \mathbf{i} + \dot{V}_y \mathbf{j}) + (V_x \dot{\mathbf{i}} + V_y \dot{\mathbf{j}})$$

The first two terms in the expression represent that part of the total derivative of $\mathbf{V}$ which is measured relative to the x-y reference system, and the second two terms represent that part of the derivative due to the rotation of the reference system. With the expressions for $\dot{\mathbf{i}}$ and $\dot{\mathbf{j}}$ from Eqs. 5/11, we may now write

$$\left(\frac{d\mathbf{V}}{dt}\right)_{XY} = \left(\frac{d\mathbf{V}}{dt}\right)_{xy} + \boldsymbol{\omega} \times \mathbf{V} \tag{5/13}$$

Here $\boldsymbol{\omega} \times \mathbf{V}$ represents the difference between the time derivative of the vector as measured in a fixed reference and its time derivative as measured in the rotating reference. As we shall see in Art. 7/2 where three-dimensional motion is introduced, Eq. 5/13 is valid in three dimensions as well as in two dimensions.

Figure 5/12

The physical significance of Eq. 5/13 is illustrated in Fig. 5/12, which shows the vector $\mathbf{V}$ at time t as observed both in the fixed axes X-Y and in the rotating axes x-y. Since we are dealing with the effects of rotation only, the vector may be shown through the origin of coordinates without loss of generality. During time dt the vector swings to position $\mathbf{V}'$, and the observer in x-y measures the components dV due to its change in magnitude and $V\,d\beta$ due to its rotation $d\beta$ relative to x-y. To the rotating observer, then, the derivative $(d\mathbf{V}/dt)_{xy}$ which he measures has the components dV/dt and $V\,d\beta/dt = V\dot{\beta}$. The remaining part of the total time derivative not measured by the rotating observer has the magnitude $V\,d\theta/dt$ and, expressed as a vector, is $\boldsymbol{\omega} \times \mathbf{V}$. Thus we see physically from the diagram that

$$(\dot{\mathbf{V}})_{XY} = (\dot{\mathbf{V}})_{xy} + \boldsymbol{\omega} \times \mathbf{V}$$

which is Eq. 5/13.

(*c*) *Relative acceleration.* The relative-acceleration equation may be obtained by differentiating the relative-velocity relation, Eq. 5/12. Thus

$$\mathbf{a}_A = \mathbf{a}_B + \dot{\boldsymbol{\omega}} \times \mathbf{r} + \boldsymbol{\omega} \times \dot{\mathbf{r}} + \dot{\mathbf{v}}_{\text{rel}}$$

From the derivation of Eq. 5/12 we saw that

$$\dot{\mathbf{r}} = \frac{d}{dt}(x\mathbf{i} + y\mathbf{j}) = (x\dot{\mathbf{i}} + y\dot{\mathbf{j}}) + (\dot{x}\mathbf{i} + \dot{y}\mathbf{j})$$

$$= \boldsymbol{\omega} \times \mathbf{r} + \mathbf{v}_{\text{rel}}$$

Therefore the third term on the right side of the acceleration equation becomes

$$\boldsymbol{\omega} \times \dot{\mathbf{r}} = \boldsymbol{\omega} \times (\boldsymbol{\omega} \times \mathbf{r} + \mathbf{v}_{\text{rel}}) = \boldsymbol{\omega} \times (\boldsymbol{\omega} \times \mathbf{r}) + \boldsymbol{\omega} \times \mathbf{v}_{\text{rel}}$$

With the aid of Eqs. 5/11, the last term on the right side of the equation for $\mathbf{a}_A$ becomes

$$\dot{\mathbf{v}}_{\text{rel}} = \frac{d}{dt}(\dot{x}\mathbf{i} + \dot{y}\mathbf{j}) = (\dot{x}\dot{\mathbf{i}} + \dot{y}\dot{\mathbf{j}}) + (\ddot{x}\mathbf{i} + \ddot{y}\mathbf{j})$$

$$= \boldsymbol{\omega} \times (\dot{x}\mathbf{i} + \dot{y}\mathbf{j}) + (\ddot{x}\mathbf{i} + \ddot{y}\mathbf{j})$$

$$= \boldsymbol{\omega} \times \mathbf{v}_{\text{rel}} + \mathbf{a}_{\text{rel}}$$

Substitution into the expression for $\mathbf{a}_A$ and collection of terms give

$$\boxed{\mathbf{a}_A = \mathbf{a}_B + \dot{\boldsymbol{\omega}} \times \mathbf{r} + \boldsymbol{\omega} \times (\boldsymbol{\omega} \times \mathbf{r}) + 2\boldsymbol{\omega} \times \mathbf{v}_{\text{rel}} + \mathbf{a}_{\text{rel}}} \quad (5/14)$$

Equation 5/14 is the general vector expression for the absolute acceleration of a particle A in terms of its acceleration $\mathbf{a}_{\text{rel}}$ measured relative to a moving coordinate system which rotates with an angular velocity $\boldsymbol{\omega}$ and an angular acceleration $\dot{\boldsymbol{\omega}}$. The terms $\dot{\boldsymbol{\omega}} \times \mathbf{r}$ and $\boldsymbol{\omega} \times (\boldsymbol{\omega} \times \mathbf{r})$ are shown in Fig. 5/13. They represent, respectively, the *t*- and *n*-components of the acceleration $\mathbf{a}_{P/B}$ of the coincident point P in its circular motion with respect to B. This motion would be observed from a set of nonrotating axes moving with B. The magnitude of $\dot{\boldsymbol{\omega}} \times \mathbf{r}$ is $r\ddot{\theta}$ and its direction is tangent to the circle. The magnitude of $\boldsymbol{\omega} \times (\boldsymbol{\omega} \times \mathbf{r})$ is $r\omega^2$ and its direction is from P to B along the normal to the circle. The acceleration of A relative to the plate along the path, $\mathbf{a}_{\text{rel}}$, may be expressed in rectangular, normal and tangential, or polar coordinates in the rotating system. Generally *n*- and *t*-components are most frequently used, and these components are depicted in Fig. 5/13. The tangential component would have the magnitude $(a_{\text{rel}})_t = \ddot{s}$ where s is the distance measured along the path to A. The normal component would have the magnitude v_{rel}^2/ρ where ρ is the radius of curvature of the path as measured in *x-y*. The sense of this vector would always be toward the center of curvature.

Figure 5/13

Figure 5/14

The term $2\boldsymbol{\omega} \times \mathbf{v}_{\text{rel}}$, shown in Fig. 5/13, is known as the *Coriolis* acceleration* and represents the difference between the acceleration of A relative to P as measured from nonrotating axes and from rotating axes. The direction is always normal to the term $\mathbf{v}_{\text{rel}}$, and the sense is established by the right-hand rule for the cross product.

The Coriolis acceleration is difficult to visualize because it is composed of two separate physical effects. To help with this visualization, we will consider the simplest possible motion in which this term appears. In Fig. 5/14*a* we have a rotating disk with a radial slot in which a small particle A is confined to slide. Let the disk turn with a constant angular velocity $\omega = \dot{\theta}$ and let the particle move along the slot with a constant speed $v_{\text{rel}} = \dot{x}$ relative to the slot. The velocity of A has the components $\dot{x}$ due to motion along the slot and $x\omega$ due to the rotation of the slot. The changes in these two velocity components due to the rotation of the disk are shown in the *b*-part of the figure for the interval dt during which the x-y axes rotate with the disk through the angle $d\theta$ to x'-y'. The velocity increment due to the change in direction of $\mathbf{v}_{\text{rel}}$ is $\dot{x}\,d\theta$ and that due to the change in magnitude of $x\omega$ is $\omega\,dx$, both being in the y-direction normal to the slot. Dividing each increment by dt and adding gives the sum $\omega\dot{x} + \dot{x}\omega = 2\dot{x}\omega$, which is the magnitude of the Coriolis acceleration $2\boldsymbol{\omega} \times \mathbf{v}_{\text{rel}}$. Dividing the remaining velocity increment $x\omega\,d\theta$ due to the change in direction of $x\omega$ by dt gives $x\omega\dot{\theta}$ or $x\omega^2$, which is the acceleration of a point P fixed to the slot and momentarily coincident with the particle A. We now see how Eq. 5/14 fits these results. With the origin B in that equation taken at the fixed center O, $\mathbf{a}_B = \mathbf{0}$. With constant angular velocity, $\dot{\boldsymbol{\omega}} \times \mathbf{r} = \mathbf{0}$. With $\mathbf{v}_{\text{rel}}$ constant in magnitude and no curvature to the slot, $\mathbf{a}_{\text{rel}} = \mathbf{0}$. We are left with

$$\mathbf{a}_A = \boldsymbol{\omega} \times (\boldsymbol{\omega} \times \mathbf{r}) + 2\boldsymbol{\omega} \times \mathbf{v}_{\text{rel}}$$

Replacing $\mathbf{r}$ by $x\mathbf{i}$, $\boldsymbol{\omega}$ by $\omega\mathbf{k}$, and $\mathbf{v}_{\text{rel}}$ by $\dot{x}\mathbf{i}$ gives

$$\mathbf{a}_A = -x\omega^2\mathbf{i} + 2\dot{x}\omega\mathbf{j}$$

which checks our analysis from Fig. 5/14. We also note that this same result is contained in our polar-coordinate analysis of plane curvilinear motion in Eq. 2/14 when we let $\ddot{r} = 0$ and $\ddot{\theta} = 0$ and replace r by x and $\dot{\theta}$ by ω. If the slot in the disk of Fig. 5/14 had been curved, we would have had a normal component of acceleration relative to the slot so that $\mathbf{a}_{\text{rel}}$ would not be zero.

The following comparison will help to establish the equivalence and clarify the difference between the relative-acceleration equations written for rotating and nonrotating reference axes.

---

* Named after the French military engineer G. Coriolis (1792–1843), who was the first to call attention to this term.

$$\mathbf{a}_A = \mathbf{a}_B + \underbrace{\dot{\omega} \times \mathbf{r} + \omega \times (\omega \times \mathbf{r})}_{} + \underbrace{2\omega \times \mathbf{v}_{\text{rel}} + \mathbf{a}_{\text{rel}}}_{}$$

$$\mathbf{a}_A = \mathbf{a}_B + \underbrace{\mathbf{a}_{P/B}}_{} + \mathbf{a}_{A/P}$$

$$\mathbf{a}_A = \underbrace{\mathbf{a}_P}_{} + \mathbf{a}_{A/P}$$

$$\mathbf{a}_A = \mathbf{a}_B + \underbrace{\mathbf{a}_{A/B}}_{}$$

$$(5/14a)$$

The equivalence of $\mathbf{a}_{P/B}$ and $\dot{\omega} \times \mathbf{r} + \omega \times (\omega \times \mathbf{r})$, as shown in the second equation, has already been described. From the third equation where $\mathbf{a}_B + \mathbf{a}_{P/B}$ has been combined to give $\mathbf{a}_P$, it is seen that the relative-acceleration term $\mathbf{a}_{A/P}$, unlike the corresponding relative-velocity term, is not equal to the relative acceleration $\mathbf{a}_{\text{rel}}$ measured from the rotating x-y frame of reference. The Coriolis term is, therefore, the difference between the acceleration $\mathbf{a}_{A/P}$ of A relative to P as measured in a nonrotating system and the acceleration $\mathbf{a}_{\text{rel}}$ of A relative to P as measured in a rotating system. From the fourth equation it is seen that the acceleration $\mathbf{a}_{A/B}$ of A with respect to B as measured in a nonrotating system, Eq. 2/21, is a combination of the last four terms in the first equation for the rotating system.

The results expressed by Eq. 5/14 may be visualized somewhat more simply by writing the acceleration of A in terms of the acceleration of the coincident point P. Since the acceleration of P is $\mathbf{a}_P = \mathbf{a}_B + \dot{\omega} \times \mathbf{r} + \omega \times (\omega \times \mathbf{r})$, we may rewrite Eq. 5/14 as

$$\boxed{\mathbf{a}_A = \mathbf{a}_P + 2\omega \times \mathbf{v}_{\text{rel}} + \mathbf{a}_{\text{rel}}} \qquad (5/14b)$$

When written in this form, point P may not be picked at random because it is the one point attached to the rotating reference frame coincident with A at the instant of analysis. Again, reference to Fig. 5/13 should be made to clarify the meaning of each of the terms in Eq. 5/14 and its equivalent Eq. 5/14b.

In summary, once we have chosen our rotating reference system, we must recognize the following quantities:

$\mathbf{v}_B$ = absolute velocity of the origin B of the rotating axes
$\mathbf{a}_B$ = absolute acceleration of the origin B of the rotating axes
$\mathbf{r}$ = position vector of the coincident point P measured from B
ω = angular velocity of the rotating axes
$\dot{\omega}$ = angular acceleration of the rotating axes
$\mathbf{v}_{\text{rel}}$ = velocity of A measured relative to the rotating axes
$\mathbf{a}_{\text{rel}}$ = acceleration of A measured relative to the rotating axes

Also, it is well to be reminded that our vector analysis depends on the consistent use of a right-handed set of coordinate axes. Lastly, note is made of the fact that Eqs. 5/12 and 5/14, developed here for plane motion, hold equally well for space motion, and this extension of generality will be covered in Art. 7/6 on relative motion in three dimensions.

Sample Problem 5/16

At the instant represented the disk with the radial slot is rotating about O with a counterclockwise angular velocity of 4 rad/s which is decreasing at the rate of 10 rad/s². The motion of slider A is separately controlled, and at this instant $r = 150$ mm, $\dot{r} = 125$ mm/s, and $\ddot{r} = 2025$ mm/s². Determine the absolute velocity and acceleration of A for this position.

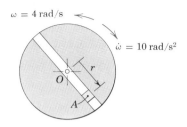

Solution. We have motion relative to a rotating path, so that a rotating coordinate system with origin at O is indicated. We attach x-y axes to the disk and use the unit vectors $\mathbf{i}$ and $\mathbf{j}$.

Velocity. With the origin at O the term $\mathbf{v}_B$ of Eq. 5/12 disappears and we have

① $$\mathbf{v}_A = \boldsymbol{\omega} \times \mathbf{r} + \mathbf{v}_{\text{rel}}$$

The angular velocity as a vector is $\boldsymbol{\omega} = 4\mathbf{k}$ rad/s where $\mathbf{k}$ is the unit
② vector normal to the x-y plane in the $+z$-direction. Our relative-velocity equation becomes

$$\mathbf{v}_A = 4\mathbf{k} \times 0.150\mathbf{i} + 0.125\mathbf{i} = 0.600\mathbf{j} + 0.125\mathbf{i} \text{ m/s} \qquad Ans.$$

in the direction indicated and has the magnitude

$$v_A = \sqrt{(0.600)^2 + (0.125)^2} = 0.613 \text{ m/s} \qquad Ans.$$

Acceleration. Equation 5/14 written for zero acceleration of the origin of the rotating coordinate system is

$$\mathbf{a}_A = \boldsymbol{\omega} \times (\boldsymbol{\omega} \times \mathbf{r}) + \dot{\boldsymbol{\omega}} \times \mathbf{r} + 2\boldsymbol{\omega} \times \mathbf{v}_{\text{rel}} + \mathbf{a}_{\text{rel}}$$

The terms become

③ $$\boldsymbol{\omega} \times (\boldsymbol{\omega} \times \mathbf{r}) = 4\mathbf{k} \times (4\mathbf{k} \times 0.150\mathbf{i}) = 4\mathbf{k} \times 0.6\mathbf{j} = -2.4\mathbf{i} \text{ m/s}^2$$

$$\dot{\boldsymbol{\omega}} \times \mathbf{r} = -10\mathbf{k} \times 0.150\mathbf{i} = -1.5\mathbf{j} \text{ m/s}^2$$

$$2\boldsymbol{\omega} \times \mathbf{v}_{\text{rel}} = 2(4\mathbf{k}) \times 0.125\mathbf{i} = 1.0\mathbf{j} \text{ m/s}^2$$

$$\mathbf{a}_{\text{rel}} = 2.025\mathbf{i} \text{ m/s}^2$$

The total acceleration is, therefore,

$$\mathbf{a}_A = (2.025 - 2.4)\mathbf{i} + (1.0 - 1.5)\mathbf{j} = -0.375\mathbf{i} - 0.5\mathbf{j} \text{ m/s}^2 \qquad Ans.$$

in the direction indicated and has the magnitude

$$a_A = \sqrt{(0.375)^2 + (0.5)^2} = 0.625 \text{ m/s}^2 \qquad Ans.$$

Vector notation is certainly not essential to the solution of this problem. The student should be able to work out the steps with scalar notation just as easily. The correct direction of the Coriolis-acceleration term may always be found by the direction in which the head of the $\mathbf{v}_{\text{rel}}$ vector would move if rotated about its tail in the sense of ω as shown.

① This equation is the same as $\mathbf{v}_A = \mathbf{v}_P + \mathbf{v}_{A/P}$ where P is a point attached to the disk coincident with A at this instant.

② Note that the x-y-z axes chosen constitute a right-handed system.

③ Be sure to recognize that $\boldsymbol{\omega} \times (\boldsymbol{\omega} \times \mathbf{r})$ and $\dot{\boldsymbol{\omega}} \times \mathbf{r}$ represent the normal and tangential components of acceleration of a point P on the disk coincident with A. This description becomes that of Eq. 5/14b.

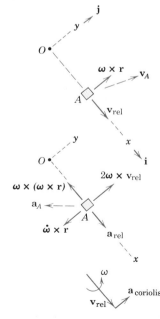

Sample Problem 5/17

The pin A of the hinged link AC is confined to move in the rotating slot of link OD. The angular velocity of OD is $\omega = 2$ rad/s clockwise and is constant for the interval of motion concerned. For the position where $\theta = 45°$ with AC horizontal, determine the velocity of pin A and the velocity of A relative to the rotating slot in OD.

Solution. Motion of a point (pin A) along a rotating path (the slot) suggests the use of rotating coordinate axes x-y attached to arm OD. With the origin at the fixed point O, the term $\mathbf{v}_B$ of Eq. 5/12 vanishes, and we have $\mathbf{v}_A = \boldsymbol{\omega} \times \mathbf{r} + \mathbf{v}_{\text{rel}}$.

The velocity of A in its circular motion about C is

$$\mathbf{v}_A = \boldsymbol{\omega}_{CA} \times \mathbf{r}_{CA} = \omega_{CA}\mathbf{k} \times (225/\sqrt{2})(-\mathbf{i} - \mathbf{j}) = (225/\sqrt{2})\omega_{CA}(\mathbf{i} - \mathbf{j})$$

① where the angular velocity ω_{CA} is arbitrarily assigned in a clockwise sense in the positive z-direction ($+\mathbf{k}$).

The angular velocity ω of the rotating axes is that of the arm OD and, by the right-hand rule, is $\boldsymbol{\omega} = \omega\mathbf{k} = 2\mathbf{k}$ rad/s. The vector from the origin to the point P on OD coincident with A is $\mathbf{r} = \overline{OP}\mathbf{i} = \sqrt{(450 - 225)^2 + (225)^2}\,\mathbf{i} = 225\sqrt{2}\mathbf{i}$ mm. Thus

$$\boldsymbol{\omega} \times \mathbf{r} = 2\mathbf{k} \times 225\sqrt{2}\mathbf{i} = 450\sqrt{2}\mathbf{j} \text{ mm/s}$$

Lastly, the relative-velocity term $\mathbf{v}_{\text{rel}}$ is the velocity measured by an observer attached to the rotating reference frame and is $\mathbf{v}_{\text{rel}} = \dot{x}\mathbf{i}$. Substitution into the relative-velocity equation gives

$$(225/\sqrt{2})\omega_{CA}(\mathbf{i} - \mathbf{j}) = 450\sqrt{2}\mathbf{j} + \dot{x}\mathbf{i}$$

Equating separately the coefficients of the $\mathbf{i}$ and $\mathbf{j}$ terms gives

$$(225/\sqrt{2})\omega_{CA} = \dot{x} \quad \text{and} \quad -(225/\sqrt{2})\omega_{CA} = 450\sqrt{2}$$

giving $\quad \omega_{CA} = -4$ rad/s $\quad$ and $\quad \dot{x} = v_{\text{rel}} = -450\sqrt{2}$ mm/s $\quad$ *Ans.*

With a negative value for ω_{CA} the actual angular velocity of CA is counterclockwise, so the velocity of A is up with a magnitude of

② $$v_A = 225(4) = 900 \text{ mm/s} \quad\quad\quad \textit{Ans.}$$

Geometric clarification of the terms is helpful and is easily shown. Using the equivalence between the third and the first of Eqs. 5/12a with $\mathbf{v}_B = \mathbf{0}$ enables us to write $\mathbf{v}_A = \mathbf{v}_P + \mathbf{v}_{A/P}$ where P is the point on the rotating arm OD coincident with A. Clearly, $v_P = \overline{OP}\omega = 225\sqrt{2}(2) = 450\sqrt{2}$ mm/s and its direction is normal to OD. The relative velocity $\mathbf{v}_{A/P}$, which is the same as $\mathbf{v}_{\text{rel}}$, is seen from the figure to be along the slot toward O. This conclusion becomes clear when it is observed that A is approaching P along the slot from below before coincidence and is receding from P upward along the slot following coincidence. The velocity of A is tangent to its circular arc about C. The vector equation may now be satisfied since there are only two remaining scalar unknowns, namely, the magnitude of $\mathbf{v}_{A/P}$ and the magnitude of $\mathbf{v}_A$. For the 45° position the figure requires $v_{A/P} = 450\sqrt{2}$ mm/s and $v_A = 900$ mm/s each in its direction shown. The angular velocity of AC is

$$[\omega = v/r] \quad\quad \omega_{AC} = v_A/\overline{AC} = 900/225 = 4 \text{ rad/s counterclockwise}$$

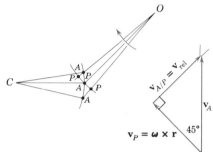

① It is clear enough physically that CA will have a counterclockwise angular velocity for the conditions specified, so we anticipate a negative value for ω_{CA}.

② Solution of the problem is not restricted to the reference axes used. Alternatively, the origin of the x-y axes, still attached to OD, could be chosen at the coincident point P on CD. This choice would merely replace the $\boldsymbol{\omega} \times \mathbf{r}$ term by its equal, $\mathbf{v}_P$. As a further selection, all vector quantities could be expressed in terms of X-Y components using unit vectors $\mathbf{I}$ and $\mathbf{J}$.

A direct conversion between the two reference systems is obtained from the geometry of the unit circle and gives

$$\mathbf{i} = \mathbf{I} \cos\theta - \mathbf{J} \sin\theta$$

and $\quad \mathbf{j} = \mathbf{I} \sin\theta + \mathbf{J} \cos\theta$

Sample Problem 5/18

For the conditions of Sample Problem 5/17 determine the angular acceleration of AC and the acceleration of A relative to the rotating slot in arm OD.

Solution. We attach the rotating coordinate system x-y to arm OD and use Eq. 5/14. With the origin at the fixed point O, the term $\mathbf{a}_B$ becomes zero so that

$$\mathbf{a}_A = \dot{\boldsymbol{\omega}} \times \mathbf{r} + \boldsymbol{\omega} \times (\boldsymbol{\omega} \times \mathbf{r}) + 2\boldsymbol{\omega} \times \mathbf{v}_{\text{rel}} + \mathbf{a}_{\text{rel}}$$

From the solution to Sample Problem 5/17 we use the values $\boldsymbol{\omega} = 2\mathbf{k}$ rad/s, $\boldsymbol{\omega}_{CA} = -4\mathbf{k}$ rad/s, $\mathbf{v}_{\text{rel}} = -450\sqrt{2}\mathbf{i}$ mm/s and write

$$\mathbf{a}_A = \dot{\boldsymbol{\omega}}_{CA} \times \mathbf{r}_{CA} + \boldsymbol{\omega}_{CA} \times (\boldsymbol{\omega}_{CA} \times \mathbf{r}_{CA})$$

$$= \dot{\omega}_{CA}\mathbf{k} \times \frac{225}{\sqrt{2}}(-\mathbf{i} - \mathbf{j}) - 4\mathbf{k} \times \left(-4\mathbf{k} \times \frac{225}{\sqrt{2}}[-\mathbf{i} - \mathbf{j}]\right)$$

$$\dot{\boldsymbol{\omega}} \times \mathbf{r} = 0 \text{ since } \omega = \text{constant}$$

$$\boldsymbol{\omega} \times (\boldsymbol{\omega} \times \mathbf{r}) = 2\mathbf{k} \times (2\mathbf{k} \times 225\sqrt{2}\mathbf{i}) = -900\sqrt{2}\mathbf{i} \text{ mm/s}^2$$

$$2\boldsymbol{\omega} \times \mathbf{v}_{\text{rel}} = 2(2\mathbf{k}) \times (-450\sqrt{2}\mathbf{i}) = -1800\sqrt{2}\mathbf{j} \text{ mm/s}^2$$

$$\mathbf{a}_{\text{rel}} = \ddot{x}\mathbf{i}$$

Substitution into the relative-acceleration equation yields

$$\frac{1}{\sqrt{2}}(225\dot{\omega}_{CA} + 3600)\mathbf{i} + \frac{1}{\sqrt{2}}(-225\dot{\omega}_{CA} + 3600)\mathbf{j}$$
$$= -900\sqrt{2}\mathbf{i} - 1800\sqrt{2}\mathbf{j} + \ddot{x}\mathbf{i}$$

Equating separately the $\mathbf{i}$ and $\mathbf{j}$ terms gives

$$(225\dot{\omega}_{CA} + 3600)/\sqrt{2} = -900\sqrt{2} + \ddot{x}$$

and $\qquad (-225\dot{\omega}_{CA} + 3600)/\sqrt{2} = -1800\sqrt{2}$

Solving for the two unknowns gives

$$\dot{\omega}_{CA} = 32 \text{ rad/s}^2 \qquad \text{and} \qquad \ddot{x} = a_{\text{rel}} = 8910 \text{ mm/s}^2 \qquad Ans.$$

If desired, the acceleration of A may also be written as

$$\mathbf{a}_A = (225/\sqrt{2})(32)(\mathbf{i} - \mathbf{j}) + (3600/\sqrt{2})(\mathbf{i} + \mathbf{j}) = 7637\mathbf{i} - 2546\mathbf{j} \text{ mm/s}^2$$

We make use here of the geometric representation of the relative-acceleration equation to further clarify the problem. The geometric approach may be used as an alternative solution. Again, we introduce point P on OD coincident with A. The equivalent scalar terms are

$$(a_A)_t = |\dot{\boldsymbol{\omega}}_{CA} \times \mathbf{r}_{CA}| = r\dot{\omega}_{CA} = r\alpha_{CA} \text{ normal to } CA, \text{ sense unknown}$$

$$(a_A)_n = |\boldsymbol{\omega}_{CA} \times (\boldsymbol{\omega}_{CA} \times \mathbf{r}_{CA})| = r\omega_{CA}^2 \text{ from } A \text{ to } C$$

$$(a_P)_n = |\boldsymbol{\omega} \times (\boldsymbol{\omega} \times \mathbf{r})| = \overline{OP}\omega^2 \text{ from } P \text{ to } O$$

$$(a_P)_t = |\dot{\boldsymbol{\omega}} \times \mathbf{r}| = r\dot{\omega} = 0 \text{ since } \omega = \text{constant}$$

$$|2\boldsymbol{\omega} \times \mathbf{v}_{\text{rel}}| = 2\omega v_{\text{rel}} \text{ directed as shown}$$

$$a_{\text{rel}} = \ddot{x} \text{ along } OD, \text{ sense unknown}$$

We start with the known vectors and add them head-to-tail for each side of the equation beginning at R and ending at S where the intersection of the known directions of $(\mathbf{a}_A)_t$ and $\mathbf{a}_{\text{rel}}$ establishes the solution. Closure of the polygon determines the sense of each of the two unknown vectors, and their magnitudes are easily calculated from the geometry of the figure.

(1) If the slot had been curved with a radius of curvature ρ, the term $\mathbf{a}_{\text{rel}}$ would have had a component v_{rel}^2/ρ normal to the slot and directed toward the center of curvature in addition to its component along the slot.

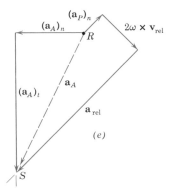

(e)

(2) It is always possible to avoid a simultaneous solution by projecting the vectors onto the perpendicular to one of the unknowns.

PROBLEMS

5/129 Car *B* is rounding the curve with a constant speed of 72 km/h, and car *A* is approaching car *B* in the intersection with a constant speed of 54 km/h. Determine the velocity which car *A* appears to have to an observer riding in and turning with car *B*. The *x*-*y* axes are attached to car *B*. Is this apparent velocity the negative of the velocity which *B* appears to have to a nonrotating observer in car *A*? The distance separating the two cars at the instant depicted is 20 m.

Ans. $\mathbf{v}_{\text{rel}} = 15\mathbf{i} - 22\mathbf{j}$ m/s, no

Problem 5/129

5/130 Cars *A* and *B* are rounding the curves with equal speeds of 72 km/h. Determine the velocity which *A* appears to have to an observer riding in and turning with car *B* for the instant represented. Does the curvature of the road for car *A* affect the result? Axes *x*-*y* are attached to car *B*.

Ans. $\mathbf{v}_{\text{rel}} = -46\mathbf{i}$ m/s

Problem 5/130

5/131 The disk with the radial slot rotates in a horizontal plane about *O* with an angular acceleration $\dot{\omega} = 20$ rad/s² while the 2-kg slider *A* moves with a constant speed $\dot{x} = 120$ mm/s relative to the slot during a certain interval of its motion. If the angular velocity of the disk is $\omega = 15$ rad/s at the instant when the slider crosses the center *O* of rotation of the disk, determine the acceleration of the slider at this instant.

Problem 5/131

5/132 In negotiating an unbanked curve at a steady speed of 25 km/h, the center *C* of the railroad car follows a circular path of radius $\rho = 60$ m. The longitudinal axis of the car remains tangent to the circle. Determine the absolute velocity **v** of a person *P* who walks at the constant speed of 1.5 m/s relative to the car when he is at points *A*, *B*, and *C*. Use the *x*-*y* axes attached to the car as shown.

Ans. $\mathbf{v}_A = -8.44\mathbf{i} + 1.157\mathbf{j}$ m/s
$\mathbf{v}_C = -8.44\mathbf{i}$ m/s
$\mathbf{v}_B = -8.44\mathbf{i} - 1.157\mathbf{j}$ m/s

Problem 5/132

Problem 5/133

Problem 5/135

Problem 5/137

5/133 The space shuttle *A* is in an equatorial circular orbit of 240-km altitude and is moving from west to east. Determine the velocity and acceleration which it appears to have to an observer *B* fixed to and rotating with the earth at the equator as the shuttle passes overhead. Use *R* = 6378 km for the radius of the earth. Also use Fig. 1/1 for the appropriate value of *g* and carry out your calculations to 4-figure accuracy.

5/134 Consider a straight and level railroad track with a 50 000-kg railroad car moving along it at a constant speed of 15 m/s. Determine the horizontal force *R* exerted by the rails on the car if the track were located hypothetically at (*a*) the north pole and (*b*) the equator, oriented in a north-south direction.

5/135 A vehicle *A* travels west at high speed on a perfectly straight road *B* which is tangent to the earth's surface at the equator. The road has no curvature whatsoever in the vertical plane. Determine the necessary speed v_{rel} of the vehicle relative to the road which will give rise to zero acceleration of the vehicle in the vertical direction. Assume that the center of the earth has no acceleration.
Ans. v_{rel} = 837 km/h

5/136 If the road of Prob. 5/135, instead of being straight with no curvature, followed the curvature of the earth's surface, determine the necessary speed v_{rel} of the vehicle to the west relative to the road which will give rise to zero vertical acceleration of the vehicle. Assume that the center of the earth has no acceleration.

5/137 The disk with the circular slot rotates about *O* with a constant counterclockwise angular velocity ω = 10 rad/s. The pin *A* moves in the slot so that its radial line *AO* rotates relative to the line *OB*, fixed to the disk, with a constant rate $\dot{\beta}$ = 5 rad/s for a certain interval of its motion. Solve for the acceleration of *A* by both Eqs. 2/14 and 5/14.
Ans. $a = a_x = -33.8$ m/s^2

5/138 The slider A moves in the slot at the same time that the disk rotates about its center O with an angular speed ω positive in the counterclockwise sense. Determine the x- and y-components of the absolute acceleration of A if, at the instant represented, $\omega = 5$ rad/s, $\dot{\omega} = -10$ rad/s^2, $x = 100$ mm, $\dot{x} = 150$ mm/s, and $\ddot{x} = 500$ mm/s^2.

Problem 5/138   **Problem 5/139**

5/139 The disk with the circular slot of 200-mm radius rotates about O with a constant angular velocity $\omega = 15$ rad/s. Determine the acceleration of the slider A at the instant when it passes the center of the disk if, at that moment, $\dot{\theta} = 12$ rad/s and $\ddot{\theta} = 0$.   *Ans.* $\mathbf{a}_A = 43.2\mathbf{j}$ m/s^2

5/140 A radar station B situated at the equator observes a satellite A in a circular equatorial orbit of 200-km altitude and moving from west to east. When the satellite is 30° above the horizon, determine the difference between the velocity of the satellite relative to the radar station, as measured from a nonrotating frame of reference, and the velocity as measured relative to the reference frame of the radar station.

Problem 5/140

5/141 Two satellites are in circular equatorial orbits of different altitudes. Satellite A is in a geosynchronous orbit (one with the same period as the earth's rotation so that it "hovers" over the same spot on the equator). Satellite B has an orbit of radius $r_B = 30\,000$ km. Calculate the velocity which A appears to have to an observer fixed in B when the elevation angle θ is (*a*) 0° and (*b*) 90°. The x-y axes are attached to B whose antenna always points toward the center of the earth ($-y$-direction). Consult Art. 3/12 and Appendix D for necessary orbital information.

Ans. (*a*) $\mathbf{v}_{\text{rel}} = 5250\mathbf{i} - 5187\mathbf{j}$ km/h
(*b*) $\mathbf{v}_{\text{rel}} = 7380\mathbf{i}$ km/h

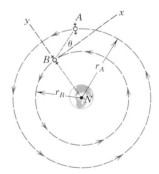

Problem 5/141

5/142 If the cars of Prob. 5/130 both have a constant speed of 72 km/h as they round the curves, determine the acceleration which A appears to have to an observer riding in and turning with car B for the instant represented. Axes x-y are attached to car B.

Problem 5/143

Problem 5/144

Problem 5/145

Problem 5/146

5/143 Each of the two cars A and B is traveling with a constant speed of 72 km/h. Determine the velocity and acceleration of car A as seen by an observer moving and rotating with car B when the cars are in the positions shown. The x-y axes are attached to car B. Sketch both relative motion vectors.

Ans. $\mathbf{v}_{rel} = -47.3\mathbf{i} + 10\mathbf{j}$ m/s
$\mathbf{a}_{rel} = -4\mathbf{i} - 12.93\mathbf{j}$ m/s^2

5/144 The fire truck is moving forward at a speed of 60 km/h and is decelerating at the rate of 3 m/s^2. Simultaneously, the ladder is being raised and extended. At the instant considered the angle θ is 30° and is increasing at the constant rate of 10 deg/s. Also at this instant the extension b of the ladder is 1.5 m, with $\dot{b} = 0.6$ m/s and $\ddot{b} = -0.3$ m/s^2. For this instant determine the acceleration of the end A of the ladder (a) with respect to the truck and (b) with respect to the ground.

5/145 For an alternative solution to Prob. 5/144 assign r-θ coordinates with origin at B as shown. Then make use of the polar coordinate relations for the acceleration of A relative to B. The r- and θ-components of the absolute acceleration should coincide with the components along and normal to the ladder as found in Prob. 5/144.

Ans. $\mathbf{a}_A = -3.13\mathbf{e}_r + 1.709\mathbf{e}_\theta$ m/s^2

5/146 Two boys A and B are sitting on opposite sides of a horizontal turntable which rotates at a constant counterclockwise angular velocity ω as seen from above. Boy A throws a ball toward B by giving it a horizontal velocity $\mathbf{u}$ relative to the turntable toward B. Assume that the ball has no horizontal acceleration once released and write an expression for the acceleration $\mathbf{a}_{rel}$ which B would observe the ball to have in the plane of the turntable just after it is thrown. Sketch the path of the ball on the turntable as observed by B.

5/147 A boy starts from the center O of a merry-go-round and walks along the radial line OM fixed to the rotating platform. When he is 2.4 m from the center O of the platform at point P, his velocity relative to the platform in the direction of M is 0.6 m/s and is decreasing at the rate of 0.3 m/s². The merry-go-round rotates in the direction shown at the constant rate of one revolution each 10 s. If $\theta = 30°$ when the boy reaches P, determine (a) his absolute velocity in terms of X-Y components in the ground-fixed axes (use unit vectors $\mathbf{I}$ and $\mathbf{J}$) and (b) his absolute acceleration in terms of x-y components.

$$Ans.\ \mathbf{v}_A = -0.234\mathbf{I} + 1.606\mathbf{J}\ \text{m/s}$$
$$\mathbf{a}_A = -1.247\mathbf{i} + 0.754\mathbf{j}\ \text{m/s}^2$$

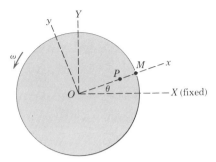

Problem 5/147

5/148 A tall building is situated on the equator. The north face of the building houses a standard 12-h clock whose center is a distance h above the ground (which is essentially at sea level). Develop expressions for the velocity and acceleration of the tip A of the hour hand at 12 o'clock as measured from an earth-centered nonrotating coordinate system. Take the positive direction of the x-axis to be from the center of the earth radially outward toward the building and that of the z-axis toward the north. The hour hand has a length l and makes two complete rotations relative to the clock during one complete rotation of the earth. The radius and angular velocity of the earth are R and ω.

5/149 Near the end of its takeoff roll, the airplane is "rotating" (nose pitching up) just prior to lift-off. The velocity and acceleration of the aircraft, expressed in terms of the motion of the wheel assembly C, are v_C and a_C, both directed horizontally forward. The pitch angle is θ, and the pitch rate $\omega = \dot{\theta}$ is increasing at the rate $\alpha = \dot{\omega}$. If a person A is walking forward in the center aisle with a velocity and an acceleration v_{rel} and a_{rel}, both measured forward relative to the cabin, derive expressions for the velocity and acceleration of A as observed by a ground-fixed observer.

$$Ans.\ \mathbf{v}_A = (v_C \cos \theta - \omega h + \dot{L})\mathbf{i} + (\omega L - v_C \sin \theta)\mathbf{j}$$
$$\mathbf{a}_A = (a_C \cos \theta - \alpha h - L\omega^2 + \ddot{L})\mathbf{i}$$
$$+ (-a_C \sin \theta - h\omega^2 + L\alpha + 2\omega \dot{L})\mathbf{j}$$

Problem 5/149

Problem 5/150

5/150 The figure shows the vanes of a centrifugal-pump impeller which turns with a constant clockwise speed of 200 rev/min. The fluid particles are observed to have an absolute velocity whose component in the *r*-direction is 3 m/s at discharge from the vane. Furthermore, the magnitude of the velocity of the particles measured relative to the vane is increasing at the rate of 24 m/s² just before they leave the vane. Determine the magnitude of the total acceleration of a fluid particle an instant before it leaves the impeller. The radius of curvature ρ of the vane at its end is 200 mm.

Problem 5/151

5/151 In the position shown the bar *DC* is rotating counterclockwise at the constant rate $N = 2$ rad/s. Determine the angular velocity ω and the angular acceleration α of *EBO* at this instant.

Ans. $\omega = 2$ rad/s CCW
$\alpha = 8$ rad/s² CW

Problem 5/152

5/152 The crank *OA* revolves clockwise with a constant angular velocity of 10 rad/s within a limited arc of its motion. For the position $\theta = 30°$ determine the angular velocity of the slotted link *CB* and the acceleration of *A* as measured relative to the slot in *CB*.

5/153 The Geneva wheel of Prob. 5/41 is shown again here. Determine the angular acceleration α_2 of wheel C for the instant when $\theta = 20°$. Wheel A has a constant clockwise angular velocity of 2 rad/s. *Ans.* $\alpha_2 = 16.53$ rad/s² CCW

Problem 5/153

5/154 The pin A in the bell crank AOD is guided by the flanges of the collar B, which slides with a constant velocity v_B of 0.9 m/s along the fixed shaft. For the position $\theta = 30°$ determine the acceleration of the plunger CE, whose upper end is positioned by the radial slot in the bell crank. *Ans.* $a_C = 24.9$ m/s² up

Problem 5/154

5/155 One wheel of an experimental vehicle F, which has a constant velocity $v = 36$ km/h, is shown. The wheel rolls without slipping and causes an oscillation of the slotted arm through the action of its pin A. Control rod DB, in turn, moves back and forth relative to the vehicle by virtue of the motion imparted to pin B. For the position shown, determine the acceleration a_B of the control rod DB. (*Suggestion:* Consider the justification and advantage of using a reference frame attached to the vehicle.) *Ans.* $a_B = 27.3$ m/s² to the right

Problem 5/155

5/8   PROBLEM FORMULATION AND REVIEW

In Chapter 5 we have applied our knowledge of basic kinematics, which we developed in Chapter 2, to the plane motion of rigid bodies. We approached the problem essentially in two ways. First we wrote an equation which described the general geometric configuration of a given problem in terms of knowns and variables. Then we differentiated the equation with respect to time to obtain velocities and accelerations both linear and angular. We called this approach an absolute-motion analysis, and it is satisfactory when the geometric expression and its time derivatives can be easily written.

In our second approach we applied the principles of relative motion to rigid bodies and found that this approach enabled us to solve many problems which would be too awkward to handle by mathematical differentiation. The relative-velocity equation, the instantaneous center of zero velocity, and the relative-acceleration equation all require that we visualize clearly and analyze correctly the case of circular motion of one point around another point as viewed from nonrotating axes. Recognition of the relationships for circular motion is certainly an essential key without which little progress can be made.

The relative-velocity and relative-acceleration relationships are vector equations which we may solve in any one of three ways: by a scalar-geometric analysis of the vector polygon, by vector algebra, or by a graphical construction of the vector polygon. Each method has its advantages, and the student is urged to gain experience with all three. If one uses a vector-algebra approach, he should sketch the geometry of the vectors in any event, because the ability to visualize physical-geometrical relationships of force and motion is probably the most important experience of all in the study of mechanics.

In Chapter 6 we shall study the kinetics of rigid bodies in plane motion. Here we find that the ability to solve for the linear and angular accelerations of rigid bodies is absolutely necessary in order to apply the force and moment equations which relate the applied forces to the resulting motions. Thus the material of Chapter 5 is essential to that in Chapter 6.

Finally, in Chapter 5 we introduced rotating coordinate systems which enable us to solve problems where the motion is observed relative to a rotating frame of reference. Whenever a point moves along a path which itself is turning, analysis by rotating axes is indicated if a relative-motion approach is used. In deriving Eq. 5/12 for relative velocity and Eq. 5/14 for relative acceleration, where the relative terms are measured from a rotating reference system, it was necessary for us to account for the time derivatives of the unit vectors $\mathbf{i}$ and $\mathbf{j}$. Equations 5/12 and 5/14 also apply to spatial motion, as will be shown in Chapter 7.

REVIEW PROBLEMS

5/156 A point P located on a rigid body rotating about the fixed z-axis has a velocity $\mathbf{v} = 3\mathbf{i} - 2\mathbf{j}$ m/s at a certain instant. If the angular velocity of the body is $\omega = 4\mathbf{k}$ rad/s at this same instant, calculate the coordinates of P.   *Ans.* $x = -\frac{1}{2}$ m, $y = -\frac{3}{4}$ m

5/157 A point P on a rigid body which rotates in the x-y plane about an axis through the origin O has coordinates $x = -75$ mm and $y = 100$ mm at a certain instant. At this same instant the angular velocity and angular acceleration of the body around the z-axis are $\omega = 2\mathbf{k}$ rad/s and $\alpha = -3\mathbf{k}$ rad/s², respectively. Determine the vector expression for the velocity $\mathbf{v}$ and acceleration $\mathbf{a}$ of P for this instant. Identify the normal and tangential components of the acceleration.

$$Ans. \ \mathbf{v} = -200\mathbf{i} - 150\mathbf{j} \text{ mm/s}$$
$$\mathbf{a}_n = 300\mathbf{i} - 400\mathbf{j} \text{ mm/s}^2$$
$$\mathbf{a}_t = 300\mathbf{i} + 225\mathbf{j} \text{ mm/s}^2$$

5/158 The equilateral triangular plate is guided by the two vertex rollers A and B which are confined to move in the perpendicular slots. The control rod gives A a constant velocity v_A to the left for an interval of its motion. Determine the value of θ for which the horizontal component of the velocity of C is zero.

Problem 5/158

5/159 The frictional resistance to the rotation of a flywheel consists of a retardation due to air friction which varies as the square of the angular velocity and a constant frictional retardation in the bearing. As a result the angular acceleration of the flywheel while it is allowed to coast is given by $\alpha = -K - k\omega^2$ where K and k are constants. Determine an expression for the time required for the flywheel to come to rest from an initial angular velocity ω_0.

$$Ans. \ t = \frac{1}{\sqrt{Kk}} \tan^{-1}\left(\omega_0\sqrt{\frac{k}{K}}\right)$$

Problem 5/160

5/160 The telephone-cable reel is rolled down the incline by the cable leading from the upper drum and wrapped around the inner hub of the reel. If the drum turns at the constant rate $\omega_1 = 2$ rad/s, calculate the time required for the center of the reel to move 30 m along the incline.

Problem 5/161

5/161 The wheel rolls to the left without slipping. If the velocity of B relative to A is 1.5 m/s, determine the magnitude of the velocity of point P for this position. *Ans.* $v_P = 1.186$ m/s

Problem 5/162

5/162 The hydraulic cylinder moves pin A to the right with a constant velocity v. Use the fact that the distance from A to B is invariant, where B is the point on AC momentarily in contact with the gear, and write an expression for the angular velocity ω of the gear and the angular velocity of the rack AC.

Problem 5/163

5/163 In the linkage shown OC has a constant clockwise angular velocity $\omega = 2$ rad/s during an interval of motion while the hydraulic cylinder gives pin A a constant velocity of 1.2 m/s to the right. For the position shown where OC is vertical and BC is horizontal calculate the angular velocity of BC. Solve by drawing the necessary velocity polygon. *Ans.* $\omega_{BC} = 2$ rad/s CW

Problem 5/164

5/164 Solve Prob. 5/163 by vector algebra starting with the position-vector equation $\mathbf{r}_C + \mathbf{r}_{B/C} = \mathbf{r}_A + \mathbf{r}_{B/A}$ as shown in the separate diagram here.

5/165 The tilting device maintains a sloshing water bath for washing vegetable produce. If the crank *OA* oscillates about the vertical and has a clockwise angular velocity of 4π rad/s when *OA* is vertical, determine the angular velocity of the basket in the position shown where $\theta = 30°$.

Ans. $\omega_{DB} = 3.24$ rad/s CW

Problem 5/165

5/166 The figure illustrates a commonly used quick-return mechanism which produces a slow cutting stroke of the tool (attached to *D*) and a rapid return stroke. If the driving crank *OA* is turning at the constant rate $\dot{\theta} = 3$ rad/s, determine the velocity of point *B* for the instant when $\theta = 30°$.

Ans. $v_B = 288$ mm/s

Problem 5/166

5/167 The right-angled link *ABC* has plane motion in the *x-y* plane. At a certain instant the acceleration of *A* with respect to *B* is $\mathbf{a}_{A/B} = -450\mathbf{i} - 600\mathbf{j}$ mm/s² and the acceleration of *C* is zero. Determine the acceleration of *A* for this instant.

Problem 5/167

5/168 The two pulleys are fastened together to form a single rigid unit, and each of the two cables is wrapped securely around its respective pulley. If point *A* on the hoisting cable has an upward acceleration of 1 m/s², determine the magnitudes of the velocity and acceleration of points *O* and *B* if point *A* has an upward velocity of 1.5 m/s at the instant depicted.

Ans. $v_O = 1$ m/s, $v_B = 1.414$ m/s
$a_O = 0.667$ m/s², $a_B = 2.75$ m/s²

Problem 5/168

Problem 5/169

Problem 5/171

Problem 5/173

Problem 5/174

5/169 At the instant represented $x = 50$ mm and $\dot{s} = 1.6$ m/s. Determine the corresponding velocity of point B. *Ans.* $v_B = 1.029$ m/s

5/170 Determine the acceleration of the mass center G of the connecting rod AB for the slider-crank mechanism of Sample Problem 5/15 for (*a*) $\theta = 0$ and (*b*) $\theta = 180°$.

5/171 To speed up the unrolling of a telephone cable the trailer with the reel of cable starts from rest and is given an initial acceleration of 0.9 m/s². Simultaneously, the tow truck pulls the free end of the cable horizontally in the opposite direction with an initial acceleration of 0.6 m/s². If both vehicles start from rest at the same instant, determine the magnitude of the total acceleration of a point A on the forward end of the horizontal reel diameter (*a*) just as the motion starts and (*b*) one second after the start of the motion.
Ans. (*a*) $a_A = 2.42$ m/s²
(*b*) $a_A = 5.23$ m/s²

5/172 Determine the angular acceleration of the basket of the vegetable washer of Prob. 5/165 for the position where OA is vertical. In this position OA has an angular velocity of 4π rad/s and no angular acceleration. *Ans.* $\alpha_{BD} = 41.2$ rad/s² CW

▶**5/173** The slotted disk sector rotates with a constant counterclockwise angular velocity $\omega = 3$ rad/s. Simultaneously the slotted arm OC oscillates about the line OB (fixed to the disk) so that θ changes at the constant rate of 2 rad/s except at the extremities of the oscillation during reversal of direction. Determine the magnitude of the total acceleration of the pin A when $\theta = 30°$ and $\dot{\theta}$ is positive (clockwise). *Ans.* $a_A = 1.413$ m/s²

▶**5/174** The crank OB revolves clockwise at the constant rate ω_0 of 5 rad/s. For the instant when $\theta = 90°$ determine the angular acceleration α of the rod BD, which slides through the pivoted collar at C. *Ans.* $\alpha = 6.25$ rad/s² CW

Computer-oriented problems

*5/175 For the Geneva wheel of Prob. 5/41, shown again here, write the expression for the angular velocity ω_2 of the slotted wheel C during engagement of pin P and plot ω_2 for the range $-45° \leq \theta \leq 45°$. The driving wheel A has a constant angular velocity $\omega_1 = 2$ rad/s.

Problem 5/175

*5/176 Link OA is given a constant counterclockwise angular velocity ω. Determine the angular velocity ω_{AB} of link AB as a function of θ. Compute and plot the ratio ω_{AB}/ω for the range $0 \leq \theta \leq 90°$. Indicate the value of θ for which the angular velocity of AB is half that of OA.   *Ans.* $\theta = 46.1°$

Problem 5/176

*5/177 For the slider-crank configuration shown, derive the expression for the velocity v_A of the piston (taken positive to the right) as a function of θ. Substitute the numerical data of Sample Problem 5/15 and calculate v_A as a function of θ for $0 \leq \theta \leq 180°$. Plot v_A versus θ and find its maximum magnitude and the corresponding value of θ. (By symmetry anticipate the results for $180° \leq \theta \leq 360°$.)
               Ans. $(v_A)_{max} = 20.9$ m/s at $\theta = 72.3°$

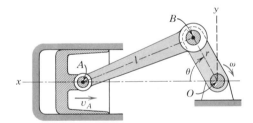

Problem 5/177

*5/178 For the slider-crank of Prob. 5/177 derive the expression for the acceleration a_A of the piston (taken positive to the right) as a function of θ for $\omega = \dot{\theta} = $ constant. Substitute the numerical data of Sample Problem 5/15 and calculate a_A as a function of θ for $0 \leq \theta \leq 180°$. Plot a_A versus θ and find the value of θ for which $a_A = 0$. (By symmetry anticipate the results for $180° \leq \theta \leq 360°$).
               Ans. $a_A = 0$ at $\theta = 72.3°$

PLANE KINETICS OF RIGID BODIES

6

6/1 INTRODUCTION

The kinetics of rigid bodies treats the relationships between the forces which act upon the bodies from sources external to their boundaries and the corresponding translational and rotational motions of the bodies. In Chapter 5 we developed the kinematical relationships for the plane motion of rigid bodies, and we will use these relationships extensively in this present chapter where the effects of forces on the two-dimensional motion of rigid bodies are examined.

For our purpose in Chapter 6 a body, which can be approximated as a thin slab with its motion confined to the plane of the slab, will be treated under the category of plane motion. The plane of motion will contain the mass center, and all forces which act on the body will be projected onto the plane of motion. A body which has appreciable dimensions normal to the plane of motion but is symmetrical about its plane of motion through the mass center may be treated as a plane-motion problem. These idealizations clearly fit a very large category of rigid-body motions.

In Chapter 3 we found that two force equations of motion were required to define the plane motion of a particle whose motion has two linear components. For the plane motion of a rigid body an additional equation is needed to specify the state of rotation of the body. Thus two force equations and one moment equation or their equivalent are required to determine the state of rigid-body plane motion.

The relationships which form the basis for most of the analysis of rigid-body motion were developed in Chapter 4 for a general system of particles. Frequent reference will be made to these equations as they are further developed in Chapter 6 and applied specifically to the plane motion of rigid bodies. It is advisable to refer to the developments of Chapter 4 frequently as Chapter 6 is studied. Also, it is inadvisable to proceed further until a firm grasp of the calculation of velocities and accelerations as developed in

Chapter 5 for rigid-body plane motion is well in hand. Without the ability to determine accelerations correctly from the principles of kinematics, it is frequently useless to attempt application of the force and moment principles of motion. Consequently, it is essential to master the necessary kinematics, including the calculation of relative accelerations, before proceeding.

Basic to the approach to kinetics is the isolation of the body or system to be analyzed. This isolation was illustrated and used in Chapter 3 for particle kinetics and will be employed consistently in the present chapter. For problems involving the instantaneous relationships among force, mass, and acceleration or momentum, the body or system should be explicitly defined by isolating it with its *free-body diagram*. When the principles of work and energy are employed, an *active-force diagram* which shows only those external forces that do work on the system may be used in lieu of the free-body diagram. *No solution of a problem should be attempted without first defining the complete external boundary of the body or system and identifying all external forces that act on it.*

Chapter 6 is organized into the same three sections in which we treated the kinetics of particles in Chapter 3. Section A relates the forces and moments to the instantaneous linear and angular accelerations. Section B treats the solution of problems by the method of work and energy. Section C covers the methods of impulse and momentum. Virtually all of the basic concepts and approaches covered in these three sections were treated in Chapter 3 on particle kinetics. This repetition will result in accelerated progress in Chapter 6 provided the kinematics of rigid-body plane motion is well in hand. In each of the three sections the three types of motion, namely, translation, fixed-axis rotation, and general plane motion, will be treated.

In the kinetics of rigid bodies which have angular motion it is necessary to introduce a property of the body which accounts for the radial distribution of its mass with respect to a particular axis of rotation normal to the plane of motion. This property is known as the *mass moment of inertia* of the body, and it is essential that we be able to calculate this property in order to solve rotational problems. Familiarity with the calculation of mass moments of inertia is assumed of the reader at this point. Appendix B treats this topic for those who need instruction or review.

SECTION A. FORCE, MASS, AND ACCELERATION

6/2 GENERAL EQUATIONS OF MOTION

In Arts. 4/2 and 4/4 we derived the force and moment vector equations of motion for a perfectly general system of mass. We now apply these results by starting, first, with a general rigid body in three dimensions. The force equation, Eq. 4/1,

$$\Sigma\mathbf{F} = m\bar{\mathbf{a}} \tag{4/1}$$

tells us that the resultant $\Sigma\mathbf{F}$ of the external forces acting on the body equals the mass m of the body times the acceleration $\bar{\mathbf{a}}$ of its mass center G. The moment equation taken about the mass center, Eq. 4/9,

$$\Sigma\mathbf{M}_G = \dot{\mathbf{H}}_G \tag{4/9}$$

tells us that the resultant moment about the mass center of the external forces on the body equals the time rate of change of the angular momentum of the body about the mass center.

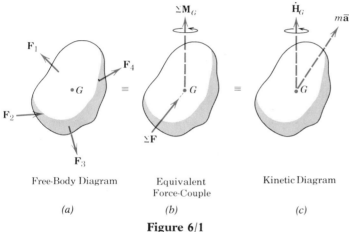

Free-Body Diagram

(a)

Equivalent
Force-Couple

(b)

Kinetic Diagram

(c)

Figure 6/1

We recall from our study of statics that a general system of forces acting on a rigid body may be replaced by a resultant force applied at a chosen point and a corresponding couple. By replacing the external forces by their equivalent force-couple system with the resultant force acting through the mass center, we may then visualize the action of the forces and the corresponding dynamic response of the body with the aid of Fig. 6/1. The *a*-part of the figure represents the free-body diagram of the body. The *b*-part of the figure shows the equivalent force-couple system with the resultant force applied through G. The *c*-part of the figure represents the resulting dynamic effects as specified by Eqs. 4/1 and 4/9 and is termed here the *kinetic diagram*. The equivalence between the free-body diagram and the kinetic diagram enables us to clearly visualize and easily remember the separate translational and rotational effects of the forces applied to a rigid body. We shall express this equivalence mathematically as we apply these results to the treatment of rigid-body plane motion which follows.

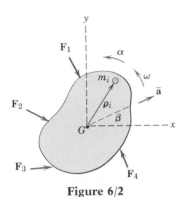

Figure 6/2

Plane motion equations.   We now apply the foregoing re-lationships to the case of plane motion. Figure 6/2 represents a rigid body moving with plane motion in the *x-y* plane. The mass center G has an acceleration $\bar{\mathbf{a}}$, and the body has an angular velocity $\boldsymbol{\omega} = \omega\mathbf{k}$ and an angular acceleration $\boldsymbol{\alpha} = \alpha\mathbf{k}$, both taken in the positive *z*-direction. Because the *z*-direction of both $\boldsymbol{\omega}$ and $\boldsymbol{\alpha}$ remains perpendicular to the plane of motion, we may use scalar notation ω and $\alpha = \dot{\omega}$ to represent the angular velocity and angular acceleration.

The angular momentum about the mass center for the gen-eral system was expressed in Eq. 4/8 as $\mathbf{H}_G = \Sigma\boldsymbol{\rho}_i \times m_i\dot{\boldsymbol{\rho}}_i$ where $\boldsymbol{\rho}_i$ is the position vector relative to G of the representative par-ticle of mass m_i. For our rigid body the velocity of m_i relative to G is $\dot{\boldsymbol{\rho}}_i = \boldsymbol{\omega} \times \boldsymbol{\rho}_i$, which has a magnitude $\rho_i\omega$ and lies in the plane of motion normal to $\boldsymbol{\rho}_i$. The product $\boldsymbol{\rho}_i \times \dot{\boldsymbol{\rho}}_i$ is, then, a vector normal to the *x-y* plane in the sense of $\boldsymbol{\omega}$, and its magnitude is $\rho_i^2\omega$. Thus the magnitude of $\mathbf{H}_G$ becomes $H_G = \Sigma\rho_i^2 m_i\omega = \omega\Sigma\rho_i^2 m_i$. The summation, which may also be written as $\int \rho^2\, dm$, is defined as the *mass moment of inertia* $\bar{I}$ of the body about the *z*-axis through G. (See Appendix B for a discussion of the calcula-tion of mass moments of inertia.) We may now write

$$H_G = \bar{I}\omega$$

where $\bar{I}$ is a constant property of the body and is a measure of the rotational inertia or resistance to change in rotational ve-locity due to the radial distribution of mass around the *z*-axis through G. With this substitution our moment equation, Eq. 4/9, becomes

$$\Sigma M_G = \dot{H}_G = \bar{I}\dot{\omega} = \bar{I}\alpha$$

where $\alpha = \dot{\omega}$ is the angular acceleration of the body.

The moment equation and the vector form of the generalized Newton's second law of motion, Eq. 4/1, are now written

$$\boxed{\begin{aligned} \Sigma\mathbf{F} &= m\bar{\mathbf{a}} \\ \Sigma M_G &= \bar{I}\alpha \end{aligned}} \qquad (6/1)$$

Equations 6/1 are the general equations of motion for a rigid body in plane motion. In applying Eqs. 6/1, the vector force equation will be expressed in terms of its two scalar components using *x-y*, *n-t*, or *r-θ* coordinates according to the description which is most convenient for the problem at hand.

As an alternative approach it is instructive to derive the mo-ment equation by referring directly to the forces which act on the representative particle of mass m_i, as shown in Fig. 6/3. The ac-celeration of m_i equals the vector sum of $\bar{a}$ and the relative

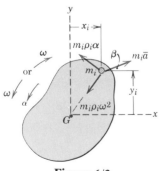

Figure 6/3

terms $\rho_i \omega^2$ and $\rho_i \alpha$ where the mass center G is used as the reference point. It follows that the resultant of all forces on m_i has the components $m_i \bar{a}$, $m_i \rho_i \omega^2$, and $m_i \rho_i \alpha$ in the directions shown. The sum of the moments of these force components about G in the sense of α becomes

$$M_{G_i} = m_i \rho_i^2 \alpha + (m_i \bar{a} \sin \beta)x_i - (m_i \bar{a} \cos \beta)y_i$$

Similar moment expressions exist for all particles in the body, and the sum of these moments about G for the resultant forces acting on all particles may be written as

$$\Sigma M_G = \Sigma m_i \rho_i^2 \alpha + \bar{a} \sin \beta \, \Sigma m_i x_i - \bar{a} \cos \beta \, \Sigma m_i y_i$$

But the origin of coordinates is taken at the mass center, so that $\Sigma m_i x_i = m\bar{x} = 0$ and $\Sigma m_i y_i = m\bar{y} = 0$. Thus the moment sum becomes

$$\Sigma M_G = \Sigma m_i \rho_i^2 \alpha = \bar{I}\alpha$$

as before. The contribution to ΣM_G of the forces internal to the body is, of course, zero since they occur in pairs of equal and opposite forces of action and reaction between neighboring particles. Thus ΣM_G, as before, represents the sum of moments about the mass center G of only the external forces acting on the body as disclosed by the free-body diagram. We note that the force component $m_i \rho_i \omega^2$ has no moment about G and conclude, therefore, that the angular velocity ω has no influence on the moment equation about the mass center G.

The results embodied in our basic equations of motion for a rigid body in plane motion, Eqs. 6/1, are represented diagramatically in Fig. 6/4, which is the two-dimensional counterpart of the *a*- and *c*-parts of Fig. 6/1 for a general three-dimensional body. The free-body diagram discloses the forces and moments appearing on the left-hand side of our equations of motion. The kinetic diagram discloses the resulting dynamic response in terms of the translational term $m\bar{\mathbf{a}}$ and the rotational term $\bar{I}\alpha$ which appear

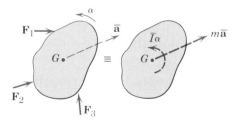

Free-Body Diagram   Kinetic Diagram

Figure 6/4

on the right-hand side of Eqs. 6/1. As previously mentioned, the translational term $m\bar{a}$ will be expressed by its x-y, n-t, or r-θ components once the appropriate inertial reference system is designated. The equivalence depicted in Fig. 6/4 is basic to our understanding of the kinetics of plane motion and will be employed frequently in the solution of problems. Representation of the resultants $m\bar{a}$ and $\bar{I}\alpha$ will help ensure that the force and moment sums determined from the free-body diagram are equated to their proper resultants.

Alternative moment equation. In Art. 4/4 of Chaper 4 on systems of particles we developed a general moment equation about an arbitrary point P, Eq. 4/12, which is

$$\Sigma\mathbf{M}_P = \dot{\mathbf{H}}_G + \bar{\rho} \times m\bar{a} \qquad [4/12]$$

where $\bar{\rho}$ is the vector from P to the mass center G and $\bar{a}$ is the mass-center acceleration. As we have shown earlier in this article, for a rigid body in plane motion $\dot{\mathbf{H}}_G$ becomes $\bar{I}\alpha$. Also, the cross product $\bar{\rho} \times m\bar{a}$ is simply the moment of magnitude $m\bar{a}d$ of $m\bar{a}$ about P. Therefore, for the two-dimensional body illustrated in Fig. 6/5 with its free-body diagram and kinetic diagram, we may rewrite Eq. 4/12 simply as

$$\boxed{\Sigma M_P = \bar{I}\alpha + m\bar{a}d} \qquad (6/2)$$

Clearly, all three terms are positive in the counterclockwise sense for the example shown, and the choice of P eliminates reference to $\mathbf{F}_1$ and $\mathbf{F}_3$. If we had wished to eliminate reference to $\mathbf{F}_2$ and $\mathbf{F}_3$, for example, by choosing their intersection as the reference point, then P would lie on the opposite side of the $m\bar{a}$ vector, and the clockwise moment of $m\bar{a}$ about P would be a negative term in the equation. Equation 6/2 is easily remembered as it is merely an expression of the familiar principle of moments where the sum of the moments about P equals the combined moment about P of their sum, expressed by the resultant couple $\Sigma M_G = \bar{I}\alpha$ and the resultant force $\Sigma\mathbf{F} = m\bar{a}$.

In Art. 4/4 we also developed an alternative moment equation about P, Eq. 4/13, which is

$$\Sigma\mathbf{M}_P = (\dot{\mathbf{H}}_P)_{\text{rel}} + \bar{\rho} \times m\mathbf{a}_P \qquad [4/13]$$

For rigid-body plane motion, if P is chosen as a point *fixed* to the body, then in scalar form $(\dot{\mathbf{H}}_P)_{\text{rel}}$ becomes $I_P\alpha$ where I_P is the mass moment of inertia about an axis through P and α is the angular acceleration of the body. So we may write the equation as

$$\boxed{\Sigma\mathbf{M}_P = I_P\alpha + \bar{\rho} \times m\mathbf{a}_P} \qquad (6/3)$$

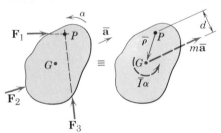

Free-Body Diagram   Kinetic Diagram

Figure 6/5

where the acceleration of P is $\mathbf{a}_P$ and the position vector from P to G is $\bar{\boldsymbol{\rho}}$. When $\bar{\boldsymbol{\rho}} = \mathbf{0}$, point P becomes the mass center G, and Eq. 6/3 reduces to the scalar form $\Sigma M_G = \bar{I}\alpha$, previously derived. When $\mathbf{a}_P = \mathbf{0}$, point P becomes point O, fixed in an inertial reference system and attached to the body (or body extended), and Eq. 6/3 in scalar form reduces to

$$\Sigma M_O = I_O \alpha \qquad\qquad (6/4)$$

Equation 6/4, then, applies to the rotation of a rigid body about a nonaccelerating point O fixed to the body and is the two-dimensional simplification of Eq. 4/7.

Unconstrained and constrained motion. The motion of a rigid body may be unconstrained or constrained. The rocket moving in a vertical plane, Fig. 6/6a, is an example of unconstrained motion as there are no physical confinements to its motion. The two components $\bar{a}_x$ and $\bar{a}_y$ of the mass-center acceleration and the angular acceleration α may be determined independently of one another by direct application of Eqs. 6/1. The bar in Fig. 6/6b, on the other hand, represents a constrained motion where the vertical and horizontal guides for the ends of the bar impose a kinematical relationship between the acceleration components of the mass center and the angular acceleration of the bar. Thus it is necessary to determine this kinematic relationship from the principles established in Chapter 5 and to combine it with the force and moment equations of motion before a solution can be carried out. In general, dynamics problems which involve physical constraints to motion require a kinematical analysis relating linear to angular acceleration before the force and moment equations of motion can be solved. It is for this reason that an under-

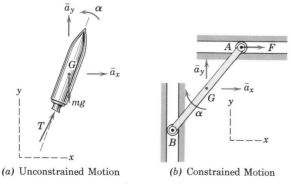

(a) Unconstrained Motion   (b) Constrained Motion

Figure 6/6

standing of the principles and methods of Chapter 5 is so vital to the work of Chapter 6.

System of interconnected bodies.   Upon occasion when dealing with two or more connected rigid bodies whose motions are related kinematically, it is convenient to analyze the bodies as an entire system. Figure 6/7 illustrates two rigid bodies hinged at A and subjected to the external forces shown. The forces in the connection at A are internal to the system and are not disclosed. The resultant of all external forces must equal the vector sum of the two resultants $m_1\bar{\mathbf{a}}_1$ and $m_2\bar{\mathbf{a}}_2$, and the sum of the moments about some arbitrary point such as P of all external forces must equal the moment of the resultants, $\bar{I}_1\alpha_1 + \bar{I}_2\alpha_2 + m_1\bar{a}_1d_1 + m_2\bar{a}_2d_2$. Thus we may state

$$\Sigma\mathbf{F} = \Sigma m\bar{\mathbf{a}}$$
$$\Sigma M_P = \Sigma\bar{I}\alpha + \Sigma m\bar{a}d$$

$$(6/5)$$

where the summations on the right-hand side of the equations represent as many terms as there are separate bodies. If there are more than three remaining unknowns in a system, however, the three independent scalar equations of motion, when applied to the system, are not sufficient to solve the problem. In this case more advanced methods such as virtual work (Art. 6/7) or Lagrange's equations (not discussed in this book*) could be employed or else the system dismembered and each part analyzed separately with the resulting equations solved simultaneously.

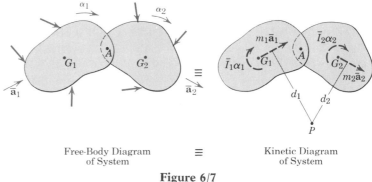

Free-Body Diagram   ≡   Kinetic Diagram
of System   of System

Figure 6/7

* When an interconnected system has more than one degree of freedom, that is, requires more than one coordinate to specify completely the configuration of the system, the more advanced equations of Lagrange are generally used. See the senior author's *Dynamics, 2nd Edition SI Version* 1975, John Wiley & Sons, for a treatment of Lagrange's equations.

Analysis procedure. In solving force-mass-acceleration problems for the plane motion of rigid bodies, the following steps should be taken after the conditions and requirements of the problem are clearly in mind.

(*1*) *Kinematics.* First identify the class of motion and then solve for any needed linear and angular accelerations which can be determined solely from given kinematic information. In the case of constrained plane motion, it is usually necessary to establish the relation between the linear acceleration of the mass center and the angular acceleration of the body by first solving the appropriate relative velocity and relative acceleration equations. Again, it is emphasized that success in working force-mass-acceleration problems in this chapter is contingent upon the ability to describe the necessary kinematics, so that frequent review of Chapter 5 is recommended.

(*2*) *Diagrams.* Always draw the complete free-body diagram of the body to be analyzed. Assign a convenient inertial coordinate system and label all known and unknown quantities. The kinetic diagram should also be constructed so as to clarify the equivalence between the applied forces and the resulting dynamic response.

(*3*) *Equations of motion.* Apply the three equations of motion from Eqs. 6/1, being consistent with the algebraic signs in relation to the choice of reference axes. Equation 6/2 or 6/3 may be employed as an alternative to one of Eqs. 6/1. Combine the results from any needed kinematic analysis. Count the number of unknowns and be certain that there are an equal number of independent equations available. For a solvable rigid-body problem in plane motion, there can be no more than five scalar unknowns which can be determined from the three scalar equations of motion, from Eqs. 6/1, and the two scalar component relations which come from the relative-acceleration equation.

The sequence of steps to be followed in any given problem will depend on the conditions of the problem. To organize a solution in a logical sequence of steps, it is recommended that the dependency of a required quantity on intermediate quantities be recognized. Tracing the dependencies back to the given information establishes the reverse order of the steps to be followed in carrying out the actual solution.

The foregoing developments will now be applied to the three cases of motion in a plane, namely, translation, fixed-axis rotation, and general plane motion in the three articles which follow.

6/3 TRANSLATION

Rigid-body translation in plane motion was described in Art. 5/1 and illustrated in Figs. 5/1*a* and 5/1*b* where we saw that every line in a translating body remains parallel to its original position at all times. In rectilinear translation all points move in straight lines, whereas in curvilinear translation all points move on congruent curved paths. In either case there can be no angular motion of the translating body, so that both ω and α are zero. Therefore, from the moment relation of Eqs. 6/1, we see that all reference to the moment of inertia is eliminated for a translating body.

For a translating body, then, our general equations for plane motion, Eqs. 6/1, may be written

$$\boxed{\begin{array}{l} \Sigma \mathbf{F} = m\bar{\mathbf{a}} \\ \Sigma M_G = \bar{I}\alpha = 0 \end{array}}$$

(6/6)

For rectilinear translation, illustrated in Fig. 6/8a, if the x-axis is chosen in the direction of the acceleration, then the two scalar force equations become $\Sigma F_x = m\bar{a}_x$ and $\Sigma F_y = m\bar{a}_y = 0$. For curvilinear translation, Fig. 6/8b, using *n-t* coordinates, the two scalar force equations become $\Sigma F_n = m\bar{a}_n$ and $\Sigma F_t = m\bar{a}_t$. In both cases $\Sigma M_G = 0$.

We may also employ the alternative moment equation, Eq. 6/2, with the aid of the kinetic diagram. For rectilinear translation we see that $\Sigma M_P = m\bar{a}d$ and $\Sigma M_A = 0$. For curvilinear translation the kinetic diagram permits us to write $\Sigma M_A = m\bar{a}_n d_A$ in the clockwise sense and $\Sigma M_B = m\bar{a}_t d_B$ in the counterclockwise sense. Thus we have complete freedom for choosing a convenient moment center.

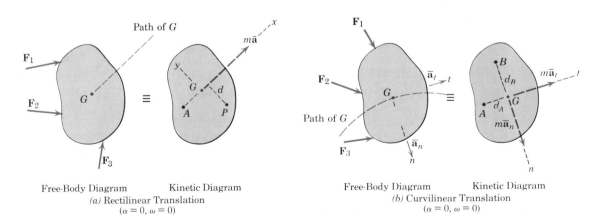

Free-Body Diagram   Kinetic Diagram
(a) Rectilinear Translation
($\alpha = 0, \omega = 0$)

Free-Body Diagram   Kinetic Diagram
(b) Curvilinear Translation
($\alpha = 0, \omega = 0$)

Figure 6/8

Sample Problem 6/1

The 1500-kg pickup truck reaches a speed of 50 km/h from rest in a distance of 60 m up the 10-percent incline with constant acceleration. Calculate the normal force under each pair of wheels and the friction force under the rear driving wheels. The effective coefficient of friction between the tires and the road is known to be at least 0.8.

Solution. We will assume that the mass of the wheels is negligible compared with the total mass of the truck. The truck may now be simulated by a single rigid body in rectilinear translation with an acceleration of

$[v^2 = 2as]$ $\qquad \bar{a} = \dfrac{(50/3.6)^2}{2(60)} = 1.608 \text{ m/s}^2$

The free-body diagram of the complete truck shows the normal forces N_1 and N_2, the friction force F in the direction to oppose the slipping of the driving wheels, and the weight mg represented by its two components. With $\theta = \tan^{-1} 1/10 = 5.71°$ these components are $mg \cos \theta = 1500(9.81) \cos 5.71° = 14.64(10^3)$ N and $mg \sin \theta = 1500(9.81) \sin 5.71° = 1.464(10^3)$ N. The kinetic diagram shows the resultant which passes through the mass center and is in the direction of its acceleration. Its magnitude is

$$m\bar{a} = 1500(1.608) = 2411 \text{ N}$$

Applying the three equations of motion, Eqs. 6/1, for the three unknowns gives

$[\Sigma F_x = m\bar{a}_x]$ $\qquad F - 1464 = 2411 \qquad F = 3875 \text{ N}$ *Ans.*

$[\Sigma F_y = m\bar{a}_y = 0]$ $\qquad N_1 + N_2 - 14.64(10^3) = 0$ $\qquad (a)$

$[\Sigma M_G = \bar{I}\alpha = 0]$ $\qquad 1.5N_1 + 3875(0.6) - N_2(1.5) = 0$ $\qquad (b)$

Solving (a) and (b) simultaneously gives

$$N_1 = 6546 \text{ N} \qquad N_2 = 8096 \text{ N} \qquad \textit{Ans.}$$

In order to support a friction force of 3875 N, a coefficient of friction of at least $F/N_2 = 3875/8096 = 0.48$ is required. Since our coefficient of friction is at least 0.8, the surfaces are rough enough to support the calculated value of F so that our result is correct.

Alternative solution. From the kinetic diagram we see that N_1 and N_2 can be obtained independently of one another by writing separate moment equations about A and B.

$[\Sigma M_A = m\bar{a}d]$ $3N_2 - 1.5(14.64)(10^3) - 0.6(1464) = 2411(0.6)$

$$N_2 = 8096 \text{ N} \qquad \textit{Ans.}$$

$[\Sigma M_B = m\bar{a}d]$ $14.64(10^3)(1.5) - 1464(0.6) - 3N_1 = 2411(0.6)$

$$N_1 = 6546 \text{ N} \qquad \textit{Ans.}$$

① Without this assumption we would be obliged to account for the relatively small additional forces which produce moments that give the wheels their angular acceleration.

② Recall that 3.6 km/h is 1 m/s.

③ We must be careful not to use the friction equation $F = \mu N$ here since we do not have a case of slipping or impending slipping. If the given coefficient of friction were less than 0.48, the friction force would be μN_2, and the car would be unable to attain the acceleration of 1.608 m/s². In this case the unknowns would be N_1, N_2, and a.

④ The left-hand side of the equation is evaluated from the free-body diagram, and the right-hand side from the kinetic diagram. The positive sense for the moment sum is arbitrary but must be the same for both sides of the equation. In this problem we have taken the clockwise sense as positive for the moment of the resultant force about B.

Sample Problem 6/2

The vertical bar AB has a mass of 150 kg with center of mass G midway between the ends. The bar is elevated from rest at $\theta = 0$ by means of the parallel links of negligible mass with a constant couple $M = 5$ kN · m applied to the lower link at C. Determine the angular acceleration α of the links as a function of θ, and find the force B in the link DB at the instant when $\theta = 30°$.

Solution. The motion of the bar is seen to be curvilinear translation since the bar itself does not rotate during the motion. With the circular motion of the mass center G, we choose n- and t-coordinates as the most convenient description. With negligible mass of the links the tangential component A_t of the force at A is obtained from the free-body diagram of AC where $\Sigma M_C \cong 0$ and $A_t = M/\overline{AC} = 5/1.5 = 3.33$ kN. The force at B is along the link. All applied forces are shown on the free-body diagram of the bar, and the kinetic diagram is also indicated where the $m\bar{a}$ resultant is shown in terms of its two components.

(1) Generally speaking the best choice of reference axes is to make them coincide with the directions in which the components of the mass-center acceleration are expressed. Examine the consequences of choosing horizontal and vertical axes in this problem.

The sequence of solution is established by noting that A_n and B depend on the n-summation of forces and, hence, on $m\bar{r}\omega^2$ at $\theta = 30°$. The value of ω depends on the variation of $\alpha = \ddot{\theta}$ with θ. This dependency is established from a force summation in the t-direction for a general value of θ where $\bar{a}_t = (\bar{a}_t)_A = \overline{AC}\alpha$. Thus we begin with

$[\Sigma F_t = m\bar{a}_t] \qquad 3.33 - 0.15(9.81) \cos \theta = 0.15(1.5\alpha)$

$\qquad\qquad \alpha = 14.81 - 6.54 \cos \theta \text{ rad/s}^2 \qquad Ans.$

With α a known function of θ, the angular velocity ω of the links is obtained from

$[\omega \, d\omega = \alpha \, d\theta] \quad \int_0^\omega \omega \, d\omega = \int_0^\theta (14.81 - 6.54 \cos \theta) \, d\theta$

$\qquad\qquad \omega^2 = 29.6\theta - 13.08 \sin \theta$

Substitution for $\theta = 30°$ gives

$\qquad (\omega^2)_{30°} = 8.97 \text{ (rad/s)}^2 \qquad \alpha_{30°} = 9.15 \text{ rad/s}^2$

and $\qquad\qquad m\bar{r}\omega^2 = 0.15(1.5)(8.97) = 2.02$ kN

$\qquad\qquad m\bar{r}\alpha = 0.15(1.5)(9.15) = 2.06$ kN

The force B may be obtained by a moment summation about A, which eliminates A_n and A_t and the weight. Or a moment summation may be taken about the intersection of A_n and the line of action of $m\bar{r}\alpha$, which eliminates A_n and $m\bar{r}\alpha$. Using A as a moment center gives

$[\Sigma M_A = m\bar{a}d] \quad 1.8 \cos 30° \, B = 2.02(1.2) \cos 30° + 2.06(0.6)$

$\qquad\qquad B = 2.14 \text{ kN} \qquad Ans.$

The component A_n could be obtained from a force summation in the n-direction or from a moment summation about G or about the intersection of B and the line of action of $m\bar{r}\alpha$.

(2) The force and moment equations for a body of negligible mass become the same as the equations of equilibrium. Link BD, therefore, acts as a two-force member in equilibrium.

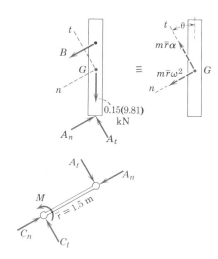

PROBLEMS

6/1 The right-angled 3-kg bar with equal legs is freely hinged to the vertical plate at C. The bar is prevented from rotating by the two pegs A and B fixed to the plate. Determine the acceleration a of the plate for which no force is exerted on the bar by either peg A or B.   *Ans.* $a = 3g$

Problem 6/1

6/2 In Prob. 6/1, if the plate is given a horizontal acceleration $a = 2g$, calculate the force exerted on the bar by either peg A or B.   *Ans.* $B = 7.36$ N

6/3 The homogeneous crate of mass m is mounted on small wheels at its front and back edges. Determine the maximum force P which can be applied without overturning the crate about (a) its front edge with $h = b$ and (b) its back edge with $h = 0$.

Problem 6/3

6/4 The uniform 100-kg pole AB is suspended in the horizontal position by the three wires shown. If wire CB breaks, calculate the tension in wire BD immediately after the break. (*Suggestion:* By a thoughtful choice of moment center solve by using only one equation of motion.)

Ans. $T_B = 347$ N

Problem 6/4

6/5 The device shown consists of a vertical frame A to which are freely pivoted a geared sector at O and a balanced gear and attached pointer at C. Under a steady horizontal acceleration a to the right, the sector undergoes a clockwise angular displacement, thus causing the pointer to register a steady counterclockwise angle θ from the zero-acceleration position at $\theta = 0$. Determine the acceleration corresponding to an angle θ.

Problem 6/5

Problem 6/6

Problem 6/7

Problem 6/9

Problem 6/10

6/6 The arm *OA* of the classifying accelerometer has a mass of 0.25 kg with center of mass at *G*. The adjusting screw and spring are preset to a force of 12 N at *B*. At what acceleration *a* would the electrical contacts at *A* be on the verge of opening? Motion is in the vertical plane of the figure.

Ans. a = 8.34g

6/7 Arm *AB* of a classifying accelerometer has a mass of 0.1 kg with mass center at *G* and is pivoted freely to the frame *F* at *A*. The torsion spring at *A* is set to preload the arm with an applied clockwise moment of 225 N·mm. Determine the downward acceleration *a* of the frame at which the contacts at *B* will separate and break the electrical circuit.

6/8 If the downward acceleration of the accelerometer of Prob. 6/7 is 2*g*, calculate the force between the contacts at *B* and the force on the pin at *A*.

Ans. A = 3.99 N, B = 3.01 N

6/9 A passenger car of an overhead monorail system is driven by one of its two small wheels *A* or *B*. Select the one for which the car can be given the greater acceleration without slipping the driving wheel and compute the maximum acceleration if the effective coefficient of friction is limited to 0.25 between the wheels and the rail. Neglect the small mass of the wheels.

6/10 A drill press with mass center at *G* is transported on the flat bed of a truck. If the driver neglects to secure the drill press to the truck, determine the maximum acceleration *a* which the truck may have before the drill press slips or tips about its base at *A*. The coefficient of static friction between the truck bed and the drill press is 0.40.

Ans. a = 4.91 m/s²

6/11 A cleated conveyor belt transports solid homogeneous cylinders up a 15° incline. The diameter of the cylinders is half their height. Determine the maximum acceleration which the belt may have without tipping the cylinders as it starts.

Problem 6/11

6/12 The loaded trailer has a mass of 900 kg with center of mass at *G* and is attached at *A* to a rear-bumper hitch. If the car and trailer reach a velocity of 60 km/h on a level road in a distance of 30 m from rest with constant acceleration, compute the vertical component of the force supported by the hitch at *A*. Neglect the small friction force exerted on the relatively light wheels.   *Ans.* $A_y = 1389$ N

Problem 6/12

6/13 The uniform 4-m boom has a mass of 60 kg and is pivoted to the back of a truck at *A* and secured by a cable at *C*. Calculate the magnitude of the total force supported by the connection at *A* if the truck starts from rest with an acceleration of 5 m/s².

Problem 6/13

6/14 A laminate roller consists of the uniform 2-kg bar *ACB* with two light rollers that apply force to the laminates on the top and bottom of a countertop along its edge. Determine the force exerted by each roller on the laminate when a 50-N force is applied normal to the bar in the position shown. Neglect all friction.
          Ans. $B = 79.6$ N, $C = 134.9$ N

Dimensions in Millimeters

Problem 6/14

Problem 6/15

Problem 6/16

Problem 6/17

Problem 6/18

6/15 The uniform steel bar *AD* is rigidly mounted in the collar, which is given a vertical harmonic oscillation as indicated. Neglect the weight of the bar compared with other forces which act and determine the maximum bending moment *M* induced in the bar during its vertical oscillation for an amplitude $y_0 = 3$ mm and a frequency $n = 6$ cycles per second.

6/16 The block *A* and attached rod have a combined mass of 60 kg and are confined to move along the 60° guide under the action of the 800-N applied force. The uniform horizontal rod has a mass of 20 kg and is welded to the block at *B*. Friction in the guide is negligible. Compute the bending moment *M* exerted by the weld on the rod at *B*.

Ans. M = 196.0 N·m

6/17 The uniform rod *AB* of mass *m* and length *l* is welded at *A* to the vertical member of the parallelogram linkage and always remains horizontal. If link *OC* is given a constant counterclockwise angular velocity ω in the vertical plane, determine expressions for the bending moment *M*, the tension *T*, and the shear *V* applied to the beam at *A* by the weld in terms of θ.

$$Ans. \ M = m\frac{l}{2}(g + r\omega^2 \cos\theta)$$
$$T = mr\omega^2 \sin\theta$$
$$V = m(g + r\omega^2 \cos\theta)$$

6/18 The coefficient of static friction at both ends of the uniform bar is 0.40. Determine the maximum horizontal acceleration *a* which the truck may have without causing the bar to slip. (*Suggestion:* The problem can be solved by using only one equation, a moment equation. The location of the moment center may be determined graphically.)

Ans. a = 2.38 m/s²

6/19 Determine the expression for the maximum velocity v which the microbus can reach in a distance s from rest without slipping its rear driving wheels if the effective coefficient of friction between the tires and the pavement is μ. Neglect the mass of the wheels.

Problem 6/19

6/20 A jet transport with a landing speed of 200 km/h reduces its speed to 60 km/h with a negative thrust R from its jet thrust reversers in a distance of 425 m along the runway with constant deceleration. The total mass of the aircraft is 140 Mg with mass center at G. Compute the reaction N under the nose wheel B toward the end of the braking interval and prior to the application of mechanical braking. At the lower speed aerodynamic forces on the aircraft are small and may be neglected.

Ans. $N = 257$ kN

Problem 6/20

6/21 The rocket sled shown is used to test human tolerance to high accelerations. The sled and rocket motors have an initial mass of 1200 kg with mass center at G_1. The mass of the man and instrumentation together is 300 kg with mass center at G_2. If the rocket motors produce a total thrust T of 100 kN and the initial frictional resistance at each runner is 30 percent of the vertical load supported by the runner, calculate the vertical reaction N_A under the front runner.

Ans. $N_A = 14.03$ kN

Problem 6/21

6/22 The riding power mower has a mass of 140 kg with center of mass at G_1. The operator has a mass of 90 kg with center of mass at G_2. Calculate the minimum effective coefficient of friction μ which will permit the front wheels of the mower to lift off the ground as the mower starts to move forward.

Problem 6/22

Problem 6/23

6/23 The figure shows the Saturn V mobile launch platform A together with the umbilical tower B, unfueled rocket C, and crawler-transporter D which carries the system to the launch site. The approximate dimensions of the structure and locations of the mass centers G are given. The approximate masses are $m_A = 3$ Gg, $m_B = 3.3$ Gg, $m_C = 0.23$ Gg, and $m_D = 3$ Gg. The minimum stopping distance from the top speed of 1.5 km/h is 0.1 m. Compute the vertical component of the reaction under the front crawler unit F during the period of maximum deceleration. *Ans.* $F = 59.5$ MN

Problem 6/24

6/24 The 25-kg bar BD is attached to the two light links AB and CD and moves in the vertical plane. The lower link is subjected to a clockwise torque $M = 200$ N·m applied through its shaft at A. If each link has an angular velocity $\omega = 5$ rad/s as it passes the horizontal position, calculate the force which the upper link exerts on the bar at D at this instant. Also find the angular acceleration of the links at this position. *Ans.* $D = 234$ N, $\alpha = 5.87$ rad/s^2

Problem 6/25

6/25 The car seen from the rear is traveling at a speed v around a turn of mean radius r banked inward at an angle θ. The effective coefficient of friction between the tires and the road is μ. Determine (a) the proper bank angle for a given v to eliminate any tendency to slip or tip, and (b) the maximum speed v before the car tips or slips for a given θ. Note that the forces and the acceleration lie in the plane of the figure so that the problem may be treated as one of plane motion even though the velocity is normal to this plane.

6/26 Determine the maximum mass m of the cylinder for which the loaded 2000-kg coal car will not overturn about the rear wheels B. Neglect the mass of all pulleys and wheels. (Note that the tension in the cable at C is not $2mg$.)

Ans. m = 3.23 Mg

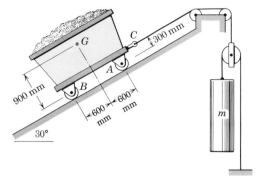

Problem 6/26

6/27 The parallelogram linkage is used to transfer crates from platform A to platform B and is hydraulically operated. The oil pressure in the cylinder is programmed to provide a smooth transition from $\theta = 0$ to $\theta = \theta_0 = \pi/3$ rad given by $\theta = \dfrac{\pi}{6}\left(1 - \cos\dfrac{\pi t}{2}\right)$

where t is in seconds. Determine the force at D on the pin (a) just after the start of the motion with θ and t essentially zero and (b) when $t = 1$ s. The crate and platform have a combined mass of 200 kg with mass center at G. The mass of each link is small and may be neglected.

Ans. (a) D = 1714 N, (b) D = 2178 N

Problem 6/27

6/28 The two wheels of the vehicle are connected by a 20-kg link AB with center of mass at G. The link is pinned to the wheel at B, and the pin at A fits into a smooth horizontal slot in the link. If the vehicle has a constant speed of 4 m/s, determine the magnitude of the force supported by the pin at B for the position $\theta = 30°$.

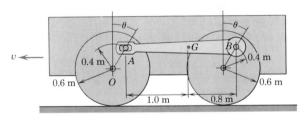

Problem 6/28

6/29 The 900-kg crate with center of mass G in its geometric center is suspended by the two cables at A and B and by a rope at C all in the vertical plane. If the rope is suddenly released with $\theta = 60°$, calculate the tension in the cable at A an instant after release. Solve, first, by writing three motion equations in the three unknowns, and, second, by writing only one motion equation. Compare the merits of the two approaches.   *Ans. T_A = 1616 N*

Problem 6/29

Problem 6/30

6/30 Model tests of the landing of the lunar excursion module (LEM) are conducted using the pendulum which consists of the model suspended by the two parallel wires A and B. If the model has a mass of 10 kg with center of mass at G, and if $\dot{\theta} = 2$ rad/s when $\theta = 60°$, calculate the tension in each of the wires at this instant.

Ans. $T_A = 147.9$ N, $T_B = 21.1$ N

Problem 6/31

6/31 The homogeneous rectangular plate has a mass of 20 kg and is supported in the vertical plane by the light parallel links shown. If a couple $M = 100$ N·m is applied to the end of link AB with the system initially at rest, calculate the force supported by the pin at C as the plate lifts off its support with $\theta = 30°$. *Ans.* $C = 207$ N

Problem 6/32

▶**6/32** The uniform 200-kg bar AB is raised in the vertical plane by the application of a constant couple $M = 3$ kN·m applied to the link at C. The mass of the links is small and may be neglected. If the bar starts from rest at $\theta = 0$, determine the magnitude of the force supported by the pin at A as the position $\theta = 60°$ is passed. *Ans.* $A = 2.02$ kN

6/4 FIXED-AXIS ROTATION

Rotation of a rigid body about a fixed axis O was described in Art. 5/2 and illustrated in Fig. 5/1c. For this motion we saw that all points in the body describe circles about the rotation axis, and all lines of the body in the plane of motion have the same angular velocity ω and angular acceleration α. The acceleration components of the mass center for circular motion are most easily expressed in n-t coordinates, so we have $\bar{a}_n = \bar{r}\omega^2$ and $\bar{a}_t = \bar{r}\alpha$, as shown in Fig. 6/9$a$ for rotation of the rigid body about the fixed axis through O. The b-part of the figure represents the free-body diagram, and the equivalent kinetic diagram in the c-part of the figure shows the $m\bar{\mathbf{a}}$ resultant in terms of its n- and t-components and the $\bar{I}\alpha$ resultant couple. Our general equations for plane motion, Eqs. 6/1, are directly applicable and are repeated here.

$$\boxed{\begin{aligned} \Sigma\mathbf{F} &= m\bar{\mathbf{a}} \\ \Sigma M_G &= \bar{I}\alpha \end{aligned}} \qquad [6/1]$$

Thus the two scalar components of the force equation become $\Sigma F_n = m\bar{r}\omega^2$ and $\Sigma F_t = m\bar{r}\alpha$. In applying the moment equation about G it is essential to account for the moment of the force applied to the body at O, so this force must not be omitted from the free-body diagram.

For fixed-axis rotation it is generally useful to apply a moment equation directly about the rotation axis O. We derived this equation previously as Eq. 6/4, which is repeated here.

$$\boxed{\Sigma M_O = I_O\alpha} \qquad [6/4]$$

From the kinetic diagram in Fig. 6/9c, we may obtain Eq. 6/4 very easily by evaluating the moment of the resultants about O, which becomes $\Sigma M_O = \bar{I}\alpha + m\bar{a}_t\bar{r}$. Substitution of the transfer-of-axis relation for mass moments of inertia, $I_O = \bar{I} + m\bar{r}^2$, gives $\Sigma M_O = (I_O - m\bar{r}^2)\alpha + m\bar{r}^2\alpha = I_O\alpha$.

For the common case of rotation of a rigid body about a fixed axis through its mass center G, clearly $\bar{\mathbf{a}} = \mathbf{0}$, and therefore $\Sigma\mathbf{F} = \mathbf{0}$. The resultant of the applied forces, then, is the couple $\bar{I}\alpha$.

We may combine the resultant force $m\bar{a}_t$ and resultant couple $\bar{I}\alpha$ by moving $m\bar{a}_t$ to a parallel position through point Q, Fig. 6/10, located by $m\bar{r}\alpha q = \bar{I}\alpha + m\bar{r}\alpha(\bar{r})$. Using the transfer-of-axis theorem and $I_O = k_O^2 m$ gives $q = k_O^2/\bar{r}$. Point Q is called the *center of percussion* and has the unique property that the resultant of all forces applied to the body must pass through it. It follows that the sum of the moments of all forces about the center of percussion is always zero, $\Sigma M_Q = 0$.

Fixed-Axis Rotation

(a)

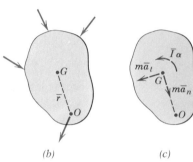

(b)   (c)

Free-Body Diagram   Kinetic Diagram

Figure 6/9

Figure 6/10

Sample Problem 6/3

The 300-kg concrete block is elevated by the hoisting mechanism shown where the cables are securely wrapped around the respective drums. The drums, which are fastened together and turn as a single unit about their mass center at O, have a combined mass of 150 kg and a radius of gyration about O of 450 mm. If a constant tension P of 1.80 kN is maintained by the power unit at A, determine the vertical acceleration of the block and the resultant force on the bearing at O.

Solution I. The free-body and kinetic diagrams of the drums and concrete block are drawn showing all forces which act including the components O_x and O_y of the bearing reaction. The resultant of the force system on the drums for centroidal rotation is the couple $\bar{I}\alpha = I_O\alpha$ where

(2) $[I = k^2 m]$ $\qquad \bar{I} = I_O = (0.450)^2 150 = 30.4 \text{ kg} \cdot \text{m}^2$

Taking moments about the mass center O for the pulley in the sense of the angular acceleration α gives

$[\Sigma M_G = \bar{I}\alpha]$ $\qquad 1800(0.600) - T(0.300) = 30.4\alpha$ $\qquad (a)$

The acceleration of the block is described by

$[\Sigma F_y = ma_y]$ $\qquad T - 300(9.81) = 300a$ $\qquad (b)$

From $a_t = r\alpha$ we have $a = 0.300\alpha$. With this substitution Eqs. (a) and (b) are combined to give

$\qquad T = 3252 \text{ N} \qquad \alpha = 3.44 \text{ rad/s}^2 \qquad a = 1.031 \text{ m/s}^2 \qquad Ans.$

The bearing reaction is computed from its components. Since $\bar{a} = 0$ we use the equilibrium equations

$[\Sigma F_x = 0]$ $\quad O_x - 1800 \cos 45° = 0 \qquad O_x = 1273 \text{ N}$

$[\Sigma F_y = 0]$ $\quad O_y - 150(9.81) - 3252 - 1800 \sin 45° = 0 \quad O_y = 5996 \text{ N}$

$\qquad O = \sqrt{(1273)^2 + (5996)^2} = 6130 \text{ N} \qquad Ans.$

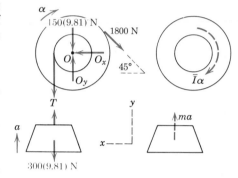

(1) Be alert to the fact that the tension T is not $300(9.81)$ N. If it were, the block would not accelerate.

(2) Don't overlook the need to express k_o in meters when using g in m/s^2.

Solution II. We may use a more condensed approach by drawing the free-body diagram of the entire system, thus eliminating reference to T, which becomes internal to the new system. From the kinetic diagram for the system we see that the moment sum about O must equal the resultant couple $\bar{I}\alpha$ for the drums plus the moment of the resultant ma for the block. Thus from the principle of Eq. 6/5 we have

$[\Sigma M_O = \bar{I}\alpha + m\bar{a}d]$

$\qquad 1800(0.600) - 300(9.81)(0.300) = 30.4\alpha + 300a(0.300)$

With $a = 0.300\alpha$ the solution gives, as before, $a = 1.031 \text{ m/s}^2$.

We may equate the force sums on the entire system to the sums of the resultants. Thus

$[\Sigma F_y = \Sigma m\bar{a}_y]$

$O_y - 150(9.81) - 300(9.81) - 1800 \sin 45° = 150(0) + 300(1.031)$

$\qquad\qquad O_y = 5996 \text{ N}$

$[\ F_x = \Sigma m\bar{a}_x]$ $\qquad O_x - 1800 \cos 45° = 0 \qquad O_x = 1273 \text{ N}$

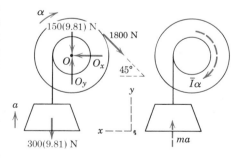

Sample Problem 6/4

The pendulum has a mass of 7.5 kg with center of mass at G and has a radius of gyration about the pivot O of 295 mm. If the pendulum is released from rest at $\theta = 0$, determine the total force supported by the bearing at the instant when $\theta = 60°$. Friction in the bearing is negligible.

Solution. The free-body diagram of the pendulum in a general position is shown along with the corresponding kinetic diagram where the components of the resultant force have been drawn through G.

The normal component O_n is found from a force equation in the n-direction which involves the normal acceleration $\bar{r}\omega^2$. Since the angular velocity ω of the pendulum is found from the integral of the angular acceleration and since O_t depends on the tangential acceleration $\bar{r}\alpha$, it follows that α should be obtained first. To this end with $I_O = k_O^2 m$, the moment equation about O gives

② $[\Sigma M_O = I_O \alpha]$ $7.5(9.81)(0.25) \cos \theta = (0.295)^2(7.5)\alpha$

$$\alpha = 28.2 \cos \theta \text{ rad/s}^2$$

and for $\theta = 60°$

$[\omega \, d\omega = \alpha \, d\theta]$ $\displaystyle\int_0^\omega \omega \, d\omega = \int_0^{\pi/3} 28.2 \cos \theta \, d\theta$

$$\omega^2 = 48.8 \text{ (rad/s)}^2$$

The remaining two equations of motion applied to the 60° position yield

$[\Sigma F_n = m\bar{r}\omega^2]$ $O_n - 7.5(9.81) \sin 60° = 7.5(0.25)(48.8)$

$$O_n = 155.2 \text{ N}$$

$[\Sigma F_t = m\bar{r}\alpha]$ $-O_t + 7.5(9.81) \cos 60° = 7.5(0.25)(28.2) \cos 60°$

$$O_t = 10.37 \text{ N}$$

$$O = \sqrt{(155.2)^2 + (10.37)^2} = 155.6 \text{ N} \qquad Ans.$$

The proper sense for O_t may be observed at the outset by applying the moment equation $\Sigma M_G = \bar{I}\alpha$ where the moment about G due to O_t must be clockwise to agree with α. The force O_t may also be obtained initially by a moment equation about the center of percussion Q, shown in the lower figure, which avoids the necessity of computing α. First, we must obtain the distance q, which is

$[q = k_O^2/\bar{r}]$ $q = \dfrac{(0.295)^2}{0.250} = 0.348 \text{ m}$

$[\Sigma M_Q = 0]$ $O_t(0.348) - 7.5(9.81)(\cos 60°)(0.348 - 0.250) = 0$

$$O_t = 10.37 \text{ N} \qquad Ans.$$

① The acceleration components of G are, of course, $\bar{a}_n = \bar{r}\omega^2$ and $\bar{a}_t = \bar{r}\alpha$.

② Review the theory again and satisfy yourself that $\Sigma M_O = I_O \alpha = \bar{I}\alpha + m\bar{r}^2\alpha = m\bar{r}\alpha q$.

③ Note especially here that the force summations are taken in the positive direction of the acceleration components of the mass center G.

Problem 6/33

Problem 6/34

Problem 6/35

Problem 6/36

Problem 6/37

Problem 6/38

PROBLEMS

6/33 Derive the moment equation of motion $\Sigma M_O = I_O \alpha$ for the plane rotation of a general rigid body about the fixed axis O. Start with the equation of motion for a representative particle of mass m_i of the body.

6/34 Explain why a baseball bat should be swung so that the impact of the ball with the bat does not occur opposite the mass center G of the bat as shown. Where should impact occur?

6/35 The uniform slender bar is suspended by a light cord from its end and swings as a pendulum. With the aid of a free-body diagram show that the bar cannot remain in line with the cord during the motion.

6/36 Each of the two drums and connected hubs of 250-mm radius has a mass of 100 kg and a radius of gyration about its center of 375 mm. Calculate the angular acceleration of each drum. Friction in each bearing is negligible.
Ans. $\alpha_a = 3.20 \text{ rad/s}^2$, $\alpha_b = 3.49 \text{ rad/s}^2$

6/37 The 750-mm slender bar has a mass of 9 kg and is mounted on a vertical shaft at O. If a torque $M = 10 \text{ N·m}$ is applied to the bar through its shaft, calculate the horizontal force R on the bearing as the bar starts to rotate. *Ans.* $R = 14.3 \text{ N}$

6/38 The uniform 20-kg slender bar is pivoted at O and swings freely in the vertical plane. If the bar is released from rest in the horizontal position, calculate the initial value of the force R exerted by the bearing on the bar an instant after release.

6/39 If the bar of Prob. 6/38 has an initial angular acceleration of 6 rad/s² when released from rest in the horizontal position, calculate the frictional moment M_f acting on the bar at O. Also find the force R exerted by the bearing on the bar at O.

6/40 The automotive dynamometer is able to simulate road conditions for an acceleration of 0.6g for the loaded pickup truck with a total mass of 2.8 Mg. Calculate the required moment of inertia of the dynamometer drum about its center O assuming that the drum turns freely during the acceleration phase of the test.

Ans. $I_O = 2.27$ Mg·m²

Problem 6/40

6/41 The uniform 5-kg bar is mounted in a bearing at swings in a vertical plane about the pivot at A. If $\dot{\theta} = 2$ rad/s when $\theta = 30°$, compute the force supported by the pin at A at that instant.

Ans. $A = 56.3$ N

Problem 6/41

6/42 The uniform 5-kg bar is mounted in a bearing at O and hangs in a vertical position. The bearing is mounted in stiff elastic supports instrumented with electrical strain gages calibrated to record the horizontal force applied to the bearing at O. If the gages record a peak value of 215 N during the sudden application of the horizontal force P applied to the end of the bar, calculate the peak value of P.

6/43 The half ring of mass m and radius r is welded to a small horizontal shaft mounted in a bearing as shown. Neglect the mass of the shaft and determine the angular acceleration of the ring resulting from a torque M applied to the shaft.

$$Ans.\ \alpha = \frac{2M}{mr^2}$$

Problem 6/42   **Problem 6/43**

6/44 The uniform slender bar of density ρ rotates about a vertical axis through its mass center with a constant angular velocity ω. Determine an expression for the tensile stress σ in the bar as a function of r.

Problem 6/44

Problem 6/45

6/45 The ring of mean radius r and small cross-sectional dimensions is rotating in a horizontal plane with a constant angular velocity ω about its center O. Determine the tension T in the ring. Solve, first, by analyzing one half of the ring as a free body and, second, by analyzing a differential element of arc as a free body.

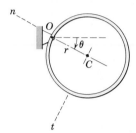

Problem 6/46

6/46 The narrow ring of mass m is free to rotate in the vertical plane about O. If the ring is released from rest at $\theta = 0$, determine expressions for the n- and t-components of the force at O in terms of θ.

$$Ans.\ O_t = -\frac{mg}{2}\cos\theta,\ O_n = 2\,mg\sin\theta$$

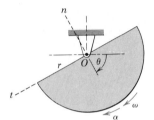

Problem 6/47

6/47 The semicircular disk of mass m and radius r is released from rest at $\theta = 0$ and rotates freely in the vertical plane about its fixed bearing at O. Derive expressions for the n- and t-components of the force F on the bearing as a function of θ.

Problem 6/48

6/48 Gear A has a mass of 24 kg and a centroidal radius of gyration of 150 mm. Gear B has a mass of 6 kg and a centroidal radius of gyration of 75 mm. Calculate the angular acceleration of gear B when a torque M of 8 N·m is applied to the shaft of gear A. (Gear teeth are not shown.)

$$Ans.\ \alpha_B = 23.7\ \text{rad/s}^2$$

Problem 6/49

6/49 The solid circular disk of mass m with center at G is attached to the vertical shaft at O and rotates in the horizontal plane under the action of an applied couple M as shown. Determine the required length of time t, starting from rest, for the n- and t-components of the force exerted by the disk on the shaft to become momentarily equal.

6/50 The solid cylindrical rotor *B* has a mass of 43 kg and is mounted on its central axis *C-C*. The frame *A* rotates about the fixed vertical axis *O-O* under the applied torque $M = 30$ N·m. The rotor may be unlocked from the frame by withdrawing the locking pin *P*. Calculate the angular acceleration α of the frame *A* if the locking pin is (*a*) in place and (*b*) withdrawn. Neglect all friction and the mass of the frame.

Ans. (*a*) $\alpha = 8.46$ rad/s^2
(*b*) $\alpha = 11.16$ rad/s^2

Problem 6/50

6/51 A reel of flexible power cable is mounted on the dolly, which is fixed in position. There are 60 m of cable having a mass of 0.65 kg per meter of length wound on the reel at a radius of 375 mm. The empty spool has a mass of 28 kg and a radius of gyration about its axis of 300 mm. A tension *T* of 90 N is required to overcome frictional resistance to turning. Calculate the angular acceleration α of the reel if a tension of 180 N is applied to the free end of the cable.

Problem 6/51

6/52 A gimbal pedestal supports a payload in the space shuttle and deploys it when the doors of the cargo bay are opened when in orbit. The payload is modeled as a homogeneous rectangular block with a mass of 6000 kg. The torque on the gimbal axis *O-O* is 30 N·m supplied by a d-c brushless motor. With the shuttle orbiting in a "weightless" condition, determine the time *t* required to bring the payload from its stowed position at $\theta = 0$ to its deployed position at $\theta = 90°$ if the torque is applied for the first 45° of angle and then reversed for the remaining 45° to bring the payload to a stop ($\dot{\theta} = 0$).   *Ans.* $t = 78.6$ s

Problem 6/52

Problem 6/53

6/53 A device for impact testing consists of a 34-kg pendulum with mass center at G and with radius of gyration about O of 620 mm. The distance b for the pendulum is selected so that the force on the bearing at O is the least possible value during impact with the specimen at the bottom of the swing. Determine b and calculate the magnitude of the total force R on the bearing O an instant after release from rest at $\theta = 60°$.

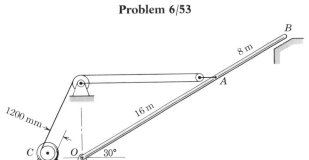

Problem 6/54

6/54 The uniform 24-m mast has a mass of 300 kg and is hinged at its lower end to a fixed support at O. If the winch C develops a starting torque of 1300 N·m, calculate the total force supported by the pin at O as the mast begins to lift off its support at B. Also find the corresponding angular acceleration α of the mast. The cable at A is horizontal and the mass of the pulleys and winch is negligible.
Ans. $\alpha = 0.071$ rad/s^2, $O = 5263$ N

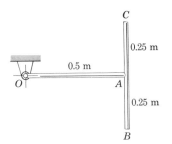

Problem 6/55

6/55 Each of the two uniform slender bars OA and BC has a mass of 8 kg. The bars are welded at A to form a T-shaped member and are rotating freely about a horizontal axis through O. If the bars have an angular velocity ω of 4 rad/s as OA passes the horizontal position shown, calculate the total force R supported by the bearing at O.
Ans. $R = 101.3$ N

Problem 6/56

6/56 The rig is used to hoist a rocket to a vertical launch position and consists of a uniform 6-m steel beam with a mass of 720 kg to which the rocket is secured. The rocket has a mass of 1600 kg with mass center at G_2. The moment of inertia of the rocket about a transverse axis through G_2 (normal to its length) is 4800 kg·m^2. If the initial tension in the piston rod of the hydraulic cylinder is 80 kN, calculate the initial angular acceleration α of the assembly as it lifts the rocket off its supporting pads. Treat the beam as a slender rod.
Ans. $\alpha = 0.239$ rad/s^2

6/57 The uniform semicircular bar of mass m and radius r is hinged freely about a horizontal axis through A. If the bar is released from rest in the position shown where AB is horizontal, determine the initial angular acceleration α of the bar and the expression for the force exerted on the bar by the pin at A. (Note carefully that the initial tangential acceleration of the mass center is not vertical.)

Problem 6/57

Problem 6/58

6/58 The semicircular ring of mass m and radius r is welded to the vertical shaft, which permits the ring to rotate in the horizontal plane about axis O-O. If a torque M is applied to the ring through its shaft, determine the expression for the resulting angular acceleration α of the ring and the force F acting in the horizontal plane on the ring at O as the ring starts from rest.

$$Ans.\ \alpha = \frac{M}{2mr^2},\ F = 0.593M/r$$

6/59 If the bar of Prob. 6/44 starts from rest with an angular acceleration $\alpha = \dot{\omega}$ under the action of a couple applied to the shaft, determine the resulting bending moment M and shear force V in the bar as a function of r as observed in the horizontal plane of rotation.

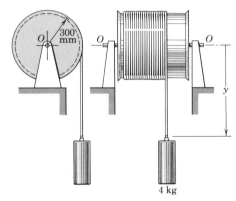

Problem 6/60

6/60 A flexible cable 60 meters long with a mass of 0.160 kg per meter of length is wound around the reel. With $y = 0$, the weight of the 4-kg cylinder is required to start turning the reel to overcome friction in its bearings. Determine the downward acceleration a in meters per second squared of the cylinder as a function of y in meters. The empty reel has a mass of 16 kg with a radius of gyration about its bearing of 200 mm. *Ans.* $a = 0.0758y$

6/61 The hoisting cable for the 50-kg cylinder is wound securely around each of the two drums attached rigidly to the large driving gear A. The drums and gear have a combined mass of 80 kg with radius of gyration about O of 260 mm. If the motor pinion applies a tangential force of 280 N to the teeth of the driving gear at B, calculate the upward acceleration a of the cylinder. Neglect friction and the mass of the cable. *Ans.* $a = 0.509$ m/s^2

Problem 6/61

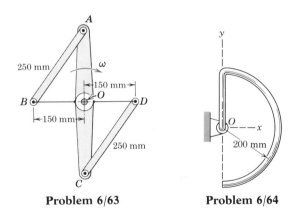

Problem 6/62

6/62 Prior to deployment of its two instrument arms *AB*, the spacecraft shown in the upper view is spinning at the constant rate of 1 revolution per second. Each instrument arm, shown in the lower view, has a mass of 10 kg with mass center at *G*. Calculate the tension *T* in the deployment cable prior to its release. Also find the magnitude of the force on the pin at *A*. Neglect any acceleration of the center *O* of the spacecraft.

6/63 Link *AC* is made to revolve with a constant angular velocity $\omega = 10$ rad/s about a fixed vertical axis through its center *O*. Each of the uniform links *AB* and *CD* has a mass of 4 kg and is held in the configuration shown by a cord leading perpendicularly to the rotating link *AC*. Calculate the tension *T* in *BO* and *DO*. *Ans.* $T = 30$ N

Problem 6/63

Problem 6/64

6/64 A uniform metal rod with a mass of 0.60 kg per meter of length is bent into the shape shown in the vertical plane and pivoted freely about a horizontal axis at *O* normal to the plane. If the rod is released from rest in the position shown with the straight section vertical, compute the initial *x*- and *y*-components of the reaction **R** supported by the bearing at *O*. *Ans.* $R_x = 0.339$ N, $R_y = 3.52$ N

6/65 Two slender bars *AB*, each of mass *m* and length *l*, are pivoted at *A* to the plate. The plate rotates in the horizontal plane about a fixed vertical axis through its center *O* and is given a constant angular acceleration α. (*a*) Determine the force *F* exerted on each of the two rollers as the assembly starts to rotate. (*b*) Find the total force on the pin at *A* and show that it remains constant as long as $F > 0$. (*c*) Determine the angular velocity ω at which contact with the rollers ceases.

Ans. (*a*) $F = ml\alpha/6$

(*b*) $A = \dfrac{\sqrt{10}}{6} ml\alpha$

(*c*) $\omega = \sqrt{\alpha/3}$

Problem 6/65

6/66 The split ring of mass m and radius r is rotating in its plane with an angular velocity ω about a fixed vertical axis through one end at O. Determine the bending moment M and shear force V in the ring at A if ω is constant.

Problem 6/66

6/67 The uniform slender rod of length l and mass m is welded at its end tangent to the rim of the circular disk of radius r which rotates about a vertical axis through O. Determine the bending moment M, the shear force V, and the axial force T which the weld exerts on the rod (a) for a constant angular velocity ω of the disk and (b) as the disk starts from rest with a counterclockwise angular acceleration α.

$$\text{Ans. } (a)\ M = \frac{mrl\omega^2}{2},\ V = mr\omega^2,\ T = \frac{ml\omega^2}{2}$$

$$(b)\ M = -\frac{ml^2\alpha}{3},\ V = -\frac{ml\alpha}{2},\ T = mr\alpha$$

Problem 6/67

6/68 The curved bar of mass m is hinged to the rotating disk at O and bears against one of the smooth pins A and B which are fastened to the disk. If the disk rotates about its vertical axis C, determine the force exerted on the bar by the hinge at O and the reaction A or B on the bar (a) if the disk has a constant angular velocity ω and (b) as the disk starts from rest with a counterclockwise angular acceleration α.

$$\text{Ans. } (a)\ O = \frac{2mr\omega^2}{\pi},\ A = \frac{2mr\omega^2}{\pi}$$

$$(b)\ O = mr\alpha\sqrt{1 + \frac{4}{\pi^2}},\ B = mr\alpha\left(1 - \frac{2}{\pi}\right)$$

Problem 6/68

▶ **6/69** The uniform slender bar of mass m and length l is released from rest in the vertical position and pivots on its square end about the corner at O. (a) If the bar is observed to slip when $\theta = 30°$, find the coefficient of static friction μ_s between the bar and the corner. (b) If the end of the bar is notched so that it cannot slip, find the angle θ at which contact between the bar and the corner ceases.

$$\text{Ans. } (a)\ \mu_s = 0.188,\ (b)\ \theta = 53.1°$$

Problem 6/69

6/5 GENERAL PLANE MOTION

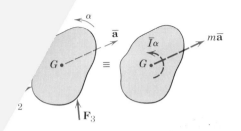

Free-Body Diagram   Kinetic Diagram

Figure 6/4
(Repeated)

The dynamics of general plane motion of a rigid body combines translation and rotation. In Art. 6/2 we represented such a body in Fig. 6/4 with its free-body diagram and its kinetic diagram, which discloses the dynamic resultants of the applied forces. Figure 6/4 and Eq. 6/1, which apply to general plane motion, are repeated again here for convenient reference.

$$\Sigma \mathbf{F} = m\bar{\mathbf{a}}$$
$$\Sigma M_G = \bar{I}\alpha$$

[6/1]

As previously, the two scalar components of the force equation are expressed in whatever coordinate system most readily describes the acceleration of the mass center. Direct application of these equations expresses the equivalence between the externally applied forces, as disclosed by the free-body diagram, and their force and moment resultants, as represented by the kinetic diagram.

In Art. 6/2 we also showed, with the aid of Fig. 6/5, the application of the alternative moment relation about any point P, Eq. 6/2. This figure and this equation are also repeated here for easy reference.

$$\Sigma M_P = \bar{I}\alpha + m\bar{a}d$$

[6/2]

In some instances it may be more convenient to use the alternative moment relation of Eq. 6/3 when moments are taken about a point P whose acceleration is known. It is also noted that the moment equation about a nonaccelerating point O on the body, Eq. 6/4, constitutes still another alternative moment relation and at times may be used to advantage.

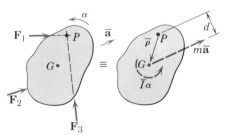

Free-Body Diagram   Kinetic Diagram

Figure 6/5
(Repeated)

In working a problem in general plane motion we first observe whether the motion is unconstrained or constrained, as illustrated in the examples of Fig. 6/6. If the motion is constrained, we must account for the kinematic relationship between the linear and the angular accelerations and incorporate it into our force and moment equations of motion. If the motion is unconstrained, the accelerations can be determined independently of one another by direct application of the three motion equations, Eqs. 6/1.

In order for a rigid-body problem to be solvable, the number of unknowns cannot exceed the number of independent equations available to describe them, and a check on the sufficiency of the relationships should always be made. At the most, for plane motion we have three scalar equations of motion and two scalar components of the vector relative-acceleration equation for constrained motion. Thus we can handle as many as five unknowns.

Strong emphasis should be placed on the importance of a clear choice of the body to be isolated and the representation of this isolation by a correct free-body diagram. Only after this vital step has been completed can we properly evaluate the equivalence between the external forces and their resultants. Of equal importance in the analysis of plane motion is a clear understanding of the kinematics involved. Very often the difficulties experienced at this point have to do with kinematics, and a thorough review of the relative acceleration relations for plane motion will be most helpful. In formulating the solution to a problem we recognize that the directions of certain forces or accelerations may not be known at the outset, so that it may be necessary to make initial assumptions whose validity will be proved or disproved when the solution is carried out. It is essential, however, that all assumptions made be consistent with the principle of action and reaction and with any kinematical requirements, which are also called conditions of constraint. Thus if a wheel is rolling on a horizontal surface, its center is constrained to move on a horizontal line. Furthermore, if the unknown linear acceleration a of the center of the wheel is assumed positive to the right, then the unknown angular acceleration α would be positive in a clockwise sense in order that $a = +r\alpha$, assuming the wheel does not slip. Also, we note that for a wheel which rolls without slipping, the static friction force between the wheel and its supporting surface is generally *less* than its maximum value, so that $F \neq \mu_s N$. But if the wheel slips as it rolls, $a \neq r\alpha$, and a kinetic friction force is generated which is given by $F = \mu_k N$. It may be necessary to test the validity of either assumption, slipping or no slipping, in a given problem. The difference between the static and kinetic coefficients of friction, μ_s and μ_k, is sometimes ignored, in which case μ is used for either or both coefficients.

Sample Problem   6/5

A metal hoop with a radius $r = 150$ mm is released from rest on the 20° incline. If the coefficients of static and kinetic friction are $\mu_s = 0.15$ and $\mu_k = 0.12$, determine the angular acceleration α of the hoop and the time t for the hoop to move a distance of 3 m down the incline.

150 mm

$\mu_s = 0.15$
$\mu_k = 0.12$

20°

Solution.   The free-body diagram shows the unspecified weight mg, the normal force N, and the friction force F acting on the hoop at the contact point C with the incline. The kinetic diagram shows the resultant force $m\bar{a}$ through G in the direction of its acceleration and the couple $\bar{I}\alpha$. The counterclockwise angular acceleration requires a counterclockwise moment about G, so F must be up the incline.

Assume that the hoop rolls without slipping, so that $\bar{a} = r\alpha$. Application of the components of Eqs. 6/1 with x- and y-axes assigned gives

$[\Sigma F_x = m\bar{a}_x]$ $mg \sin 20° - F = m\bar{a}$

$[\Sigma F_y = m\bar{a}_y = 0]$ $N - mg \cos 20° = 0$

① $[\Sigma M_G = \bar{I}\alpha]$ $Fr = mr^2\alpha$

$r = 150$ mm

Elimination of F between the first and third equations and substitution of the kinematic assumption $\bar{a} = r\alpha$ give

② $$\bar{a} = \frac{g}{2} \sin 20° = \frac{9.81}{2} (0.3420) = 1.678 \text{ m/s}^2$$

Alternatively, with our assumption of $\bar{a} = r\alpha$ for pure rolling, a moment sum about C by Eq. 6/2 gives $\bar{a}$ directly. Thus

$[\Sigma M_C = \bar{I}\alpha + m\bar{a}d]$ $mgr \sin 20° = mr^2 \dfrac{\bar{a}}{r} + m\bar{a}r$ $\bar{a} = \dfrac{g}{2} \sin 20°$

To check our assumption of no slipping, we calculate F and N and compare F with its limiting value. From the above equations

$$F = mg \sin 20° - m\frac{g}{2} \sin 20° = 0.1710mg$$

$$N = mg \cos 20° = 0.9397mg$$

But the maximum possible friction force is

$[F_{max} = \mu_s N]$ $F_{max} = 0.15(0.9397mg) = 0.1410mg$

Because our calculated value of $0.1710mg$ exceeds the limiting value of $0.1410mg$, we conclude that our assumption of pure rolling was wrong. Therefore the hoop slips as it rolls and $\bar{a} \neq r\alpha$. The friction force then becomes the kinetic value

$[F = \mu_k N]$ $F = 0.12(0.9397mg) = 0.1128mg$

The motion equations now give

$[\Sigma F_x = m\bar{a}_x]$ $mg \sin 20° - 0.1128mg = m\bar{a}$

$\bar{a} = 0.229(9.81) = 2.25 \text{ m/s}^2$

③ $[\Sigma M_G = \bar{I}\alpha]$ $0.1128mg(r) = mr^2\alpha$

$$\alpha = \frac{0.1128(9.81)}{0.150} = 7.37 \text{ rad/s}^2 \qquad Ans.$$

The time required for the center G of the hoop to move 3 m from rest with constant acceleration is

$[x = \frac{1}{2}at^2]$ $t = \sqrt{\dfrac{2x}{\bar{a}}} = \sqrt{\dfrac{2(3)}{2.25}} = 1.633 \text{ s} \qquad Ans.$

① Because all of the mass of a hoop is a distance r from its center G, its moment of inertia about G must be mr^2.

② Note that $\bar{a}$ is independent of both m and r.

③ Note that α is independent of m but dependent on r.

Sample Problem 6/6

The drum A is given a constant angular acceleration α_0 of 3 rad/s² and causes the 70-kg spool B to roll on the horizontal surface by means of the connecting cable which wraps around the inner hub of the spool. The radius of gyration $\bar{k}$ of the spool about its mass center G is 250 mm, and the coefficient of static friction between the spool and the horizontal surface is 0.25. Determine the tension T in the cable and the friction force F exerted by the horizontal surface on the spool.

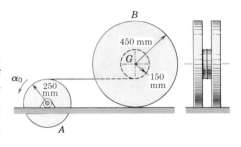

Solution. The free-body diagram and the kinetic diagram of the spool are drawn as shown. The correct direction of the friction force may be assigned in this problem by observing from both diagrams that with counterclockwise angular acceleration a moment sum about point G (and also about point D) must be counterclockwise. A point on the connecting cable has an acceleration $a_t = r\alpha = 0.25(3) = 0.75$ m/s², which is also the horizontal component of the acceleration of point D on the spool. It will be assumed initially that the spool rolls without slipping, in which case it would have a counterclockwise
① angular acceleration $\alpha = (a_D)_x/\overline{DC} = 0.75/0.30 = 2.5$ rad/s². The acceleration of the mass center G is, therefore, $\bar{a} = r\alpha = 0.45(2.5) = 1.125$ m/s².

With the kinematics determined, we now apply the three equations of motion, Eqs. 6/1.

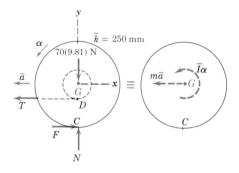

$[\Sigma F_x = m\bar{a}_x]$ $\qquad F - T = 70(-1.125)$ $\qquad\qquad (a)$

$[\Sigma F_y = m\bar{a}_y]$ $\qquad N - 70(9.81) = 0$ $\qquad N = 687$ N

② $[\Sigma M_G = \bar{I}\alpha]$ $\qquad F(0.450) - T(0.150) = 70(0.250)^2(2.5)$ $\qquad (b)$

Solving (a) and (b) simultaneously gives

$$F = 75.8 \text{ N} \qquad \text{and} \qquad T = 154.6 \text{ N} \qquad\qquad Ans.$$

To establish the validity of our assumption of no slipping, we see that the surfaces are capable of supporting a maximum friction force $F_{\max} = \mu_s N = 0.25(687) = 171.7$ N. Since only 75.8 N of friction force are required, we conclude that our assumption of rolling without slipping is valid.

If the coefficient of static friction had been 0.1, for example, then the friction force would have been limited to $0.1(687) = 68.7$ N which is less than 75.8 N, and the spool would slip. In this event the kinematical relation $\bar{a} = r\alpha$ would no longer hold. With $(a_D)_x$ known, the
③ angular acceleration would be $\alpha = [\bar{a} - (a_D)_x]/\overline{GD}$. Using this relation along with $F = \mu_k N < 68.7$ N, we would then re-solve the three equations of motion for the unknowns T, $\bar{a}$, and α.

Alternatively, with point C as a moment center, we may use Eq. 6/2 and obtain T directly. Thus

$[\Sigma M_C = \bar{I}\alpha + m\bar{a}r]$ $\qquad 0.3T = 70(0.25)^2(2.5) + 70(1.125)(0.45)$

$$T = 154.6 \text{ N} \qquad\qquad Ans.$$

where the kinematic condition of no slipping, $\bar{a} = \bar{r}\alpha$, has been incor-
④ porated. We could also write a moment equation about point D to obtain F directly.

① The relation between $\bar{a}$ and α is the kinematical requirement which accompanies the constraint to the motion imposed by the assumption that the spool rolls without slipping.

② Be careful not to make the mistake of using $\frac{1}{2}mr^2$ for $\bar{I}$ of the spool, which is not a uniform circular disk.

③ Our principles of relative acceleration are a necessity here. Hence the relation $(a_{G/D})_t = \overline{GD}\alpha$ should be recognized.

④ The flexibility in the choice of moment centers provided by the kinetic diagram can usually be employed to simplify the analysis.

Sample Problem 6/7

The slender 30-kg bar AB moves in the vertical plane with its ends constrained to follow the smooth horizontal and vertical guides. If the 150-N force is applied at A with the bar initially at rest in the position for which $\theta = 30°$, calculate the resulting angular acceleration of the bar and the forces on the small end rollers at A and B.

Solution. The bar undergoes constrained motion, so that we must establish the relationship between the mass-center acceleration and the angular acceleration. The relative-acceleration equation

(1) $\mathbf{a}_A = \mathbf{a}_B + \mathbf{a}_{A/B}$ must be solved first, and then the equation $\bar{\mathbf{a}} = \mathbf{a}_G = \mathbf{a}_B + \mathbf{a}_{G/B}$ is next solved to obtain expressions relating $\bar{a}$ and α. With α assigned in its clockwise physical sense, the acceleration polygons which represent these equations are shown, and their solution gives

$$\bar{a}_x = \bar{a} \cos 30° = 0.6\alpha \cos 30° = 0.520\alpha \text{ m/s}^2$$
$$\bar{a}_y = \bar{a} \sin 30° = 0.6\alpha \sin 30° = 0.3\alpha \text{ m/s}^2$$

Next we construct the free-body diagram and the kinetic diagram as shown. With $\bar{a}_x$ and $\bar{a}_y$ now known in terms of α, the remaining unknowns are α and the forces A and B. We now apply Eqs. 6/1 which give

(2) $[\Sigma M_G = \bar{I}\alpha]$

$$150(0.6 \cos 30°) - A(0.6 \sin 30°) + B(0.6 \cos 30°) = \frac{1}{12}30(1.2)^2\alpha$$

$[\Sigma F_x = m\bar{a}_x]$ $\qquad 150 - B = 30(0.520\alpha)$

$[\Sigma F_y = m\bar{a}_y]$ $\qquad A - 30(9.81) = 30(0.3\alpha)$

Solving the three equations simultaneously gives us the results

$$A = 337 \text{ N} \qquad B = 76.8 \text{ N} \qquad \alpha = 4.69 \text{ rad/s}^2 \qquad Ans.$$

As an alternative solution we can use Eq. 6/2 with point C as the moment center and avoid the necessity of solving three equations simultaneously. This choice eliminates reference to forces A and B and gives α directly. Thus

(3) $[\Sigma M_C = \bar{I}\alpha + \Sigma m\bar{a}d]$

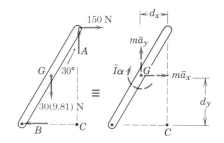

$$150(1.2 \cos 30°) - 30(9.81)(0.6 \sin 30°) = \frac{1}{12}30(1.2)^2\alpha$$
$$+ 30(0.520\alpha)(0.6 \cos 30°) + 30(0.3\alpha)(0.6 \sin 30°)$$
$$67.6 = 14.40\alpha \qquad \alpha = 4.69 \text{ rad/s}^2 \qquad Ans.$$

With α determined we can now apply the force equations independently and get

$[\Sigma F_y = m\bar{a}_y]$ $\qquad A - 30(9.81) = 30(0.3)(4.69)$ $\qquad A = 337 \text{ N} \qquad Ans.$
$[\Sigma F_x = m\bar{a}_x]$ $\qquad 150 - B = 30(0.520)(4.69)$ $\qquad B = 76.8 \text{ N} \qquad Ans.$

(1) If the application of the relative-acceleration equations is not perfectly clear at this point, then review Art. 5/6. Note that the relative normal acceleration term is absent since there is no angular velocity of the bar.

(2) Recall that the moment of inertia of a slender rod about its center is $\frac{1}{12}ml^2$.

(3) From the kinetic diagram the terms $\Sigma m\bar{a}d$ are $m\bar{a}_x d_y + m\bar{a}_y d_x$. Since both are clockwise in the sense of $\bar{I}\alpha$, they are positive.

Sample Problem 6/8

A car door is inadvertently left slightly open when the brakes are applied to give the car a constant rearward acceleration a. Derive expressions for the angular velocity of the door as it swings past the 90° position and the components of the hinge reactions for any value of θ. The mass of the door is m, its mass center is a distance $\bar{r}$ from the hinge axis O, and the radius of gyration about O is k_O.

Solution. Because the angular velocity ω increases with θ, we need to find how the angular acceleration α varies with θ so that we may integrate it over the interval to obtain ω. We obtain α from a moment equation about O. First we draw the free-body diagram of the door in the horizontal plane for a general position θ. The only forces in this plane are the components of the hinge reaction shown here in the x- and y-directions. On the kinetic diagram, in addition to the resultant couple $\bar{I}\alpha$ shown in the sense of α, we represent the resultant force $m\bar{a}$ in terms of its components by using an equation ① of relative acceleration with respect to O. This equation becomes the kinematical equation of constraint and is

$$\bar{a} = a_G = a_O + (a_{G/O})_n + (a_{G/O})_t$$

The magnitudes of the $m\bar{a}$ components are, then,

② $\qquad ma_O = ma \qquad m(a_{G/O})_n = m\bar{r}\omega^2 \qquad m(a_{G/O})_t = m\bar{r}\alpha$

where $\omega = \dot{\theta}$ and $\alpha = \ddot{\theta}$.

For a given angle θ the three unknowns are α, O_x, and O_y. We can eliminate O_x and O_y by a moment equation about O, which gives

③ $\quad [\Sigma M_O = \bar{I}\alpha + \Sigma m\bar{a}d] \qquad 0 = m(k_O{}^2 - \bar{r}^2)\alpha + m\bar{r}\alpha(\bar{r}) - ma(\bar{r}\sin\theta)$

④ Solving for α gives $\qquad \alpha = \dfrac{a\bar{r}}{k_O{}^2}\sin\theta$

Now we integrate α first to a general position and get

$$[\omega\, d\omega = \alpha\, d\theta] \qquad \int_0^\omega \omega\, d\omega = \int_0^\theta \frac{a\bar{r}}{k_O{}^2}\sin\theta\, d\theta$$

$$\omega^2 = \frac{2a\bar{r}}{k_O{}^2}(1 - \cos\theta)$$

For $\theta = \pi/2$ $\qquad\qquad \omega = \dfrac{1}{k_O}\sqrt{2a\bar{r}}$ $\qquad$ *Ans.*

⑤ To find O_x and O_y for any given value of θ the force equations give

$[\Sigma F_x = m\bar{a}_x] \qquad O_x = ma - m\bar{r}\omega^2\cos\theta - m\bar{r}\alpha\sin\theta$

$$= m\left(a - \frac{2a\bar{r}^2}{k_O{}^2}[1 - \cos\theta]\cos\theta - \frac{a\bar{r}^2}{k_O{}^2}\sin^2\theta\right)$$

$$= ma\left(1 - \frac{\bar{r}^2}{k_O{}^2}[1 + 2\cos\theta - 3\cos^2\theta]\right) \qquad \textit{Ans.}$$

$[\Sigma F_y = m\bar{a}_y] \qquad O_y = m\bar{r}\alpha\cos\theta - m\bar{r}\omega^2\sin\theta$

$$= m\bar{r}\frac{a\bar{r}}{k_O{}^2}\sin\theta\cos\theta - m\bar{r}\frac{2a\bar{r}}{k_O{}^2}(1 - \cos\theta)\sin\theta$$

$$= \frac{ma\bar{r}^2}{k_O{}^2}(3\cos\theta - 2)\sin\theta \qquad \textit{Ans.}$$

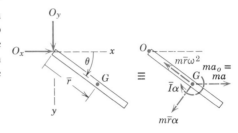

① Point O is chosen because it is the only point on the door whose acceleration is known.

② Be careful to place $m\bar{r}\alpha$ in the sense of positive α with respect to rotation about O.

③ The free-body diagram shows that there is zero moment about O. We use the transfer-of-axis theorem here and substitute $k_O{}^2 = \bar{k}^2 + \bar{r}^2$. If this relation is not totally familiar, review Art. B/1 in Appendix B.

④ We may also use Eq. 6/3 with O as a moment center

$$\Sigma M_O = I_O\alpha + \bar{\rho} \times ma_O$$

where the scalar values of the terms are $I_O\alpha = mk_O{}^2\alpha$ and $\bar{\rho} \times ma_O$ becomes $-\bar{r}ma\sin\theta$.

⑤ The kinetic diagram shows clearly the terms which make up $m\bar{a}_x$ and $m\bar{a}_y$.

Problem 6/70

Problem 6/71

Problem 6/73

Problem 6/74

Problem 6/75

PROBLEMS

6/70 The 30-kg wheel has a centroidal radius of gyration of 150 mm. If friction is sufficient to prevent slipping, compute the friction force F acting on the wheel during its downhill roll. What is the minimum coefficient of static friction μ_s to prevent slipping? *Ans.* $F = 34.8$ N, $\mu_s = 0.128$

6/71 For a given coefficient of static friction μ_s determine the maximum angle θ which will permit the solid circular disk to roll down the incline without slipping. *Ans.* $\theta = \tan^{-1} 3\mu_s$

6/72 If the coefficient of static friction between the wheel of Prob. 6/70 and the incline is 0.10, calculate the acceleration of the center of the wheel and its angular acceleration. Do the results depend on the weight of the wheel?
Ans. $a = 2.87$ m/s^2, $\alpha = 9.06$ rad/s^2, no

6/73 A long cable of length L and mass ρ per unit length is wrapped around the periphery of a spool of negligible mass. One end of the cable is fixed, and the spool is released from rest in the position shown. Show that the center of the spool has a constant acceleration during the motion.

6/74 The 30-kg solid circular disk is initially at rest on the horizontal surface when a 12-N force P, constant in magnitude and direction, is applied to the cord wrapped securely around its periphery. The disk is supported by an air cushion supplied by air jets through many small holes in the surface and, consequently, slides over the surface with essentially no frictional resistance. Calculate the angular velocity ω of the disk after the 12-N force has been applied for 2 seconds and find the linear velocity v of the center of the disk after it has moved 1.2 meters from rest.
Ans. $\omega = 6.4$ rad/s, $v = 0.980$ m/s

6/75 The circular disk of mass m and radius r is rolling through the bottom of the circular path of radius R. If the disk has an angular velocity ω, determine the force N exerted by the path on the disk.

6/76 The spacecraft is spinning with a constant angular velocity ω about the z-axis at the same time that its mass center O is traveling with a velocity v_O in the y-direction. If a tangential hydrogen-peroxide jet is fired when the craft is in the position shown, determine the expression for the absolute acceleration of point A on the spacecraft rim at the instant the jet force is F. The radius of gyration of the craft about the z-axis is k, and its mass is m.

Problem 6/76

6/77 Above the earth's atmosphere at an altitude of 400 km where the acceleration due to gravity is 8.69 m/s² a certain rocket has a total remaining mass of 300 kg and is directed 30° from the vertical. If the thrust T from the rocket motor is 4 kN and if the rocket nozzle is tilted through an angle of 1° as shown, calculate the angular acceleration α and the x- and y-components of the acceleration of the mass center G. The rocket has a centroidal radius of gyration of 1.5 m.

$$Ans. \quad \alpha = 0.310 \text{ rad/s}^2$$
$$\bar{a}_x = 6.87 \text{ m/s}^2$$
$$\bar{a}_y = 2.74 \text{ m/s}^2$$

Problem 6/77

6/78 The unbalanced wheel with mass center at G rolls down the incline with increasing speed v of its geometric center O. What is the least speed v for which the wheel could lose contact momentarily with the incline? Identify the position of G where this condition occurs.

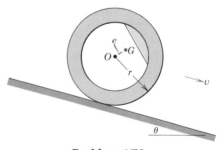

Problem 6/78

6/79 The mass center G of the 10-kg wheel is off center by 10 mm. If G is in the position shown as the wheel rolls without slipping through the bottom of the circular path of 1.2-m radius with an angular velocity ω of 10 rad/s, compute the force P exerted by the path on the wheel. (Be careful to use the correct mass-center acceleration.)

$$Ans. \quad P = 101.4 \text{ N}$$

Problem 6/79

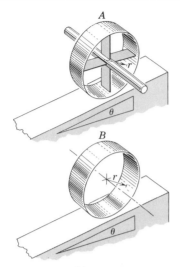

Problem 6/80

6/80 Determine the angular acceleration of each of the two wheels as they roll without slipping down the inclines. For wheel *A* investigate the case where the mass of the rim and spokes is negligible and the mass of the bar is concentrated along its centerline. For wheel *B* assume that the thickness of the rim is negligible compared with its radius so that all of the mass is concentrated in the rim. Also specify the minimum coefficient of static friction μ_s required to prevent each wheel from slipping.

Problem 6/81

6/81 The hoop of radius *r* and negligible mass has four equally spaced spokes, each of mass *m* and length *r*. If the hoop is released on the incline and rolls without slipping, determine the acceleration *a* of the center of the hoop and the friction force *F* acting on the hoop. For a given angle θ, what is the minimum value μ_s of the static coefficient of friction for which no slipping will occur?

Ans. $a = \frac{3}{4}g \sin \theta, \ \mu_s = \frac{1}{4}\tan \theta$

Problem 6/82

6/82 The circular disk of 200-mm radius has a mass of 25 kg with centroidal radius of gyration of 175 mm and has a concentric circular groove of 75 mm radius cut into it. If a steady horizontal force of 20 N is applied to a cord wrapped around the groove as shown, calculate the angular acceleration α of the disk as it starts from rest. The coefficient of static friction between the disk and the horizontal surface is 0.10. (Be sure to recognize that the wheel rolls clockwise and *not* counterclockwise. Assume first that the wheel does not slip and then verify your assumption with the results.) *Ans.* $\alpha = 1.416$ rad/s^2

6/83　In the design of a space station, remote from gravitational attraction, a long tubular antenna is to be deployed in the following way. Thin spring-steel tape of length L and mass ρ per unit of length is preformed into a single split tube, after which it is flattened and wound around a spool of radius r. The bending moment in the tape as it leaves the spool tangentially is M. This moment acts to unroll the tape, and the spool is pushed away from the clamped end A of the tube. The guiding bracket contains the tape and keeps it tangent to the spool as it unwinds. Neglect the mass of the spool and bracket compared with the mass of the tape and determine the acceleration a of the center of the spool and the compressive force P in the tube as the roll accelerates away from A. Neglect any variation in r and assume that the clamp A is fixed in space. The roll may be analyzed with the free-body diagram shown.

Problem 6/83

6/84　The wheel and its hub have a mass of 30 kg with a radius of gyration about the center of 450 mm. A cord wrapped securely around its hub is attached to the fixed support, and the wheel is released from rest on the incline. If the static and kinetic coefficients of friction between the wheel and the incline are 0.40 and 0.30, respectively, calculate the acceleration a of the center of the wheel. First prove that the wheel slips.　　*Ans. $a = 1.256$ m/s^2*

Problem 6/84

6/85　The uniform slender bar of length l and mass m is resting on a smooth horizontal surface. If a horizontal force F is applied normal to the bar at end A, determine the expression for the acceleration a_B of end B during application of F.

Problem 6/85

6/86　A 15-kg roll of wrapping paper in the form of a solid circular cylinder with a diameter of 300 mm is resting on a horizontal surface. If a force of 40 N is applied to the paper at the angle $\theta = 30°$, determine the initial acceleration a of the center of the roll and the corresponding angular acceleration α. The coefficient of kinetic friction between the paper and the horizontal surface is 0.2.
　　　　Ans. $a = 0.614$ m/s^2, $\alpha = 12.95$ rad/s^2

Problem 6/86

Problem 6/87

Problem 6/88

Problem 6/89

Problem 6/90

Problem 6/91

6/87 Small ball-bearing rollers mounted on the ends of the slender bar of mass m and length l constrain the motion of the bar in the horizontal x-y slots. If a couple M is applied to the bar initially at rest at $\theta = 45°$, determine the forces exerted on the rollers at A and B as the bar starts to move.

6/88 End A of the uniform 5-kg bar is pinned freely to the collar which has an acceleration $a = 4$ m/s² along the fixed horizontal shaft. If the bar has a clockwise angular velocity $\omega = 2$ rad/s as it swings past the vertical, determine the components of the force on the bar at A for this instant.

Ans. $A_x = 5$ N, $A_y = 57.1$ N

6/89 The uniform heavy bar AB of mass m is moving on its light end rollers along the horizontal with a velocity v when end A passes point C and begins to move on the curved portion of the path with radius r. Determine the force exerted by the path on the roller A immediately after it passes C.

6/90 The uniform steel beam of mass m and length l is suspended by the two cables at A and B. If the cable at B suddenly breaks, determine the tension T in the cable at A immediately after the break occurs. Treat the beam as a slender rod and show that the result is independent of the length of the beam.

Ans. $T = \dfrac{2\sqrt{3}}{13}\, mg$

6/91 The uniform 12-kg square panel is suspended from point C by the two wires at A and B. If the wire at B suddenly breaks, calculate the tension T in the wire at A an instant after the break occurs.

6/92 The unbalanced wheel has a mass of 10 kg and rolls without slipping on the horizontal surface. When the mass center G passes the horizontal line through O as shown, the angular velocity of the wheel is 2 rad/s. For this instant compute the normal force N and friction force F acting on the wheel at its point of contact with the horizontal surface. The wheel has a radius of gyration about its mass center of 64 mm.

Ans. $N = 87.7$ N, $F = 24.4$ N

Problem 6/92

6/93 The truck, initially at rest with a solid cylindrical roll of paper in the position shown, moves forward with a constant acceleration a. Find the distance s which the truck goes before the paper rolls off the edge of its horizontal bed. Friction is sufficient to prevent slipping.

Ans. $s = 3d/2$

Problem 6/93

6/94 A uniform rod of length l and mass m is secured to a circular hoop of radius l as shown. The mass of the hoop is negligible. If the rod and hoop are released from rest on a horizontal surface in the position illustrated, determine the initial values of the friction force F and normal force N under the hoop if friction is sufficient to prevent slipping.

Ans. $F = \frac{3}{8}mg$, $N = \frac{13}{16}mg$

Problem 6/94

6/95 From the answers to Prob. 6/94 we see that the minimum coefficient of static friction between the hoop and the horizontal surface to prevent slipping is $F/N = \frac{3}{8}/\frac{13}{16} = 0.46$. If the static coefficient is less than this value and if the kinetic coefficient of friction is 0.30, determine the initial angular acceleration α of the hoop and bar as they are released from rest in the horizontal position of the bar.

Problem 6/96

Problem 6/97

Problem 6/98

Problem 6/99

6/96 The fairing which covers the spacecraft package in the nose of the booster rocket is jettisoned when the rocket is in space where gravitational attraction is negligible. A mechanical actuator moves the two halves slowly from the closed position I to position II at which point the fairings are released to rotate freely about their hinges at O under the influence of a constant acceleration a of the rocket. When position III is reached, the hinge at O is released and the fairings drift away from the rocket. Determine the angular velocity ω of the fairings at the 90° position. The mass of each fairing is m with center of mass at G and radius of gyration k_O about O.

6/97 The uniform 3.6-m pole is hinged to the truck bed and released from the vertical position as the truck starts from rest with an acceleration of 0.9 m/s². If the acceleration remains constant during the motion of the pole, calculate the angular velocity ω of the pole as it reaches the horizontal position.
Ans. $\omega = 2.99$ rad/s

6/98 The uniform slender bar AB has a mass of 0.8 kg and is driven by crank OA and constrained by link CB of negligible mass. If OA has an angular acceleration $\alpha_0 = 4$ rad/s² and an angular velocity $\omega_0 = 2$ rad/s when both OA and CB are normal to AB, calculate the force in CB for this instant. (*Suggestion:* Consider the use of Eq. 6/3 with A as a moment center.) *Ans.* $B = 4.03$ N (tension)

6/99 The slender rod of mass m and length l is released from rest in the vertical position with the small roller at end A resting on the incline. Determine the initial acceleration of A.

6/100 The uniform 15-kg bar is supported on the horizontal surface at A by a small roller of negligible mass. If the coefficient of kinetic friction between end B and the vertical surface is 0.30, calculate the initial acceleration of end A as the bar is released from rest in the position shown.

Ans. $a_A = 5.93$ m/s^2

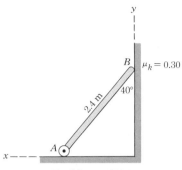

Problem 6/100

6/101 The rectangular block, which is solid and homogeneous, is supported at its corners by small rollers resting on horizontal surfaces. If the supporting surface at B is suddenly removed, determine the expression for the initial acceleration of corner A.

Ans. $a_A = \dfrac{3g}{4\dfrac{b}{h} + \dfrac{h}{b}}$

Problem 6/101

6/102 The hydraulic cylinder BC of the dump truck is broken and is disconnected. The driver (who has passed a course in dynamics) decides to calculate the minimum acceleration a of the truck required to tilt the dump about its pivot at A. He then proceeds to calculate the initial angular acceleration α of the dump if the truck is given an acceleration of $1.2a$. What are his correct answers for a and α and would he be able to carry out the experiment? The dump container may be modeled as a homogeneous and solid rectangular block with mass center at G.

Problem 6/102

6/103 Cable is unwound from the reel as its trailer is moved to the right. The combined mass of the reel and cable wound on it is 50 kg with radius of gyration about O_1 of 420 mm. The trailer body has a mass of 100 kg. Each of the trailer's two wheels has a mass of 15 kg with a radius of gyration about its center of 300 mm. Calculate the acceleration a of the trailer if the horizontal component of the force acting on the hitch at A is $P = 400$ N. Friction between the wheels and the road is sufficient to prevent slipping. Ans. $a = 1.723$ m/s^2

Problem 6/103

Problem 6/104

6/104 The 0.6-kg connecting rod AB of a certain internal combustion engine has a mass center at G and a radius of gyration about G of 28 mm. The piston and piston pin A have a combined mass of 0.82 kg. The engine is running at a constant speed of 3000 rev/min, so that the angular velocity of the crank is $3000(2\pi)/60 = 100\pi$ rad/s. Neglect the weights of the components and the force exerted by the gas in the cylinder compared with the dynamic forces generated and calculate the magnitude of the force on the piston pin A for the crank angle $\theta = 90°$. (*Suggestion:* Use the alternative moment relation, Eq. 6/3, with B as the moment center.)
Ans. $A = 1522$ N

6/105 Solve for the magnitude of the total force on the piston pin A in Prob. 6/104 for the crank angle $\theta = 180°$.

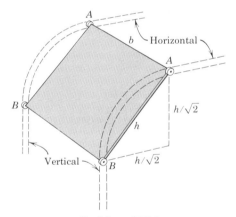

Problem 6/106

6/106 The overhead garage door is a homogeneous rectangular panel of mass m and is guided by its corner rollers that run in the tracks shown (dotted). If the door is released from rest in the position shown, determine the force exerted on the door by each of the rollers at A and B. Neglect any friction.
Ans. $A = \frac{5}{16}mg$, $B = \frac{3}{16}mg$

6/107 In a study of head injury against the instrument panel of a car during sudden or crash stops where lap belts without shoulder straps are used, the segmented human model shown in the figure is analyzed. The hip joint O is assumed to remain fixed relative to the car, and the torso above the hip is treated as a rigid body of mass m freely pivoted at O. The center of mass of the torso is at G with the initial position of OG taken as vertical. The radius of gyration of the torso about O is k_O. If the car is brought to a sudden stop with a constant deceleration a, determine the velocity v relative to the car with which the model's head strikes the instrument panel. Substitute the values $m = 50$ kg, $\bar{r} = 450$ mm, $r = 800$ mm, $k_O = 550$ mm, $\theta = 45°$, and $a = 10g$ and compute v.
Ans. $v = 11.73$ m/s

Problem 6/107

▶6/108 The 4-kg pendulum *OA* with center of mass at *G* has a radius of gyration about *O* of 300 mm. The carriage, to which the pendulum is pivoted, has a mass of 2 kg and moves freely along the horizontal rails. If the pendulum is released from rest in the 45° position, calculate the initial acceleration *a* of the carriage.　　*Ans. a* = 2.95 m/s²

Problem 6/108

▶6/109 A rocket is given a thrust *T* at an angle *θ* with its axis in order to change the direction of its motion. Treat the rocket as a uniform slender bar of length *l* and mass *m*, and write the expression for the bending moment *M* in the rocket as a function of *x* when the rocket is in the vertical position shown.

$$Ans.\ M = \left(\frac{l-x}{l}\right)^2 xT\sin\theta$$

Problem 6/109

▶6/110 The uniform square plate of mass *m* is suspended in the vertical plane by its corner rollers *A* and *B* as it moves horizontally to the right with a velocity *v*. Determine the force which each roller exerts on the plate an instant after corner *B* has entered the circular portion of its guide.

$$Ans.\ A = \frac{m}{2}\left(g - \frac{v^2}{6r}\right),\ B = \frac{m}{2}\left(g - \frac{5v^2}{6r}\right)$$

Problem 6/110

▶6/111 If corners *A* of the overhead door of Prob. 6/106 have a velocity *v* and the door is moving under the action of its own weight in the position shown, determine the angular acceleration *α* of the door and the forces exerted on the door by each of the four corner rollers. Solve by considering that corners *A* are still moving horizontally but approaching the circular portion of their path while corners *B* are still moving in the circular path of radius *h*/√2 but approaching the straight vertical section.

$$Ans.\ A = \frac{m}{16}(5g + \sqrt{2}v^2/h),\ B = \frac{m}{16}(3g - \sqrt{2}v^2/h)$$

SECTION B. WORK AND ENERGY

6/6 *WORK-ENERGY RELATIONS*

In our study of the kinetics of particles in Arts. 3/6 and 3/7 we developed and applied the principles of work and energy to the motion of a particle and to selected cases of connected particles. We found that these principles were especially useful in describing motion resulting from the cumulative effect of forces acting through distances. Furthermore, when the forces were conservative, we were able to determine velocity changes by analyzing the energy conditions at the beginning and end of the motion interval. For finite displacements the work-energy method eliminates the necessity for determining the acceleration and integrating it over the interval to obtain the velocity change. These same advantages are realized when we extend the work-energy principles to describe rigid-body motion. Before carrying out this extension it is strongly recommended that the definitions and concepts of work, kinetic energy, gravitational and elastic potential energy, conservative forces, and power treated in Arts. 3/6 and 3/7 be carefully reviewed. We shall assume familiarity with these quantities as we apply them to rigid-body problems. In Arts. 4/3 and 4/4 on the kinetics of systems of particles we extended the principles of Arts. 3/6 and 3/7 to encompass any general system of mass particles, which includes rigid bodies, and this discussion should also be reviewed.

(*a*) *Work of forces and couples.*   The work done by a force **F** has been treated in detail in Art. 3/6 and is given by

$$U = \int \mathbf{F} \cdot d\mathbf{r} \qquad \text{or} \qquad U = \int (F \cos \alpha) \, ds$$

where $d\mathbf{r}$ is the infinitesimal vector displacement of the point of application of **F** during time dt. In the equivalent scalar form of the integral α is the angle between **F** and the direction of the displacement and ds is the magnitude of the vector displacement $d\mathbf{r}$. We frequently have occasion to evaluate the work done by a couple M which acts on a rigid body during its motion. Figure 6/11 shows a couple $M = Fb$ acting on a rigid body which moves in the plane of the couple. During time dt the body rotates through an angle $d\theta$, and line AB moves to $A'B'$. We may consider this motion in two parts, first a translation to $A'B''$ and then a rotation $d\theta$ about A'. We see immediately that during the translation the work done by one of the forces cancels that due to the other force, so that the net work done is $dU = F(b \, d\theta) = M \, d\theta$ due to the rotational part of the motion. If the couple acts in the opposite sense to the rotation, the work done is negative. During a

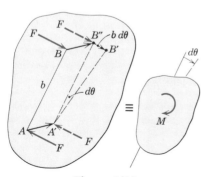

Figure 6/11

finite rotation the work done by a couple M whose plane is parallel to the plane of motion is, therefore,

$$U = \int M \, d\theta$$

(*b*) **Kinetic energy.**   We now use the familiar expression for the kinetic energy of a particle to develop expressions for the kinetic energy of a rigid body for each of the three classes of rigid-body plane motion illustrated in Fig. 6/12.

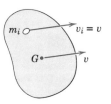

(*a*) Translation

(*Translation*) The translating rigid body of Fig. 6/12*a* has a mass m and all of its particles have a common velocity v. The kinetic energy of any particle of mass m_i of the body is $T_i = \frac{1}{2}m_i v^2$, so for the entire body $T = \Sigma \frac{1}{2}m_i v^2 = \frac{1}{2}v^2 \Sigma m_i$ or

$$\boxed{T = \tfrac{1}{2}mv^2} \qquad (6/7)$$

This expression holds for both rectilinear and curvilinear translation.

(*b*) Fixed–Axis
     Rotation

(*Fixed-axis rotation*) The rigid body in Fig. 6/12*b* rotates with an angular velocity ω about the fixed axis through O. The kinetic energy of a representative particle of mass m_i is $T_i = \frac{1}{2}m_i(r_i\omega)^2$. Thus for the entire body $T = \frac{1}{2}\omega^2 \Sigma m_i r_i^2$. But the moment of inertia of the body about O is $I_O = \Sigma m_i r_i^2$, so

$$\boxed{T = \tfrac{1}{2}I_O\omega^2} \qquad (6/8)$$

The similarity in form of the kinetic energy expressions for translation and rotation should be noted. The reader should verify that the dimensions of the two expressions are identical.

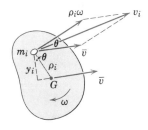

(*c*) General Plane
     Motion

Figure 6/12

(*General plane motion*) The rigid body in Fig. 6/12*c* executes plane motion where, at the instant considered, the velocity of its mass center G is $\bar{v}$ and the angular velocity is ω. The velocity v_i of a representative particle of mass m_i may be expressed in terms of the mass-center velocity $\bar{v}$ and the velocity $\rho_i\omega$ relative to the mass center as shown. With the aid of the law of cosines we write the kinetic energy of the body as the sum ΣT_i of the kinetic energies of all its particles. Thus

$$T = \Sigma \tfrac{1}{2}m_i v_i^2 = \Sigma \tfrac{1}{2}m_i(\bar{v}^2 + \rho_i^2\omega^2 + 2\bar{v}\rho_i\omega \cos \theta)$$

Since ω and $\bar{v}$ are common to all terms in the third summation, we may factor them out. Thus the third term in the expression for T becomes

$$\omega\bar{v}\Sigma m_i\rho_i \cos \theta = \omega\bar{v}\Sigma m_i y_i = 0$$

since $\Sigma m_i y_i = m\bar{y} = 0$. The kinetic energy of the body is, then,

$$T = \tfrac{1}{2}\bar{v}^2 \Sigma m_i + \tfrac{1}{2}\omega^2 \Sigma m_i \rho_i^2 \text{ or}$$

$$\boxed{T = \tfrac{1}{2}m\bar{v}^2 + \tfrac{1}{2}\bar{I}\omega^2} \tag{6/9}$$

where $\bar{I}$ is the moment of inertia of the body about its mass center. This expression for kinetic energy clearly shows the separate contributions to the total kinetic energy resulting from the translational velocity $\bar{v}$ of the mass center and the rotational velocity ω about the mass center.

The kinetic energy of plane motion may also be expressed in terms of the rotational velocity about the instantaneous center C of zero velocity. Since C momentarily has zero velocity, the proof leading to Eq. 6/8 for the fixed point O holds equally well for point C, so that, alternatively, we may write the kinetic energy of a rigid body in plane motion as

$$\boxed{T = \tfrac{1}{2}I_C\omega^2} \tag{6/10}$$

In Art. 4/3 we derived Eq. 4/4 for the kinetic energy of any system of mass. We now see that this expression becomes equivalent to Eq. 6/9 when the mass system is rigid. For a rigid body the quantity $\dot{\rho}_i$ in Eq. 4/4 is the velocity of the representative particle relative to the mass center and is the vector $\boldsymbol{\omega} \times \boldsymbol{\rho}_i$ which has the magnitude $\rho_i\omega$. The summation term in Eq. 4/4 becomes $\Sigma\tfrac{1}{2}m_i(\rho_i\omega)^2 = \tfrac{1}{2}\omega^2\Sigma m_i\rho_i^2 = \tfrac{1}{2}\bar{I}\omega^2$, which brings Eq. 4/4 into agreement with Eq. 6/9.

(*c*) *Potential energy.* Gravitational potential energy V_g and elastic potential energy V_e were covered in detail in Art. 3/7. The only additional comment is a reminder that, if the potential energy term ΔV_g is used, then the work of the weight $W = mg$ must *not* be included also in the expression for U.

Work-energy equation. The work-energy relation, Eq. 3/17, was introduced in Art. 3/7 for particle motion and was generalized in Art. 4/3 to include the motion of a general system of particles. This equation,

$$\boxed{U = \Delta T + \Delta V_g + \Delta V_e} \tag{[4/3]}$$

applies to any conservative mechanical system. For application to the motion of a single rigid body the term ΔV_e disappears, and the total work U done on the body by the external forces (other than gravity forces) equals the corresponding change ΔT in the kinetic energy of the body plus the change ΔV_g in its potential energy of position in the gravitational field. Alternatively the equation may be written $U = \Delta T$ provided that the work of gravitational forces is included in the expression for U.

When applied to an interconnected conservative system of rigid bodies, Eq. 4/3 will include the change ΔV_e in the stored elastic energy in the connections. The term U will include the work of all forces external to the system (other than gravitational forces) including the negative work of internal friction forces if any. The term ΔT is the sum of the changes in kinetic energy of all moving parts during the interval of motion in question, and ΔV_g is the sum of the changes in gravitational potential energy for the various members. Alternatively, if the work of gravitational forces is included in the U-term, then the ΔV_g-term must be omitted.

When the work-energy principle is applied to a single rigid body, either a *free-body diagram* or an *active-force diagram* should be used. In the case of an interconnected system of rigid bodies, an active-force diagram of the entire system should be drawn in order to isolate the system and disclose all forces which do work on the system. Diagrams should also be drawn which disclose the initial and final positions of the system for the given interval of motion.

The work-energy equation provides a direct relationship between the forces which do work and the corresponding changes in the motion of a mechanical system. However, if there is appreciable internal mechanical friction, then the system must be dismembered in order to disclose the kinetic-friction forces and account for the negative work that they do. When the system is dismembered, one of the primary advantages of the work-energy approach is automatically lost. The work-energy method is most useful for analyzing conservative systems of interconnected bodies where energy loss due to the negative work of friction forces is negligible.

(*d*) *Power.* The concept of power was discussed in Art. 3/6 on work-energy for particle motion. It is recalled that power is the time rate at which work is performed. For a force $\mathbf{F}$ acting on a rigid body in plane motion, the power developed by that force at a given instant was given by Eq. 3/12 and is the rate at which it is doing work or

$$P = \frac{dU}{dt} = \frac{\mathbf{F} \cdot d\mathbf{r}}{dt} = \mathbf{F} \cdot \mathbf{v}$$

where $d\mathbf{r}$ and $\mathbf{v}$ are, respectively, the differential displacement and the velocity of the point of application of the force. Similarly, for a couple M acting on the body, the power developed by the couple at a given instant is the rate at which it is doing work or

$$P = \frac{dU}{dt} = \frac{M\,d\theta}{dt} = M\omega$$

where $d\theta$ and ω are, respectively, the differential angular displacement and the angular velocity of the body. If the senses of M and ω are the same, the power is positive and energy is supplied to the body. Conversely, if M and ω have opposite senses, the power is negative and energy is removed from the body. If the force $\mathbf{F}$ and the couple M act simultaneously, the total instantaneous power would be

$$P = \mathbf{F} \cdot \mathbf{v} + M\omega$$

We may also express power by evaluating the rate at which the total mechanical energy of a rigid body or a system of rigid bodies is changing. The work-energy relation, Eq. 4/3, for an infinitesimal displacement is

$$dU = dT + dV_g + dV_e$$

where dU is the work of the active forces and couples applied to the body or to the system of bodies. Excluded from dU are the works of gravitational forces and spring forces which are accounted for in the dV_g and dV_e terms. Dividing by dt gives the total power of the active forces and couples as

$$P = \frac{dU}{dt} = \dot{T} + \dot{V}_g + \dot{V}_e = \frac{d}{dt}(T + V)$$

Thus we see that the power developed by the active forces and couples equals the rate of change of the total mechanical energy of the body or system of bodies. We note from Eq. 6/9 that, for a given body, the first term may be written

$$\dot{T} = \frac{dT}{dt} = \frac{d}{dt}\left(\frac{1}{2}m\bar{\mathbf{v}} \cdot \bar{\mathbf{v}} + \frac{1}{2}\bar{I}\omega^2\right)$$

$$= \frac{1}{2}m(\bar{\mathbf{a}} \cdot \bar{\mathbf{v}} + \bar{\mathbf{v}} \cdot \bar{\mathbf{a}}) + \bar{I}\omega\dot{\omega}$$

$$= m\bar{\mathbf{a}} \cdot \bar{\mathbf{v}} + \bar{I}\alpha(\omega) = \mathbf{R} \cdot \bar{\mathbf{v}} + \bar{M}\omega$$

where $\mathbf{R}$ is the resultant of *all* forces acting on the body and $\bar{M}$ is the resultant moment about the mass center G of all forces. The dot product accounts for the case of curvilinear motion of the mass center where $\bar{\mathbf{a}}$ and $\bar{\mathbf{v}}$ are not in the same direction.

Sample Problem 6/9

The wheel rolls up the incline on its hubs without slipping and is pulled by the 100-N force applied to the cord wrapped around its outer rim. If the wheel starts from rest, compute its angular velocity ω after its center has moved a distance of 3 m up the incline. The wheel has a mass of 40 kg with center of mass at O and has a centroidal radius of gyration of 150 mm. Determine the power input from the 100-N force at the end of the 3-m motion interval.

Solution. Of the four forces shown on the free-body diagram of the wheel only the 100-N pull and the weight of $40(9.81) = 392$ N do work. The friction force does no work as long as the wheel does not slip. By use of the concept of the instantaneous center C of zero velocity we see that a point A on the cord to which the 100-N force is applied has a velocity $v_A = [(200 + 100)/100]v$. Hence point A on the cord moves a distance of $(200 + 100)/100 = 3$ times as far as the center O. Thus, with the effect of the weight included in the U-term, the work done on the wheel becomes

$$U = 100 \frac{200 + 100}{100}(3) - (392 \sin 15°)(3) = 595 \text{ J}$$

The wheel has general plane motion, so that the change in its kinetic energy is

$$[T = \tfrac{1}{2}m\bar{v}^2 + \tfrac{1}{2}\bar{I}\omega^2] \quad \Delta T = \left[\frac{1}{2}40(0.10\omega)^2 + \frac{1}{2}40(0.15)^2\omega^2\right] - 0$$
$$= 0.650\omega^2$$

The work-energy equation gives

$[U = \Delta T]$ $595 = 0.650\omega^2$ $\omega = 30.3$ rad/s

Alternatively the kinetic energy of the wheel may be written

$$[T = \tfrac{1}{2}I_C\omega^2] \quad T = \frac{1}{2}40[(0.15)^2 + (0.10)^2]\omega^2 = 0.650\omega^2$$

The power input from the 100-N force when $\omega = 30.3$ rad/s is

$[P = \mathbf{F} \cdot \mathbf{v}]$ $P_{100} = 100(0.3)(30.3) = 908$ W

① Since the velocity of the instantaneous center C on the wheel is zero, it follows that the rate at which the friction force does work is continuously zero. Hence F does no work as long as the wheel does not slip. If the wheel were rolling on a moving platform, however, the friction force would do work even though the wheel were not slipping.

② Note that the component of the weight down the plane does negative work.

③ Be careful to use the correct radius in the expression $v = r\omega$ for the velocity of the center of the wheel.

④ Recall that $I_C = \bar{I} + m\overline{OC}^2$ where $\bar{I} = I_O = mk_O{}^2$.

⑤ The velocity here is that of the application point of the 100-N force.

Sample Problem 6/10

The 1200-mm slender bar has a mass of 20 kg with mass center at B and is released from rest in the position for which θ is essentially zero. Point B is confined to move in the smooth vertical guide, while end A moves in the smooth horizontal guide and compresses the spring as the bar falls. Determine (a) the angular velocity of the bar as the position $\theta = 30°$ is passed and (b) the velocity with which B strikes the horizontal surface if the stiffness of the spring is 5 kN/m.

Solution. With the friction and mass of the small rollers at A and B neglected, the system may be treated as being conservative.

Part (a). For the first interval of motion from $\theta = 0$ to $\theta = 30°$ the spring is not engaged, so that there is no V_e term in the energy equation. If we adopt the alternative of treating the work of the weight in the V_g term, then there are no other forces which do work and $U = 0$.

Since we have a constrained plane motion, there is a kinematic relation between the velocity v_B of the center of mass and the angular velocity ω of the bar. This relation is easily obtained by using the instantaneous center C of zero velocity and noting that $v_B = \overline{CB}\omega$. Thus the kinetic energy of the bar in the 30° position becomes

$$[T = \tfrac{1}{2}m\bar{v}^2 + \tfrac{1}{2}\bar{I}\omega^2]$$

$$T = \frac{1}{2}\,20(0.300\omega)^2 + \frac{1}{2}\left(\frac{1}{12}\,20[1.2]^2\right)\omega^2 = 2.10\omega^2$$

The change in gravitational potential energy is the weight times the change in height of the mass center and equals

$$[\Delta V_g = W\,\Delta h] \quad \Delta V_g = 20(9.81)(0.600\cos 30° - 0.600) = -15.77 \text{ J}$$

We now substitute into the energy equation and get

$$[U = \Delta T + \Delta V_g] \quad 0 = 2.10\omega^2 - 15.77 \quad \omega = 2.74 \text{ rad/s} \qquad Ans.$$

Part (b). For the entire interval of motion we include the spring as a part of the system where

$$[V_e = \tfrac{1}{2}kx^2] \quad \Delta V_e = \frac{1}{2}\,(5000)(0.600 - 0.450)^2 - 0 = 56.25 \text{ J}$$

In the final horizontal position point A has no velocity, so that the bar is, in effect, rotating about A. Hence its kinetic energy is

$$[T = \tfrac{1}{2}I_A\omega^2] \quad T = \frac{1}{2}\left(\frac{1}{3}\,20[1.2]^2\right)\left(\frac{v_B}{0.600}\right)^2 = 13.33v_B^2$$

The change in gravitational potential energy is

$$[\Delta V_g = W\,\Delta h] \quad \Delta V_g = 20(9.81)(-0.600) = -117.7 \text{ J}$$

Substituting into the energy equation gives

$$[U = \Delta T + \Delta V_g + \Delta V_e] \quad 0 = (13.33v_B^2 - 0) - 117.7 + 56.25$$

$$v_B = 2.15 \text{ m/s} \qquad Ans.$$

Alternatively, if the bar alone constitutes the system, the active-force diagram shows the weight which does positive work and the spring force kx which does negative work. We would then write

$$[U = \Delta T] \quad 117.7 - 56.25 = 13.33v_B^2$$

which is identical with the previous result.

① We recognize that the forces acting on the bar at A and B are normal to the respective directions of motion and, hence, do no work.

② Note that we have kept all quantities in base units.

(Alternative Kinetic Diagram)

Sample Problem 6/11

In the mechanism shown each of the two wheels has a mass of 30 kg and a centroidal radius of gyration of 100 mm. Each link OB has a mass of 10 kg and may be treated as a slender bar. The 7-kg collar at B slides on the fixed vertical shaft with negligible friction. The spring has a stiffness $k = 30$ kN/m and is contacted by the bottom of the collar when the links reach the horizontal position. If the collar is released from rest at the position $\theta = 45°$ and if friction is sufficient to prevent the wheels from slipping, determine (a) the velocity v_B of the collar as it first strikes the spring, and (b) the maximum deformation x of the spring.

Solution. The mechanism executes plane motion and is conservative with the neglect of kinetic friction losses. The datum for zero gravitational potential energy V_g is conveniently taken through O as shown.

(a) For the interval from $\theta = 45°$ to $\theta = 0$, it is noted that ΔT_{wheels} is zero since each wheel starts from rest and momentarily comes to rest at $\theta = 0$. Also, at the lower position each link is merely rotating about its point O so that

$$\Delta T = [2(\tfrac{1}{2}I_O\omega^2) - 0]_{\text{links}} + [\tfrac{1}{2}mv^2 - 0]_{\text{collar}}$$

$$= \frac{1}{3}\,10(0.375)^2\left(\frac{v_B}{0.375}\right)^2 + \frac{1}{2}\,7v_B{}^2 = 6.83v_B{}^2$$

The collar at B drops a distance $0.375/\sqrt{2} = 0.265$ m so that

$$\Delta V = \Delta V_g = 0 - 2(10)(9.81)\frac{0.265}{2} - 7(9.81)(0.265) = -44.2 \text{ J}$$

① Also, $U = 0$. Hence,

$$[U = \Delta T + \Delta V] \qquad 0 = 6.83v_B{}^2 - 44.2 \qquad v_B = 2.54 \text{ m/s} \qquad Ans.$$

(b) At the condition of maximum deformation x of the spring, all parts are momentarily at rest which makes $\Delta T = 0$. Thus

$$[U = \Delta T + \Delta V_g + \Delta V_e]$$

$$0 = 0 - 2(10)(9.81)\left(\frac{0.265}{2} + \frac{x}{2}\right) - 7(9.81)(0.265 + x) + \tfrac{1}{2}(30)(10^3)x^2$$

Solution for the positive value of x gives

$$x = 60.1 \text{ mm} \qquad Ans.$$

It should be noted that the results of parts (a) and (b) involve a very simple net energy change despite the fact that the mechanism has undergone a fairly complex sequence of motions. Solution of this and similar problems by other than a work-energy approach is not an inviting prospect.

① With the work of the weight of the collar B included in the ΔV_g term, there are no other forces external to the system which do work. The friction force acting under each wheel does no work since the wheel does not slip, and, of course, the normal force does no work here. Hence $U = 0$.

PROBLEMS

(In the following problems neglect any energy loss due to kinetic friction unless otherwise indicated.)

Problem 6/112

6/112 The log is suspended by the two parallel 5-m cables and used as a battering ram. At what angle θ should the log be released from rest in order to strike the object to be smashed with a velocity of 5 m/s?
Ans. $\theta = 41.8°$

6/113 The 830-kg steel beam is held by the force P in the equilibrium position shown. If P is suddenly released, calculate the velocity v of the dolly when it reaches the dotted position with A' directly under point C. Point A is the center of mass of the beam.
Ans. $v = 7.00$ m/s

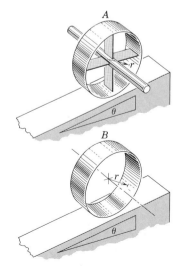

Problem 6/113

6/114 The two wheels of Prob. 6/80, shown again here, represent two extreme conditions of distribution of mass. For case A all of the mass m is assumed to be concentrated in the center of the hoop in the axial bar of negligible diameter. For case B all of the mass m is assumed to be concentrated in the rim. Determine the velocity of the center of each hoop after it has traveled a distance x down the incline from rest. The hoops roll without slipping.

6/115 The 15-kg wheel is released from rest and rolls on its hubs without slipping. Calculate the velocity v of the center O of the wheel after it has moved a distance $x = 3$ m down the incline. The radius of gyration of the wheel about O is 125 mm.
Ans. $v = 0.899$ m/s

Problem 6/114

Problem 6/115

6/116 A 1200-kg flywheel with a radius of gyration of 400 mm has its speed reduced from 5000 to 3000 rev/min during a 2-min interval. Calculate the average power supplied by the flywheel. Express your answer both in kilowatts and in horsepower.

6/117 The two identical links, each of length b and mass m, may be treated as uniform slender bars. If they are released from rest in the position shown with end A constrained by the smooth vertical guide, determine the velocity v with which A reaches O with θ essentially zero.   *Ans.* $v = \sqrt{6gb} \sin (\theta/2)$

Problem 6/117

6/118 The drum of 375-mm radius and its shaft have a mass of 41 kg and a radius of gyration of 300 mm about the axis of rotation. A total of 18 m of flexible steel cable with a mass of 3.08 kg per meter of length is wrapped around the drum with one end secured to the surface of the drum. The free end of the cable has an initial overhang $x = 0.6$ m as the drum is released from rest. Determine the angular velocity ω of the drum for the instant when $x = 6$ m. Assume that the center of mass of the portion of cable remaining on the drum at any time lies on the shaft axis. Neglect friction.
Ans. $\omega = 9.68$ rad/s

Problem 6/118

6/119 The 13-m steel I-beam of Prob. 6/113 has a mass of 830 kg and is being hoisted into position by the force $P = 5.4$ kN applied to the dolly on which end B is secured. The cable AC is attached to the center of the beam. If the system starts from rest in the dotted position, determine the velocity v of the dolly when it reaches the position B directly under C with AC horizontal.

6/120 The uniform 3-kg bar ABC is initially at rest with end A bearing against the stop in the horizontal guide. When a constant couple $M = 8$ N·m is applied to end C, the bar rotates causing end A to strike the side of the vertical guide with a velocity of 3 m/s. Calculate the loss of energy ΔE due to friction in the guides and rollers. The mass of the rollers may be neglected.   *Ans.* $\Delta E = 0.059$ J

Problem 6/120

Problem 6/121

Problem 6/122

Problem 6/123

6/121 A small press used for specialty forming of metals is shown. The flywheel and attached shaft have a combined mass of 40 kg and a radius of gyration of 180 mm about its vertical axis. The screw has a double thread with a lead (advancement per revolution) of 40 mm. The forming head and die have a mass of 30 kg and slide in the vertical guide. If the flywheel and shaft have an angular speed of 100 rev/min in the position shown and an angular speed of 120 rev/min when the die strikes the work, calculate the loss ΔE of energy due to friction.

6/122 Each of the hinged bars has a mass ρ per unit length, and the assembly is suspended at O in the vertical plane. If the bars are released from rest with θ essentially zero, determine the angular velocity ω common to all bars when A and B and C and D come together. *Ans.* $\omega = \sqrt{6g/b}$

6/123 A facility for testing the performance of motorized golf carts consists of an endless belt where the angle θ can be adjusted. The cart of mass m is slowly brought up to its rated ground speed v with the braking torque M on the upper pulley constantly adjusted so that the cart remains in a fixed position A on the test stand. With no cart on the belt, a torque M_0 is required on the pulley to overcome friction and turn the pulleys regardless of speed. Friction is sufficient to prevent the wheels from slipping on the belt. Determine an expression for the power P absorbed by the braking torque M. Do the static friction forces between the wheels and the belt do work?

6/124 The 8-kg crank *OA*, with mass center at *G* and radius of gyration about *O* of 0.22 m, is connected to the 12-kg uniform slender bar *AB*. If the linkage is released from rest in the position shown, compute the velocity *v* of end *B* as *OA* swings through the vertical.   *Ans. v = 2.29 m/s*

Problem 6/124

6/125 The electric motor shown is delivering 4 kW at 1725 rev/min to a pump which it drives. Calculate the angle δ through which the motor deflects under load if the stiffness of each of its four spring mounts is 15 kN/m. In what direction does the motor turn?
                    Ans. δ = 2.11°, motor turns clockwise

Problem 6/125

6/126 A constant force *F* is applied in the vertical direction to the symmetrical linkage starting from the rest position shown. Determine the angular velocity ω which the links acquire as they reach the position $\theta = 0$. Each link has a mass m_0. The wheel is a solid circular disk of mass *m* and rolls on the horizontal surface without slipping.

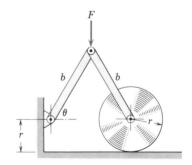

Problem 6/126

6/127 The bar *ACB* has a mass of 20 kg with radius of gyration about its mass center *C* of 160 mm. The bar is secured to the face of the uniform circular disk, which has a mass of 130 kg. If the disk and bar are released from rest in the position shown, calculate the maximum angular velocity of the disk during its subsequent motion. The disk rolls without slipping.   *Ans. ω = 0.990 rad/s*

Problem 6/127

Problem 6/128

Problem 6/129

Problem 6/130

Problem 6/131

6/128 The sheave of 400-mm radius has a mass of 50 kg and a radius of gyration of 300 mm. The sheave and its 100-kg load are suspended by the cable and the spring, which has a stiffness of 1.5 kN/m. If the system is released from rest with the spring initially stretched 100 mm, determine the velocity of O after it has dropped 50 mm.

6/129 The horizontal platform A is elevated by the action of the worm-gear motor unit B, which supplies a torque of 1500 N·m to the upper shaft that is welded to the two legs. The platform and motor unit have a combined mass of 150 kg, and each of the four identical legs has a mass of 12 kg with mass center at G and a radius of gyration about its lower end of 700 mm. Calculate the angular velocity ω of the legs when $\theta = 90°$ if the platform starts from rest with θ essentially zero.

Ans. $\omega = 3.21$ rad/s

6/130 Specify the unstretched length l_0 of the spring of stiffness $k = 1400$ N/m which will result in a velocity of 0.25 m/s for the contact at A if the toggle is given a slight nudge from its null position at $\theta = 0$. The toggle has a mass of 1.5 kg and a radius of gyration about O of 55 mm. Motion occurs in the horizontal plane.

6/131 The uniform slender link OA has a mass of 3 kg and is attached to the shaft of a gear motor at O. Link AB is a uniform slender bar with a mass of 4 kg. The motor is activated with the links at rest at $\theta = 0$ and delivers a constant counterclockwise torque of 20 N·m to the link at O. Calculate the angular velocity of OA as θ passes the 90° position.

Ans. $\omega = 16.16$ rad/s

6/132 Motive power for the experimental 10-Mg bus comes from the energy stored in a rotating flywheel which it carries. The flywheel has a mass of 1500 kg and a radius of gyration of 500 mm and is brought up to a maximum speed of 4000 rev/min. If the bus starts from rest and acquires a speed of 72 km/h at the top of a hill 20 m above the starting position, compute the reduced speed N of the flywheel. Assume that 10 percent of the energy taken from the flywheel is lost. Neglect the rotational energy of the wheels of the bus.

Problem 6/132

Problem 6/133

6/133 The two uniform bars pivoted as shown have a mass ρ per unit length and move in the vertical plane. If they are released from rest in the position $\theta = 0$, determine the velocity v with which end B approaches the support at O with θ essentially 90°.

Ans. $v = 2\sqrt{gb}$

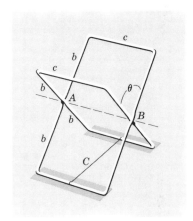

6/134 The two identical steel frames with the dimensions shown are fabricated from the same bar stock and are hinged at the midpoints A and B of their sides. If the frame is resting in the position shown on a horizontal surface with negligible friction, determine the velocity v with which each of the upper ends of the frame hits the horizontal surface if the cord at C is cut.

Problem 6/134

6/135 The figure shows the cross section of a 100-kg garage door which is a uniform rectangular panel 2.4 by 2.4 m. The door carries two spring assemblies, one on each side of the door, like the one shown. Each spring has a stiffness of 700 N/m and is unstretched when the door is in the open position shown. If the door is released from rest in this position, calculate the velocity of the edge at A as it strikes the garage floor.

Ans. $v = 3.04$ m/s

Problem 6/135

Problem 6/136

Problem 6/137

Problem 6/138

6/136 The figure shows the cross section *AB* of a 100-kg door which is a 1.2 by 1.8-m panel of uniform thickness. The door is supported by a framework of negligible mass hinged about a horizontal shaft at *O*. In the position shown, the spring, which has a stiffness $k = 450$ N/m, is unstretched. If a constant torque $M = 880$ N·m is applied to the frame through its shaft at *O* starting from the rest position shown, determine the angular velocity of the door when it reaches the horizontal position.

6/137 The figure shows the cross section *AB* of a garage door which is a rectangular 2.5-m by 5-m panel of uniform thickness with a mass of 200 kg. The door is supported by the struts having negligible mass and hinged at *O*. Two spring-and-cable assemblies, one on each side of the door, control the movement. When the door is in the horizontal open position, each spring is unextended. If the door is given a slight unbalance from the open position and allowed to fall, determine the value of the spring constant *k* for each spring which will limit the angular velocity of the door to 1.5 rad/s when edge *B·* strikes the floor.

Ans. $k = 1.270$ kN/m

6/138 The small vehicle is designed for high-speed travel over the snow. The endless tread for each side of the vehicle has a mass ρ per unit length and is driven by the front wheels. Determine that portion *M* of the constant front-axle torque required to give both vehicle treads their motion corresponding to a vehicle velocity *v* achieved with constant acceleration in a distance *s* from rest on level terrain.

6/139 The solid circular cylinder of mass *m* and radius *r* is resting on a horizontal surface which is given a constant acceleration *a* to the right from rest. Determine the work done on the cylinder during the interval in which it has rotated through 360°. The cylinder rolls without slipping. (*Hint:* The cylinder rolls in the direction shown relative to the moving surface.)

Ans. $U = \pi mra$

Problem 6/139

6/140 The 10-kg double wheel with radius of gyration of 125 mm about O is connected to the spring of stiffness $k = 600$ N/m by a cord which is wrapped securely around the inner hub. If the wheel is released from rest on the incline with the spring stretched 225 mm, calculate the maximum velocity v of its center O during the ensuing motion. The wheel rolls without slipping.

Problem 6/140

6/141 The figure shows the cross section of a uniform 100-kg ventilator door hinged about its upper horizontal edge at O. The door is controlled by the spring-loaded cable which passes over the small pulley at A. The spring has a stiffness of 200 N per meter of stretch and is undeformed when $\theta = 0$. If the door is released from rest in the horizontal position, determine the maximum angular velocity ω reached by the door and the corresponding angle θ.   *Ans.* $\omega = 3.32$ rad/s, $\theta = 58.5°$

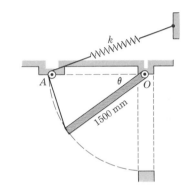

Problem 6/141

6/142 A small experimental vehicle has a total mass m of 500 kg including wheels and driver. Each of the four wheels has a mass of 40 kg and a centroidal radius of gyration of 400 mm. Total frictional resistance R to motion is 400 N and is measured by towing the vehicle at a constant speed on a level road with engine disengaged. Determine the power output of the engine for a speed of 72 km/h when going up the 10-percent grade (*a*) with zero acceleration and (*b*) with an acceleration of 3 m/s^2. (*Hint:* Power equals the time rate of increase of the total energy of the vehicle plus the rate at which frictional work is overcome.)

Problem 6/142

6/143 Each of the two links has a mass of 2 kg and a centroidal radius of gyration of 60 mm. The slider at B has a mass of 3 kg and moves freely in the vertical guide. The spring has a stiffness of 6 kN/m. If a constant torque $M = 20$ N·m is applied to link OA through its shaft at O starting from the rest position at $\theta = 45°$, determine the angular velocity ω of OA when $\theta = 0$.   *Ans.* $\omega = 9.51$ rad/s

Problem 6/143

Problem 6/144

Uncompressed
position

Problem 6/145

Problem 6/146

Problem 6/147

6/144 A solid roll of wrapping paper with an initial radius r_0 is released from rest on the incline and allowed to unroll with the free end clamped at the top. Determine the velocity of the roll in terms of the distance x which it has moved down the slope. The total length of the paper in the roll is L. Can you reconcile the difference between the initial energy of the roll and the final energy of the paper after all motion has ceased?

6/145 Slider A of mass m is placed in the smooth horizontal tube and released from rest in the position for which the spring of stiffness k has been compressed a distance x_0. Determine the maximum power P developed by the spring in moving the slider and the displacement x at which this condition occurs.

$$Ans. \ P_{max} = \frac{k}{2}\sqrt{\frac{k}{m}}\,x_0{}^2 \ \text{at} \ x = x_0/\sqrt{2}$$

6/146 The front-end suspension of a small truck is shown. The frame F must be jacked up to a height $y = 350$ mm to release the compression in the springs. If the jack is suddenly removed, allowing the frame to drop, determine the maximum downward velocity v_{max} of the frame and the corresponding value of y. Also find the minimum value of y during the motion. The frame and that portion of the truck supported by it have a mass of 800 kg, and each spring has a stiffness of 120 kN/m. The shock absorbers have been disconnected for the test.

$$Ans. \ v_{max} = 0.724 \text{ m/s at } y = 297 \text{ mm}$$
$$y_{min} = 243 \text{ mm}$$

▶**6/147** Calculate the constant force P required to give the center of the pulley a velocity of 1.2 m/s in an upward movement of the center of 0.9 m from the rest position shown. The pulley has a mass of 15 kg with a radius of gyration of 250 mm, and the cable has a total length of 4.5 m with a mass of 3 kg/m. $Ans. \ P = 178.3$ N

▶ **6/148** The uniform bar ABC has a mass m and starts from rest with $\theta = 180°$ where A, B, C, and O are collinear. If the applied force P is constant in magnitude, determine the angular velocity ω of the bar as B reaches O with $\theta = 0$. The mass of the roller at B and the mass of the strut OC are negligible. (*Hint:* Replace P by a force P at B and a couple.)

$$Ans. \; \omega = \sqrt{\frac{6P\pi}{13mb}}$$

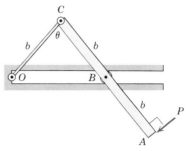

Problem 6/148

▶ **6/149** The two corners A and B of the uniform square plate of mass m are confined to move in the vertical and horizontal slots, respectively. If the plate is subjected to a force P greater than the weight, constant in magnitude, and always perpendicular to the edge of the plate, determine the velocity v_A with which corner A strikes the top of the vertical slot. Friction in the small guide rollers is negligible. (*Hint:* Replace P by an equivalent force at A and a couple.)

$$Ans. \; v_A = \sqrt{3b}\sqrt{\frac{P}{2m}(1 + \pi) - g}$$

Problem 6/149

6/7 ACCELERATION FROM WORK-ENERGY; VIRTUAL WORK

In addition to the determination of velocities resulting from the action of forces acting over finite displacements, we may use the work-energy equation to establish the instantaneous accelerations of the members of a system of interconnected bodies as a result of the active forces applied. We may also modify the equation to establish the configuration of such a system when it undergoes a constant acceleration.

For an infinitesimal interval of motion Eq. 4/3 becomes

$$dU = dT + dV$$

The term dU represents the total work done by all active nonpotential forces acting on the system under consideration during the infinitesimal displacement of the system. The work of potential forces is included in the dV-term. If we use the subscript i to denote a representative body of the interconnected system, the differential change in kinetic energy T for the entire system becomes

$$dT = d(\Sigma \tfrac{1}{2} m_i \bar{v}_i{}^2 + \Sigma \tfrac{1}{2} \bar{I}_i \omega_i{}^2) = \Sigma m_i \bar{v}_i \, d\bar{v}_i + \Sigma \bar{I}_i \omega_i \, d\omega_i$$

where $d\bar{v}_i$ and $d\omega_i$ are the respective changes in the magnitudes of the velocities and where the summation is taken over all bodies of the system. But for each body $m_i \bar{v}_i \, d\bar{v}_i = m_i \bar{\mathbf{a}}_i \cdot d\bar{\mathbf{s}}_i$ and $\bar{I}_i \omega_i \, d\omega_i = \bar{I}_i \alpha_i \, d\theta_i$, where $d\bar{\mathbf{s}}_i$ represents the infinitesimal linear displacement of the center of mass and where $d\theta_i$ represents the infinitesimal angular displacement of the body in the plane of motion. We note that $\bar{\mathbf{a}}_i \cdot d\bar{\mathbf{s}}_i$ is identical to $(\bar{a}_i)_t \, d\bar{s}_i$ where $(\bar{a}_i)_t$ is the component of $\bar{\mathbf{a}}_i$ along the tangent to the curve described by the mass center of the body in question. Also α_i represents $\ddot{\theta}_i$, the angular acceleration of the representative body. Consequently for the entire system

$$dT = \Sigma m_i \bar{\mathbf{a}}_i \cdot d\bar{\mathbf{s}}_i + \Sigma \bar{I}_i \alpha_i \, d\theta_i$$

This change may also be written as

$$dT = \Sigma \mathbf{R}_i \cdot d\bar{\mathbf{s}}_i + \Sigma \mathbf{M}_{G_i} \cdot d\boldsymbol{\theta}_i$$

where $\mathbf{R}_i$ and $\mathbf{M}_{G_i}$ are the resultant force and resultant couple acting on body i and where $d\boldsymbol{\theta}_i = d\theta_i \mathbf{k}$. These last two equations merely show us that the differential change in kinetic energy equals the differential work done on the system by the resultant forces and resultant couples acting on all the bodies of the system.

The term dV represents the differential change in the total gravitational potential energy V_g and the total elastic potential energy V_e and has the form

$$dV = d(\Sigma m_i g h_i + \Sigma \tfrac{1}{2} k_j x_j{}^2) = \Sigma m_i g \, dh_i + \Sigma k_j x_j \, dx_j$$

where h_i represents the vertical distance of the center of mass of the representative body of mass m_i above any convenient datum plane and where x_j stands for the deformation, tensile or compressive, of a representative elastic member of the system (spring) whose stiffness is k_j.

The complete expression for dU may now be written as

$$dU = \Sigma m_i \bar{\mathbf{a}}_i \cdot d\bar{\mathbf{s}}_i + \Sigma \bar{I}_i \alpha_i\, d\theta_i + \Sigma m_i g\, dh_i + \Sigma k_j x_j\, dx_j \qquad (6/11)$$

In applying Eq. 6/11 to a system of one degree of freedom, the terms $m_i \bar{\mathbf{a}}_i \cdot d\bar{\mathbf{s}}_i$ and $\bar{I}_i \alpha_i\, d\theta_i$ will be positive if the accelerations are in the same direction as the respective displacements and negative if in the opposite direction. Equation 6/11 has the advantage of relating the accelerations to the active forces directly, which eliminates the need for dismembering the system and then eliminating the internal forces and reactive forces by simultaneous solution of the force-mass-acceleration equations for each separate member.

Virtual work. In Eq. 6/11 the differential motions are differential changes in the real or actual displacements which occur. For a mechanical system which assumes a steady-state configuration during constant acceleration we often find it convenient to introduce the concept of *virtual work*. The concepts of virtual work and virtual displacement have been introduced and used to establish equilibrium configurations for static systems of interconnected bodies (see Chapter 7 of *Vol. 1 Statics*). A virtual displacement is any assumed and arbitrary displacement, linear or angular, away from the natural or actual position. For a system of connected bodies the virtual displacements must be consistent with the constraints of the system. For example, when one end of a link is hinged about a fixed pivot, the virtual displacement of the other end must be normal to the line joining the two ends. Such requirements for displacements consistent with the constraints are purely kinematical and provide what are known as the *equations of constraint*. If a set of virtual displacements satisfying the equations of constraint and therefore consistent with the constraints is given to a mechanical system, the proper relationship between the coordinates which specify the configuration of the system will be established by applying the work-energy relationship of Eq. 6/11 expressed in terms of virtual changes. Thus

$$\delta U = \Sigma m_i \bar{\mathbf{a}}_i \cdot \delta \bar{\mathbf{s}}_i + \Sigma \bar{I}_i \alpha_i\, \delta\theta_i + \Sigma m_i g\, \delta h_i + \Sigma k_j x_j\, \delta x_j \qquad (6/11a)$$

It is customary to use the differential symbol d to refer to differential changes in the *real* displacements, whereas the symbol δ is used to signify differential changes which are assumed or *virtual* changes.

Sample Problem 6/12

The movable rack A has a mass of 3 kg, and rack B is fixed. The gear has a mass of 2 kg and a radius of gyration of 60 mm. In the position shown the spring, which has a stiffness of 1.2 kN/m, is stretched a distance of 40 mm. For the instant represented determine the acceleration a of rack A under the action of the 80-N force. The plane of the figure is vertical.

Solution. The given figure represents the active-force diagram for the entire system, which is conservative.

During an infinitesimal upward displacement dx of rack A the work dU done on the system is 80 dx, where x is in meters, and this work equals the sum of the corresponding changes in the total energy of the system. These changes which appear in Eq. 6/11 are as follows:

$$[dT = \Sigma m_i \bar{\mathbf{a}}_i \cdot d\bar{\mathbf{s}}_i + \Sigma \bar{I}_i \alpha_i \, d\theta_i]$$

$$dT_{\text{rack}} = 3a \, dx$$

$$dT_{\text{gear}} = 2 \frac{a}{2} \frac{dx}{2} + 2(0.06)^2 \frac{a/2}{0.08} \frac{dx/2}{0.08} = 0.781a \, dx$$

The change in potential energies of the system from Eq. 6/11 becomes

$$[dV = \Sigma m_i g \, dh_i + \Sigma k_j x_j \, dx_j]$$

$$dV_{\text{rack}} = 3g \, dx = 3(9.81) \, dx = 29.4 \, dx$$

$$dV_{\text{gear}} = 2g(dx/2) = g \, dx = 9.81 \, dx$$

$$dV_{\text{spring}} = k_j x_j \, dx_j = 1200(0.04) \, dx/2 = 24 \, dx$$

Substitution into Eq. 6/11 gives us

$$80 \, dx = 3a \, dx + 0.781a \, dx + 29.4 \, dx + 9.81 \, dx + 24 \, dx$$

Canceling dx and solving for a give

$$a = 16.76/3.781 = 4.43 \text{ m/s}^2 \qquad \qquad Ans.$$

We see that using the work-energy method for an infinitesimal displacement has given us the direct relation between the applied force and the resulting acceleration. It was unnecessary to dismember the system, draw two free-body diagrams, apply $\Sigma F = m\bar{a}$ twice, apply $\Sigma M_G = \bar{I}\alpha$ and $F = kx$, eliminate unwanted terms, and finally solve for a.

① Note that none of the remaining forces external to the system do any work. The work done by the weight and by the spring is accounted for in the potential energy terms.

② Note that $\bar{a}_i$ for the gear is its mass-center acceleration, which is half that for the rack A. Also its displacement is $dx/2$. For the rolling gear the angular acceleration from $a = r\alpha$ becomes $\alpha_1 = (a/2)/0.08$, and the angular displacement from $ds = r \, d\theta$ is $d\theta_i = (dx/2)/0.08$.

③ Note here that the displacement of the spring is one-half that of the rack. Hence $x_i = x/2$.

Sample Problem 6/13

A constant force P is applied to end A of the two identical and uniform links and causes them to move to the right in their vertical plane with a horizontal acceleration a. Determine the steady-state angle θ made by the bars with one another.

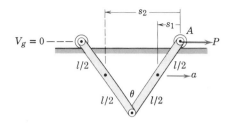

Solution. The figure constitutes the active-force diagram for the system. To find the steady-state configuration, consider a virtual displacement of each bar from the natural position assumed during the acceleration. Measurement of the displacement with respect to end A eliminates any work done by force P during the virtual displacement. Thus

①

$$\delta U = 0$$

② The virtual change in kinetic energy from Eq. 6/11a is

$$\delta T = \Sigma m\bar{a} \cdot \delta\bar{s} = ma(-\delta s_1) + ma(-\delta s_2)$$

$$= -ma\left[\delta\left(\frac{l}{2}\sin\frac{\theta}{2}\right) + \delta\left(\frac{3l}{2}\sin\frac{\theta}{2}\right)\right]$$

③

$$= -ma\left(l\cos\frac{\theta}{2}\,\delta\theta\right)$$

We choose the horizontal line through A as the datum for zero potential energy. Thus the potential energy of the links is

$$V_g = 2mg\left(-\frac{l}{2}\cos\frac{\theta}{2}\right)$$

and the virtual change in potential energy becomes

$$\delta V_g = \delta\left(-2mg\frac{l}{2}\cos\frac{\theta}{2}\right) = \frac{mgl}{2}\sin\frac{\theta}{2}\,\delta\theta$$

Substitution into the work-energy equation for virtual changes, Eq. 6/11a, gives

$$[\delta U = \delta T + \delta V_g] \qquad 0 = -mal\cos\frac{\theta}{2}\,\delta\theta + \frac{mgl}{2}\sin\frac{\theta}{2}\,\delta\theta$$

from which

$$\theta = 2\tan^{-1}\frac{2a}{g} \qquad\qquad Ans.$$

Again, in this problem we see that the work-energy approach obviated the necessity for dismembering the system, drawing separate free-body diagrams, applying motion equations, eliminating unwanted terms, and solving for θ.

① Note that we use the symbol δ to refer to an assumed or virtual differential change rather than the symbol d, which refers to an infinitesimal change in the real displacement.

② In evaluating δT we are finding the variation of T resulting from a virtual change in displacement. Therefore we must use the expression $m\mathbf{a} \cdot \delta\mathbf{s}$ and not the expression $\delta(\frac{1}{2}mv^2)$ which represents the variation of T resulting from a variation in v.

③ We have chosen to use the angle θ to describe the configuration of the links, although we could have used the distance between the two ends of the links just as well.

PROBLEMS

Problem 6/150

6/150 The uniform slender links are released from rest in the vertical plane. The light rollers at A and B are free to move in the horizontal guide. Determine the initial angular acceleration α of the links.

$$Ans.\ \alpha = \frac{3}{2}\frac{g}{b}\cos\theta$$

Problem 6/151

6/151 Each of the links A and B has a mass of 4 kg, and bar C has a mass of 6 kg. Calculate the angle θ assumed by the links if the body to which they are pinned is given a steady horizontal acceleration a of 1.2 m/s².
$$Ans.\ \theta = 7.0°$$

Problem 6/152

6/152 A torque M_0 applied to O_1A through its shaft at O_1 is just sufficient to reduce the force between the 120-kg platform C and the stop at D to zero. If a torque $M = 2M_0$ is applied to link O_1A, determine the initial angular acceleration α of the links without dismembering the system. Each link may be treated as a uniform slender bar. Motion is in the vertical plane.

Problem 6/153

6/153 If the force P required to maintain equilibrium of the 25-kg sheave and attached 50-kg load is doubled, calculate the acceleration a of the center O of the sheave. The radius of gyration of the sheave about its center is 150 mm. $Ans.\ a = 8.26$ m/s²

6/154 The sliding block is given a horizontal acceleration to the right which is slowly increased to a steady value a. The attached pendulum of mass m and mass center G assumes a steady angular deflection θ. The torsion spring at O exerts a moment $M = K\theta$ on the pendulum to oppose the angular deflection. Determine the torsional stiffness K which will allow a steady deflection θ.

Problem 6/154

6/155 The slotted disk revolves in a horizontal plane about its center O. Each spring has a stiffness $k = 300$ N/m and is unstretched with the disk at rest and with $x = 20$ mm. Determine the steady-state position x of each 0.3-kg slider A for a constant disk speed of 250 rev/min.   *Ans.* $x = 63.6$ mm

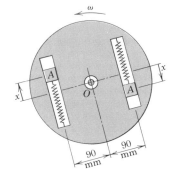

Problem 6/155

6/156 Pinion A of the electric motor turns gear B and its attached pinion C. This unit meshes with gear D and its attached hoisting drum E. The motor rotor and pinion A have a combined mass of 30 kg and a radius of gyration of 150 mm. The unit B and C, as well as the unit D and E, has a mass of 95 kg with a radius of gyration of 300 mm. The motor receives 2 kW of electrical power, 94 percent of which is converted into mechanical power. For an instant when the upward velocity of the 80-kg mass is 0.9 m/s, calculate the upward acceleration a of the load.

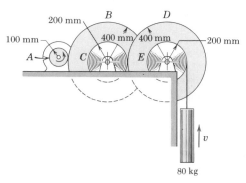

Problem 6/156

6/157 The cargo box of the food-delivery truck for aircraft servicing has a loaded mass m and is elevated by the application of a couple M on the lower end of the link which is hinged to the truck frame. The horizontal slots allow the linkage to unfold as the cargo box is elevated. Determine the upward acceleration of the box in terms of h for a given value of M. Neglect the mass of the links.

$$Ans. \ a = \frac{M}{2mb\sqrt{1 - (h/2b)^2}} - g$$

Problem 6/157

Problem 6/158

Problem 6/159

Problem 6/160

Problem 6/161

6/158 The hydraulic cylinder A exerts an upward force P on pin B of the uniform bar BC of mass m. Neglect the mass of link OG and determine the angular acceleration α of BC as the bar starts from rest in the position shown. Motion occurs in the vertical plane.

$$Ans. \ \alpha = \frac{3}{4b}\left(\frac{2P}{m} - g\right)\cos\theta$$

6/159 The two uniform slender bars, each of mass m, are suspended in the vertical plane by the small rollers which bear on the horizontal surface. The bars are prevented from collapse by the cord A. If the cord is cut, determine the initial downward acceleration of the pin at O which joins the bars.

$$Ans. \ a = \frac{3(1 + \cos\theta)}{5 + 3\cos\theta}g$$

6/160 The small vehicle has a total mass m, and each of its four wheels has a moment of inertia I about its center. The vehicle is propelled by an electric motor which supplies a torque M to the frictional-drive pinion of radius r_1. The rotor of the electric motor and the attached pinion have a combined moment of inertia I_O about the shaft axis. Determine the torque M required to give the vehicle an acceleration a up the incline. No slipping occurs between any of the frictional surfaces at the wheels or pinion.

6/161 The sector and attached wheels are released from rest in the position shown in the vertical plane. Each wheel is a solid 5-kg circular disk and rolls on the fixed circular path without slipping. The sector has a mass of 8 kg and is closely approximated by one-fourth of a solid circular disk of 400-mm radius. Determine the initial angular acceleration α of the sector.

$$Ans. \ \alpha = 10.84 \ \text{rad/s}^2$$

6/162 Each of the uniform bars *OA* and *OB* has a mass of 2 kg and is freely hinged at *O* to the vertical shaft, which is given an upward acceleration *a* = *g*/2. The links which connect the light collar *C* to the bars have negligible weight, and the collar slides freely on the shaft. The spring has a stiffness *k* = 130 N/m and is uncompressed for the position equivalent to $\theta = 0$. Calculate the angle θ assumed by the bars under conditions of steady acceleration.

Dimensions in Millimeters

Problem 6/162

6/163 The mechanical tachometer measures the rotational speed *N* of the shaft by the horizontal motion of the collar *B* along the rotating shaft. This movement is caused by the centrifugal action of the two 350-g masses *A* which rotate with the shaft. Collar *C* is fixed to the shaft. Determine the rotational speed *N* of the shaft for a reading $\beta = 15°$. The stiffness of the spring is 900 N/m, and it is uncompressed when $\theta = 0$ and $\beta = 0$. Neglect the weights of the links. 　　*Ans. N = 132.8 rev/min*

Problem 6/163

6/164 A spring-controlled governor for a steam turbine is shown where the axis *O-O* of rotation is horizontal. As the speed increases, the rotational radius of the 0.25-kg balls tends to increase thus increasing the angle β. This change activates a control which throttles the steam inlet pressure to limit the turbine speed. When the rotational speed is zero, the spring is uncompressed and $\beta = 0$. Determine the steady-state value of β for a shaft speed *N* = 1800 rev/min. Neglect the small mass of the right-angled arms. 　　*Ans. β = 14.48°*

Problem 6/164

Problem 6/165

Problem 6/166

Problem 6/167

Problem 6/168

6/165 The aerial tower shown is designed to elevate a workman in a vertical direction. An internal mechanism at *B* maintains the angle between *AB* and *BC* at twice the angle θ between *BC* and the ground. If the combined mass of the man and the cab is 200 kg and if all other masses are neglected, determine the torque *M* applied to *BC* at *C* and the torque M_B in the joint at *B* required to give the cab an initial vertical acceleration of 1.2 m/s² when started from rest in the position $\theta = 30°$.
Ans. $M_B = 11.44$ kN·m
$M = 0$

▶**6/166** The sliding bar of mass *m* in Prob. 6/148 is repeated here. Determine the angular acceleration of *AC* due to the action of the force *P* as the bar starts from rest with any value θ. Link *OC* has negligible mass, and the smooth guide is horizontal.
Ans. $\alpha = \left(\dfrac{3P}{mb}\right)\dfrac{2 + \cos\theta}{7 + 6\cos\theta}$

▶**6/167** Each of the two uniform links has a mass of 3 kg. The rod *AC* has a mass of 2.5 kg and is free to slide through the pivoted collar at *B*. The spring has negligible mass, so that the total mass of the assembly is 8.5 kg. Determine the steady-state angle θ assumed by the links under a constant accelerating force *F* = 50 N. The links are suspended in the vertical plane by the rollers, and the spring has a stiffness *k* = 350 N/m. Also, the spring is uncompressed when θ is essentially zero for *F* = 0.
Ans. $\theta = 27.0°$

▶**6/168** The 36-kg pulley has a radius of gyration about its center of 250 mm. The chain has a mass of 7.5 kg/m and has a length of 2.4 m plus the half-circumferential portion over the pulley. For a very slight unbalance starting from rest, the pulley turns clockwise and the chain piles up on the platform. As each link strikes the platform and comes to rest, it is unable to transmit any force to the links above it. Compute the angular velocity ω of the pulley at the instant when *x* = 1.2 m. (*Hint:* Analyze the energy change during a differential interval of motion noting the energy loss of each increment of chain which comes to rest on the platform.)
Ans. $\omega = 4.53$ rad/s

SECTION C. IMPULSE AND MOMENTUM

6/8 IMPULSE-MOMENTUM EQUATIONS

The principles of impulse and momentum were developed and used in Art. 3/8 in the description of particle motion. In this treatment we observed that these principles were of particular importance when the applied forces were expressible as functions of the time and when interactions between particles occurred during short periods of time, such as with impact. Similar advantages result when the impulse-momentum principles are applied to the motion of rigid bodies.

In Art. 4/2 the impulse-momentum principles were extended to cover any defined system of mass particles without restriction as to the connections between the particles of the system. These extended relations all apply to the motion of a rigid body, which is merely a special case of a general system of mass. We will now apply these equations directly to rigid-body motion in two dimensions.

(*a*) *Linear momentum.* In Art. 4/4 we defined the linear momentum of a mass system as the vector sum of the linear momenta of all of its particles and wrote $\mathbf{G} = \Sigma m_i \mathbf{v}_i$. With $\mathbf{r}_i$ representing the position vector to m_i we have $\mathbf{v}_i = \dot{\mathbf{r}}_i$ and $\mathbf{G} = \Sigma m_i \dot{\mathbf{r}}_i$ which, for a system whose total mass is constant, may be written as $\mathbf{G} = d(\Sigma m_i \mathbf{r}_i)/dt$. When we substitute the principle of moments $m\bar{\mathbf{r}} = \Sigma m_i \mathbf{r}_i$ to locate the mass center, the momentum becomes $\mathbf{G} = d(m\bar{\mathbf{r}})/dt = m\dot{\bar{\mathbf{r}}}$ where $\dot{\bar{\mathbf{r}}}$ is the velocity $\bar{\mathbf{v}}$ of the mass center. Therefore, as before, we find that the linear momentum of any mass system, rigid or nonrigid, is

$$\mathbf{G} = m\bar{\mathbf{v}} \qquad\qquad [4/5]$$

In the derivation of Eq. 4/5 we note that it was unnecessary to employ the kinematical condition for a rigid body, Fig. 6/13, which is $\mathbf{v}_i = \bar{\mathbf{v}} + \boldsymbol{\omega} \times \boldsymbol{\rho}_i$. In that case we obtain the same result by writing $\mathbf{G} = \Sigma m_i(\bar{\mathbf{v}} + \boldsymbol{\omega} \times \boldsymbol{\rho}_i)$. The first sum is $\bar{\mathbf{v}}\Sigma m_i = m\bar{\mathbf{v}}$, and the second sum becomes $\boldsymbol{\omega} \times \Sigma m_i \boldsymbol{\rho}_i = \boldsymbol{\omega} \times m\bar{\boldsymbol{\rho}} = \mathbf{0}$ since $\boldsymbol{\rho}_i$ is measured from the mass center making $\bar{\boldsymbol{\rho}}$ zero.

Next in Art. 4/4 we rewrote Newton's generalized second law as Eq. 4/6. This equation and its integrated form are

$$\Sigma \mathbf{F} = \dot{\mathbf{G}} \qquad \text{and} \qquad \int_{t_1}^{t_2} \Sigma \mathbf{F}\, dt = \mathbf{G}_2 - \mathbf{G}_1 \qquad (6/12)$$

Equation 6/12 may be written in its scalar component form which,

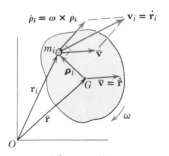

Figure 6/13

for plane motion in the *x-y* plane, gives

$$\boxed{\begin{aligned}\Sigma F_x &= \dot{G}_x \\[2mm] \Sigma F_y &= \dot{G}_y\end{aligned}} \qquad \text{and} \qquad \boxed{\begin{aligned}\int_{t_1}^{t_2} \Sigma F_x \; dt &= G_{x_2} - G_{x_1} \\[2mm] \int_{t_1}^{t_2} \Sigma F_y \; dt &= G_{y_2} - G_{y_1}\end{aligned}} \qquad (6/12a)$$

In words the first of Eqs. 6/12 and 6/12*a* states that the resultant force equals the time rate of change of momentum. The integrated form of Eqs. 6/12 and 6/12*a* states that the linear impulse on the body during the interval $t_2 - t_1$ equals the corresponding change in linear momentum. As in the force-mass-acceleration formulation, the force summations in Eqs. 6/12 and 6/12*a* must include *all* forces acting externally on the body considered. We emphasize, therefore, that when using the impulse-momentum equations, it is essential to construct the complete free-body diagram so as to disclose all forces which appear in the force summation. In contrast to the method of work and energy, all forces exert impulses whether they do work or not.

(*b*) *Angular momentum.* Angular momentum is defined as the moment of linear momentum. In Art. 4/4 we expressed the angular momentum about the mass center of any prescribed system of mass as $\mathbf{H}_G = \Sigma \boldsymbol{\rho}_i \times m_i \mathbf{v}_i$, which is merely the vector sum of the moments about *G* of the linear momenta of all particles. We showed in Art. 4/4 that this sum could also be written as $\mathbf{H}_G = \Sigma \boldsymbol{\rho}_i \times m_i \dot{\boldsymbol{\rho}}_i$ where $\dot{\boldsymbol{\rho}}_i$ is the velocity of m_i with respect to *G*. Although we have simplified this expression in Art. 6/2 in the course of deriving the moment equation of motion, we will pursue this same expression again for sake of emphasis by using the rigid body in plane motion represented in Fig. 6/13. The relative velocity becomes $\dot{\boldsymbol{\rho}}_i = \boldsymbol{\omega} \times \boldsymbol{\rho}_i$ where the angular velocity of the body is $\boldsymbol{\omega} = \omega \mathbf{k}$. The unit vector $\mathbf{k}$ is directed into the paper for the sense of $\boldsymbol{\omega}$ shown. Since $\boldsymbol{\rho}_i$, $\dot{\boldsymbol{\rho}}_i$, and $\boldsymbol{\omega}$ are at right angles to one another, the magnitude of $\dot{\boldsymbol{\rho}}_i$ is $\rho_i \omega$, and the magnitude of $\boldsymbol{\rho}_i \times m_i \dot{\boldsymbol{\rho}}_i$ is $\rho_i^2 \omega m_i$. Thus we may write $\mathbf{H}_G = \Sigma \rho_i^2 m_i \omega \mathbf{k} = \bar{I} \omega \mathbf{k}$ where $\bar{I} = \Sigma m_i \rho_i^2$ is the mass moment of inertia of the body about its mass center. Because the angular-momentum vector is always normal to the plane of motion, vector notation is generally unnecessary, and we may write the angular momentum about the mass center as the scalar

$$\boxed{H_G = \bar{I}\omega} \qquad (6/13)$$

This angular momentum appears in the moment-angular-momentum relation, Eq. 4/9, which in scalar notation for plane motion, along with its integrated form, is

$$\boxed{\Sigma M_G = \dot{H}_G} \qquad \text{and} \qquad \boxed{\int_{t_1}^{t_2} \Sigma M_G \, dt = H_{G_2} - H_{G_1}} \quad \textbf{(6/14)}$$

In words the first of Eqs. 6/14 states that the sum of the moments about the mass center of *all* forces acting on the body equals the time rate of change of angular momentum about the mass center. The integrated form of Eq. 6/14 states that the angular impulse about the mass center of all forces acting on the body during the interval $t_2 - t_1$ equals the corresponding change in the angular momentum about G. The sense for positive rotation must be clearly established, and the algebraic signs of ΣM_G, H_{G_2}, and H_{G_1} must be consistent with this choice. Again, a free-body diagram is essential.

With the moments about G of the linear momenta of all particles accounted for by $H_G = \bar{I}\omega$, it follows that we may represent the linear momentum $\mathbf{G} = m\bar{\mathbf{v}}$ as a vector through the mass center G, as shown in Fig. 6/14a. Thus $\mathbf{G}$ and $\mathbf{H}_G$ have vector properties analogous to those of the resultant force and couple.

With the establishment of the linear and angular momentum resultants in Fig. 6/14a, which represents the momentum diagram, the angular momentum H_O about any point O is easily written as

$$\boxed{H_O = \bar{I}\omega + m\bar{v}d} \qquad \textbf{(6/15)}$$

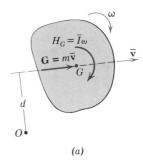

(a)

This expression holds at any particular instant of time about O, which may be a fixed or moving point on or off the body.

When a body rotates about a fixed point O on the body or body extended, as shown in Fig. 6/14b, the relations $\bar{v} = \bar{r}\omega$ and $d = \bar{r}$ may be substituted into the expression for H_O, giving $H_O = (\bar{I}\omega + m\bar{r}^2\omega)$. But $\bar{I} + m\bar{r}^2 = I_O$ so that

$$\boxed{H_O = I_O\omega} \qquad \textbf{(6/16)}$$

In Art. 4/2 we derived Eq. 4/7, which is the moment-angular-momentum equation about a fixed point O. This equation, written in scalar notation for plane motion, along with its integrated form is

$$\boxed{\Sigma M_O = I_O\dot{\omega}} \qquad \text{and} \qquad \boxed{\int_{t_1}^{t_2} \Sigma M_O \, dt = I_O(\omega_2 - \omega_1)} \quad \textbf{(6/17)}$$

In addition to drawing a complete free-body diagram in order that the force and moment summations may be correctly evalu-

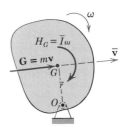

(b)

Figure 6/14

ated when applying the linear and angular impulse-momentum equations for rigid-body motion, we find it useful to draw the momentum diagram which indicates the resultant linear-momentum vector and the angular-momentum couple. We caution the reader here not to add linear momentum and angular momentum for the same reason that force and moment cannot be added directly.

(*c*) *Interconnected rigid bodies.* The equations of impulse and momentum may also be used for a system of interconnected rigid bodies since the momentum principles are applicable to any general system of constant mass. In Fig. 6/15 are shown the combined free-body diagram and momentum diagram for two interconnected bodies. Equations 4/6 and 4/7, which are $\Sigma \mathbf{F} = \dot{\mathbf{G}}$ and $\Sigma \mathbf{M}_O = \dot{\mathbf{H}}_O$ where O is a fixed reference point, may be written for each member of the system and added. The sums would be

$$\Sigma \mathbf{F} = \dot{\mathbf{G}}_1 + \dot{\mathbf{G}}_2 + \cdots$$
$$\Sigma \mathbf{M}_O = \dot{\mathbf{H}}_{O_1} + \dot{\mathbf{H}}_{O_2} + \cdots \qquad (6/18)$$

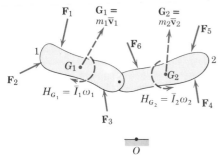

Figure 6/15

In integrated form for a finite time interval, these expressions are

$$\int_{t_1}^{t_2} \Sigma \mathbf{F} \, dt = (\Delta \mathbf{G})_{\text{system}} \qquad \int_{t_1}^{t_2} \Sigma \mathbf{M}_O \, dt = (\Delta \mathbf{H}_O)_{\text{system}} \quad (6/19)$$

We note that the equal and opposite actions and reactions in the connections are internal to the system and cancel one another so will not be involved in the force and moment summations. Also point O is one fixed reference point for the entire system.

(*d*) *Conservation of momentum.* In Part (*b*) of Art. 4/5 the principles of conservation of momentum for a general mass system were expressed by Eqs. 4/11 and 4/12. These principles are applicable to either a single rigid body or to a system of interconnected rigid bodies. Thus, if $\Sigma \mathbf{F} = \mathbf{0}$ for a given interval of time, then

$$\boxed{\Delta \mathbf{G} = \mathbf{0}} \qquad\qquad [4/15]$$

which says that the linear momentum vector undergoes no change in the absence of a resultant linear impulse. For the system of interconnected rigid bodies there may be linear momentum changes of individual parts of the system during the interval, but there will be no resultant momentum change for the system if there is no resultant linear impulse.

Similarly, if the resultant moment about a given fixed point O or about the mass center is zero during a particular interval of time for a single rigid body or for a system of interconnected rigid bodies, then

$$\boxed{\Delta \mathbf{H}_O = 0} \quad \text{or} \quad \boxed{\Delta \mathbf{H}_G = 0} \qquad [4/16]$$

which says that the angular momentum either about the fixed point or about the mass center undergoes no change in the absence of a corresponding resultant angular impulse. Again, in the case of the interconnected system, there may be angular-momentum changes of individual components during the interval, but there will be no resultant angular-momentum change for the system if there is no resultant angular impulse about the fixed point or the mass center. Either of Eqs. 4/16 may hold without the other. In the case of an interconnected system the use of the center of mass for the system is generally inconvenient. As was illustrated previously in Art. 3/9 in the chapter on particle motion, the use of momentum principles greatly facilitates the analysis of situations where forces and couples act for very short periods of time.

(*e*) *Impact of rigid bodies.*   Impact phenomena involve a fairly complex interrelationship of energy and momentum transfer, energy dissipation, elastic and plastic deformation, relative impact velocity, and body geometry. In Art. 3/11 we treated the impact of bodies modeled as particles and considered only the case of central impact where the contact forces of impact passed through the mass centers of the bodies, as would always happen with colliding smooth spheres, for example. To relate the conditions after impact to those before impact required the introduction of the so-called coefficient of restitution e or impact coefficient, which compares the relative separation velocity to the relative approach velocity measured along the direction of the contact forces. Although in the classical theory of impact e was considered a constant for given materials, more modern investigations show that e is highly dependent on geometry and impact velocity as well as on materials. At best, even for spheres and rods under direct central and longitudinal impact, the coefficient of restitution is a complex and variable factor of limited use.

Any attempt to extend this simplified theory of impact utilizing a coefficient of restitution for the noncentral impact of rigid bodies of varying shape is a gross oversimplification which has little practical value. For this reason we shall not include such an exercise in this book, even though such a theory is easily developed and appears in certain references. We can and do, however, make full use of the principles of conservation of linear and angular momentum where applicable in discussing the impact and other interactions of rigid bodies.

Sample Problem   6/14

The force P, which is applied to the cable wrapped around the central hub of the symmetrical wheel, is increased slowly according to $P = 6.5t$, where P is in newtons and t is the time in seconds after P is first applied. Determine the angular velocity ω of the wheel 10 s after P is applied if the wheel is rolling to the left with a velocity of its center of 0.9 m/s at time $t = 0$. The wheel, which has a mass of 60 kg and a radius of gyration about its center of 250 mm, rolls without slipping.

Solution. The free-body diagram of the wheel for any position within the interval is shown. Also indicated are the initial linear and angular momenta at time $t = 0$ and the final linear and angular momenta at time $t = 10$ s. The correct direction of the friction force F is that to oppose the slipping which would occur without friction.

Application of the linear impulse-momentum equation and the angular impulse-momentum equation over the *entire* interval gives

② $\left[\int_{t_1}^{t_2} \Sigma F_x\, dt = G_{x_2} - G_{x_1} \right]$ $\int_0^{10} (6.5t - F)\, dt = 60[0.450\omega - (-0.9)]$

③ $\left[\int_{t_1}^{t_2} \Sigma M_G\, dt = H_{G_2} - H_{G_1} \right]$

$\int_0^{10} (0.450F - 0.225[6.5t])\, dt = 60(0.250)^2 \left[\omega - \left(-\dfrac{0.9}{0.450} \right) \right]$

Since the force F is variable, it must remain under the integral sign. We eliminate F between the two equations by dividing the second one by 0.450 and adding to the first one. Integrating and solving for ω give

$$\omega = 2.60 \text{ rad/s clockwise} \qquad Ans.$$

Alternative solution. We could avoid the necessity of a simultaneous solution by applying the second of Eqs. 6/17 about a fixed point O on the horizontal surface. The moments of the 60(9.81)-N weight and the equal and opposite force N cancel one another, and F is eliminated since its moment about O is zero. Thus the angular momentum about O becomes $H_O = \bar{I}\omega + m\bar{v}r = m\bar{k}^2\omega + mr^2\omega = m(\bar{k}^2 + r^2)\omega$ where $\bar{k}$ is the centroidal radius of gyration and r is the 450-mm rolling radius. Thus we see that $H_O = H_C$ since $\bar{k}^2 + r^2 = k_C^2$ and $H_C = I_C\omega = mk_C^2\omega$. Equation 6/17 now gives

$\left[\int_{t_1}^{t_2} \Sigma M_O\, dt = H_{O_2} - H_{O_1} \right]$

$\int_0^{10} 6.5t(0.450 - 0.225)\, dt = 60[(0.250)^2 + (0.450)^2] \left(\omega - \left[-\dfrac{0.9}{0.450} \right] \right)$

Solution of this one equation is equivalent to the simultaneous solution of the two previous equations.

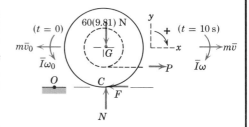

① Also, we note the clockwise unbalance of moments about C which causes a clockwise angular acceleration as the wheel rolls without slipping. Since the moment sum about G must also be in the clockwise sense of α, the friction force must act to the left to provide it.

② Note carefully the signs of the momentum terms. The final linear velocity is assumed in the positive x-direction, so G_{x_2} is positive. The initial linear velocity is negative, so G_{x_1} is negative. When we subtract the negative term, we get the double minus sign.

③ Since the wheel rolls without slipping, a positive x-velocity requires a clockwise angular velocity, and vice versa. Again we subtract a minus quantity.

Sample Problem 6/15

The sheave E of the hoisting rig shown has a mass of 30 kg and a centroidal radius of gyration of 250 mm. The 40-kg load D which is carried by the sheave has an initial downward velocity $v_0 = 1.2$ m/s at the instant when a clockwise torque is applied to the hoisting drum A to maintain essentially a constant force $F = 380$ N in the cable at B. Compute the angular velocity ω of the sheave 5 s after the torque is applied to the drum and find the tension T in the cable at O during the interval. Neglect all friction.

Solution.   The load and the sheave taken together constitute the system, and its free-body diagram is shown. The tension T in the cable at O and the final angular velocity ω of the sheave are the two unknowns. We eliminate T initially by applying the moment-angular-momentum equation about the fixed point O, taking counterclockwise as positive.

$$\left[\int_{t_1}^{t_2} \Sigma M_O \, dt = H_{O_2} - H_{O_1} \right]$$

$$\int_{t_1}^{t_2} \Sigma M_O \, dt = \int_0^5 [380(0.750) - (30 + 40)(9.81)(0.375)] \, dt$$

$$= 137.4 \text{ N·m·s}$$

① $(H_{O_2} - H_{O_1})_D = mv_2 d - mv_1 d = md(v_2 - v_1)$

$$= 40(0.375)(v - [-1.2]) = 15(0.375\omega + 1.2)$$

$$= 5.63\omega + 18 \text{ N·m·s}$$

$(H_{O_2} - H_{O_1})_E = \bar{I}(\omega_2 - \omega_1) + md(\bar{v}_2 - \bar{v}_1)$

$$= 30(0.250)^2(\omega - [-1.2/0.375])$$

$$+ 30(0.375)(0.375\omega - [-1.2])$$

$$= 6.09\omega + 19.50 \text{ N·m·s}$$

Substituting into the momentum equation gives

$$137.4 = 5.63\omega + 18 + 6.09\omega + 19.50$$

$$\omega = 8.53 \text{ rad/s counterclockwise} \qquad \textit{Ans.}$$

The force-linear-momentum equation is now applied to the system to determine T. With the positive direction up we have

$$\left[\int_{t_1}^{t_2} \Sigma F \, dt = G_2 - G_1 \right]$$

$$\int_0^5 [T + 380 - 70(9.81)] \, dt = 70[0.375(8.53) - (-1.2)]$$

$$5T = 1841 \qquad T = 368 \text{ N} \qquad \textit{Ans.}$$

If we had taken our moment equation around the center C of the sheave instead of point O, it would contain both unknowns T and ω, and we would be obliged to solve it simultaneously with the foregoing force equation which would also contain the same two unknowns.

① Watch for the double minus sign in subtracting a negative quantity. Also, the units of angular momentum, which are those of angular impulse, may be written as kg·m^2·s^{-1}.

Sample Problem 6/16

The uniform rectangular block of dimensions shown is sliding to the left on the horizontal surface with a velocity v when it strikes the small step in the surface. Assume negligible rebound at the step, and compute the minimum value of v which will permit the block to pivot about the edge of the step and just reach the standing position A with no velocity. Compute the percentage energy loss $\Delta E/E$ for $b = c$.

Solution. It will be assumed that the edge of the step O acts as a latch on the corner of the block, so that the block pivots about O. Furthermore, the height of the step is assumed negligible compared with the dimensions of the block. During impact the only force which exerts a moment about O is the weight mg, but the angular impulse

① due to the weight is extremely small since the time of impact is negligible. Thus we may assume that the angular momentum about O is conserved.

The initial angular momentum of the block about O just before impact is the moment of its linear momentum and is $H_O = mv(b/2)$.

② The velocity of the center of mass G immediately after impact is $\bar{v}$, and the angular velocity is $\omega = \bar{v}/\bar{r}$. The angular momentum about O just after impact when the block is starting its rotation about O is

③ $[H_O = I_O\omega]$ $\qquad H_O = \left[\frac{1}{12} m(b^2 + c^2) + m\left(\left[\frac{c}{2}\right]^2 + \left[\frac{b}{2}\right]^2 \right) \right]\omega$

$$= \frac{m}{3}(b^2 + c^2)\omega$$

Conservation of angular momentum gives

$[\Delta H_O = 0]$ $\qquad \frac{m}{3}(b^2 + c^2)\omega = mv\frac{b}{2}$ $\qquad \omega = \frac{3vb}{2(b^2 + c^2)}$

This angular velocity will be sufficient to raise the block just past position A if the kinetic energy of rotation equals the increase in potential energy. Thus

$[\Delta T + \Delta V_g = 0]$ $\qquad \frac{1}{2}I_O\omega^2 - mg\left(\sqrt{\left(\frac{b}{2}\right)^2 + \left(\frac{c}{2}\right)^2} - \frac{b}{2} \right) = 0$

$$\frac{1}{2}\frac{m}{3}(b^2 + c^2)\left[\frac{3vb}{2(b^2 + c^2)} \right]^2 - \frac{mg}{2}(\sqrt{b^2 + c^2} - b) = 0$$

$$v = 2\sqrt{\frac{g}{3}\left(1 + \frac{c^2}{b^2}\right)(\sqrt{b^2 + c^2} - b)} \qquad\qquad Ans.$$

The percentage loss of energy is

$$\frac{\Delta E}{E} = \frac{\frac{1}{2}mv^2 - \frac{1}{2}I_O\omega^2}{\frac{1}{2}mv^2} = 1 - \frac{k_O^2\omega^2}{v^2} = 1 - \left(\frac{b^2 + c^2}{3}\right)\left[\frac{3b}{2(b^2 + c^2)} \right]^2$$

$$= 1 - \frac{3}{4\left(1 + \frac{c^2}{b^2}\right)}, \qquad \Delta E/E = 62.5\% \quad \text{for } b = c \qquad Ans.$$

① If the corner of the block struck a spring instead of the rigid step, then the time of the interaction during compression of the spring could become appreciable, and the angular impulse about the fixed point at the end of the spring due to the moment of the weight would have to be accounted for.

② Notice the abrupt change in direction and magnitude of the velocity of G during the impact.

③ Be sure to use the transfer theorem $I_O = \bar{I} + m\bar{r}^2$ correctly here.

PROBLEMS

6/169 A person who walks through the revolving door exerts a 90-N horizontal force on one of the four door panels and keeps the 15° angle constant relative to a line which is normal to the panel. If each panel is modeled by a 60-kg uniform rectangular plate which is 1.2 m in length as viewed from above, determine the final angular velocity ω of the door if the person exerts the force for 3 seconds. The door is initially at rest and friction may be neglected. *Ans.* $\omega = 2.26$ rad/s

Problem 6/169

6/170 The center of the homogeneous solid cylinder is given an initial velocity of 0.6 m/s up the incline. Determine the time t required for it to reach a velocity of 1.2 m/s down the incline if it rolls without slipping. *Ans.* $t = 2.77$ s

Problem 6/170

6/171 The 75-kg flywheel has a radius of gyration about its shaft axis of $\bar{k} = 0.50$ m and is subjected to the torque $M = 10(1 - e^{-t})$ N·m, where t is in seconds. If the flywheel is at rest at time $t = 0$, determine its angular velocity ω at $t = 3$ s. *Ans.* $\omega = 1.093$ rad/s

Problem 6/171

6/172 The constant tensions of 200 N and 160 N are applied to the hoisting cable as shown. If the velocity v of the load is 2 m/s down and the angular velocity ω of the pulley is 8 rad/s counterclockwise at time $t = 0$, determine v and ω after the cable tensions have been applied for 5 s. Note the independence of the results.

Problem 6/172

Problem 6/173

Problem 6/174

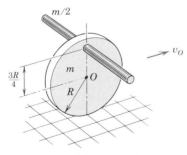

Problem 6/177

6/173 The constant 40-N force is applied to the 36-kg stepped cylinder as shown. The centroidal radius of gyration of the cylinder is $\bar{k} = 200$ mm and it rolls on the incline without slipping. If the cylinder is at rest when the force is first applied, determine its angular velocity ω eight seconds later.

Ans. $\omega = 24.2$ rad/s

6/174 The 30-g bullet has a horizontal velocity of 500 m/s as it strikes the 25-kg compound pendulum, which has a radius of gyration $k_O = 0.95$ m. If $h = 1.1$ m, calculate the angular velocity ω of the pendulum with its embedded bullet immediately after the impact.

Ans. $\omega = 0.730$ rad/s

6/175 If the bullet of Prob. 6/174 takes 10^{-4} s to embed itself in the pendulum, calculate the time average of the horizontal force O_x exerted by the pin on the bar at O during the impact. Use the results cited for Prob. 6/174.

6/176 For what value of h will the bullet of Prob. 6/174 cause no horizontal force reaction at O?

6/177 The homogeneous circular cylinder of mass m and radius R carries a slender rod of mass $m/2$ attached to it as shown. If the cylinder rolls on the surface without slipping with a velocity v_O of its center O, determine the angular momenta H_G and H_O of the system for the instant shown.

Ans. $H_G = \frac{11}{16}mRv_O$, $H_O = \frac{37}{32}mRv_O$

6/178 Repeat the calculations of H_G and H_O for the cylinder and rod of Prob. 6/177 if the cylinder is rotated 90° clockwise from the position shown in the figure.

6/179 The uniform circular disk of 200-mm radius has a mass of 25 kg and is mounted on the rotating bar *OA* in three different ways. In each case the bar rotates about its vertical shaft at *O* with a clockwise angular velocity $\omega = 4$ rad/s. In case (*a*) the disk is welded to the bar. In case (*b*) the disk, which is pinned freely at *A*, moves with curvilinear translation and therefore has no rigid-body rotation. In case (*c*) the relative angle between the disk and the bar is increasing at the rate $\dot\theta = 8$ rad/s. Calculate the angular momentum of the disk about point *O* for each case.

Ans. (*a*) $H_O = 18$ kg·m²/s
(*b*) $H_O = 16$ kg·m²/s
(*c*) $H_O = 14$ kg·m²/s

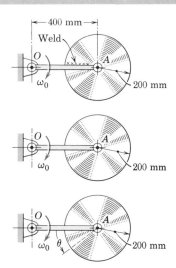

Problem 6/179

6/180 Two small variable-thrust jets are actuated to keep the spacecraft angular velocity about the *z*-axis constant at $\omega_0 = 1.25$ rad/s as the two telescopic booms are extended from $r_1 = 1.2$ m to $r_2 = 4.5$ m at a constant rate over a two-minute period. Determine the necessary thrust *T* for each jet as a function of time where $t = 0$ is the time when the telescoping action is begun. The small 10-kg experiment modules at the ends of the booms may be treated as particles, and the mass of the rigid booms is negligible.

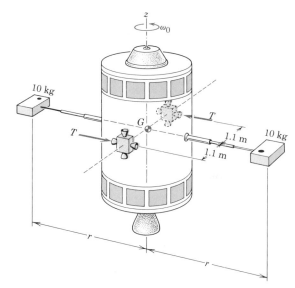

Problem 6/180

6/181 With the gears initially at rest and the couple *M* equal to zero, the forces exerted by the frame on the shafts of the gears at *A* and *B* are 120 and 64 N, respectively, both upward to support the weights of the two gears. A couple $M = 6.8$ N·m is now applied to the larger gear through its shaft at *A*. After 4 s the larger gear has a clockwise angular momentum of 15 kg·m²/s, and the smaller gear has a counterclockwise angular momentum of 5 kg·m²/s. Calculate the new values of the forces R_A and R_B exerted by the frame on the shafts during the 4-s interval. Isolate the two gears together as the system.

Ans. $R_A = 106.8$ N, $R_B = 77.2$ N

Problem 6/181

Problem 6/182

6/182 The elements of a spacecraft with axial mass symmetry and a reaction-wheel control system are shown in the figure. When the motor exerts a torque on the reaction wheel, an equal and opposite torque is exerted on the spacecraft, thereby changing its angular momentum in the z-direction. If all system elements start from rest and the motor exerts a constant torque M for a time period t, determine the final angular velocity of (a) the spacecraft and (b) the wheel relative to the spacecraft. The mass moment of inertia about the z-axis of the entire spacecraft, including the wheel, is I and that of the wheel alone is I_w. The spin axis of the wheel is coincident with the z-axis of symmetry of the spacecraft.

Problem 6/183

6/183 If the system shown is released from rest, how much time passes before the velocity of the 10-kg block reaches 2 m/s? The radius of gyration of the 12-kg pulley at O is $k_O = 150$ mm and the mass of the lower pulley is negligible. Neglect all friction.
Ans. $t = 1.586$ s

Problem 6/184

6/184 A cylindrical shell of 400-mm diameter and mass m is rotating about its central horizontal axis with an angular velocity $\omega_0 = 30$ rad/s when it is released onto a horizontal surface with no velocity of its center ($v_0 = 0$). If slipping between the shell and the surface occurs for 1.5 s, calculate the coefficient of kinetic friction μ_k and the maximum velocity v reached by the center of the shell.

6/185 The cylindrical shell of Prob. 6/184 has a horizontal velocity $v_0 = 6$ m/s of its center and no angular velocity ($\omega_0 = 0$) when it is released onto the horizontal surface. If slipping between the shell and the surface occurs for 1.5 seconds, calculate the coefficient of kinetic friction μ_k and the final velocity v reached by the center of the shell.
Ans. $\mu_k = 0.204$, $v = 3$ m/s

6/186 The large rotor has a mass of 60 kg and a radius of gyration about its vertical axis of 200 mm. The small rotor is a solid circular disk with a mass of 8 kg and is initially rotating with an angular velocity $\omega_1 = 80$ rad/s with the large rotor at rest. A spring-loaded pin P which rotates with the large rotor is released and bears against the periphery of the small disk bringing it to a stop relative to the large rotor. Neglect any bearing friction and calculate the final angular velocity of the assembly.

Problem 6/186

6/187 The uniform circular disk D has a mass of 4 kg and is free to turn about the bearing axis C-C. The 2.5-kg arm B is fastened to the vertical shaft O-O. The arm may be approximated as a slender bar 300 mm in length, and the moment of inertia of the vertical shaft O-O may be neglected. The 3-kg rod A is fastened securely to the arm. If the initial angular velocity of D is $\omega_0 = 7$ rad/s in the direction shown and the arm B is at rest, determine the angular velocity ω_B of the arm after a torque $M = 1.5$ N·m has been applied to the shaft for 4 s.

Ans. $\omega_B = 34.0$ rad/s

Problem 6/187

6/188 Each of the two 300-mm uniform rods A has a mass of 1.5 kg and is hinged at its end to the rotating base B. The 4-kg base has a radius of gyration of 40 mm and is initially rotating freely about its vertical axis with a speed of 300 rev/min and with the rods latched in the vertical positions. If the latches are released and the rods assume the horizontal dotted positions, calculate the new rotational speed N of the assembly.

Problem 6/188

Problem 6/189

6/189 The slender bar of mass m and length l is released from rest in the horizontal position shown. If point A of the bar becomes attached to the pivot at B upon impact, determine the angular velocity ω of the bar immediately after impact in terms of the distance x. Evaluate your expression for $x = 0$, $l/2$, and l.

$$Ans. \ \omega = \left(\frac{l}{2} - x\right) \sqrt{2gh}/(\tfrac{1}{3}l^2 - lx + x^2)$$

Problem 6/190

6/190 The 300-mm slender bar has a mass of 1.6 kg and is initially centered in the radial slot of the circular disk which is rotating freely about its vertical axis with a speed of 400 rev/min. The 16-kg disk has a radius of gyration of 225 mm. If the bar is given a slight nudge, it will slide radially from its central position. Calculate the rotational velocity of the disk when the end of the bar reaches the periphery of the disk.

Problem 6/191

6/191 A man of mass m stands at point A on the horizontal turntable initially at rest. The turntable is free to rotate about O with negligible friction and has a moment of inertia I about O. If the man starts to walk clockwise in a circle of radius r, determine the angle θ through which the turntable rotates until the man reaches his starting point in the new position A'. If the man walks at the speed v_r relative to the surface of the turntable, what is the corresponding angular velocity ω of the turntable?

$$Ans. \ \theta = \frac{2\pi}{1 + (I/mr^2)}, \ \omega = \frac{v_r/r}{1 + (I/mr^2)}$$

Problem 6/192

6/192 The slotted circular disk whose mass is 6 kg has a radius of gyration about O of 175 mm. The disk carries the four steel balls, each of mass 0.15 kg and located as shown, and rotates freely about a vertical axis through O with an angular speed of 120 rev/min. Each of the small balls is held in place by a latching device not shown. If the balls are released while the disk is rotating and come to rest in the dotted positions relative to the slots, compute the new angular velocity ω of the disk. Also find the magnitude $|\Delta E|$ of the energy loss due to the impact of the balls with the ends of the slots. Neglect the diameter of the balls and discuss this approximation.

6/193 The 17.5-Mg lunar landing module with center of mass at G has a radius of gyration of 1.8 m about G. The module is designed to contact the lunar surface with a vertical free-fall velocity of 8 km/h. If one of the four legs hits the lunar surface on a small incline and suffers no rebound, compute the angular velocity ω of the module immediately after impact as it pivots about the contact point. The 9-m dimension is the distance across the diagonal of the square formed by the four feet as corners.

Ans. $\omega = 0.308$ rad/s

Problem 6/193

6/194 A uniform pole of length L is dropped at an angle θ with the vertical, and both ends have a velocity v as end A hits the ground. If end A pivots about its contact point during the remainder of the motion, determine the velocity v' with which end B hits the ground.

Problem 6/194

6/195 The small gear is made to rotate in a horizontal plane about the large stationary gear by means of the torque M applied to the arm OA. The small gear has a mass of 3 kg and may be treated as a circular disk. The 2-kg arm OA has a radius of gyration about the fixed bearing at O of 150 mm. Determine the constant torque M required to give the arm OA an absolute angular velocity of 20 rad/s in 3 s, starting from rest. Neglect friction.

Ans. $M = 1.819$ N·m

Problem 6/195

6/196 Determine the minimum velocity v which the wheel may have and just roll over the obstruction. The centroidal radius of gyration of the wheel is k, and it is assumed that the wheel does not slip.

Problem 6/196

Problem 6/197

6/197 The uniform slender bar of mass m and length l has no angular velocity as end A strikes the ground against the stop with no rebound. If $\alpha = 15°$, what is the minimum magnitude of the initial velocity $\mathbf{v}_1$ for which the bar will rotate about A to the vertical position? *Ans.* $v_1 = 2.41\sqrt{gl}$

6/198 The bar of Prob. 6/197 has no angular velocity as end A strikes the ground with no rebound. For a given value of the speed v_1 just prior to impact, determine the minimum angle α for which the bar will rotate to the vertical position after impact.

Problem 6/199

▶ **6/199** The motor M drives disk A which turns disk B with no slipping. Disk A and its attached shaft and motor armature have a mass of 18 kg and a combined radius of gyration of 85 mm. Disk B has a mass of 5 kg and a radius of gyration of 140 mm. The motor housing and attached arm C have a combined mass of 24 kg and a radius of gyration about the axis O-O of 450 mm. Before the motor is turned on, the entire assembly is rotating as a unit about O-O with a rotational speed $\omega_0 = 30$ rev/min in the direction shown. The motor M has an operating speed of 1720 rev/min in the direction shown as measured with C fixed. Determine the new rotational speed of arm C if the motor is turned on.

Ans. $N = 26.2$ rev/min clockwise

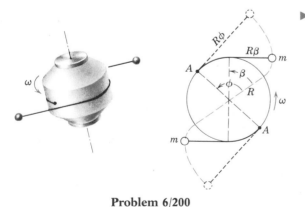

Problem 6/200

▶ **6/200** A mechanical device for reducing the spin of a satellite consists of two masses m attached to wires or thin metal tapes which unwind when released and acquire additional angular momentum at the expense of the angular momentum of the satellite. The masses are then released when unwound. Each tape has a length $R\phi$ and is fastened to its point A. If the satellite is spinning at an initial speed ω_0 when the masses are released with $\beta = 0$, determine the rate $\dot{\beta}$ of unwinding and find the new angular velocity ω_ϕ of the satellite when $\beta = \phi$. The moment of inertia of the satellite exclusive of the m's is I. (*Hint:* Equate the angular momentum and the kinetic energy of the system to their respective initial values.)

Ans. $\dot{\beta} = \omega_0 = $ constant

$$\omega_\phi = \frac{I + 2mR^2(1 - \phi^2)}{I + 2mR^2(1 + \phi^2)}\,\omega_0$$

6/9 *PROBLEM FORMULATION AND REVIEW*

In Chapter 6 we have made use of essentially all of the elements of dynamics studied so far. We have found that a knowledge of kinematics, using both absolute- and relative-motion analysis, is an essential part of the solution to problems in rigid-body kinetics. Our approach in Chapter 6 paralleled Chapter 3 where we developed the kinetics of particles using force-mass-acceleration, work-energy, and impulse-momentum methods.

The following outline will help to summarize the most important considerations in solving rigid-body problems in plane motion.

(*a*) *Identification of body or system.* It is essential that we make an unambiguous decision as to which body or system of bodies we are analyzing and then isolate the body or system selected by drawing its free-body diagram, kinetic diagram, or active-force diagram, whichever is appropriate.

(*b*) *Type of motion.* Next identify the category of motion as rectilinear translation, curvilinear translation, fixed-axis rotation, or general plane motion. Always see that the kinematics of the problem is properly described before attempting to solve the kinetic equations.

(*c*) *Coordinate system.* Choose an appropriate coordinate system. The geometry of the particular kinematics involved is usually the deciding factor. Designate the positive sense for moment and force summations and be consistent with the choice.

(*d*) *Principle and method.* If the instantaneous relationship between the applied forces and the acceleration is desired, then the equivalence between the forces and their $m\bar{a}$ and $\bar{I}\alpha$ resultants, as disclosed by the free-body and kinetic diagrams, will indicate the most direct approach to a solution.

When motion occurs over an interval of displacement, the work-energy approach is indicated, and we relate initial to final velocities without calculating the acceleration. We have seen the advantage of this approach for interconnected mechanical systems with negligible internal friction.

If the interval of motion is specified in terms of time rather than displacement, the impulse-momentum approach is indicated. When the angular motion of a rigid body is suddenly generated or changed, we use the principle of conservation of angular momentum.

(*e*) *Assumptions and approximations.* By now we should have acquired a feel for the practical significance of certain assumptions and approximations such as treating a rod as an ideal slender bar and neglecting friction where minimal. These and other idealizations are important to the process of obtaining solutions to real problems.

REVIEW PROBLEMS

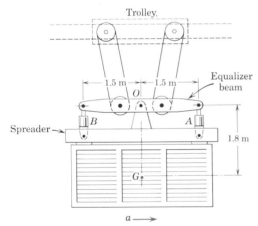

Problem 6/201

6/201 The dump truck carries 5 m³ of dirt with a density of 1600 kg/m³, and the elevating mechanism rotates the dump about the pivot *A* at a constant angular rate of 4 deg/s. The mass center of the dump and load is at *G*. Determine the maximum power *P* required during the tilting of the load.
Ans. P = 8.22 kW

6/202 An antiswing system for unloading cargo containers from ships consists of the moving trolley and its suspended equalizer beam and spreader, to which the loaded container is secured. Rotation of the spreader relative to the equalizer beam is controlled by the hydraulic cylinders *A* and *B* which have a piston diameter of 100 mm. During the horizontal acceleration of the system with a load of 4 Mg the oil pressure in cylinder *A* is 400 kPa and that in cylinder *B* is 500 kPa. Compute the acceleration *a* of the system. The center of mass of the load is at *G*. *Ans. a* = 0.1636 m/s²

Problem 6/202

6/203 The force *P* is applied to the homogeneous crate of mass *m*. If the coefficient of kinetic friction between the crate and the horizontal platform is μ_k, determine the limiting values of *h* so that the crate will slide without tipping about either the front edge or the rear edge.

Problem 6/203

6/204 The forklift truck with center of mass at G_1 has a mass of 1600 kg including the vertical mast. The fork and load have a combined mass of 900 kg with center of mass at G_2. The roller guide at *B* is capable of supporting horizontal force only, whereas the connection at *C*, in addition to supporting horizontal force, also transmits the vertical elevating force. If the fork is given an upward acceleration which is sufficient to reduce the force under the rear wheels at *A* to zero, calculate the corresponding reaction at *B*. *Ans. B = 10.46 kN*

Problem 6/204

6/205 A type of speed governor for a flywheel consists of the "spider" *A* which is able to undergo an angular deflection δ from its equilibrium position with respect to the flywheel during periods of angular acceleration. This relative movement is used to activate the speed-control device. Specify the stiffness *k* of each of the two restraining springs so that the relative angular deflection δ is limited to 6° during a period of steady acceleration $\alpha = 4$ rad/s². The mass of the spider is 3 kg, and its radius of gyration about its center *O* is 90 mm.

Problem 6/205

6/206 A space telescope is shown in the figure. One of the reaction wheels of its attitude-control system is spinning as shown at 10 rad/s, and at this speed the friction in the wheel bearing causes an internal moment of 10^{-6} N·m. Both the wheel speed and the friction moment may be considered constant over a time span of several hours. If the mass moment of inertia of the entire spacecraft about the *x*-axis is $150(10^3)$ kg·m², determine how much time passes before the line of sight of the initially stationary spacecraft drifts by 1 arc-second, which is 1/3600 degree. All other elements are fixed relative to the spacecraft, and no torquing of the reaction wheel shown is performed to correct the attitude drift. Neglect external torques.

Ans. t = 1206 s

Problem 6/206

Problem 6/207

Problem 6/208

Problem 6/209

Problem 6/210

6/207 The angle bar of Prob. 6/1 is replaced by two identical 3-kg bars spot welded together at D and hinged to the plate at C. Find the magnitudes of the force D and the moment M supported by the weld at D if the vertical plate is given an acceleration $a = 4g$. Show that the moment is independent of the acceleration.

6/208 The uniform slender bar has a mass of 30 kg and is released from rest in the near-vertical position shown where the spring of stiffness 150 N/m is unstretched. Calculate the velocity with which end A strikes the horizontal surface.

Ans. $v_A = 3.70$ m/s

6/209 Each of the solid circular disk wheels has a mass of 2 kg, and the inner solid cylinder has a mass of 3 kg. The disks and cylinder are mounted on the small central shaft so that each can rotate independently of the other with negligible friction in the bearings. Calculate the acceleration of the center of the wheels when the 20-N force is applied as shown. The coefficients of friction between the wheels and the horizontal surface are $\mu_k = 0.30$ and $\mu_s = 0.40$.

6/210 A couple $M = 12$ N·m is applied at C to the spring toggle mechanism which is released from rest in the position $\theta = 45°$. In this position the spring, which has a stiffness of 140 N/m, is stretched 150 mm. Bar AB has a mass of 3 kg and BC a mass of 6 kg. Calculate the angular velocity ω of BC as it crosses the position $\theta = 0$. Motion is in the vertical plane, and friction is negligible.

Ans. $\omega = 4.15$ rad/s

6/211 The dump truck of Prob. 6/102 is shown again here. The loaded container has a mass of 8 Mg and may be modeled as a solid homogeneous block. The hydraulic cylinder applies a force of 300 kN to the container in the starting position shown. Calculate the initial angular acceleration α of the container (*a*) if the truck is not moving and (*b*) if the truck has a forward acceleration of 3 m/s². *Ans.* (*a*) $\alpha = 0.504$ rad/s², (*b*) $\alpha = 1.166$ rad/s²

1.5 m   1.5 m
0.9 m
0.9 m
0.5 m
1 m   1 m

Problem 6/211

6/212 The uniform 30-kg pole *AC* is balanced in the vertical position when a horizontal force of 240 N is suddenly applied to the rope at *B*. In the absence of friction between the pole and the horizontal surface determine the initial acceleration of end *C*.

C
3 m
240 N
B
1 m
A

Problem 6/212

6/213 In Prob. 6/212 if the coefficients of static and kinetic friction between the vertical pole and the horizontal surface are both 0.30, calculate the initial acceleration of end *C*. *Ans.* $a_C = 1.89$ m/s² to the left

6/214 In an acrobatic stunt, man *A* of mass m_A drops from a raised platform onto the end of the light but strong beam with a velocity v_0. The boy of mass m_B is propelled upward with a velocity v_B. For a given ratio $n = m_B/m_A$ determine *b* in terms of *L* which will maximize the upward velocity of the boy. Assume that both man and boy act as rigid bodies.

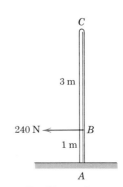

A
v_0
B
O
b
L

Problem 6/214

6/215 A uniform slender bar of mass *M* and length *L* is translating on the smooth horizontal *x-y* plane with a velocity v_M when a particle of mass *m* traveling with a velocity v_m as shown strikes and becomes embedded in the bar. Determine the final linear velocity of the mass center of the system and its angular velocity.

$\frac{3L}{4}$ $\frac{L}{4}$
M
y
v_M
• m
x
v_m

Problem 6/215

Problem 6/216

6/216 The car shown has a mass of 1500 kg with center of mass at *G*. Each of its four 620-mm diameter wheels has a mass of 32 kg and a radius of gyration of 210 mm. If the engine transmits a torque of 2100 N·m to the two rear wheels combined, calculate the resulting acceleration *a* of the car and find the least value of the coefficient of static friction between the wheels and the road which will enable the car to accelerate without slipping its rear wheels. Also find the friction force under the front wheels.

Ans. a = 4.35 m/s², F_B = 127.6 N, μ_s = 0.772

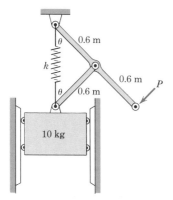

Problem 6/217

6/217 The elements of a hand-operated toggle press are shown. With no force on the handle the spring, whose unstretched length is 0.6 m, maintains equilibrium at $\theta = 45°$. Without dismembering the system, determine the downward acceleration of the 10-kg mass resulting from the sudden application of (*a*) a force *P* = 100 N perpendicular to the handle and (*b*) a clockwise couple *M* = 120 N·m applied to the handle at its end.

Problem 6/218

6/218 The link *OA* and pivoted circular disk are released from rest in the position shown and swing in the vertical plane about the fixed bearing at *O*. The 6-kg link *OA* has a radius of gyration about *O* of 375 mm. The disk has a mass of 8 kg. The two bearings are assumed to be frictionless. Find the force F_O exerted at *O* on the link (*a*) just after release and (*b*) as *OA* swings through the vertical position *OA'*.

Ans. (*a*) F_O = 32.8 N, (*b*) F_O = 347 N

Problem 6/219

6/219 The light circular hoop of radius *r* carries a heavy uniform band of mass *m* around half of its circumference and is released from rest on the incline in the upper position shown. After the hoop has rolled one-half of a revolution, (*a*) determine its angular velocity ω and (*b*) find the normal force *N* under the hoop if $\theta = 10°$.

Ans. (*a*) $\omega = \sqrt{\dfrac{g}{r} \dfrac{4 \cos \theta + \pi^2 \sin \theta}{\pi - 2}}$

(*b*) *N* = 4.14*mg*

6/220 The small block of mass m slides in the smooth radial slot of the disk, which turns freely in its bearing. If the block is displaced slightly from the center position when the angular velocity of the disk is ω_0, determine its radial velocity v_r as a function of the radial distance r. The mass moment of inertia of the disk about its axis of rotation is I_0.

Problem 6/220

6/221 Each of the solid square blocks is allowed to fall by rotating clockwise from the rest positions shown. The support at O in case (a) is a hinge and in case (b) is a small roller. Determine the angular velocity ω of each block as edge OC becomes horizontal just before striking the supporting surface.

Ans. $(a)\ \omega = 4.94$ rad/s, $(b)\ \omega = 6.25$ rad/s

Problem 6/221

6/222 The mechanical flyball governor operates with a vertical shaft O-O. As the shaft speed N is increased, the rotational radius of the two 1.5-kg balls tends to increase, and the 9-kg mass A is lifted up by the collar B. Determine the steady-state value of β for a rotational speed of 150 rev/min. Neglect the mass of the arms and collar.

Ans. $\beta = 22.5°$

Problem 6/222

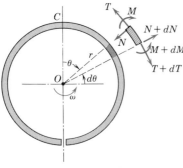

Problem 6/223

▶ **6/223** The split ring of radius *r* is rotating about a vertical axis through its center *O* with a constant angular velocity ω. Use a differential element of the ring and derive expressions for the shear force *N* and rim tension *T* in the ring in terms of the angle θ. Determine the bending moment M_C at point *C* by using one-half of the ring as a free body. The mass of the ring per unit length of rim is ρ.

$$Ans. \quad N = \rho r^2 \omega^2 \sin \theta$$
$$T = \rho r^2 \omega^2 (1 + \cos \theta)$$
$$M_C = 2\rho r^3 \omega^2$$

Problem 6/224

▶ **6/224** The two slender bars each having a mass of 4 kg are hinged at *B* and pivoted at *C*. If a horizontal impulse $\int F\, dt = 14\ \text{N} \cdot \text{s}$ is applied to the end *A* of the lower bar during an interval of 0.1 s during which the bars are still essentially in their vertical rest positions, compute the angular velocity ω_2 of the upper bar immediately after the impulse.

$$Ans. \quad \omega_2 = 2.50\ \text{rad/s}$$

Problem 6/225

▶ **6/225** Before it hits the ground a falling chimney, such as the one shown, will usually crack at the point where the bending moment is greatest. Show that the position of maximum moment occurs at the center of percussion relative to the upper end for a slender chimney of constant cross section. Neglect any restraining moment at the base.

Computer-oriented problems

*6/226 For the slotted disk and small slider of Prob. 6/220, the system properties are $\omega_0 = 2$ rad/s, $m = 4$ kg, and $r_d = 0.3$ m. If the radial velocity of the slider is to be limited to 0.25 m/s as it reaches the edge of the disk, determine the maximum permissible moment of inertia I_O of the disk. Plot the variation of the radial velocity v_r with the radial displacement for this value of I_O.

$$\text{Ans. } I_O = 0.0756 \text{ kg·m}^2, \; v_r = 2r\sqrt{\frac{0.0756}{0.0756 + 4r^2}}$$

*6/227 The homogeneous square block of mass m is released from rest at θ essentially zero and pivots at the midpoint of its base about the fixed corner at O. Determine and plot the normal and tangential forces, expressed in dimensionless form N/mg and F/mg, exerted on the block by the corner as functions of θ. (a) If a small notch at O prevents the block from slipping, determine the angle θ at which contact with the corner ceases. (b) In the absence of a notch and with a coefficient of static friction of 0.8, determine the angle θ at which slipping first occurs.   *Ans. (a) $\theta = 56.9°$, (b) $\theta = 45.1°$*

Problem 6/227

*6/228 The crate slides down the incline with velocity v_1 and its corner strikes a small obstacle at A. Determine the minimum required velocity v_1 if the crate is to rotate about A so that it travels on the conveyor belt on its side as indicated in the figure. Plot the variation of v_1 with α for $0 \leq \alpha \leq 45°$.
$$\text{Ans. } v_1 = 4.71\sqrt{1 - \cos(45° - \alpha)} \text{ m/s}$$

Problem 6/228

*6/229 The 30-kg slender bar has an initial angular velocity $\omega_0 = 4$ rad/s in the vertical position where the spring is unstretched. Determine the minimum angular velocity ω_{min} reached by the bar and the corresponding angle θ. Also find the angular velocity of the bar as it strikes the horizontal surface.

$$\text{Ans. } \omega_{min} = 0.910 \text{ rad/s at } \theta = 74.0°$$
$$\omega = 1.59 \text{ rad/s at } \theta = 90°$$

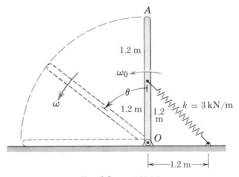

Problem 6/229

INTRODUCTION TO THREE-DIMENSIONAL DYNAMICS OF RIGID BODIES

7

7/1 INTRODUCTION

Although a large percentage of dynamics problems in engineering lend themselves to solution by means of the principles of plane motion, modern developments have focused increasing attention upon problems which call for the analysis of motion in three dimensions. Inclusion of the third dimension adds considerable complexity to the kinematic and kinetic relationships. Not only does the added dimension introduce a third component to vectors which represent force, linear velocity, linear acceleration, and linear momentum, but the introduction of the third dimension adds the possibility of two additional components for vectors representing angular quantities including moments of forces, angular velocity, angular acceleration, and angular momentum. It is in three-dimensional motion that the full power of vector analysis is utilized.

A good background in the dynamics of plane motion is extremely useful in the study of three-dimensional dynamics, as the approach to problems and many of the terms are the same as or analogous to those in two dimensions. If the study of three-dimensional dynamics is undertaken without the benefit of prior study of plane-motion dynamics, more time will be required to master the principles and to become familiar with the approach to problems.

The treatment presented in Chapter 7 is not intended as a complete development of the three-dimensional motion of rigid bodies but merely as a basic introduction to the subject. This introduction should, however, be sufficient to solve many of the more common problems in three-dimensional motion and also to lay the foundation for more advanced study. We shall proceed as we did for particle motion and for rigid-body plane motion by first examining the necessary kinematics and then proceeding to the kinetics.

SECTION A. KINEMATICS

7/2 TRANSLATION

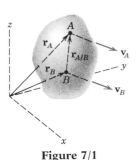

Figure 7/1

Figure 7/1 shows a rigid body translating in three-dimensional space. Any two points in the body, such as *A* and *B*, will move along parallel straight lines if the motion is one of *rectilinear translation* or will move along congruent curves if the motion is one of *curvilinear translation*. In either case every line in the body, such as *AB*, remains parallel to its original position. The position vectors and their first and second time derivatives are

$$\mathbf{r}_A = \mathbf{r}_B + \mathbf{r}_{A/B} \qquad \mathbf{v}_A = \mathbf{v}_B \qquad \mathbf{a}_A = \mathbf{a}_B$$

where $\mathbf{r}_{A/B}$ remains constant and therefore has no time derivative. Thus all points in the body have the same velocity and the same acceleration. The kinematics of translation presents no special difficulty, and further elaboration is unnecessary.

7/3 FIXED-AXIS ROTATION

Consider now the *rotation* of a rigid body about a fixed axis *n-n* in space with an angular velocity ω, as shown in Fig. 7/2. The angular velocity is a vector in the direction of the rotation axis with a sense established by the familiar right-hand rule. For fixed-axis rotation ω does not change its direction since it lies along the axis. We choose the origin *O* of the fixed coordinate system on the rotation axis for convenience. Any point such as *A* that is not on the axis moves in a circular arc in a plane normal to the axis and has a velocity

$$\boxed{\mathbf{v} = \omega \times \mathbf{r}} \tag{7/1}$$

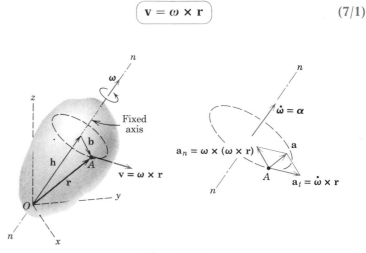

Figure 7/2

which may be seen by replacing **r** by **h** + **b** and noting that $\omega \times \mathbf{h} = \mathbf{0}$. The acceleration of A is given by the time derivative of Eq. 7/1. Thus,

$$\mathbf{a} = \dot{\omega} \times \mathbf{r} + \omega \times (\omega \times \mathbf{r}) \tag{7/2}$$

where $\dot{\mathbf{r}}$ has been replaced by its equal, $\mathbf{v} = \omega \times \mathbf{r}$. The normal and tangential components of **a** for the circular motion have the familiar magnitudes $a_n = |\omega \times (\omega \times \mathbf{r})| = b\omega^2$ and $a_t = |\dot{\omega} \times \mathbf{r}| = b\alpha$, where $\alpha = \dot{\omega}$. Inasmuch as both **v** and **a** are perpendicular to ω and $\dot{\omega}$, it follows that $\mathbf{v} \cdot \omega = 0$, $\mathbf{v} \cdot \dot{\omega} = 0$, $\mathbf{a} \cdot \omega = 0$, and $\mathbf{a} \cdot \dot{\omega} = 0$ for fixed-axis rotation.

7/4 PARALLEL-PLANE MOTION

When all points in a rigid body move in planes which are parallel to a fixed plane P, Fig. 7/3, we have a general form of plane motion. The reference plane is customarily taken through the mass center G and is referred to as the plane of motion. Since each point in the body, such as A', has a motion identical with the motion of point A in plane P, it follows that the kinematics of plane motion covered in Chapter 5 provides a complete description of the motion when applied to the reference plane.

7/5 ROTATION ABOUT A FIXED POINT

When a body rotates about a fixed point, the angular velocity vector no longer remains fixed in direction, and this change calls for a more general concept of rotation.

We must first examine the conditions under which rotation vectors obey the parallelogram law of addition and may, therefore, be treated as proper vectors. Consider a solid sphere, Fig. 7/4, which is cut from a rigid body confined to rotate about the fixed point O.

Figure 7/3

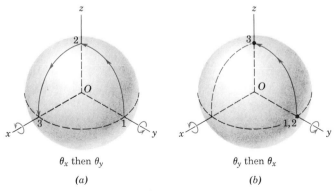

θ_x then θ_y　　　　　　θ_y then θ_x

(a)　　　　　　　　　　(b)

Figure 7/4

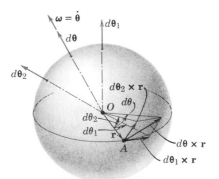

Figure 7/5

The x-y-z axes here are taken fixed in space and do not rotate with the body. In the a-part of the figure two successive 90° rotations of the sphere about, first, the x-axis and, second, the y-axis result in the motion of a point which is initially on the y-axis in position 1, to positions 2 and 3, successively. On the other hand, if the order of the rotations is reversed, the point suffers no motion during the y-rotation but moves to point 3 during the 90° rotation about the x-axis. Thus the two cases do not yield the same final position, and it is evident from this one special example that finite rotations do not generally obey the parallelogram law of vector addition and are not commutative. Thus finite rotations may *not* be treated as proper vectors.

Infinitesimal rotations, however, do obey the parallelogram law of vector addition. This fact is shown in Fig. 7/5 which represents the combined effect of two infinitesimal rotations $d\boldsymbol{\theta}_1$ and $d\boldsymbol{\theta}_2$ of a rigid body about the respective axes through the fixed point O. As a result of $d\boldsymbol{\theta}_1$ point A has a displacement $d\boldsymbol{\theta}_1 \times \mathbf{r}$, and likewise $d\boldsymbol{\theta}_2$ causes a displacement $d\boldsymbol{\theta}_2 \times \mathbf{r}$ of point A. Either order of addition of these infinitesimal displacements clearly produces the same resultant displacement, which is $d\boldsymbol{\theta}_1 \times \mathbf{r} + d\boldsymbol{\theta}_2 \times \mathbf{r} = (d\boldsymbol{\theta}_1 + d\boldsymbol{\theta}_2) \times \mathbf{r}$. Hence the two rotations are equivalent to the single rotation $d\boldsymbol{\theta} = d\boldsymbol{\theta}_1 + d\boldsymbol{\theta}_2$. It follows that the angular velocities $\boldsymbol{\omega}_1 = \dot{\boldsymbol{\theta}}_1$ and $\boldsymbol{\omega}_2 = \dot{\boldsymbol{\theta}}_2$ may be added vectorially to give $\boldsymbol{\omega} = \dot{\boldsymbol{\theta}} = \boldsymbol{\omega}_1 + \boldsymbol{\omega}_2$. We conclude, therefore, that at any instant of time a body with one fixed point is rotating instantaneously about a particular axis passing through the fixed point.

To aid in visualizing the concept of the instantaneous axis of rotation we will cite a specific example. Figure 7/6 represents a solid cylindrical rotor made of clear plastic containing many black particles embedded in the plastic. The rotor is spinning about its shaft axis at the steady rate ω_1, and its shaft, in turn, is rotating about the fixed vertical axis at the steady rate ω_2, with rotations in the directions indicated. If the rotor is photographed at a cer-

Figure 7/6

tain instant during its motion, the resulting picture would show one line of black dots in sharp focus indicating that, momentarily, their velocity was zero. This line of points with no velocity establishes the instantaneous position of the axis of rotation *O-n*. Any dot on this line, such as *A*, would have equal and opposite velocity components, v_1 due to ω_1 and v_2 due to ω_2. All other dots, such as the one at *P*, would appear out of focus, and their movements would show as short blurred streaks in the form of small circular arcs in planes normal to the axis *O-n*. Thus all particles of the body, except those on line *O-n*, are momentarily rotating in circular arcs about the instantaneous axis of rotation. If a succession of photographs were taken, we would observe that the rotation axis would be defined by a new series of dots in focus and that the axis would change position both in space and relative to the body. For rotation of a rigid body about a fixed point, then, it is seen that the rotation axis is in general not a line fixed in the body.

(*a*) *Body and space cones.* Relative to the plastic cylinder of Fig. 7/6 the instantaneous axis of rotation *O-A-n* generates a right-circular cone about the cylinder axis called the *body cone*. As the two rotations continue and the cylinder swings around the vertical axis, the instantaneous axis of rotation also generates a right-circular cone about the vertical axis called the *space cone*. These cones are shown in Fig. 7/7 for this particular example, and we see that the body cone rolls on the space cone. For a more general case where the rotations are not steady, the space and body cones will not be right-circular cones, Fig. 7/8, but the body cone still rolls on the space cone.

Figure 7/7

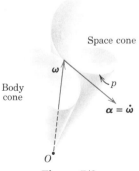

Figure 7/8

(*b*) *Angular acceleration.* The angular acceleration $\boldsymbol{\alpha}$ of a rigid body in three-dimensional motion is the time derivative of its angular velocity, $\boldsymbol{\alpha} = \dot{\boldsymbol{\omega}}$. In contrast to the case of rotation in a single plane where the scalar α measures only the change in magnitude of the angular velocity, in three-dimensional motion the vector $\boldsymbol{\alpha}$ reflects the change in direction of $\boldsymbol{\omega}$ as well as its change in magnitude. Thus in Fig. 7/8 where the tip of the angular velocity vector ω follows the space curve p and changes both in magnitude and direction, the angular acceleration α becomes a vector tangent to this curve in the direction of the change in ω.

When the magnitude of ω remains constant, the angular acceleration α becomes normal to ω. For this case if we let Ω stand for the angular velocity with which the vector ω itself rotates (*precesses*) as it forms the space cone, the angular acceleration may be written

$$\alpha = \Omega \times \omega \tag{7/3}$$

This relation is easily seen from Fig. 7/9 where the vectors α, ω, and Ω in the lower figure bear exactly the same relationship to each other as do the vectors $\mathbf{v}$, $\mathbf{r}$, and ω in the upper figure for relating the velocity of a point A on a rigid body to its position vector from O and the angular velocity of the body.

If we use Fig. 7/2 to represent a rigid body rotating about a fixed point O with the instantaneous axis of rotation n-n, we see that the velocity $\mathbf{v}$ and acceleration $\mathbf{a} = \dot{\mathbf{v}}$ of any point A in the body are given by the same expressions as apply to the case where the axis is fixed, namely,

$$\mathbf{v} = \omega \times \mathbf{r} \tag{7/1}$$
$$\mathbf{a} = \dot{\omega} \times \mathbf{r} + \omega \times (\omega \times \mathbf{r}) \tag{7/2}$$

The one difference between the case of rotation about a fixed axis and rotation about a fixed point lies in the fact that for rotation about a fixed point the angular acceleration $\alpha = \dot{\omega}$ will have a component normal to ω due to the change in direction of ω as well as a component in the direction of ω to reflect any change in the magnitude of ω. Although any point on the rotation axis n-n momentarily will have zero velocity, it will *not* have zero acceleration as long as ω is changing its direction. On the other hand, for rotation about a fixed axis $\alpha = \dot{\omega}$ has only the one component along the fixed axis to reflect the change in the magnitude of ω. Furthermore, points which lie on the fixed rotation axis clearly have no velocity or acceleration.

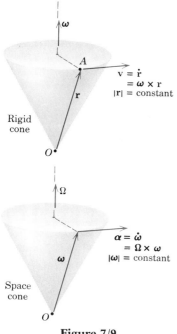

ω

A

$v = \dot{r}$
$= \omega \times r$
$|r| = $ constant

r

Rigid
cone

O

Ω

$\alpha = \dot{\omega}$
$= \Omega \times \omega$
$|\omega| = $ constant

ω

Space
cone

O

Figure 7/9

Sample Problem 7/1

The 0.8-m arm *OA* for a remote-control mechanism is pivoted about the horizontal *x*-axis of the clevis, and the entire assembly rotates about the *z*-axis with a constant speed $N = 60$ rev/min. Simultaneously the arm is being raised at the constant rate $\dot{\beta} = 4$ rad/s. For the position where $\beta = 30°$ determine (*a*) the angular velocity of *OA*, (*b*) the angular acceleration of *OA*, (*c*) the velocity of point *A*, and (*d*) the acceleration of point *A*.

Solution. (*a*) Since the arm *OA* is rotating about both the *x*- and the *z*-axes, it has the components $\omega_x = \dot{\beta} = 4$ rad/s and $\omega_z = 2\pi N/60 = 2\pi(60)/60 = 6.283$ rad/s. The angular velocity is

$$\omega = \omega_x + \omega_z = 4\mathbf{i} + 6.283\mathbf{k} \text{ rad/s} \qquad Ans.$$

(*b*) The angular acceleration of *OA* is

$$\alpha = \dot{\omega} = \dot{\omega}_x + \dot{\omega}_z$$

Since ω_z is not changing in magnitude or direction, $\dot{\omega}_z = \mathbf{0}$. But ω_x is changing direction and thus has a derivative which, from Eq. 7/3, is

$$\dot{\omega}_x = \omega_z \times \omega_x = 6.283\mathbf{k} \times 4\mathbf{i} = 25.13\mathbf{j} \text{ rad/s}^2$$

① Therefore

$$\alpha = 25.13\mathbf{j} + \mathbf{0} = 25.13\mathbf{j} \text{ rad/s}^2 \qquad Ans.$$

(*c*) With the position vector of *A* given by $\mathbf{r} = 0.693\mathbf{j} + 0.4\mathbf{k}$ m, the velocity of *A* from Eq. 7/1 becomes

$$\mathbf{v} = \omega \times \mathbf{r} = \begin{vmatrix} \mathbf{i} & \mathbf{j} & \mathbf{k} \\ 4 & 0 & 6.283 \\ 0 & 0.693 & 0.4 \end{vmatrix} = -4.35\mathbf{i} - 1.60\mathbf{j} + 2.77\mathbf{k} \text{ m/s} \quad Ans.$$

(*d*) The acceleration of *A* from Eq. 7/2 is

$$\mathbf{a} = \dot{\omega} \times \mathbf{r} + \omega \times (\omega \times \mathbf{r})$$
$$= \alpha \times \mathbf{r} + \omega \times \mathbf{v}$$
$$= \begin{vmatrix} \mathbf{i} & \mathbf{j} & \mathbf{k} \\ 0 & 25.13 & 0 \\ 0 & 0.693 & 0.4 \end{vmatrix} + \begin{vmatrix} \mathbf{i} & \mathbf{j} & \mathbf{k} \\ 4 & 0 & 6.283 \\ -4.35 & -1.60 & 2.77 \end{vmatrix}$$
$$= (10.05\mathbf{i}) + (10.05\mathbf{i} - 38.44\mathbf{j} - 6.40\mathbf{k})$$
② $$= 20.11\mathbf{i} - 38.44\mathbf{j} - 6.40\mathbf{k} \text{ m/s}^2 \qquad Ans.$$

① Alternatively consider axes *x-y-z* to be attached to the vertical shaft and clevis so that they rotate. The derivative of ω_x becomes $\dot{\omega}_x = 4\dot{\mathbf{i}}$. But from Eq. 5/11 we have $\dot{\mathbf{i}} = \omega_z \times \mathbf{i} = 6.283\mathbf{k} \times \mathbf{i} = 6.283\mathbf{j}$. Thus $\alpha = \dot{\omega}_x = 4(6.283)\mathbf{j} = 25.13\mathbf{j}$ rad/s² as before.

② To compare methods it is suggested that these results for **v** and **a** be obtained by applying Eqs. 2/18 and 2/19 for particle motion in spherical coordinates, changing symbols as necessary.

Sample Problem 7/2

The electric motor with attached disk is running at a constant low speed of 120 rev/min in the direction shown. Its housing and mounting base are initially at rest. The entire assembly is next set in rotation about the vertical Z-axis at the constant rate $N = 60$ rev/min with a fixed angle γ of 30°. Determine (a) the angular velocity and angular acceleration of the disk, (b) the space and body cones, and (c) the velocity and acceleration of point A at the top of the disk for the instant shown.

$\overline{OC} = 250$ mm
$\overline{CA} = 125$ mm

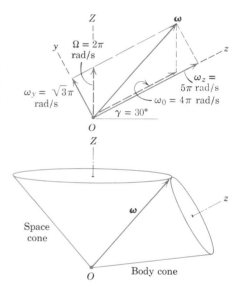

Space cone

Body cone

Solution. The axes x-y-z with unit vectors $\mathbf{i}$, $\mathbf{j}$, $\mathbf{k}$ are attached to the motor frame with the z-axis coinciding with the rotor axis and the x-axis coinciding with the horizontal axis through O about which the motor tilts. The Z-axis is vertical and carries the unit vector $\mathbf{K} = \mathbf{j} \cos \gamma + \mathbf{k} \sin \gamma$.

(a) The rotor and disk have two components of angular velocity, $\omega_0 = 120(2\pi)/60 = 4\pi$ rad/s about the z-axis and $\Omega = 60(2\pi)/60 = 2\pi$ rad/s about the Z-axis. Thus the angular velocity becomes

(1) $\qquad \boldsymbol{\omega} = \boldsymbol{\omega}_0 + \boldsymbol{\Omega} = \omega_0 \mathbf{k} + \Omega \mathbf{K}$

$\qquad = \omega_0 \mathbf{k} + \Omega(\mathbf{j} \cos \gamma + \mathbf{k} \sin \gamma) = (\Omega \cos \gamma)\mathbf{j} + (\omega_0 + \Omega \sin \gamma)\mathbf{k}$

$\qquad = (2\pi \cos 30°)\mathbf{j} + (4\pi + 2\pi \sin 30°)\mathbf{k} = \pi(\sqrt{3}\mathbf{j} + 5.0\mathbf{k})$ rad/s

$\qquad\qquad\qquad\qquad\qquad\qquad\qquad\qquad\qquad\qquad$ *Ans.*

The angular acceleration of the motor and disk from Eq. 7/3 is

(2) $\qquad \boldsymbol{\alpha} = \dot{\boldsymbol{\omega}} = \boldsymbol{\Omega} \times \boldsymbol{\omega}$

$\qquad = \Omega(\mathbf{j} \cos \gamma + \mathbf{k} \sin \gamma) \times [(\Omega \cos \gamma)\mathbf{j} + (\omega_0 + \Omega \sin \gamma)\mathbf{k}]$

$\qquad = \Omega(\omega_0 \cos \gamma + \Omega \sin \gamma \cos \gamma)\mathbf{i} - (\Omega^2 \sin \gamma \cos \gamma)\mathbf{i}$

(3) $\qquad = (\Omega \omega_0 \cos \gamma)\mathbf{i} = \mathbf{i}(2\pi)(4\pi) \cos 30° = 68.4\mathbf{i}$ rad/s² $\qquad$ *Ans.*

(b) The angular velocity vector $\boldsymbol{\omega}$ is the common element of the space and body cones which may now be constructed as shown.

(c) The position vector of point A for the instant considered is

$\qquad\qquad \mathbf{r} = 0.125\mathbf{j} + 0.250\mathbf{k}$ m

From Eq. 7/1 the velocity of A is

$$\mathbf{v} = \boldsymbol{\omega} \times \mathbf{r} = \begin{vmatrix} \mathbf{i} & \mathbf{j} & \mathbf{k} \\ 0 & \sqrt{3}\pi & 5\pi \\ 0 & 0.125 & 0.250 \end{vmatrix} = -0.1920\pi\mathbf{i} \text{ m/s} \qquad \textit{Ans.}$$

From Eq. 7/2 the acceleration of point A is

$\qquad \mathbf{a} = \dot{\boldsymbol{\omega}} \times \mathbf{r} + \boldsymbol{\omega} \times (\boldsymbol{\omega} \times \mathbf{r}) = \boldsymbol{\alpha} \times \mathbf{r} + \boldsymbol{\omega} \times \mathbf{v}$

$\qquad = 68.4\mathbf{i} \times (0.125\mathbf{j} + 0.250\mathbf{k}) + \pi(\sqrt{3}\mathbf{j} + 5\mathbf{k}) \times (-0.1920\pi\mathbf{i})$

$\qquad = -26.6\mathbf{j} + 11.83\mathbf{k}$ m/s² $\qquad\qquad\qquad$ *Ans.*

(1) Note that $\boldsymbol{\omega}_0 + \boldsymbol{\Omega} = \boldsymbol{\omega} = \boldsymbol{\omega}_y + \boldsymbol{\omega}_z$ as shown on the vector diagram.

(2) Remember that Eq. 7/3 gives the complete expression for $\boldsymbol{\alpha}$ only for steady precession where $|\boldsymbol{\omega}|$ is constant, which applies to this problem.

(3) Since the magnitude of $\boldsymbol{\omega}$ is constant, $\boldsymbol{\alpha}$ must be tangent to the base circle of the space cone which puts it in the plus x-direction in agreement with our calculated conclusion.

PROBLEMS

7/1 The cylinder rotates about the fixed axis *OA* with an angular velocity of 1200 rev/min. At a certain instant point *P* on the cylinder has a velocity whose *x*- and *z*-components are 3.86 m/s and −4.20 m/s, respectively. Find the magnitude *v* of the velocity of *P* and the radius *r* from the rotation axis to point *P*.   *Ans.* $v = 5.90$ m/s, $r = 47.0$ mm

Problem 7/1

7/2 A rigid body rotates about a fixed axis with an angular acceleration $\alpha = 2\mathbf{i} + \mathbf{j} - 3\mathbf{k}$ rad/s². At a certain instant the tangential acceleration of a point *A* on the body has *y*- and *z*-components equal to −2 m/s² and 1.5 m/s², respectively. Determine the magnitude of the tangential acceleration of *A* and the distance *r* from *A* to the rotation axis.
Ans. $a_t = 4.10$ m/s², $r = 1.096$ m

7/3 A timing mechanism consists of the rotating distributor arm *AB* and the fixed contact *C*. If the arm rotates about the fixed axis *OA* with a constant angular velocity $\omega = 3(3\mathbf{i} + 2\mathbf{j} + 6\mathbf{k})$ rad/s, and if the coordinates of the contact *C* expressed in millimeters are (75, 75, 150), determine the magnitude of the acceleration of the tip *B* of the distributor arm as it passes point *C*.

Problem 7/3

7/4 A rigid body rotates about a fixed axis with a constant angular velocity of 14 rad/s. A point *O* on the rotation axis is selected as the origin of coordinates, and a point *B* on the rotation axis has *x-y-z* coordinates of 0.2, 0.3, 0.6 meters. Determine the magnitudes of the velocity and acceleration of a point *A* on the body whose *x-y-z* coordinates are 200, 200, 100 millimeters.
Ans. $v = 2.72$ m/s, $a = 38.1$ m/s²

7/5 The disk rotates with a spin velocity of 15 rad/s about its horizontal *z*-axis first in the direction (*a*) and second in the direction (*b*). The assembly rotates with the velocity $N = 10$ rad/s about the vertical axis. Construct the space and body cones for each case.

Problem 7/5

Problem 7/6

7/6 The vertical shaft and attached clevis rotate about the z-axis at the constant rate $\Omega = 4$ rad/s. Simultaneously the shaft B revolves about its axis OA at the constant rate $\omega_0 = 3$ rad/s, and the angle γ is decreasing at the constant rate of $\pi/4$ rad/s. Determine the angular velocity ω and the magnitude of the angular acceleration α of shaft B when $\gamma = 30°$. The x-y-z axes are attached to the clevis and rotate with it.

Ans. $\omega = -0.785\mathbf{i} - 2.598\mathbf{j} + 2.5\mathbf{k}$ rad/s
$\alpha = 11.44$ rad/s^2

$\overline{OC} = 250$ mm
$\overline{CA} = 125$ mm

Problem 7/7

7/7 The electric motor and attached disk of Sample Problem 7/2 are shown again here. If the motor reaches a speed of 3000 rev/min in 2 seconds from rest with constant acceleration, determine the total angular acceleration of the rotor and disk $\frac{1}{3}$ second after it is turned on if the turntable is rotating at a constant rate $N = 30$ rev/min. The angle $\gamma = 30°$ is constant.

Ans. $\alpha = 50\pi \left(\dfrac{\pi}{2\sqrt{3}} \mathbf{i} + \mathbf{k} \right)$ rad/s^2

7/8 If the motor of Prob. 7/7 pivots about the x-axis at the constant rate $\dot{\gamma} = 3\pi$ rad/s with no rotation about the Z-axis ($N = 0$), determine the angular acceleration α of the rotor and disk as the position $\gamma = 30°$ is passed. The constant speed of the motor is 120 rev/min. Also find the velocity and acceleration of a point A, which is on the top of the disk for this position.

7/9 For the figure for Prob. 7/7 assume that the motor shaft and attached disk are rotating about the z-axis with a spin ω_0 of 120 rev/min which is increasing at the rate of 10 rad/s^2 at the position $\gamma = 30°$. At the same time $\dot{\gamma}$ is 12 rad/s and is increasing at the rate of 15 rad/s^2. Determine the vector expression for the angular acceleration α of the rotor at this instant. There is no rotation about the Z-axis.

Ans. $\alpha = -15\mathbf{i} + 48\pi\mathbf{j} + 10\mathbf{k}$ rad/s^2

7/10 The robot shown has five degrees of rotational freedom. The x-y-z axes are attached to the base ring, which rotates about the z-axis at the rate ω_1. The arm O_1O_2 rotates about the x-axis at the rate $\omega_2 = \dot{\theta}$. The control arm O_2A rotates about axis O_1-O_2 at the rate ω_3 and about a perpendicular axis through O_2 that is momentarily parallel to the x-axis at the rate $\omega_4 = \dot{\beta}$. Finally, the jaws rotate about axis O_2-A at the rate ω_5. The magnitudes of all angular rates are constant. For the configuration shown, determine the magnitude ω of the total angular velocity of the jaws for $\theta = 60°$ and $\beta = 45°$ if $\omega_1 = 2$ rad/s, $\dot{\theta} = 1.5$ rad/s, and $\omega_3 = \omega_4 = \omega_5 = 0$. Also express the angular acceleration $\boldsymbol{\alpha}$ of arm O_1O_2 as a vector.

$\textit{Ans.}\ \omega = 2.5$ rad/s, $\boldsymbol{\alpha} = 3\mathbf{j}$ rad/s^2

Problem 7/10

7/11 For the robot of Prob. 7/10 determine the angular velocity $\boldsymbol{\omega}$ of the jaws for $\theta = 60°$ and $\beta = 45°$ if $\omega_3 = 3$ rad/s, $\omega_5 = 2$ rad/s, and if $\omega_1 = \omega_2 = \omega_4 = 0$. Also find the angular acceleration $\boldsymbol{\alpha}$ of the jaws.

7/12 For the robot of Prob. 7/10 determine the total angular velocity $\boldsymbol{\omega}$ of the jaws for $\theta = 60°$ and $\beta = 45°$ if $\omega_1 = 2$ rad/s, $\omega_2 = 1.5$ rad/s, $\omega_4 = 3$ rad/s, and if $\omega_3 = \omega_5 = 0$. Also find the angular acceleration $\boldsymbol{\alpha}$ of the arm O_1O_2.

$\textit{Ans.}\ \boldsymbol{\omega} = -1.5\mathbf{i} + 2\mathbf{k}$ rad/s, $\boldsymbol{\alpha} = 3\mathbf{j}$ rad/s^2

7/13 For the robot of Prob. 7/10 determine the magnitude ω of the angular velocity of the jaws for $\theta = 60°$ and $\beta = 45°$ if $\omega_2 = 1.5$ rad/s, $\omega_3 = 2$ rad/s, $\omega_4 = 0.5$ rad/s, and if $\omega_1 = \omega_5 = 0$. Also find the angular acceleration $\boldsymbol{\alpha}$ of the jaws.

$\textit{Ans.}\ \omega = 2.24$ rad/s, $\boldsymbol{\alpha} = -3.46\mathbf{j} + 2\mathbf{k}$ rad/s^2

7/14 For the robot of Prob. 7/10 determine the magnitude ω of the angular velocity of the jaws for $\theta = 60°$ and $\beta = 45°$ if $\omega_1 = 0.5$ rad/s, $\omega_2 = 0.25$ rad/s, $\omega_3 = 1.5$ rad/s, $\omega_4 = 1.5$ rad/s, and $\omega_5 = 0.4$ rad/s.

$\textit{Ans.}\ \omega = 2.54$ rad/s

Problem 7/15

Problem 7/16

Problem 7/18

7/15 The revolving crane of Prob. 2/170 is shown again here. The boom *OP* has a length of 24 m, and the crane is revolving about the vertical axis at the constant rate of 2 rev/min in the direction shown. Simultaneously the boom is being lowered at the constant rate $\dot\beta = 0.10$ rad/s. Calculate the magnitudes of the velocity and acceleration of the end *P* of the boom for the instant when it passes the position $\beta = 30°$.

7/16 The wheel rolls without slipping in a circular arc of radius *R* and makes one complete turn about the vertical *y*-axis with constant speed in time τ. Determine the vector expression for the angular acceleration α of the wheel and construct the space and body cones.

$$Ans.\ \alpha = -\left(\frac{2\pi}{\tau}\right)^2 \frac{R}{r}\mathbf{i}$$

7/17 Determine the velocity **v** and acceleration **a** of point *A* on the wheel of Prob. 7/16 for the position shown where *A* crosses the horizontal line through the center of the wheel.

$$Ans.\ \mathbf{v} = \frac{2\pi R}{\tau}\left(\mathbf{i} - \mathbf{j} - \frac{r}{R}\mathbf{k}\right)$$
$$\mathbf{a} = -\left(\frac{2\pi}{\tau}\right)^2 R\left[\left(\frac{R}{r} + \frac{r}{R}\right)\mathbf{i} + \mathbf{k}\right]$$

7/18 The shaft *OA* of the bevel gear *B* rotates about the fixed axis *O-x* with a constant speed $N = 60$ rev/min in the direction shown. Gear *B* meshes with the bevel gear *C* along its pitch cone of semivertex angle $\gamma = \tan^{-1}\frac{3}{2}$ as shown. Determine the angular velocity ω and angular acceleration α of gear *B* if gear *C* is fixed and does not rotate. The *y*-axis revolves with the shaft *OA*.

$$Ans.\ \omega = \pi(-2\mathbf{i} + 3\mathbf{j}) \text{ rad/s},\ \alpha = -6\pi^2\mathbf{k} \text{ rad/s}^2$$

7/19 If gear *C* of Prob. 7/18 has a constant rotational velocity of 20 rev/min about the axis *O-x* in the same sense as *N* while *OA* maintains its constant rotational speed *N* = 60 rev/min, calculate the angular velocity *ω* and angular acceleration *α* of gear *B*.

▶ **7/20** The right-circular cone *A* rolls on the fixed right-circular cone *B* at a constant rate and makes one complete trip around *B* every 4 s. Compute the magnitude of the angular acceleration *α* of cone *A* during its motion. *Ans.* $\alpha = 6.32$ rad/s²

Problem 7/20

▶ **7/21** The solid right-circular cone of base radius *r* and height *h* rolls on a flat surface without slipping. The center *B* of the circular base moves in a circular path around the *z*-axis with a constant speed *v*. Determine the angular velocity *ω* and the angular acceleration *α* of the solid cone.

$$Ans.\ \omega = v\sqrt{\frac{1}{r^2} + \frac{1}{h^2}}\,\mathbf{i}$$

$$\alpha = -\frac{v^2}{h^2}\left(\frac{r}{h} + \frac{h}{r}\right)\mathbf{j}$$

Problem 7/21

7/6 GENERAL MOTION

The kinematic analysis of a rigid body which has general three-dimensional motion is best accomplished with the aid of our principles of relative motion. These principles have been applied to problems in plane motion and will now be extended to space motion. We will make use of both translating reference axes and rotating reference axes.

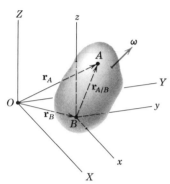

Figure 7/10

(a) Translating reference axes. In Fig. 7/10 is shown a rigid body which has an angular velocity ω. We choose any convenient point B as the origin of a translating reference system x-y-z. The velocity $\mathbf{v}$ and acceleration $\mathbf{a}$ of any other point A in the body is given by the relative-velocity and relative-acceleration expressions

$$\mathbf{v}_A = \mathbf{v}_B + \mathbf{v}_{A/B} \tag{5/4}$$

$$\mathbf{a}_A = \mathbf{a}_B + \mathbf{a}_{A/B} \tag{5/7}$$

which were developed in Arts. 5/4 and 5/6 for the plane motion of rigid bodies and which hold equally well for any plane in space in which the three vectors for each of the equations must lie.

In applying these relations to rigid-body motion in space we note from Fig. 7/10 that the distance $\overline{AB}$ remains constant. Thus, from an observer's position on x-y-z, the body appears to rotate about the point B and point A appears to lie on a spherical surface with B as the center. Consequently, we may view the general motion as a translation of the body with the motion of B plus a rotation of the body about B.

The relative motion terms represent the effect of the rotation about B and are identical to the velocity and acceleration expressions discussed in the previous article for rotation of a rigid body about a fixed point. Therefore the relative-velocity and relative-acceleration equations may be written

$$\begin{array}{l} \mathbf{v}_A = \mathbf{v}_B + \omega \times \mathbf{r}_{A/B} \\ \mathbf{a}_A = \mathbf{a}_B + \dot{\omega} \times \mathbf{r}_{A/B} + \omega \times (\omega \times \mathbf{r}_{A/B}) \end{array} \tag{7/4}$$

where ω is the instantaneous angular velocity of the body.

The selection of the reference point B is quite arbitrary in theory. In practice point B is chosen for convenience as some point in the body whose motion is known in whole or in part. If point A is chosen as the reference point, the relative motion equations become

$$\mathbf{v}_B = \mathbf{v}_A + \omega \times \mathbf{r}_{B/A}$$

$$\mathbf{a}_B = \mathbf{a}_A + \dot{\omega} \times \mathbf{r}_{B/A} + \omega \times (\omega \times \mathbf{r}_{B/A})$$

where $\mathbf{r}_{B/A} = -\mathbf{r}_{A/B}$. It should be clear that $\boldsymbol{\omega}$ and, hence, $\dot{\boldsymbol{\omega}}$ are the same vectors for either formulation since the absolute angular motion of the body is independent of the choice of reference point. When we come to the kinetic equations for general motion, we will see that the mass center of a body is frequently the most convenient reference point to choose.

If points A and B in Fig. 7/10 represent the ends of a rigid control link in a spatial mechanism where the end connections act as ball-and-socket joints (as in Sample Problem 7/3), it is necessary to impose certain kinematical requirements. Clearly any rotation of the link about its own axis AB does not affect the action of the link. Thus the angular velocity $\boldsymbol{\omega}_n$ whose vector is normal to the link describes its action. It is necessary, therefore, that $\boldsymbol{\omega}_n$ and $\mathbf{r}_{A/B}$ be at right angles which is satisfied if $\boldsymbol{\omega}_n \cdot \mathbf{r}_{A/B} = 0$. Similarly, it is only the component $\boldsymbol{\alpha}_n{}^*$ of the angular acceleration of the link normal to AB which affects its action, so that $\boldsymbol{\alpha}_n \cdot \mathbf{r}_{A/B} = 0$ must also hold.

(*b*) *Rotating reference axes.* A more general formulation of the motion of a rigid body in space calls for the use of reference axes which rotate as well as translate. The description of Fig. 7/10 is modified in Fig. 7/11 to show reference axes with origin attached to the reference point B as before but which rotate with an absolute angular velocity $\boldsymbol{\Omega}$ which may be different from the absolute angular velocity $\boldsymbol{\omega}$ of the body.

We now make use of Eqs. 5/11, 5/12, 5/13, and 5/14 developed in Art. 5/7 for describing the plane motion of a rigid body using rotating axes. The extension of these relations from two to three dimensions is easily accomplished by merely including the z-component of the vectors, and this step is left to the student to carry out. Replacing $\boldsymbol{\omega}$ in these equations by the angular velocity $\boldsymbol{\Omega}$ of our rotating x-y-z axes gives us

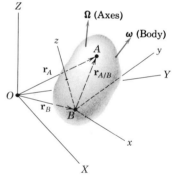

Z

$\boldsymbol{\Omega}$ (Axes)

$\boldsymbol{\omega}$ (Body)

z

A

y

$\mathbf{r}_A$

$\mathbf{r}_{A/B}$

Y

O

$\mathbf{r}_B$

B

x

X

Figure 7/11

$$\boxed{\dot{\mathbf{i}} = \boldsymbol{\Omega} \times \mathbf{i} \qquad \dot{\mathbf{j}} = \boldsymbol{\Omega} \times \mathbf{j} \qquad \dot{\mathbf{k}} = \boldsymbol{\Omega} \times \mathbf{k}} \qquad (7/5)$$

for the time derivatives of the rotating unit vectors attached to x-y-z. The expressions for the velocity and acceleration of point A become

$$\boxed{\begin{aligned} \mathbf{v}_A &= \mathbf{v}_B + \boldsymbol{\Omega} \times \mathbf{r}_{A/B} + \mathbf{v}_{\text{rel}} \\ \mathbf{a}_A &= \mathbf{a}_B + \dot{\boldsymbol{\Omega}} \times \mathbf{r}_{A/B} + \boldsymbol{\Omega} \times (\boldsymbol{\Omega} \times \mathbf{r}_{A/B}) + 2\boldsymbol{\Omega} \times \mathbf{v}_{\text{rel}} + \mathbf{a}_{\text{rel}} \end{aligned}} \qquad (7/6)$$

---

* It may be shown that $\boldsymbol{\alpha}_n = \dot{\boldsymbol{\omega}}_n$ if the angular velocity of the link about its own axis is not changing. See the senior author's *Dynamics, 2nd Edition SI Version,* John Wiley & Sons, 1975, Art. 37.

where $\mathbf{v}_{rel} = \dot{x}\mathbf{i} + \dot{y}\mathbf{j} + \dot{z}\mathbf{k}$ and $\mathbf{a}_{rel} = \ddot{x}\mathbf{i} + \ddot{y}\mathbf{j} + \ddot{z}\mathbf{k}$ are, respectively, the velocity and acceleration of point A measured relative to x-y-z by an observer attached to x-y-z. Again we note that $\boldsymbol{\Omega}$ is the angular velocity of the axes which may be different from the angular velocity $\boldsymbol{\omega}$ of the body. Also we note that $\mathbf{r}_{A/B}$ remains constant in magnitude for a rigid body but it will change direction with respect to x-y-z when the angular velocity $\boldsymbol{\Omega}$ of the axes is different from the angular velocity $\boldsymbol{\omega}$ of the body. We observe further that, if x-y-z are rigidly attached to the body, $\boldsymbol{\Omega} = \boldsymbol{\omega}$ and $\mathbf{v}_{rel}$ and $\mathbf{a}_{rel}$ are both zero, which makes the equations identical with Eqs. 7/4.

In Art. 5/7 we also developed the relationship (Eq. 5/13) between the time derivative of a vector $\mathbf{V}$ as measured in the fixed X-Y system and the time derivative of $\mathbf{V}$ as measured relative to the rotating x-y system. For our three-dimensional case this relation becomes

$$\left(\frac{d\mathbf{V}}{dt}\right)_{XYZ} = \left(\frac{d\mathbf{V}}{dt}\right)_{xyz} + \boldsymbol{\Omega} \times \mathbf{V} \qquad (7/7)$$

When we apply this transformation to the relative position vector $\mathbf{r}_{A/B} = \mathbf{r}_A - \mathbf{r}_B$ for our rigid body of Fig. 7/11, we get

$$\left(\frac{d\mathbf{r}_A}{dt}\right)_{XYZ} = \left(\frac{d\mathbf{r}_B}{dt}\right)_{XYZ} + \left(\frac{d\mathbf{r}_{A/B}}{dt}\right)_{xyz} + \boldsymbol{\Omega} \times \mathbf{r}_{A/B}$$

or

$$\mathbf{v}_A = \mathbf{v}_B + \mathbf{v}_{rel} + \boldsymbol{\Omega} \times \mathbf{r}_{A/B}$$

which gives us the first of Eqs. 7/6.

Equations 7/6 are particularly useful when the reference axes are attached to a moving body within which relative motion occurs.

Equation 7/7 may be recast as the vector operator

$$\left(\frac{d[\ ]}{dt}\right)_{XYZ} = \left(\frac{d[\ ]}{dt}\right)_{xyz} + \boldsymbol{\Omega} \times [\ ] \qquad (7/7a)$$

where $[\ ]$ stands for any vector $\mathbf{V}$ expressible both in X-Y-Z and in x-y-z. If we apply the operator to itself, we obtain the second time derivative which becomes

$$\left(\frac{d^2[\ ]}{dt^2}\right)_{XYZ} = \left(\frac{d^2[\ ]}{dt^2}\right)_{xyz} + \dot{\boldsymbol{\Omega}} \times [\ ] + \boldsymbol{\Omega} \times (\boldsymbol{\Omega} \times [\ ])$$

$$+ 2\boldsymbol{\Omega} \times \left(\frac{d[\ ]}{dt}\right)_{xyz} \qquad (7/7b)$$

This exercise is left to the student, and the form of Eq. 7/7b is observed to be the same as that of the second of Eqs. 7/6 expressed for $\mathbf{a}_{A/B} = \mathbf{a}_A - \mathbf{a}_B$.

Sample Problem 7/3

Crank CB rotates about the horizontal axis with an angular velocity $\omega_1 = 6$ rad/s which is constant for a short interval of motion which includes the position shown. The link AB has a ball-and-socket fitting on each end and connects crank DA with CB. For the instant shown, determine the angular velocity ω_2 of crank DA and the angular velocity ω_n of link AB.

Solution. The relative-velocity relation, Eq. 7/4, will be solved
① first using translating reference axes attached to B. The equation is

$$\mathbf{v}_A = \mathbf{v}_B + \boldsymbol{\omega}_n \times \mathbf{r}_{A/B}$$

② where $\boldsymbol{\omega}_n$ is the angular velocity of link AB taken normal to AB. The velocities of A and B are

$$[v = r\omega] \qquad \mathbf{v}_A = 50\omega_2\mathbf{j} \qquad \mathbf{v}_B = 100(6)\mathbf{i} = 600\mathbf{i} \text{ mm/s}$$

Also $\mathbf{r}_{A/B} = 50\mathbf{i} + 100\mathbf{j} + 100\mathbf{k}$ mm. Substitution into the velocity relation gives

$$50\omega_2\mathbf{j} = 600\mathbf{i} + \begin{vmatrix} \mathbf{i} & \mathbf{j} & \mathbf{k} \\ \omega_{n_x} & \omega_{n_y} & \omega_{n_z} \\ 50 & 100 & 100 \end{vmatrix}$$

Expanding the determinant and equating the coefficients of the $\mathbf{i}, \mathbf{j}, \mathbf{k}$ terms give

$$-6 = \qquad + \omega_{n_y} - \omega_{n_z}$$
$$\omega_2 = -2\omega_{n_x} \qquad + \omega_{n_z}$$
$$0 = 2\omega_{n_x} - \omega_{n_y}$$

These equations may be solved for ω_2, which becomes

$$\omega_2 = 6 \text{ rad/s} \qquad\qquad Ans.$$

As they stand the three equations incorporate the fact that $\boldsymbol{\omega}_n$ is normal to $\mathbf{v}_{A/B}$, but they cannot be solved until the requirement that $\boldsymbol{\omega}_n$ be
③ normal to $\mathbf{r}_{A/B}$ is included. Thus,

$$[\boldsymbol{\omega}_n \cdot \mathbf{r}_{A/B} = 0] \qquad 50\omega_{n_x} + 100\omega_{n_y} + 100\omega_{n_z} = 0$$

Combination with two of the three previous equations yields the solutions

$$\omega_{n_x} = -\tfrac{4}{3} \text{ rad/s} \qquad \omega_{n_y} = -\tfrac{8}{3} \text{ rad/s} \qquad \omega_{n_z} = \tfrac{10}{3} \text{ rad/s}$$

Thus

$$\boldsymbol{\omega}_n = \tfrac{2}{3}(-2\mathbf{i} - 4\mathbf{j} + 5\mathbf{k}) \text{ rad/s}$$

with

$$\omega_n = \tfrac{2}{3}\sqrt{2^2 + 4^2 + 5^2} = 2\sqrt{5} \text{ rad/s} \qquad\qquad Ans.$$

① We select B as the reference point since its motion can easily be determined from the given angular velocity ω_1 of CB.

② The angular velocity ω of AB is taken as a vector $\boldsymbol{\omega}_n$ normal to AB since any rotation of the link about its own axis AB has no influence on the behavior of the linkage.

③ The relative-velocity equation may be written as $\mathbf{v}_A - \mathbf{v}_B = \mathbf{v}_{A/B} = \boldsymbol{\omega}_n \times \mathbf{r}_{A/B}$ which requires that $\mathbf{v}_{A/B}$ be perpendicular to both $\boldsymbol{\omega}_n$ and $\mathbf{r}_{A/B}$. This equation alone does not incorporate the additional requirement that $\boldsymbol{\omega}_n$ be perpendicular to $\mathbf{r}_{A/B}$. Thus we must also satisfy $\boldsymbol{\omega}_n \cdot \mathbf{r}_{A/B} = 0$.

Sample Problem 7/4

Determine the angular acceleration $\dot{\omega}_2$ of crank AD in Sample Problem 7/3 for the conditions cited. Also find the angular acceleration $\dot{\omega}_n$ of link AB.

Solution. The accelerations of the links may be found from the second of Eqs. 7/4, which may be written

$$\mathbf{a}_A = \mathbf{a}_B + \dot{\omega}_n \times \mathbf{r}_{A/B} + \omega_n \times (\omega_n \times \mathbf{r}_{A/B})$$

where ω_n, as in Sample Problem 7/3, is the angular velocity of AB taken normal to AB. The angular acceleration of AB is written as $\dot{\omega}_n$. ①

In terms of their normal and tangential components, the accelerations of A and B are

$$\mathbf{a}_A = 50\omega_2{}^2\mathbf{i} + 50\dot{\omega}_2\mathbf{j} = 1800\mathbf{i} + 50\dot{\omega}_2\mathbf{j} \text{ mm/s}^2$$

$$\mathbf{a}_B = 100\omega_1{}^2\mathbf{k} + (0)\mathbf{i} = 3600\mathbf{k} \text{ mm/s}^2$$

Also

$$\omega_n \times (\omega_n \times \mathbf{r}_{A/B}) = -\omega_n{}^2\mathbf{r}_{A/B} = -20(50\mathbf{i} + 100\mathbf{j} + 100\mathbf{k}) \text{ mm/s}^2$$

$$\dot{\omega}_n \times \mathbf{r}_{A/B} = (100\dot{\omega}_{n_y} - 100\dot{\omega}_{n_z})\mathbf{i} + (50\dot{\omega}_{n_z} - 100\dot{\omega}_{n_x})\mathbf{j} + (100\dot{\omega}_{n_x} - 50\dot{\omega}_{n_y})\mathbf{k}$$

Substitution into the relative acceleration equation and equating respective coefficients of $\mathbf{i}$, $\mathbf{j}$, $\mathbf{k}$ give

$$28 = \qquad \dot{\omega}_{n_y} - \dot{\omega}_{n_z}$$
$$\dot{\omega}_2 + 40 = -2\dot{\omega}_{n_x} \qquad + \dot{\omega}_{n_z}$$
$$-32 = \quad 2\dot{\omega}_{n_x} - \dot{\omega}_{n_y}$$

Solution of these equations for $\dot{\omega}_2$ gives

$$\dot{\omega}_2 = -36 \text{ rad/s}^2 \qquad\qquad Ans.$$

The vector $\dot{\omega}_n$ is normal to $\mathbf{r}_{A/B}$ but is not normal to $\mathbf{v}_{A/B}$ as was the case with ω_n. ②

$$[\dot{\omega}_n \cdot \mathbf{r}_{A/B} = 0] \qquad 2\dot{\omega}_{n_x} + 4\dot{\omega}_{n_y} + 4\dot{\omega}_{n_z} = 0$$

which, when combined with the preceding relations for these same quantities, gives

$$\dot{\omega}_{n_x} = -8 \text{ rad/s}^2 \qquad \dot{\omega}_{n_y} = 16 \text{ rad/s}^2 \qquad \dot{\omega}_{n_z} = -12 \text{ rad/s}^2$$

Thus

$$\dot{\omega}_n = 4(-2\mathbf{i} + 4\mathbf{j} - 3\mathbf{k}) \text{ rad/s}^2 \qquad\qquad Ans.$$

and

$$|\dot{\omega}_n| = 4\sqrt{2^2 + 4^2 + 3^2} = 4\sqrt{29} \text{ rad/s}^2 \qquad\qquad Ans.$$

① If the link AB had an angular velocity component along AB, then a change in both magnitude and direction of this component could occur which would contribute to the actual angular acceleration of the link as a rigid body. However, since any rotation about its own axis AB has no influence on the motion of the cranks at C and D, we shall concern ourselves only with $\dot{\omega}_n$.

② The component of $\dot{\omega}_n$ which is not normal to $\mathbf{v}_{A/B}$ gives rise to the change in direction of $\mathbf{v}_{A/B}$.

Sample Problem 7/5

The motor housing and its bracket rotate about the Z-axis at the constant rate $\Omega = 3$ rad/s. The motor shaft and disk have a constant angular velocity of spin $p = 8$ rad/s with respect to the motor housing in the direction shown. If γ is constant at $30°$, determine the velocity and acceleration of point A at the top of the disk and the angular acceleration $\boldsymbol{\alpha}$ of the disk.

Solution. The rotating reference axes x-y-z are attached to the motor housing, and the rotating base for the motor has the momentary orientation shown with respect to the fixed axes X-Y-Z. We will use ① both X-Y-Z components with unit vectors $\mathbf{I}$, $\mathbf{J}$, $\mathbf{K}$ and x-y-z components with unit vectors $\mathbf{i}$, $\mathbf{j}$, $\mathbf{k}$. The angular velocity of the x-y-z axes becomes $\Omega = \Omega\mathbf{K} = 3\mathbf{K}$ rad/s.

① This choice for the reference axes provides a simple description for the motion of the disk relative to these axes.

Velocity. The velocity of A is given by the first of Eqs. 7/6

$$\mathbf{v}_A = \mathbf{v}_B + \boldsymbol{\Omega} \times \mathbf{r}_{A/B} + \mathbf{v}_{\text{rel}}$$

where

$$\mathbf{v}_B = \boldsymbol{\Omega} \times \mathbf{r}_B = 3\mathbf{K} \times 0.350\mathbf{J} = -1.05\mathbf{I} = -1.05\mathbf{i} \text{ m/s}$$

② $$\boldsymbol{\Omega} \times \mathbf{r}_{A/B} = 3\mathbf{K} \times (0.300\mathbf{j} + 0.120\mathbf{k})$$
$$= (-0.9 \cos 30°)\mathbf{i} + (0.36 \sin 30°)\mathbf{i} = -0.599\mathbf{i} \text{ m/s}$$

$$\mathbf{v}_{\text{rel}} = \mathbf{p} \times \mathbf{r}_{A/B} = 8\mathbf{j} \times (0.300\mathbf{j} + 0.120\mathbf{k}) = 0.960\mathbf{i} \text{ m/s}$$

Thus

$$\mathbf{v}_A = -1.05\mathbf{i} - 0.599\mathbf{i} + 0.960\mathbf{i} = -0.689\mathbf{i} \text{ m/s} \qquad Ans.$$

Acceleration. The acceleration of A is given by the second of Eqs. 7/6

$$\mathbf{a}_A = \mathbf{a}_B + \dot{\boldsymbol{\Omega}} \times \mathbf{r}_{A/B} + \boldsymbol{\Omega} \times (\boldsymbol{\Omega} \times \mathbf{r}_{A/B}) + 2\boldsymbol{\Omega} \times \mathbf{v}_{\text{rel}} + \mathbf{a}_{\text{rel}}$$

where

$$\mathbf{a}_B = \boldsymbol{\Omega} \times (\boldsymbol{\Omega} \times \mathbf{r}_B) = 3\mathbf{K} \times (3\mathbf{K} \times 0.350\mathbf{J}) = -3.15\mathbf{J}$$
$$= 3.15(-\mathbf{j} \cos 30° + \mathbf{k} \sin 30°) = -2.73\mathbf{j} + 1.58\mathbf{k} \text{ m/s}^2$$

$$\dot{\boldsymbol{\Omega}} = 0$$

② $$\boldsymbol{\Omega} \times (\boldsymbol{\Omega} \times \mathbf{r}_{A/B}) = 3\mathbf{K} \times (3\mathbf{K} \times [0.300\mathbf{j} + 0.120\mathbf{k}])$$
$$= 3\mathbf{K} \times (-0.599\mathbf{i}) = -1.557\mathbf{j} + 0.899\mathbf{k} \text{ m/s}^2$$

$$2\boldsymbol{\Omega} \times \mathbf{v}_{\text{rel}} = 2(3\mathbf{K}) \times 0.960\mathbf{i} = 5.76\mathbf{J}$$
$$= 5.76(\mathbf{j} \cos 30° - \mathbf{k} \sin 30°) = 4.99\mathbf{j} - 2.88\mathbf{k} \text{ m/s}^2$$

$$\mathbf{a}_{\text{rel}} = \mathbf{p} \times (\mathbf{p} \times \mathbf{r}_{A/B}) = 8\mathbf{j} \times (8\mathbf{j} \times [0.300\mathbf{j} + 0.120\mathbf{k}]) = -7.68\mathbf{k} \text{ m/s}^2$$

Substitution into the expression for $\mathbf{a}_A$ and collecting terms give us

$$\mathbf{a}_A = 0.703\mathbf{j} - 8.086\mathbf{k} \text{ m/s}^2$$

and

$$a_A = \sqrt{(0.703)^2 + (8.086)^2} = 8.12 \text{ m/s}^2 \qquad Ans.$$

② Note that $\mathbf{K} \times \mathbf{i} = \mathbf{J} = \mathbf{j} \cos \gamma - \mathbf{k} \sin \gamma$, $\mathbf{K} \times \mathbf{j} = -\mathbf{i} \cos \gamma$, and $\mathbf{K} \times \mathbf{k} = \mathbf{i} \sin \gamma$.

Angular acceleration. Since the precession is steady, we may use Eq. 7/3 to give us

$$\boldsymbol{\alpha} = \dot{\boldsymbol{\omega}} = \boldsymbol{\Omega} \times \boldsymbol{\omega} = 3\mathbf{K} \times (3\mathbf{K} + 8\mathbf{j})$$
$$= 0 + (-24 \cos 30°)\mathbf{i} = -20.8\mathbf{i} \text{ rad/s}^2 \qquad Ans.$$

Problem 7/22

PROBLEMS

7/22 The solid cylinder has a body cone with a semi-vertex angle of 20°. Momentarily the angular velocity ω has a magnitude of 30 rad/s and lies in the y-z plane. Determine the rate p at which the cylinder is spinning about its z-axis and write the vector expression for the velocity of B with respect to A. *Ans.* $p = 28.2$ rad/s, $\mathbf{v}_{B/A} = 4.10\mathbf{i}$ m/s

7/23 The circular disk rotates about the x-axis with an angular velocity p. At the same time the supporting arm and yoke rotate about the Z-axis with an angular velocity Ω. The x-y-z axes are attached to the arm. Determine the velocities of points A and B for the instant depicted.
$$Ans. \ \mathbf{v}_A = -b\Omega\mathbf{i} - rp\mathbf{j}, \ \mathbf{v}_B = -(b+r)\Omega\mathbf{i} + rp\mathbf{k}$$

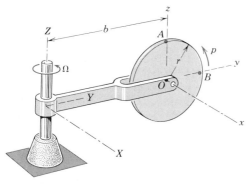

Problem 7/23

7/24 The helicopter is nosing over at the constant rate q rad/s. If the rotor blades revolve at the constant speed p rad/s, write the expression for the angular acceleration $\boldsymbol{\alpha}$ of the rotor. Take the y-axis to be attached to the fuselage and pointing forward perpendicular to the rotor axis.

Problem 7/24

7/25 If the angular rates Ω and p are both constant for Prob. 7/23, determine the acceleration of points A and B for the instant depicted. The x-y-z axes are attached to the rotating arm.
$$Ans. \ \mathbf{a}_A = 2r\Omega p\mathbf{i} - b\Omega^2\mathbf{j} - rp^2\mathbf{k}$$
$$\mathbf{a}_B = -[(r+b)\Omega^2 + rp^2]\mathbf{j}$$

7/26 The circular disk is spinning about its own axis (y-axis) at the constant rate $p = 10\pi$ rad/s. Simultaneously the frame is rotating about the Z-axis at the constant rate $\Omega = 4\pi$ rad/s. Calculate the angular acceleration $\boldsymbol{\alpha}$ of the disk and the acceleration of point A at the top of the disk. Axes x-y-z are attached to the frame, which has the momentary orientation shown with respect to the fixed axes X-Y-Z.

Problem 7/26

7/27 Determine the acceleration of point B on the disk of Prob. 7/26 for the conditions stated.

Ans. $\mathbf{a}_B = -16.4\pi^2 \mathbf{i}$ m/s^2

7/28 For the instant represented collar B is moving along the fixed shaft in the X-direction with a constant velocity $v_B = 4$ m/s. Also at this instant $X = 0.3$ m and $Y = 0.2$ m. Calculate the velocity of collar A which moves along the fixed shaft parallel to the Y-axis. Solve, first, by differentiating the relation $X^2 + Y^2 + Z^2 = L^2$ with respect to time and, second, by using the first of Eqs. 7/4 with translating axes attached to B. Each clevis is free to rotate about the axis of the rod.

Ans. $\mathbf{v}_A = -6\mathbf{j}$ m/s

Problem 7/28

7/29 The spacecraft is revolving about its z-axis, which has a fixed space orientation, at the constant rate $p = \frac{1}{10}$ rad/s. Simultaneously its solar panels are unfolding at the rate $\dot{\beta}$ which is programmed to produce the variation with β shown in the graph. Determine the angular acceleration $\boldsymbol{\alpha}$ of panel A an instant (a) before and an instant (b) after it reaches the position $\beta = 18°$.

Problem 7/29

7/30 The collar and clevis A are given a constant upward velocity of 0.2 m/s for an interval of motion and cause the ball end of the bar to slide in the radial slot in the rotating disk. Determine the angular acceleration of the bar when the bar passes the position for which $z = 75$ mm. The disk turns at the constant rate of 2 rad/s.

Ans. $\boldsymbol{\alpha} = -3\mathbf{i} - 4\mathbf{j}$ rad/s^2

Problem 7/30

Problem 7/31

7/31 The disk has a constant angular velocity p about its z-axis, and the yoke A has a constant angular velocity ω_2 about its shaft as shown. Simultaneously the entire assembly revolves about the fixed X-axis with a constant angular velocity ω_1. Determine the expression for the angular acceleration of the disk as the yoke brings it into the vertical plane in the position shown. Solve by picturing the vector changes in the angular-velocity components. *Ans.* $\boldsymbol{\alpha} = p\omega_2\mathbf{i} - p\omega_1\mathbf{j} + \omega_1\omega_2\mathbf{k}$

Problem 7/32

7/32 End A of the rigid link is confined to move in the $-x$-direction while end B is confined to move along the z-axis. Determine the component $\boldsymbol{\omega}_n$ normal to AB of the angular velocity of the link as it passes the position shown with $v_A = 0.3$ m/s.

Problem 7/33

7/33 The robot of Prob. 7/10 is shown again here where the coordinate system x-y-z with origin at O_2 rotates about the X-axis at the rate $\dot{\theta}$. Nonrotating axes X-Y-Z oriented as shown have their origin at O_1. If $\omega_2 = \dot{\theta} = 3$ rad/s constant, $\omega_3 = 1.5$ rad/s constant, $\omega_1 = \omega_5 = 0$, $\overline{O_1O_2} = 1.2$ m, $\overline{O_2A} = 0.6$ m, and $\beta = 45°$, determine the velocity of the center A of the jaws for the instant when $\theta = 60°$.
Ans. $\mathbf{v}_A = -0.636\mathbf{i} - 4.873\mathbf{j} + 1.273\mathbf{k}$ m/s

7/34 For the conditions of Prob. 7/33 determine the acceleration of the center *A* of the jaws of the robot.

7/35 For the robot shown with Prob. 7/33, determine the velocity of the center *A* of the jaws for the following conditions: $\omega_1 = 2$ rad/s constant, $\omega_2 = \dot{\theta} = -3$ rad/s constant, $\omega_4 = \dot{\beta} = 1.5$ rad/s constant, $\omega_3 = 0$, $\overline{O_1O_2} = 1.2$ m, $\overline{O_2A} = 0.6$ m, $\theta = 60°$, and $\beta = 45°$.

7/36 For the conditions of Prob. 7/33 determine the acceleration $\mathbf{a}_A$ of the center *A* of the jaws of the robot.   *Ans.* $\mathbf{a}_A = -4.77\mathbf{j} - 14.62\mathbf{k}$ m/s^2

7/37 A simulator for perfecting the docking procedure for spacecraft consists of the frame *A* which is mounted on four air-bearing pads so that it can translate and rotate freely on the horizontal surface. Mounted in the frame is a drum *B* which can rotate about the horizontal axis of the frame *A*. The coordinate axes *x-y-z* are attached to the drum *B*, and the *z*-axis of the drum makes an angle β with the horizontal. Inside the drum is the simulated command module *C* which can rotate within the drum about its *z*-axis at a rate *p*. For a certain test run frame *A* is rotating on the horizontal surface in the direction shown with a constant angular velocity of 0.2 rad/s. Simultaneously the drum *B* is rotating about the *x*-axis at the constant rate $\dot{\beta} = 0.15$ rad/s, and the module *C* is turning inside the drum at the constant rate $p = 0.9$ rad/s in the direction indicated. For these conditions determine the angular velocity ω and the angular acceleration α of the simulator *C* as it passes the position $\beta = 0$.

Ans. $\omega = 0.15\mathbf{i} + 0.2\mathbf{j} + 0.9\mathbf{k}$ rad/s
$\alpha = 0.18\mathbf{i} - 0.135\mathbf{j} - 0.03\mathbf{k}$ rad/s^2

Problem 7/37

Problem 7/38

7/38 The test chamber of a flight simulator consists of a drum which spins about the *a-a* axis at a constant angular rate p relative to the cylindrical housing A. The housing in turn is mounted on transverse horizontal bearings and rotates about the axis *b-b* at the constant rate $\dot{\theta}$. The entire assembly is made to rotate about the fixed vertical z-axis at the constant rate Ω. Write the expression for the angular acceleration $\boldsymbol{\alpha}$ of the spinning drum during the compounded motion for a given value of θ.

Ans. $\boldsymbol{\alpha} = \dot{\theta}(p \sin \theta - \Omega)\mathbf{i} - (p \, \Omega \cos \theta)\mathbf{j} + (p \, \dot{\theta} \cos \theta)\mathbf{k}$

Problem 7/39

7/39 The center O of the spacecraft is moving through space with a constant velocity. During the period of motion prior to stabilization, the spacecraft has a constant rotational rate $\Omega = \frac{1}{2}$ rad/s about its z-axis. The x-y-z axes are attached to the body of the craft, and the solar panels rotate about the y-axis at the constant rate $\dot{\theta} = \frac{1}{4}$ rad/s with respect to the spacecraft. If $\boldsymbol{\omega}$ is the absolute angular velocity of the solar panels, determine $\dot{\boldsymbol{\omega}}$. Also find the acceleration of point A when $\theta = 30°$.

Ans. $\dot{\boldsymbol{\omega}} = \frac{1}{8}\mathbf{i}$ rad/s^2
$\mathbf{a}_A = 0.0938\mathbf{i} - 0.730\mathbf{j} - 0.0325\mathbf{k}$ m/s^2

▶**7/40** For the conditions specified with Sample Problem 7/2, except that γ is increasing at the steady rate of 3π rad/s, determine the angular velocity $\boldsymbol{\omega}$ and the angular acceleration $\boldsymbol{\alpha}$ of the rotor when the position $\gamma = 30°$ is passed. (*Suggestion:* Apply Eq. 7/7 to the vector $\boldsymbol{\omega}$ to find $\boldsymbol{\alpha}$. Note that Ω in Sample Problem 7/2 is no longer the complete angular velocity of the axes.)

Ans. $\boldsymbol{\omega} = \pi(-3\mathbf{i} + \sqrt{3}\mathbf{j} + 5\mathbf{k})$ rad/s
$\boldsymbol{\alpha} = \pi^2(4\sqrt{3}\mathbf{i} + 9\mathbf{j} + 3\sqrt{3}\mathbf{k})$ rad/s^2

▶**7/41** The crank with a radius of 80 mm turns with a constant angular velocity $\omega_0 = 4$ rad/s and causes the collar A to oscillate along the fixed shaft. Determine the velocity of the collar A and the angular velocity of the rigid-body link AB as the crank crosses the vertical position shown. (*Hint:* Link AB can have no angular velocity about an axis (unit vector **n**) which is normal to both the Y-axis and the axis of the clevis pin. Thus $\boldsymbol{\omega} \cdot \mathbf{n} = 0$ where **n** has the direction of the triple vector product $\mathbf{J} \times (\mathbf{r}_{AB} \times \mathbf{J})$.

$$Ans.\ \mathbf{v}_A = 0.160\mathbf{j}\ \text{m/s}$$
$$\boldsymbol{\omega} = 0.32(-2\mathbf{i} + 4\mathbf{j} - \mathbf{k})\ \text{rad/s}$$

Problem 7/41

▶**7/42** The gyro rotor shown is spinning at the constant rate of 100 rev/min relative to the x-y-z axes in the direction indicated. If the angle γ between the gimbal ring and the horizontal X-Y plane is made to increase at the constant rate of 4 rad/s and if the unit is forced to precess about the vertical at the constant rate $N = 20$ rev/min, calculate the magnitude of the angular acceleration $\boldsymbol{\alpha}$ of the rotor when $\gamma = 30°$. Solve by using Eq. 7/7 applied to the angular velocity of the rotor.

$$Ans.\ \alpha = 42.8\ \text{rad/s}^2$$

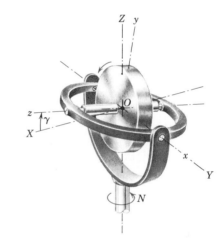

Problem 7/42

▶**7/43** The wheel of radius r is free to rotate about the bent axle CO which turns about the vertical axis at the constant rate p rad/s. If the wheel rolls without slipping on the horizontal circle of radius R, determine the expressions for the angular velocity $\boldsymbol{\omega}$ and angular acceleration $\boldsymbol{\alpha}$ of the wheel. The x-axis is always horizontal.

$$Ans.\ \boldsymbol{\omega} = p\left[\mathbf{j}\cos\theta + \mathbf{k}\left(\sin\theta + \frac{R}{r}\right)\right]$$
$$\boldsymbol{\alpha} = \left(\frac{Rp^2}{r}\cos\theta\right)\mathbf{i}$$

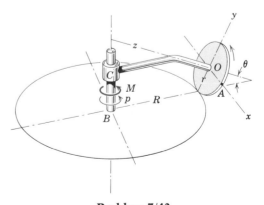

Problem 7/43

SECTION B. KINETICS

7/7 ANGULAR MOMENTUM

The force equation for a mass system, rigid or nonrigid, Eq. 4/1 or 4/6, is the generalization of Newton's second law for the motion of a particle and should require no further explanation. The moment equation for three-dimensional motion, however, is not nearly as simple as the third of Eqs. 6/1 for plane motion since the change of angular momentum has a number of additional components which are absent in plane motion.

We consider now a rigid body moving with any general motion in space, Fig. 7/12a. Axes x-y-z are *attached* to the body with origin at the mass center G. Thus the angular velocity $\boldsymbol{\omega}$ of the body becomes the angular velocity of the x-y-z axes as observed from the fixed reference axes X-Y-Z. The absolute angular momentum $\mathbf{H}_G$ of the body about its mass center G is the sum of the moments about G of the linear momenta of all elements of the body and was expressed in part (b) of Art. 4/4 as $\mathbf{H}_G = \Sigma(\boldsymbol{\rho}_i \times m_i\mathbf{v}_i)$, where $\mathbf{v}_i$ is the absolute velocity of the mass element m_i. But for the rigid body $\mathbf{v}_i = \bar{\mathbf{v}} + \boldsymbol{\omega} \times \boldsymbol{\rho}_i$ where $\boldsymbol{\omega} \times \boldsymbol{\rho}_i$ is the relative velocity of m_i with respect to G as seen from nonrotating axes. Thus we may write

$$\mathbf{H}_G = -\bar{\mathbf{v}} \times \Sigma m_i\boldsymbol{\rho}_i + \Sigma(\boldsymbol{\rho}_i \times m_i[\boldsymbol{\omega} \times \boldsymbol{\rho}_i])$$

where we have factored out $\bar{\mathbf{v}}$ from the first summation terms by reversing the order of the cross product and changing the sign. With origin at the mass center G the first term in $\mathbf{H}_G$ is zero since $\Sigma m_i\boldsymbol{\rho}_i = m\bar{\boldsymbol{\rho}} = 0$. The second term with the substitution of dm for m_i and $\boldsymbol{\rho}$ for $\boldsymbol{\rho}_i$ gives

$$\mathbf{H}_G = \int (\boldsymbol{\rho} \times [\boldsymbol{\omega} \times \boldsymbol{\rho}]) \, dm \tag{7/8}$$

Before expanding the integrand of Eq. 7/8, we consider also the case of a rigid body rotating about a fixed point O, Fig. 7/12b. The x-y-z axes are attached to the body, and both body and axes have an angular velocity $\boldsymbol{\omega}$. The angular momentum about O was expressed in part (b) of Art. 4/4 and is $\mathbf{H}_O = \Sigma(\mathbf{r}_i \times m_i\mathbf{v}_i)$ where, for the rigid body, $\mathbf{v}_i = \boldsymbol{\omega} \times \mathbf{r}_i$. Thus, with the substitution of dm for m_i and $\mathbf{r}$ for $\mathbf{r}_i$, the angular momentum is

$$\mathbf{H}_O = \int (\mathbf{r} \times [\boldsymbol{\omega} \times \mathbf{r}]) \, dm \tag{7/9}$$

We observe now that for the two cases of Figs. 7/12a and 7/12b, the position vector $\boldsymbol{\rho}_i$ or $\mathbf{r}_i$ is given by the same expression $x\mathbf{i} + y\mathbf{j} + z\mathbf{k}$. Thus Eqs. 7/8 and 7/9 are identical in form, and the symbol $\mathbf{H}$ will be used here for either case. We now carry

(a)

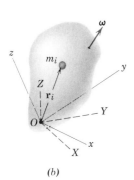

(b)

Figure 7/12

out the expansion of the integrand in the two expressions for angular momentum with recognition of the fact that the components of ω are invariant with respect to the integrals over the body and, hence, become constant multipliers of the integrals. The cross-product expansion applied to the triple vector product upon collection of terms gives

$$
\begin{aligned}
d\mathbf{H} = \mathbf{i}[(y^2 + z^2)\omega_x &\quad - xy\omega_y &\quad - xz\omega_z]\, dm\\
+\,\mathbf{j}[\quad - yx\omega_x + (z^2 + x^2)\omega_y &\quad &\quad - yz\omega_z]\, dm\\
+\,\mathbf{k}[\quad - zx\omega_x &\quad - zy\omega_y + (x^2 + y^2)\omega_z]\, dm
\end{aligned}
$$

Now let

$$
\begin{array}{ll}
I_{xx} = \displaystyle\int (y^2 + z^2)\, dm & \quad I_{xy} = \displaystyle\int xy\, dm\\[2mm]
I_{yy} = \displaystyle\int (z^2 + x^2)\, dm & \quad I_{xz} = \displaystyle\int xz\, dm\\[2mm]
I_{zz} = \displaystyle\int (x^2 + y^2)\, dm & \quad I_{yz} = \displaystyle\int yz\, dm
\end{array}
\tag{7/10}
$$

The quantities I_{xx}, I_{yy}, I_{zz} are known as the *moments of inertia* of the body about the respective axes, and I_{xy}, I_{xz}, I_{yz} are known as the *products of inertia* with respect to the coordinate axes. These quantities describe the manner in which the mass of a rigid body is distributed with respect to the chosen axes. The calculation of moments and products of inertia is explained fully in Appendix B. The double subscripts for the moments and products of inertia preserve a symmetry of notation which has special meaning in their description by tensor notation.* It is observed that $I_{xy} = I_{yx}$, $I_{xz} = I_{zx}$, $I_{yz} = I_{zy}$. With the substitutions of Eqs. 7/10, the expression for $\mathbf{H}$ becomes

$$
\begin{aligned}
\mathbf{H} = &\quad (\quad I_{xx}\omega_x - I_{xy}\omega_y - I_{xz}\omega_z)\mathbf{i}\\
&+(-I_{yx}\omega_x + I_{yy}\omega_y - I_{yz}\omega_z)\mathbf{j}\\
&+(-I_{zx}\omega_x - I_{zy}\omega_y + I_{zz}\omega_z)\mathbf{k}
\end{aligned}
\tag{7/11}
$$

and the components of $\mathbf{H}$ are clearly

$$
\begin{aligned}
H_x &= \quad I_{xx}\omega_x - I_{xy}\omega_y - I_{xz}\omega_z\\
H_y &= -I_{yx}\omega_x + I_{yy}\omega_y - I_{yz}\omega_z\\
H_z &= -I_{zx}\omega_x - I_{zy}\omega_y + I_{zz}\omega_z
\end{aligned}
\tag{7/12}
$$

* See, for example, the senior author's *Dynamics, 2nd Edition SI Version*, John Wiley & Sons, 1975, Art. 41.

Equation 7/11 is the general expression for the angular momentum about either the mass center *G* or about a fixed point *O* for a rigid body rotating with an instantaneous angular velocity *ω*.

Attention is called to the fact that in each of the two cases represented, the reference axes *x-y-z* are *attached* to the rigid body. This attachment makes the moment-of-inertia integrals and the product-of-inertia integrals of Eqs. 7/10 invariant with time. If the *x-y-z* axes were to rotate with respect to an irregular body, then these inertia integrals would be functions of the time which would introduce an undesirable complexity into the angular-momentum relations. An important exception occurs when a rigid body is spinning about an axis of symmetry, in which case the inertia integrals are not affected by the angular position of the body about its spin axis. Thus it is frequently convenient to permit a body with axial symmetry to rotate relative to the reference system about one of the coordinate axes. In addition to the momentum components due to the angular velocity **Ω** of the reference axes, then, an added angular momentum component along the spin axis due to the relative spin about the axis would have to be accounted for.

In Eq. 7/12 the array of moments and products of inertia

$$\begin{bmatrix} I_{xx} & -I_{xy} & -I_{xz} \\ -I_{yx} & I_{yy} & -I_{yz} \\ -I_{zx} & -I_{zy} & I_{zz} \end{bmatrix}$$

is known as the *inertia matrix* or *inertia tensor*. As we change the orientation of the axes the moments and products of inertia will also change in value. It can be shown* that there is one unique orientation of axes *x-y-z* for a given origin for which the products of inertia vanish and the moments of inertia I_{xx}, I_{yy}, I_{zz} take on stationary values. For this orientation the inertia matrix takes the form

$$\begin{bmatrix} I_{xx} & 0 & 0 \\ 0 & I_{yy} & 0 \\ 0 & 0 & I_{zz} \end{bmatrix}$$

and is said to be diagonalized. The axes *x-y-z* for which the products of inertia vanish are called the *principal axes of inertia*, and I_{xx}, I_{yy}, and I_{zz} are called the *principal moments of inertia*. The principal moments of inertia for a given origin represent the maximum, minimum, and an intermediate value of the moments of inertia.

If the coordinate axes coincide with the principal axes of inertia, Eq. 7/11 for the angular momentum about the mass center

---

* See, for example, the senior author's *Dynamics, 2nd Edition SI Version*, John Wiley & Sons 1975, Art. 41.

or about a fixed point becomes

$$\mathbf{H} = I_{xx}\omega_x\mathbf{i} + I_{yy}\omega_y\mathbf{j} + I_{zz}\omega_z\mathbf{k} \qquad (7/13)$$

It is always possible to locate the principal axes of inertia for a general three-dimensional rigid body. Thus we can express its angular momentum by Eq. 7/13, although it may not always be convenient to do so for geometric reasons. Except when the body rotates about one of the principal axes of inertia or when $I_{xx} = I_{yy} = I_{zz}$, the vectors $\mathbf{H}$ and $\boldsymbol{\omega}$ have different directions.

The momentum properties of a rigid body may be represented by the resultant linear momentum vector $\mathbf{G} = m\bar{\mathbf{v}}$ through the mass center and the resultant angular momentum vector $\mathbf{H}_G$ about the mass center, as shown in Fig. 7/13. Although $\mathbf{H}_G$ has the properties of a free vector, we represent it through G for convenience. These vectors have properties analogous to those of a force and a couple. Thus the angular momentum about any point A which is *not fixed* to the body equals the free vector $\mathbf{H}_G$ plus the moment of the linear momentum vector about A. Therefore we may write

$$\mathbf{H}_A = \mathbf{H}_G + \bar{\mathbf{r}} \times \mathbf{G} = \mathbf{H}_G + \bar{\mathbf{r}} \times m\bar{\mathbf{v}} \qquad (7/14)$$

Figure 7/13

Equation 7/14 constitutes a transfer theorem for angular momentum. Attention is called to the fact that we can use Eq. 7/11 to calculate $\mathbf{H}_A$ only when A is either the mass center or a fixed point in space.

7/8 KINETIC ENERGY

In Art. 4/3 of the chapter on the dynamics of systems of particles we developed the expression for the kinetic energy T of any general system of mass, rigid or nonrigid, and obtained the result

$$T = \tfrac{1}{2}m\bar{v}^2 + \Sigma\tfrac{1}{2}m_i|\dot{\boldsymbol{\rho}}_i|^2 \qquad [4/4]$$

where $\bar{v}$ is the velocity of the mass center and $\boldsymbol{\rho}_i$ is the position vector of a representative element of mass m_i with respect to the mass center. We identified the first term as the kinetic energy due to the translation of the system and the second term as the kinetic energy associated with the motion relative to the mass center. The translational term may be written alternatively as

$$\tfrac{1}{2}m\bar{v}^2 = \tfrac{1}{2}m\dot{\bar{\mathbf{r}}} \cdot \dot{\bar{\mathbf{r}}} = \tfrac{1}{2}\bar{\mathbf{v}} \cdot \mathbf{G}$$

where $\dot{\bar{\mathbf{r}}}$ is the velocity $\bar{\mathbf{v}}$ of the mass center and $\mathbf{G}$ is the linear momentum of the body.

For a rigid body the relative term becomes the kinetic energy due to rotation about the mass center. Since $\dot{\rho}_i$ is the velocity of the representative particle with respect to the mass center, then for the rigid body we may write it as $\dot{\rho}_i = \omega \times \rho_i$ where ω is the angular velocity of the body. With this substitution the relative term in the kinetic energy expression becomes

$$\Sigma \tfrac{1}{2}m_i|\dot{\rho}_i|^2 = \Sigma \tfrac{1}{2}m_i(\omega \times \rho_i) \cdot (\omega \times \rho_i)$$

If we use the fact that the dot and the cross may be interchanged in the triple scalar product, that is, $\mathbf{P} \times \mathbf{Q} \cdot \mathbf{R} = \mathbf{P} \cdot \mathbf{Q} \times \mathbf{R}$, we may write

$$(\omega \times \rho_i) \cdot (\omega \times \rho_i) = \omega \cdot \rho_i \times (\omega \times \rho_i)$$

Since ω is the same factor in all terms of the summation, it may be factored out to give

$$\Sigma \tfrac{1}{2}m_i|\dot{\rho}_i|^2 = \tfrac{1}{2}\omega \cdot \Sigma \rho_i \times m_i(\omega \times \rho_i) = \tfrac{1}{2}\omega \cdot \mathbf{H}_G$$

where $\mathbf{H}_G$ is the same as the integral expressed by Eq. 7/8. Thus the general expression for the kinetic energy of a rigid body moving with mass-center velocity $\bar{\mathbf{v}}$ and angular velocity ω is

$$\boxed{T = \tfrac{1}{2}\bar{\mathbf{v}} \cdot \mathbf{G} + \tfrac{1}{2}\omega \cdot \mathbf{H}_G} \tag{7/15}$$

Expansion of this vector equation by substitution of the expression for $\mathbf{H}_G$ written from Eq. 7/11 yields

$$T = \tfrac{1}{2}m\bar{v}^2 + \tfrac{1}{2}(\bar{I}_{xx}\omega_x{}^2 + \bar{I}_{yy}\omega_y{}^2 + \bar{I}_{zz}\omega_z{}^2)$$
$$- (\bar{I}_{xy}\omega_x\omega_y + \bar{I}_{xz}\omega_x\omega_z + \bar{I}_{yz}\omega_y\omega_z) \tag{7/16}$$

If the axes coincide with the principal axes of inertia, the kinetic energy is merely

$$T = \tfrac{1}{2}m\bar{v}^2 + \tfrac{1}{2}(\bar{I}_{xx}\omega_x{}^2 + \bar{I}_{yy}\omega_y{}^2 + \bar{I}_{zz}\omega_z{}^2) \tag{7/17}$$

When a rigid body is pivoted about a fixed point O or when there is a point O in the body which momentarily has zero velocity, the kinetic energy is $T = \Sigma \tfrac{1}{2}m_i\dot{\mathbf{r}}_i \cdot \dot{\mathbf{r}}_i$. This expression reduces to

$$\boxed{T = \tfrac{1}{2}\omega \cdot \mathbf{H}_O} \tag{7/18}$$

where $\mathbf{H}_O$ is the angular momentum about O, as may be seen by replacing ρ_i in the previous derivation by $\mathbf{r}_i$, the position vector from O. Equations 7/15 and 7/18 are the three-dimensional counterparts of Eqs. 6/9 and 6/8 for plane motion.

Sample Problem 7/6

The bent plate has a mass of 70 kg per square meter of surface area and revolves about the z-axis at the rate $\omega = 30$ rad/s. Determine (a) the angular momentum **H** of the plate about point O and (b) the kinetic energy T of the plate. Neglect the mass of the hub and the thickness of the plate compared with its surface dimensions.

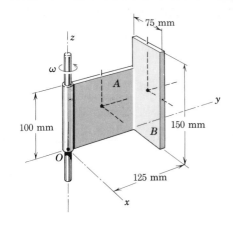

Solution. The moments and products of inertia are written with the aid of Eqs. B/3 and B/9 in Appendix B by transfer from the parallel centroidal axes for each part. First, the mass of each part is $m_A = (0.100)(0.125)(70) = 0.875$ kg, $m_B = (0.075)(0.150)(70) = 0.788$ kg.

Part A

$$[I_{xx} = \bar{I}_{xx} + md^2] \quad I_{xx} = \frac{0.875}{12}[(0.100)^2 + (0.125)^2]$$

$$+ 0.875[(0.050)^2 + (0.0625)^2] = 0.007\ 47 \text{ kg} \cdot \text{m}^2$$

$$[I_{yy} = \tfrac{1}{3}ml^2] \quad I_{yy} = \frac{0.875}{3}(0.100)^2 = 0.002\ 92 \text{ kg} \cdot \text{m}^2$$

$$[I_{zz} = \tfrac{1}{3}ml^2] \quad I_{zz} = \frac{0.875}{3}(0.125)^2 = 0.004\ 56 \text{ kg} \cdot \text{m}^2$$

$$[I_{xy} = \int xy\ dm,\ I_{xz} = \int xz\ dm] \quad I_{xy} = 0 \quad I_{xz} = 0$$

$$[I_{yz} = \bar{I}_{yz} + md_y\ d_z] \quad I_{yz} = 0 + 0.875(0.0625)(0.050) = 0.002\ 73 \text{ kg} \cdot \text{m}^2$$

Part B

$$[I_{xx} = \bar{I}_{xx} + md^2] \quad I_{xx} = \frac{0.788}{12}(0.150)^2 + 0.788[(0.125)^2 + (0.075)^2$$

$$= 0.018\ 21 \text{ kg} \cdot \text{m}^2$$

$$[I_{yy} = \bar{I}_{yy} + md^2] \quad I_{yy} = \frac{0.788}{12}[(0.075)^2 + (0.150)^2]$$

$$+ 0.788[(0.0375)^2 + (0.075)^2] = 0.007\ 38 \text{ kg} \cdot \text{m}^2$$

$$[I_{zz} = \bar{I}_{zz} + md^2] \quad I_{zz} = \frac{0.788}{12}(0.075)^2 + 0.788[(0.125)^2 + (0.0375)^2]$$

$$= 0.013\ 78 \text{ kg} \cdot \text{m}^2$$

$$[I_{xy} = \bar{I}_{xy} + md_x\ d_y] \quad I_{xy} = 0 + 0.788(0.0375)(0.125) = 0.003\ 69 \text{ kg} \cdot \text{m}^2$$
$$[I_{xz} = \bar{I}_{xz} + md_x\ d_z] \quad I_{xz} = 0 + 0.788(0.0375)(0.075) = 0.002\ 21 \text{ kg} \cdot \text{m}^2$$
$$[I_{yz} = \bar{I}_{yz} + md_y\ d_z] \quad I_{yz} = 0 + 0.788(0.125)(0.075) = 0.007\ 38 \text{ kg} \cdot \text{m}^2$$

The sum of the respective inertia terms gives for the two plates together

$$I_{xx} = 0.0257 \text{ kg} \cdot \text{m}^2 \qquad I_{xy} = 0.003\ 69 \text{ kg} \cdot \text{m}^2$$
$$I_{yy} = 0.010\ 30 \text{ kg} \cdot \text{m}^2 \qquad I_{xz} = 0.002\ 21 \text{ kg} \cdot \text{m}^2$$
$$I_{zz} = 0.018\ 34 \text{ kg} \cdot \text{m}^2 \qquad I_{yz} = 0.010\ 12 \text{ kg} \cdot \text{m}^2$$

(a) The angular momentum of the body is given by Eq. 7/11 where $\omega_z = 30$ rad/s and ω_x and ω_y are zero. Thus

$$\mathbf{H}_O = 30(-0.002\ 21\mathbf{i} - 0.010\ 12\mathbf{j} + 0.018\ 34\mathbf{k}) \text{ N} \cdot \text{m} \cdot \text{s} \quad Ans.$$

(b) The kinetic energy from Eq. 7/18 becomes

$$T = \tfrac{1}{2}\boldsymbol{\omega} \cdot \mathbf{H}_O = \tfrac{1}{2}(30\mathbf{k}) \cdot 30(-0.002\ 21\mathbf{i} - 0.010\ 12\mathbf{j} + 0.018\ 34\mathbf{k}) = 8.25 \text{ J}$$

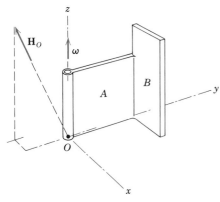

① The parallel-axis theorems for transferring moments and products of inertia from centroidal axes to parallel axes are explained in Appendix B and are most useful relations.

② Recall that the units of angular momentum may also be written in the base units as kg·m²/s.

Problem 7/44

Problem 7/45

Problem 7/46

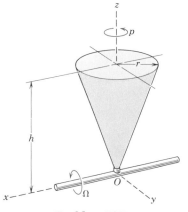

Problem 7/47

PROBLEMS

7/44 The slender shaft carries two offset particles each of mass m and rotates about the z-axis with an angular velocity ω as indicated. Write an expression for the angular momentum **H** for the system about the origin O of the x-y-z axes for the position shown. Write the kinetic energy T of the system by inspection and verify your result by applying Eq. 7/18.

$$Ans. \; \mathbf{H} = mR\omega\left[\frac{L}{3}\mathbf{j} + 2R\mathbf{k}\right]$$
$$T = mR^2\omega^2$$

7/45 The slender rod of mass m and length l is attached to the disk which rotates about the y-axis with an angular velocity ω. By inspection write the expression for the angular momentum **H** of the rod about the origin O of the x-y-z axes for the position shown. Verify the result by applying Eq. 7/11.

7/46 The slender rod of mass m and length l rotates about the y-axis as the element of a right-circular cone. If the angular velocity about the y-axis is ω, determine the expression for the angular momentum of the rod with respect to the x-y-z axes for the particular position shown.

$$Ans. \; \mathbf{H} = \tfrac{1}{3}ml^2\omega \sin\theta(\mathbf{j}\sin\theta - \mathbf{k}\cos\theta)$$

7/47 The right-circular cone of height h and base radius r spins about its axis of symmetry with an angular rate p. Simultaneously, the entire cone revolves about the x-axis with angular rate Ω. Determine the angular momentum **H** of the cone about the origin O of the x-y-z axes and the kinetic energy T for the position shown. The mass of the cone is m.

7/48 The uniform slender bar of length l and mass m is welded to the shaft, which rotates in bearings A and B with an angular velocity ω as shown. Determine the angular momentum $\mathbf{H}_O$ of the bar with respect to the x-y-z axes in terms of θ.

Problem 7/48

7/49 The circular disk of mass m and radius r is mounted on the vertical shaft with an angle α between its plane and the plane of rotation of the shaft. Determine an expression for the angular momentum $\mathbf{H}$ of the disk about O. Find the angle β which the angular momentum $\mathbf{H}$ makes with the shaft if $\alpha = 10°$.

Ans. $\mathbf{H} = \frac{1}{4}mr^2\omega[(-\sin \alpha \cos \alpha)\mathbf{i}$
$+ (\sin^2 \alpha + 2 \cos^2 \alpha)\mathbf{k}]$
$\beta = 4.96°$

Problem 7/49

7/50 The solid cube of mass m and side a revolves about an axis M-M through a diagonal with an angular velocity ω. Write the expression for the angular momentum $\mathbf{H}$ of the cube with respect to the axes indicated.

Problem 7/50

7/51 The bent rod has a mass m and revolves about the z-axis with an angular velocity ω. Determine the angular momentum of the rod about the origin O of the coordinates for the position shown. Also find the kinetic energy of the rod.

Ans. $\mathbf{H}_O = \dfrac{m\omega b^2}{3}\left(-\tfrac{1}{2}\mathbf{i} - \tfrac{1}{2}\mathbf{j} + \tfrac{11}{3}\mathbf{k}\right)$

$T = \tfrac{11}{18}mb^2\omega^2$

Problem 7/51

Problem 7/52

7/52 The 100-mm-radius wheel has a mass of 3 kg and turns about its y'-axis with an angular velocity $p = 40\pi$ rad/s in the direction shown. Simultaneously the fork rotates about its x-axis shaft with an angular velocity $\omega = 10\pi$ rad/s as indicated. Calculate the angular momentum of the wheel about its center O'. Also compute the kinetic energy of the wheel.

$$Ans. \mathbf{H}_{O'} = 0.236(\mathbf{i} + 8\mathbf{j}) \text{ kg·m}^2/\text{s}$$
$$T = 214.7 \text{ J}$$

7/53 Calculate the angular momentum of the wheel of Prob. 7/52 about the origin O of the x-y-z axes.

$$Ans. \mathbf{H}_O = 6.13\mathbf{i} + 1.88\mathbf{j} \text{ kg·m}^2/\text{s}$$

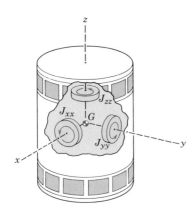

Problem 7/54

7/54 The elements of a reaction-wheel attitude-control system for a spacecraft are shown in the figure. Point G is the center of mass for the system of the spacecraft and wheels, and x, y, z are principal axes for the system. The system moments of inertia are I_{xx}, I_{yy}, and I_{zz} and the wheel axial moments of inertia are J_{xx}, J_{yy}, and J_{zz}. Determine the system angular momentum $\mathbf{H}_G$ if the spacecraft angular-velocity components are ω_x, ω_y, and ω_z and the wheel angular rates relative to the spacecraft are Ω_x, Ω_y, and Ω_z.

Problem 7/55

7/55 The solid right-circular cone of mass m, length b, and base radius r spins at an angular rate p about its axis of symmetry. Simultaneously, the bracket and attached shaft axis revolve at the rate ω about the x-axis. Determine the angular momentum $\mathbf{H}_O$ of the cone about point O and its kinetic energy T.

$$Ans. \mathbf{H}_O = m\omega(\tfrac{3}{20}r^2 + \tfrac{1}{10}b^2 + h^2)\mathbf{i} + \tfrac{3}{10}mr^2p\mathbf{j}$$
$$T = \tfrac{1}{2}m\omega^2(\tfrac{3}{20}r^2 + \tfrac{1}{10}b^2 + h^2) + \tfrac{3}{20}mr^2p^2$$

7/56 The solid circular disk of mass $m = 2$ kg and radius $r = 100$ mm rolls in a circle of radius $b = 200$ mm on the horizontal plane without slipping. If the centerline OC of the axle of the wheel rotates about the z-axis with an angular velocity $\omega = 4\pi$ rad/s, determine the expression for the angular momentum of the disk with respect to the fixed point O. Also compute the kinetic energy of the wheel.

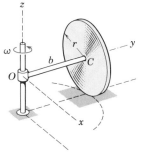

Problem 7/56

7/57 The gyro rotor of Prob. 7/42, repeated here, is spinning at the constant rate $p = 100$ rev/min relative to the x-y-z axes in the direction indicated. If the angle γ between the gimbal ring and horizontal X-Y plane is made to increase at the rate of 4 rad/s and if the unit is forced to precess about the vertical at the constant rate $N = 20$ rev/min, calculate the angular momentum $\mathbf{H}_O$ of the rotor when $\gamma = 30°$. The axial and transverse moments of inertia are $I_{zz} = 6(10^{-3})$ kg·m² and $I_{xx} = I_{yy} = 3(10^{-3})$ kg·m².
Ans. $\mathbf{H}_O = (-12\mathbf{i} + 5.44\mathbf{j} + 69.1\mathbf{k})(10^{-3})$ kg·m²/s

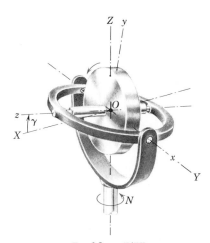

Problem 7/57

7/58 The rectangular plate, with a mass of 3 kg and a uniform small thickness, is welded at the 45° angle to the vertical shaft, which rotates with the angular velocity of 20π rad/s. Determine the angular momentum $\mathbf{H}$ of the plate about O and find the kinetic energy of the plate.

Problem 7/58

Problem 7/59

7/59 In manipulating the dumbbell-shaped object, the robotic device rotates about its base at the angular rate Ω. Simultaneously, the cylindrical arm rotates upward at the rate $\dot\gamma$ and the grips rotate about the line OG at the rate ω_p. For given values of the constant length $\overline{OG} = l$ and the angle γ, determine the angular momentum $\mathbf{H}_G$ of the dumbbell about its mass center G and its angular momentum $\mathbf{H}_O$ about point O. Express $\mathbf{H}_G$ in terms of components along the x-y-z axes attached to the dumbbell and $\mathbf{H}_O$ in terms of components along the X-Y-Z axes. The moments of inertia of the dumbbell about the y- and z-axes are $2ma^2$ and that about the x-axis is negligible.

Ans. $H_{G_x} = 0$
$$H_{G_y} = -2ma^2\dot\gamma$$
$$H_{G_z} = 2ma^2(\omega_p + \Omega\sin\gamma)$$
$$H_{O_X} = 2m\cos\gamma[a^2(\omega_p + \Omega\sin\gamma) - l^2\Omega\sin\gamma]$$
$$H_{O_Y} = -2m\dot\gamma[a^2 + l^2]$$
$$H_{O_Z} = 2m[a^2\sin\gamma(\omega_p + \Omega\sin\gamma) + l^2\Omega\cos^2\gamma]$$

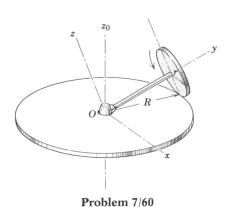

Problem 7/60

▶**7/60** The uniform circular disk of mass m and radius r is mounted on its shaft which is pivoted at O about the vertical z_0-axis. If the disk rolls without slipping and makes one complete trip around the large fixed disk in time τ, write the expression for the angular momentum of the disk with respect to the x-y-z axes through O. (*Hint:* The spin relative to x-y-z is $R\omega/r$ where $\omega = 2\pi/\tau$.)

Ans. $\mathbf{H}_O = \dfrac{2\pi mr^2}{\tau}\left[\dfrac{1}{2}\left(\dfrac{r}{R} - \dfrac{R}{r}\right)\mathbf{j}\right.$

$$\left. + \left(\dfrac{R^2}{r^2} - \dfrac{3}{4}\right)\sqrt{1 - \dfrac{r^2}{R^2}}\,\mathbf{k}\right]$$

Problem 7/61

▶**7/61** The half-cylindrical shell of mass m, radius r, and length b revolves about one edge along the z-axis with a constant rate ω as shown. Determine the angular momentum $\mathbf{H}$ of the shell with respect to the x-y-z axes.

Ans. $\mathbf{H} = mr\omega\left(\dfrac{b}{2}\mathbf{i} + \dfrac{b}{\pi}\mathbf{j} + 2r\mathbf{k}\right)$

7/9 MOMENTUM AND ENERGY EQUATIONS OF MOTION

With the description of angular momentum, inertial properties, and kinetic energy of a rigid body established in the previous two articles we are ready to apply the general momentum and energy equations of motion.

(*a*) *Momentum equations.* In Art. 4/4 of Chapter 4 we established the general linear and angular momentum equations for a system of constant mass. These equations are

$$\Sigma \mathbf{F} = \dot{\mathbf{G}} \qquad\qquad [4/6]$$
$$\Sigma \mathbf{M} = \dot{\mathbf{H}} \qquad\qquad [4/7] \text{ or } [4/9]$$

The general moment relation, Eq. 4/7 or 4/9, is expressed here by the single equation $\Sigma \mathbf{M} = \dot{\mathbf{H}}$ where the terms are taken about either a fixed point O or about the mass center G. In the derivation of the moment principle the derivative of $\mathbf{H}$ was taken with respect to an absolute coordinate system. When $\mathbf{H}$ is expressed in terms of components measured relative to a moving coordinate system x-y-z which has an angular velocity $\mathbf{\Omega}$, then by Eq. 7/7 the moment relation becomes

$$\Sigma \mathbf{M} = \left(\frac{d\mathbf{H}}{dt}\right)_{xyz} + \mathbf{\Omega} \times \mathbf{H}$$
$$= (\dot{H}_x \mathbf{i} + \dot{H}_y \mathbf{j} + \dot{H}_z \mathbf{k}) + \mathbf{\Omega} \times \mathbf{H}$$

The terms in parentheses represent that part of $\dot{\mathbf{H}}$ due to the change in magnitude of the components of $\mathbf{H}$, and the cross-product term represents that part due to the changes in direction of the components of $\mathbf{H}$. Expansion of the cross product and rearrangement of terms give

$$\begin{aligned}
\Sigma \mathbf{M} = &(\dot{H}_x - H_y\Omega_z + H_z\Omega_y)\mathbf{i} \\
&+ (\dot{H}_y - H_z\Omega_x + H_x\Omega_z)\mathbf{j} \\
&+ (\dot{H}_z - H_x\Omega_y + H_y\Omega_x)\mathbf{k}
\end{aligned} \qquad (7/19)$$

Equation 7/19 is the most general form of the moment equation about a fixed point O or about the mass center G. The Ω's are the angular velocity components of rotation of the reference axes, and the H-components in the case of a rigid body are as defined in Eq. 7/12 where the ω's are the components of the angular velocity of the body. We now apply Eq. 7/19 to a rigid body where the coordinate axes are *attached* to the body. Under these conditions when expressed in the x-y-z coordinates, the *moments and products of inertia are invariant with time*, and

$\Omega = \omega$. Thus for axes attached to the body, the three scalar components of Eq. 7/19 become

$$
\begin{aligned}
\Sigma M_x &= \dot{H}_x - H_y \omega_z + H_z \omega_y \\
\Sigma M_y &= \dot{H}_y - H_z \omega_x + H_x \omega_z \\
\Sigma M_z &= \dot{H}_z - H_x \omega_y + H_y \omega_x
\end{aligned}
\qquad (7/20)
$$

Equations 7/20 are the general moment equations for rigid-body motion with axes *attached to the body*. They hold with respect to axes through a fixed point *O* or through the mass center *G*.

In Art. 7/7 it was mentioned that, in general, for any origin fixed to a rigid body, there are three principal axes of inertia with respect to which the products of inertia vanish. If the reference axes coincide with the principal axes of inertia with origin at the mass center *G* or at a point *O* fixed to the body and fixed in space, the factors I_{xy}, I_{yz}, I_{xz} will be zero, and Eqs. 7/20 become

$$
\begin{aligned}
\Sigma M_x &= I_{xx} \dot{\omega}_x - (I_{yy} - I_{zz}) \omega_y \omega_z \\
\Sigma M_y &= I_{yy} \dot{\omega}_y - (I_{zz} - I_{xx}) \omega_z \omega_x \\
\Sigma M_z &= I_{zz} \dot{\omega}_z - (I_{xx} - I_{yy}) \omega_x \omega_y
\end{aligned}
\qquad (7/21)
$$

These relations, known as *Euler's equations**, are extremely useful in the study of rigid-body motion.

(*b*) *Energy equations.* The resultant of all external forces acting on a rigid body may be replaced by the resultant force $\Sigma \mathbf{F}$ acting through the mass center and a resultant couple $\Sigma \mathbf{M}_G$ acting about the mass center. Work is done by the resultant force and the resultant couple at the respective rates $\Sigma \mathbf{F} \cdot \bar{\mathbf{v}}$ and $\Sigma \mathbf{M}_G \cdot \boldsymbol{\omega}$ where $\bar{\mathbf{v}}$ is the linear velocity of the mass center and $\boldsymbol{\omega}$ is the angular velocity of the body. Integration over the time from condition 1 to condition 2 gives the total work done during the time interval. Equating the works done to the respective changes in kinetic energy as expressed in Eq. 7/15 gives

$$
\int_{t_1}^{t_2} \Sigma \mathbf{F} \cdot \bar{\mathbf{v}} \, dt = \tfrac{1}{2} \bar{\mathbf{v}} \cdot \mathbf{G} \Big]_1^2, \qquad \int_{t_1}^{t_2} \Sigma \mathbf{M}_G \cdot \boldsymbol{\omega} \, dt = \tfrac{1}{2} \boldsymbol{\omega} \cdot \mathbf{H}_G \Big]_1^2 \quad (7/22)
$$

These equations express the change in translational kinetic energy and the change in rotational kinetic energy, respectively, for the interval during which $\Sigma \mathbf{F}$ or $\Sigma \mathbf{M}_G$ acts, and the sum of the two expressions equals ΔT.

The work-energy relationship, developed in Chapter 4 for a

---

* Named after Leonhard Euler (1707–1783), a Swiss mathematician.

general system of particles and given by

$$U = \Delta T + \Delta V_e + \Delta V_g \qquad [4/3]$$

was used in Chapter 6 for rigid bodies in plane motion. The equation is equally applicable to rigid-body motion in three dimensions. As we have seen previously, the work-energy approach is of great advantage when analyzing end-point conditions of motion (initial and final). Here the work U done during the interval by all active forces external to the body or system is equated to the sum of the corresponding changes in kinetic energy ΔT, elastic potential energy ΔV_e, and gravitational potential energy ΔV_g. The potential-energy changes are determined in the usual way, as described previously in Art. 3/7.

We shall limit our application of the equations developed in this article to two problems of special interest, parallel-plane motion and gyroscopic motion, discussed in the next two articles.

7/10 PARALLEL-PLANE MOTION

When all particles of a rigid body move in planes which are parallel to a fixed plane, the body has a general form of plane motion, as described in Art. 7/4 and pictured in Fig. 7/3. Every line in such a body which is normal to the fixed plane remains parallel to itself at all times. We take the mass center G as the origin of coordinates x-y-z which are attached to the body with the x-y plane coinciding with the plane of motion P. The components of the angular velocity of both body and attached axes become $\omega_x = \omega_y = 0$, $\omega_z \neq 0$. For this case the angular momentum components from Eq. 7/12 become

$$H_x = -I_{xz}\omega_z \qquad H_y = -I_{yz}\omega_z \qquad H_z = I_{zz}\omega_z$$

and the moment relations of Eqs. 7/20 reduce to

$$\boxed{\begin{aligned} \Sigma M_x &= -I_{xz}\dot{\omega}_z + I_{yz}\omega_z{}^2 \\ \Sigma M_y &= -I_{yz}\dot{\omega}_z - I_{xz}\omega_z{}^2 \\ \Sigma M_z &= I_{zz}\dot{\omega}_z \end{aligned}} \qquad (7/23)$$

It is seen that the third moment equation is equivalent to the second of Eqs. 6/1 where the z-axis passes through the mass center or to Eq. 6/4 if the z-axis passes through a fixed point.

Equations 7/23 hold for an origin of coordinates at the mass center, as shown in Fig. 7/3, or for any origin on a fixed axis of rotation. The three independent force equations of motion which also apply to parallel-plane motion are clearly

$$\Sigma F_x = m\bar{a}_x \qquad \Sigma F_y = m\bar{a}_y \qquad \Sigma F_z = 0$$

Equations 7/23 find special use in describing the effect of dynamic unbalance in rotating shafts and in rolling bodies.

Sample Problem 7/7

The two circular disks each of mass m_1 are connected by the curved bar bent into quarter-circular arcs and welded to the disks. The bar has a mass m_2. The total mass of the assembly is $m = 2m_1 + m_2$. If the disks roll without slipping on a horizontal plane with a constant velocity v of the disk centers, determine the value of the friction force under each disk at the instant represented when the plane of the curved bar is horizontal.

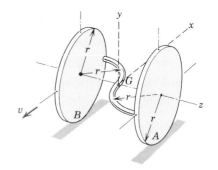

Solution. The motion is identified as parallel-plane motion since the planes of motion of all parts of the system are parallel. The free-body diagram shows the normal forces and friction forces at A and B and the total weight mg acting through the mass center G, which we take as the origin of coordinates which rotate with the body.

We now apply Eqs. 7/23 where $I_{yz} = 0$ and $\dot{\omega}_z = 0$. The moment equation about the y-axis requires determination of I_{xz}. From the diagram showing the geometry of the curved rod and with ρ standing for the mass of the rod per unit length, we have

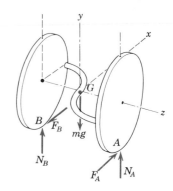

① $[I_{xz} = \int xz\, dm]$ $\qquad I_{xz} = \int_0^{\pi/2} (r\sin\theta)(-r + r\cos\theta)\rho r\, d\theta$

$$+ \int_0^{\pi/2} (-r\sin\theta)(r - r\cos\theta)\rho r\, d\theta$$

Evaluating the integrals gives

$$I_{xz} = -\rho r^3/2 - \rho r^3/2 = -\rho r^3 = -\frac{m_2 r^2}{\pi}$$

The second of Eqs. 7/23 with $\omega_z = v/r$ and $\dot{\omega}_z = 0$ gives

$[\Sigma M_y = -I_{xz}\omega_z^{\,2}]$ $\qquad F_A r + F_B r = -\left(-\frac{m_2 r^2}{\pi}\right)\frac{v^2}{r^2}$

$$F_A + F_B = \frac{m_2 v^2}{\pi r}$$

But with $\bar{v} = v$ constant, $\bar{a}_x = 0$ so that

$[\Sigma F_x = 0]$ $\qquad F_A - F_B = 0 \qquad F_A = F_B$

Thus

$$F_A = F_B = \frac{m_2 v^2}{2\pi r} \qquad\qquad Ans.$$

We also note for the given position that with $I_{yz} = 0$ and $\dot{\omega}_z = 0$ the moment equation about the x-axis gives

② $[\Sigma M_x = 0]$ $\qquad -N_A r + N_B r = 0 \qquad N_A = N_B = mg/2$

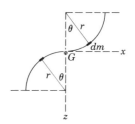

① We must be very careful to observe the correct signs for each of the coordinates of the mass element dm which make up the product xz.

② When the plane of the curved bar is not horizontal, the normal forces under the disks are no longer equal.

PROBLEMS

7/62 Each of the two rods of mass m is welded to the face of the disk which rotates about the vertical axis with a constant angular velocity ω. Determine the bending moment M acting on each rod at its base.   *Ans.* $M = \frac{1}{2}mbl\omega^2$

Problem 7/62

7/63 The shaft with two offset particles of Prob. 7/44, repeated here, rotates with the constant angular rate ω as indicated. Determine the x- and y-components of the bearing reactions at A and B due to the dynamic unbalance of the shaft for the position shown.

$$Ans.\ A_y = -mR\omega^2/3,\ B_y = mR\omega^2/3$$
$$A_x = B_x = 0$$

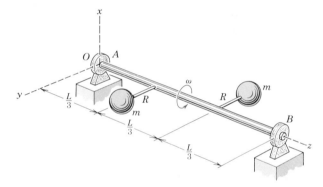

Problem 7/63

7/64 The 6-kg circular disk and attached shaft rotate at a constant speed $\omega = 10\ 000$ rev/min. If the center of mass of the disk is 0.05 mm off center, determine the magnitudes of the horizontal forces A and B supported by the bearings because of the rotational unbalance.

Problem 7/64

Problem 7/65

7/65 The rotating slender bar of Prob. 7/48 is shown again here. The bar has mass m and length l and is welded to the rotating shaft. If the angular velocity ω of the shaft is constant, determine the expression for the force supported by the bearing at B as a function of θ. Consider only the force due to the dynamic unbalance and assume that the bearings can support radial forces only.

Ans. $B = mbl\omega^2/(2c)$

7/66 If a torque $\mathbf{M} = M\mathbf{k}$ is applied to the shaft in Prob. 7/65, determine the x- and y-components of the force supported by the bearing B as the bar and shaft start from rest in the position shown. Neglect the mass of the shaft and consider dynamic forces only.

Problem 7/67

7/67 The rectangular plate of Prob. 7/58 is shown again here. The plate has a mass of 3 kg and is welded to the fixed vertical shaft which rotates at the constant speed of 20π rad/s. Compute the moment $\mathbf{M}$ applied *to* the shaft *by* the plate due to dynamic unbalance.

Ans. $\mathbf{M} = -79.0\mathbf{i}$ N·m

Problem 7/68

7/68 The circular disk of mass m and radius r, repeated from Prob. 7/49, is mounted on the vertical shaft with a small angle α between its plane and the plane of rotation of the shaft. Determine the expression for the bending moment $\mathbf{M}$ acting *on* the shaft due to the wobble of the disk at a shaft speed of ω rad/s.

7/69 The slender rod *OA* of mass *m* and length *l* is pivoted freely about a horizontal axis through *O*. The pivot at *O* and attached shaft rotate with a constant angular speed ω about the vertical *z*-axis. Write the expression for the angle θ assumed by the rod. What minimum value must ω attain before the rod will assume other than a vertical position?

$$Ans.\ \theta = \cos^{-1}\frac{3g}{2l\omega^2},\ \omega_{min} = \sqrt{\frac{3g}{2l}}$$

Problem 7/69

7/70 If the rod of Prob. 7/69 is welded to *O* at an angle θ with the vertical, determine the expression for the bending moment *M* in the rod at *O* due to the combined effect of the angular velocity ω of the shaft and the weight of the rod.

$$Ans.\ M_y = \frac{mgl}{2}\sin\theta\left(1 - \frac{2l}{3g}\omega^2\cos\theta\right)$$

7/71 The uniform slender rod of length *l* is freely hinged to the bracket *A* on the underside of the disk *B*. The disk rotates about a vertical axis with a constant angular velocity ω. Determine the value of ω which will permit the rod to maintain the position $\theta = 60°$ with $b = l/4$.

$$Ans.\ \omega = 2\sqrt{\frac{\sqrt{3}}{l}\frac{g}{l}}$$

Problem 7/71

7/72 The partial ring has a mass ρ per unit length of rim and is welded to the shaft at *A*. Determine the bending moment *M* in the ring at *A* due to the rotation of the ring for an angular velocity ω.

7/73 For the ring section of Prob. 7/72 determine the torsional moment *T* in the ring at *B* (moment about an axis tangent to the rim) as the ring starts from rest under the action of a torque *M* transmitted by the shaft to the ring at *A*. The mass per unit length of rim is ρ. *Ans.* $T = \frac{2}{3}M$

Problem 7/72

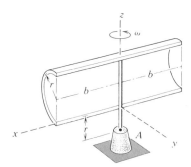

Problem 7/74

7/74 The half-cylindrical shell of radius r, length $2b$, and mass m revolves about the vertical z-axis with a constant angular velocity ω as indicated. Determine the magnitude M of the bending moment in the shaft at A due to both the weight and the rotational motion of the shell.

$$Ans.\ M = \frac{2mr}{\pi}(g + 2r\omega^2)$$

▶**7/75** The homogeneous thin triangular plate of mass m is welded to the horizontal shaft which rotates freely in the bearings at A and B. If the plate is released from rest in the horizontal position shown, determine the magnitude of the bearing reaction at A for the instant just after release.

$$Ans.\ A = mg/6$$

Problem 7/75

▶**7/76** If the homogeneous triangular plate of Prob. 7/75 is released from rest in the position shown, determine the magnitude of the bearing reaction at A after the plate has rotated 90°.

▶**7/77** Each of the two circular disks has a mass m and is welded to the end of the rigid rod of mass m_0 so that the disks have a common z-axis and are separated by a distance b. A couple M, applied to one of the disks with the assembly initially at rest, gives the centers of the disks an acceleration $\mathbf{a} = +a\mathbf{i}$. Friction is sufficient to prevent slipping. Derive expressions for the normal forces N_A and N_B exerted by the horizontal surface on the disks as they begin to roll. Express the results in terms of the acceleration a rather than the moment M.

$$Ans.\ N_A = mg + \frac{m_0 g}{2}\left(1 + \frac{a}{3g}\right)$$
$$N_B = mg + \frac{m_0 g}{2}\left(1 - \frac{a}{3g}\right)$$

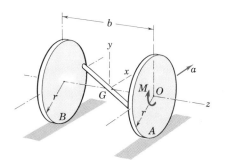

Problem 7/77

▶**7/78** If a couple $\mathbf{M} = M\mathbf{k}$ is applied to one of the disks of Sample Problem 7/7 as they start from the rest position shown, determine the momentary values of the reactions N_A and N_B.

$$Ans.\ N_A = \frac{mg}{2} - \frac{Mm_2}{3\pi mr}$$
$$N_B = \frac{mg}{2} + \frac{Mm_2}{3\pi mr}$$

7/11 *GYROSCOPIC MOTION: STEADY PRECESSION*

One of the most interesting of all problems in dynamics is that of gyroscopic motion. This motion occurs whenever the axis about which a body is spinning is itself rotating about another axis. Whereas the complete description of this motion involves considerable complexity, the most common and useful examples of gyroscopic motion occur when the axis of a rotor spinning at constant speed turns (precesses) about another axis at a steady rate. Our discussion in this article will focus on this special case.

The gyroscope has important engineering applications. With a mounting in gimbal rings (see Fig. 7/19*b*) the gyro is free from external moments, and its axis will retain a fixed direction in space irrespective of the rotation of the structure to which it is attached. In this way the gyro is used for inertial guidance systems and other directional control devices. With addition of a pendulous mass to the inner gimbal ring the earth's rotation causes the gyro to precess so that the spin axis will always point north, and this action forms the basis of the gyro compass. The gyroscope has also found important use as a stabilizing device. The controlled precession of a large gyro mounted in a ship is used to produce a gyroscopic moment to counteract the rolling of a ship at sea. The gyroscopic effect is also an extremely important consideration in the design of bearings for the shafts of rotors which are subjected to forced precessions.

We shall describe gyroscopic action, first, with a simple physical approach which relies on our previous experience with the vector changes encountered in plane dynamics. This approach will help us gain a direct physical insight into gyroscopic action. Next we will make use of the general momentum relation, Eq. 7/19, for a more complete description.

(*a*) *Simplified approach.* Figure 7/14 shows a symmetrical rotor spinning about the *z*-axis with a large angular velocity **p**, known as the *spin velocity*. If we should apply two forces *F* to the rotor axle to form a couple **M** whose vector is directed along the *x*-axis, we will find that the rotor shaft will rotate in the *x-z* plane about the *y*-axis in the sense indicated with a relatively slow angular velocity $\Omega = \dot{\psi}$ known as the *precession velocity*. Thus we identify the spin axis (**p**), the torque axis (**M**), and the precession axis ($\boldsymbol{\Omega}$) where the usual right-hand rule identifies the sense of the rotation vectors. The rotor shaft does *not* rotate about the *x*-axis in the sense of **M** as would be the case if the rotor were not spinning. To help understand this phenomenon a direct analogy may be made between the rotation vectors and the familiar vectors which describe the curvilinear motion of a particle.

Figure 7/15*a* shows a particle of mass *m* moving in the *x-z* plane. The application of a force **F** normal to its linear momen-

Figure 7/14

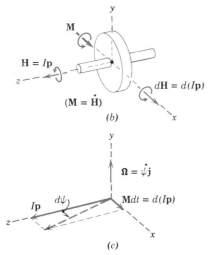

(a)

(b)

(c)

Figure 7/15

tum $\mathbf{G} = m\mathbf{v}$ causes a change $d\mathbf{G} = d(m\mathbf{v})$ in its momentum. We see that $d\mathbf{G}$, and hence $d\mathbf{v}$, is a vector in the direction of the force $\mathbf{F}$ according to Newton's second law $\mathbf{F} = \dot{\mathbf{G}}$ which may be written as $\mathbf{F}\,dt = d\mathbf{G}$. We have covered these relations extensively in Chapter 3. Recall now the analogous equation $\mathbf{M} = \dot{\mathbf{H}}$ which we developed for any prescribed mass system, rigid or nonrigid, about its mass center (Eq. 4/9) or about a fixed point O (Eq. 4/7). We now apply this relation to our symmetrical rotor as shown in Fig. 7/15b. For a high rate of spin $\mathbf{p}$ and a low precession rate $\mathbf{\Omega}$ about the y-axis the angular momentum is represented by the vector $\mathbf{H} = I\mathbf{p}$ where $I = I_{zz}$ is the moment of inertia of the rotor about the spin axis. Initially we are neglecting the small component of angular momentum about the y-axis which accompanies the slow precession. The application of the couple $\mathbf{M}$ normal to $\mathbf{H}$ causes a change $d\mathbf{H} = d(I\mathbf{p})$ in the angular momentum. We see that $d\mathbf{H}$, and hence $d\mathbf{p}$, is a vector in the direction of the couple $\mathbf{M}$ since $\mathbf{M} = \dot{\mathbf{H}}$ which may also be written $\mathbf{M}\,dt = d\mathbf{H}$. Just as the change in the linear momentum vector of the particle is in the direction of the applied force, so is the change in the angular momentum vector of the gyro in the direction of the couple. Thus we see that the vectors $\mathbf{M}$, $\mathbf{H}$, and $d\mathbf{H}$ are analogous to the vectors $\mathbf{F}$, $\mathbf{G}$, and $d\mathbf{G}$. With this insight it is no longer strange to see the rotation vector undergo a change in the direction of $\mathbf{M}$, thereby causing the axis of the rotor to precess about the y-axis.

In Fig. 7/15c we see that during time dt the angular momentum vector $I\mathbf{p}$ has swung through the angle $d\psi$, so that in the limit with $\tan d\psi = d\psi$ we have

$$d\psi = \frac{M\,dt}{Ip} \qquad \text{or} \qquad M = I\frac{d\psi}{dt}\,p$$

Substituting $\Omega = d\psi/dt$ for the magnitude of the precession velocity gives us

$$\boxed{M = I\Omega p} \qquad (7/24)$$

We note that $\mathbf{M}$, $\mathbf{\Omega}$, and $\mathbf{p}$ as vectors are mutually perpendicular, and that their vector relationship may be represented by writing the equation in the cross-product form

$$\boxed{\mathbf{M} = I\mathbf{\Omega} \times \mathbf{p}} \qquad (7/24a)$$

Equations 7/24 and 7/24a apply to moments taken about the mass center or about a fixed point on the axis of rotation.

The correct spatial relationship among the three vectors may be remembered from the fact that $d\mathbf{H}$, and hence $d\mathbf{p}$, is in the direction of $\mathbf{M}$, which establishes the correct sense for the precession Ω. Thus the spin vector **p** always tends to rotate toward the torque vector **M**. Figure 7/16 represents three orientations of the three vectors which are consistent with their correct order. Unless we establish this order correctly in a given problem, we are likely to arrive at a conclusion directly opposite to the correct one. It is emphasized that Eq. 7/24, like $\mathbf{F} = m\mathbf{a}$ and $M = I\alpha$, is an equation of motion, so that the couple **M** represents the couple due to *all* forces acting *on* the rotor as disclosed by a correct *free-body diagram of the rotor*. Also it is pointed out that, when a rotor is forced to precess, as occurs with the turbine in a ship which is executing a turn, the motion will generate a *gyroscopic couple* **M** which obeys Eq. 7/24a both in magnitude and sense.

In the foregoing discussion of gyroscopic motion it was assumed that the spin was large and the precession was small. Although we can see from Eq. 7/24 that for given values of I and M the precession Ω must be small if p is large, let us now examine the influence of Ω on the momentum relations. Again, we restrict our attention to steady precession where Ω has a constant magnitude. Figure 7/17 shows our same rotor again. Because it has a moment of inertia about the y-axis and an angular velocity of precession about this axis, there will be an additional component of angular momentum about the y-axis. Thus we have the two components $H_z = Ip$ and $H_y = I_0\Omega$ where I_0 stands for I_{yy} and, again, I stands for I_{zz}. The total angular momentum is **H** as shown. The change in **H** remains $d\mathbf{H} = \mathbf{M}\,dt$ as previously, and the precession during time dt is the angle $d\psi = M\,dt/H_z = M\,dt/(Ip)$ as before. Thus Eq. 7/24 is still valid and for steady precession is an exact description of the motion as long as the spin axis is perpendicular to the axis around which precession occurs.

Consider now the steady precession of a symmetrical top, Fig. 7/18, spinning about its axis with a high angular velocity **p** and supported at its point O. Here the spin axis makes an angle θ with the vertical Z-axis around which precession occurs. Again, we will neglect the small angular-momentum component due to the precession and consider **H** equal to $I\mathbf{p}$, the angular momentum about the axis of the top associated with the spin only. The moment about O is due to the weight and is $mg\bar{r}\sin\theta$ where $\bar{r}$ is the distance from O to the mass center G. From the diagram we see that the angular momentum vector $\mathbf{H}_O$ has a change $d\mathbf{H}_O = \mathbf{M}_O\,dt$ in the direction of **M** during time dt and that θ is unchanged. The increment in precessional angle around the Z-axis is

$$d\psi = \frac{M_O\,dt}{Ip\sin\theta}$$

Substituting the value of M_O and $\Omega = d\psi/dt$ gives

$$mg\bar{r}\sin\theta = I\Omega p\sin\theta \qquad \text{or} \qquad mg\bar{r} = I\Omega p$$

Figure 7/16

Figure 7/17

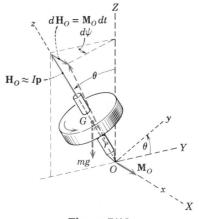

Figure 7/18

which is independent of θ. Introducing the radius of gyration so that $I = mk^2$ and solving for the precessional velocity give

$$\Omega = \frac{g\bar{r}}{k^2 p} \qquad (7/25)$$

Unlike Eq. 7/24 which is an exact description for the rotor of Fig. 7/17 with precession confined to the x-z plane, Eq. 7/25 is an approximation based on the assumption that the angular momentum associated with Ω is negligible compared with that associated with p. We shall see the amount of the error in part (*b*) of this article. On the basis of our analysis the top will have a steady precession at the constant angle θ only if it is set in motion with a value of Ω which satisfies Eq. 7/25. When these conditions are not met, the precession becomes unsteady, and θ may oscillate with an amplitude which increases as the spin velocity decreases. The corresponding rise and fall of the rotation axis is known as *nutation.*

(a)

(b)

(*b*) *More complete analysis.*   We will now make direct use of Eq. 7/19, which is the general angular-momentum equation for a rigid body, by applying it to a body spinning about its axis of rotational symmetry. This equation is valid for rotation about a fixed point or for rotation about the mass center. A spinning top, the rotor of a gyroscope, and a space capsule are examples of bodies whose motions can be described by the equations for rotation about a point. The general moment equations for this class of problems are fairly complex, and their complete solutions involve the use of elliptic integrals and somewhat lengthy computations. A large fraction of engineering problems where the motion is one of rotation about a point involve the steady precession of bodies of revolution which are spinning about their axis of symmetry. These conditions introduce simplifications which greatly facilitate solution of the equations.

Consider a body with axial symmetry, Fig. 7/19*a*, rotating about a fixed point O on its axis which is taken to be the z-direction. With O as origin, the x- and y-axes automatically become principal axes of inertia along with the z-axis. This same description may be used for the rotation of a similar symmetrical body about its center of mass G which is taken as the origin of coordinates, as shown with the gimbaled gyroscope rotor of Fig. 7/19*b*. Again the x- and y-axes are principal axes of inertia for point G. The same description may also be used to represent the rotation about its mass center of an axially symmetric body in space, such as the space capsule in Fig. 7/19*c*. In each case it is noted that, regardless of the rotation of the axes or of the rotation of the body relative to the axes (spin about the z-axis),

(c)

Figure 7/19

Article 7/11   **GYROSCOPIC MOTION: STEADY PRECESSION**   **517**

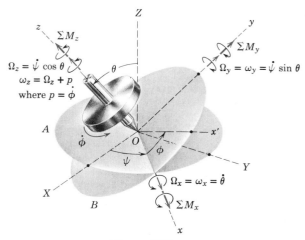

Figure 7/20

the moments of inertia about the x- and y-axes remain constant with time. The principal moments of inertia are again designated $I_{zz} = I$ and $I_{xx} = I_{yy} = I_0$. The products of inertia are, of course, zero.

Before applying Eq. 7/19 we introduce a set of coordinates which provide a natural description for our problem. These coordinates are shown in Fig. 7/20 for the example of rotation about a fixed point O. The axes X-Y-Z are fixed in space, and plane A contains the X-Y axes and the fixed point O on the rotor axis. Plane B contains point O and is always normal to the rotor axis. Angle θ measures the inclination of the rotor axis from the vertical Z-axis and is also a measure of the angle between planes A and B. The intersection of the two planes is the x-axis which is located by the angle ψ from the X-axis. The y-axis lies in plane B, and the z-axis coincides with the rotor axis. The angles θ and ψ completely specify the position of the rotor axis. The angular displacement of the rotor with respect to axes x-y-z is specified by the angle ϕ measured from the x-axis to the x'-axis which is attached to the rotor. The spin velocity becomes $p = \dot{\phi}$.

The components of the angular velocity $\boldsymbol{\omega}$ of the rotor and the angular velocity $\boldsymbol{\Omega}$ of the axes x-y-z from Fig. 7/20 become

$$\Omega_x = \dot{\theta} \qquad\qquad \omega_x = \dot{\theta}$$
$$\Omega_y = \dot{\psi} \sin \theta \qquad \omega_y = \dot{\psi} \sin \theta$$
$$\Omega_z = \dot{\psi} \cos \theta \qquad \omega_z = \dot{\psi} \cos \theta + p$$

It is important to note that the axes and the body have identical x- and y-components of angular velocity but that the z-components differ by the relative angular velocity p.

The angular-momentum components from Eq. 7/12 become

$$H_x = I_{xx}\omega_x = I_0\dot{\theta}$$

$$H_y = I_{yy}\omega_y = I_0\dot{\psi}\sin\theta$$

$$H_z = I_{zz}\omega_z = I(\dot{\psi}\cos\theta + p)$$

Substitution of the angular-velocity and angular-momentum components into Eq. 7/19 yields

$$\Sigma M_x = I_0(\ddot{\theta} - \dot{\psi}^2\sin\theta\cos\theta) + I\dot{\psi}(\dot{\psi}\cos\theta + p)\sin\theta$$

$$\Sigma M_y = I_0(\ddot{\psi}\sin\theta + 2\dot{\psi}\dot{\theta}\cos\theta) - I\dot{\theta}(\dot{\psi}\cos\theta + p) \qquad (7/26)$$

$$\Sigma M_z = I\frac{d}{dt}(\dot{\psi}\cos\theta + p)$$

Equations 7/26 are the general equations of rotation of a symmetrical body about either a fixed point O or the mass center G. In a given problem the solution to the equations will depend on the moment sums applied to the body about the three coordinate axes. We will confine our use of these equations to two particular cases of rotation about a point which are described in the following sections.

(*c*) *Steady-state precession.* We now examine the conditions under which the rotor precesses at a steady rate $\dot{\psi}$ at a constant angle θ and with constant spin velocity p. Thus

$$\dot{\psi} = \text{constant}, \quad \ddot{\psi} = 0$$

$$\theta = \text{constant}, \quad \dot{\theta} = \ddot{\theta} = 0$$

$$p = \text{constant}, \quad \dot{p} = 0$$

and Eqs. 7/26 become

$$\Sigma M_x = \dot{\psi}\sin\theta[I(\dot{\psi}\cos\theta + p) - I_0\dot{\psi}\cos\theta]$$

$$\Sigma M_y = 0 \qquad (7/27)$$

$$\Sigma M_z = 0$$

From these results it is seen that the required moment acting on the rotor about O (or about G) must be in the x-direction since the y- and z-components are zero. Furthermore, with the constant values of θ, $\dot{\psi}$, and p the moment is seen to be of constant magnitude. It is also important to observe that the moment axis is perpendicular to the plane defined by the precession axis (Z-axis) and the spin axis (z-axis).

We may also obtain Eqs. 7/27 by observing that the components of $\mathbf{H}$ remain constant as observed in x-y-z so that $(\dot{\mathbf{H}})_{xyz} =$

0. Since in general $\Sigma\mathbf{M} = (\dot{\mathbf{H}})_{xyz} + \mathbf{\Omega} \times \mathbf{H}$, we have for the case of steady precession

$$\boxed{\Sigma\mathbf{M} = \mathbf{\Omega} \times \mathbf{H}} \qquad (7/28)$$

which reduces to Eqs. 7/27 upon substitution of the values of $\mathbf{\Omega}$ and $\mathbf{H}$.

By far the most common engineering examples of gyroscopic motion occur when precession takes place about an axis which is normal to the rotor axis, as in Fig. 7/14. Thus with the substitution $\theta = \pi/2$, $\omega_z = p$, $\dot{\psi} = \Omega$, and $\Sigma M_x = M$, we have from Eqs. 7/27

$$M = I\Omega p \qquad [7/24]$$

which we derived initially in this article from a direct analysis of this special case.

Now let us examine the steady precession of the rotor (symmetrical top) of Fig. 7/20 for any constant value of θ other than $\pi/2$. The moment ΣM_x about the x-axis is due to the weight of the rotor and is $mg\bar{r}\sin\theta$. Substitution into Eqs. 7/27 and rearrangement of terms give us

$$mg\bar{r} = I\dot{\psi}p - (I_0 - I)\dot{\psi}^2\cos\theta$$

We see that $\dot{\psi}$ is small when p is large, so that the second term on the right-hand side of the equation becomes very small compared with $I\dot{\psi}p$. If we neglect this smaller term we have $\dot{\psi} = mg\bar{r}/(Ip)$ which, upon making the previous substitutions $\Omega = \dot{\psi}$ and $mk^2 = I$, becomes

$$\Omega = \frac{g\bar{r}}{k^2p} \qquad [7/25]$$

We derived this same relation earlier by assuming that the angular momentum was entirely along the spin axis.

(d) Steady precession with zero moment. Consider now the motion of a symmetrical rotor with no external moment about its mass center. Such motion is encountered with spacecraft and projectiles which both spin and precess during flight. Figure 7/21 represents such a body. Here the Z-axis, which has a fixed direction in space, is chosen to coincide with the direction of the angular momentum $\mathbf{H}_G$ which is constant since $\Sigma\mathbf{M}_G = \mathbf{0}$. The x-y-z axes are attached in the manner described in Fig. 7/20. From Fig. 7/21 the three components of momentum are $H_{G_x} = 0$, $H_{G_y} = H_G\sin\theta$, $H_{G_z} = H_G\cos\theta$. From the defining relations, Eqs. 7/12, with the notation of this article these components are also given by $H_{G_x} = I_0\omega_x$, $H_{G_y} = I_0\omega_y$, $H_{G_z} = I\omega_z$. Thus $\omega_x = \Omega_x = 0$ so that θ is constant. This result means that the motion is

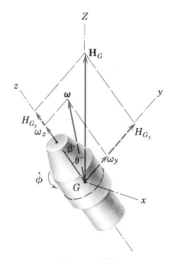

Figure 7/21

one of steady precession about the constant $\mathbf{H}_G$ vector. With no x-component the angular velocity $\boldsymbol{\omega}$ of the rotor lies in the y-z plane along with the Z-axis and makes an angle β with the z-axis. The relationship between β and θ is obtained from $\tan \theta = H_{G_y}/H_{G_z} = I_0\omega_y/(I\omega_z)$, which is

$$\tan \theta = \frac{I_0}{I} \tan \beta \qquad (7/29)$$

Thus the angular velocity $\boldsymbol{\omega}$ makes a constant angle β with the spin axis.

The rate of precession is easily obtained from Eq. 7/27 with $M = 0$ which gives

$$\dot{\psi} = \frac{Ip}{(I_0 - I)\cos \theta} \qquad (7/30)$$

It is clear from this relation that the direction of the precession depends on the relative magnitudes of the two moments of inertia.

If $I_0 > I$, then $\beta < \theta$, as indicated in Fig. 7/22*a*, and the precession is said to be *direct*. Here the body cone rolls on the outside of the space cone.

If $I > I_0$, then $\theta < \beta$, as indicated in Fig. 7/22*b*, and the precession is said to be *retrograde*. In this instance the space cone is internal to the body cone, and $\dot{\psi}$ and p have opposite signs.

If $I = I_0$, then $\theta = \beta$ from Eq. 7/29, and Fig. 7/22 shows that both angles must be zero to be equal. For this case the body has no precession and merely rotates with an angular velocity $\mathbf{p}$. This condition occurs for a body with point symmetry, such as with a homogeneous sphere.

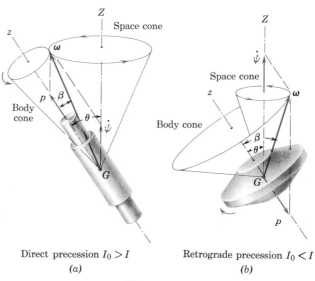

Direct precession $I_0 > I$

(a)

Retrograde precession $I_0 < I$

(b)

Figure 7/22

Sample Problem 7/8

The turbine rotor in a ship's power plant has a mass of 1000 kg with center of mass at G and a radius of gyration of 200 mm. The rotor shaft is mounted in bearings A and B with its axis in the horizontal fore-and-aft direction and turns counterclockwise at a speed of 5000 rev/min when viewed from the stern. Determine the vertical components of the bearing reactions at A and B if the ship is making a turn to port (left) of 400-m radius at a speed of 25 knots (1 knot = 0.514 m/s). Does the bow of the ship tend to rise or fall because of the gyroscopic action?

Solution. The vertical component of the bearing reactions will equal the static reactions R_1 and R_2 due to the weight of the rotor plus or minus the increment ΔR due to the gyroscopic effect. The moment principle from statics easily gives $R_1 = 5886$ N and $R_2 = 3924$ N. The ① given directions of the spin velocity $\mathbf{p}$ and the precession velocity $\mathbf{\Omega}$ are shown with the free-body diagram of the rotor. Since the spin axis always tends to rotate toward the torque axis, we see that the torque axis $\mathbf{M}$ points in the starboard direction as shown. The sense of the ΔR's is, therefore, up at B and down at A to produce the couple $\mathbf{M}$. Thus the bearing reactions at A and B are

$$R_A = R_1 - \Delta R \quad \text{and} \quad R_B = R_2 + \Delta R$$

The precession velocity Ω is the speed of the ship divided by the radius of its turn.

$$[v = \rho\Omega] \qquad \Omega = \frac{25(0.514)}{400} = 0.0321 \text{ rad/s}$$

Equation 7/24 is now applied around the mass center G of the rotor to give

$$[M = I\Omega p] \quad 1.500\,\Delta R = 1000(0.200)^2(0.0321)\left(\frac{5000[2\pi]}{60}\right)$$

$$\Delta R = 449 \text{ N}$$

The required bearing reactions become

$$R_A = 5886 - 449 = 5437 \text{ N} \quad \text{and} \quad R_B = 3924 + 449 = 4373 \text{ N}$$

Ans.

We now observe from the principle of action and reaction that ② the forces just computed are those exerted *on* the rotor shaft *by* the structure of the ship. Consequently, the equal and opposite reactions are applied to the ship *by* the rotor shaft, as shown in the bottom sketch. Therefore the effect of the gyroscopic couple is to generate the increments ΔR shown, and the bow will tend to fall and the stern to rise (but only slightly).

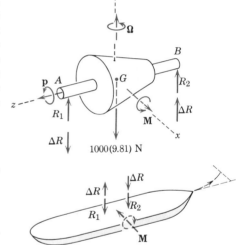

① If the ship is making a left turn, the rotation is counterclockwise when viewed from above, and the precession vector Ω is up by the right-hand rule.

② After figuring the correct sense of $\mathbf{M}$ on the rotor, the common mistake is to apply it to the ship in the same sense, forgetting the action-and-reaction principle. Clearly the results are then reversed. (Be certain not to make this mistake when operating a vertical gyro stabilizer in your yacht to counteract its roll!)

Sample Problem 7/9

A proposed space station is closely approximated by four uniform spherical shells each of mass m and radius r. The mass of the connecting structure and internal equipment may be neglected as a first approximation. If the station is designed to rotate about its z-axis at the rate of one revolution every 4 seconds, determine (a) the number n of complete cycles of precession for each revolution about the z-axis if the plane of rotation deviates only slightly from a fixed orientation, and (b) find the period τ of precession if the spin axis z makes an angle of 20° with respect to the axis of fixed orientation about which precession occurs. Draw the space and body cones for this latter condition.

Solution. (a) The number of precession cycles or wobbles for each revolution of the station about the z-axis would be the ratio of the precessional velocity $\dot\psi$ to the spin velocity p which, from Eq. 7/30, is

$$\frac{\dot\psi}{p} = \frac{I}{(I_0 - I)\cos\theta}$$

The moments of inertia are

$$I_{zz} = I = 4[\tfrac{2}{3}mr^2 + m(2r)^2] = \tfrac{56}{3}mr^2$$

$$I_{xx} = I_0 = 2(\tfrac{2}{3})mr^2 + 2[\tfrac{2}{3}mr^2 + m(2r)^2] = \tfrac{32}{3}mr^2$$

With θ very small, $\cos\theta \cong 1$, and the ratio of angular rates becomes

$$n = \frac{\dot\psi}{p} = \frac{\frac{56}{3}}{\frac{32}{3} - \frac{56}{3}} = -\frac{7}{3} \qquad Ans.$$

The minus sign indicates retrograde precession where, in the present case, $\dot\psi$ and p are essentially of opposite sense. Thus the station will make 7 wobbles for every 3 revolutions.

(b) For $\theta = 20°$ and $p = 2\pi/4$ rad/s, the period of precession or wobble is $\tau = 2\pi/|\dot\psi|$, so that from Eq. 7/30

$$\tau = \frac{2\pi}{2\pi/4}\left|\frac{I_0 - I}{I}\right|\cos\theta = 4(\tfrac{3}{7})\cos 20° = 1.61 \text{ s} \qquad Ans.$$

The precession is retrograde, and the body cone is external to the space cone as shown in the illustration where the body-cone angle, from Eq. 7/29, is

$$\tan\beta = \frac{I}{I_0}\tan\theta = \frac{56/3}{32/3}(0.3640) = 0.6370 \qquad \beta = 32.50°$$

① Our theory is based on the assumption that $I_{xx} = I_{yy}$ = the moment of inertia about any axis through G perpendicular to the z-axis. Such is the case here, and students should prove it to their own satisfaction.

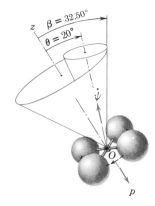

PROBLEMS

7/79 If the ship of Sample Problem 7/8 is on a straight course but its bow is rising as a wave passes under it, determine the direction of the gyroscopic moment exerted *by* the turbine rotor *on* the hull structure and its effect on the motion of the ship.
Ans. Bow tends to swing to port

7/80 The 225-kg rotor for a turbojet engine has a radius of gyration of 250 mm and rotates counterclockwise at 18 000 rev/min when viewed from the front of the airplane. If the airplane is traveling at 1000 km/h and making a turn to the left of 3-km radius, compute the gyroscopic moment M which the rotor bearings must support. Does the nose of the airplane tend to rise or fall as a result of the gyroscopic action?
Ans. $M = 2450$ N·m, nose tends to rise

7/81 A car makes a turn to the right on a level road. Determine whether the normal reaction under the right rear wheel is increased or decreased as a result of the gyroscopic effect of the precessing wheels.

7/82 The special-purpose fan is mounted as shown. The motor armature, shaft, and blades have a combined mass of 2.2 kg with radius of gyration of 60 mm. The axial position b of the 0.8-kg block A can be adjusted. With the fan turned off, the unit is balanced about the x-axis when $b = 180$ mm. The motor and fan operate at 1725 rev/min in the direction shown. Determine the value of b which will produce a steady precession of 0.2 rad/s about the positive y-axis. *Ans.* $b = 216$ mm

Problem 7/82

7/83 An airplane has just cleared the runway with a takeoff speed v. Each of its freely spinning wheels has a mass m with a radius of gyration k about its axle. As seen from the front of the airplane the wheel precesses at the angular rate Ω as the landing strut is folded into the wing about its pivot O. As a result of the gyroscopic action the supporting member A exerts a torsional moment M on B to prevent the tubular member from rotating in the sleeve at B. Determine M and identify whether it is in the sense of M_1 or M_2.

Problem 7/83

7/84 The two identical disks are rotating freely on the shaft with angular velocities equal in magnitude and opposite in direction as shown. The shaft in turn is caused to rotate about the vertical axis in the sense indicated. Prove whether the shaft bends as in *A* or as in *B* because of gyroscopic action.

Ans. A

Problem 7/84

7/85 A small air compressor for an aircraft cabin consists of the 3.50-kg turbine *A* which drives the 2.40-kg blower *B* at a speed of 20 000 rev/min. The shaft of the assembly is mounted transversely to the direction of flight and is viewed from the rear of the aircraft in the figure. The radii of gyration of *A* and *B* are 79.0 and 71.0 mm, respectively. Calculate the radial forces exerted on the shaft by the bearings at *C* and *D* if the aircraft executes a clockwise roll (rotation about the longitudinal flight axis) of 2 rad/s viewed from the rear of the aircraft. Neglect the small moments caused by the weights of the rotors.

Problem 7/85

7/86 The figure shows a gyro mounted with vertical axis and used to stabilize a hospital ship against rolling. The motor *A* turns the pinion which precesses the gyro by rotating the large precession gear *B* and attached rotor assembly about a horizontal transverse axis in the ship. The rotor turns inside the housing at a clockwise speed of 960 rev/min when viewed from the top and has a mass of 80 Mg with radius of gyration of 1.45 m. Calculate the moment exerted on the hull structure by the gyro if the motor turns the precession gear *B* at the rate of 0.320 rad/s. In which of the two directions, (*a*) or (*b*), should the motor turn in order to counteract a roll of the ship to port?

Ans. M = 5410 kN·m, direction (b)

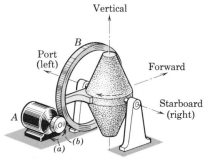

Problem 7/86

7/87 The electric motor has a total mass of 10 kg and is supported by the mounting brackets *A* and *B* attached to the rotating disk. The armature of the motor has a mass of 2.5 kg and a radius of gyration of 35 mm and turns counterclockwise at a speed of 1725 rev/min as viewed from *A* to *B*. The turntable revolves about its vertical axis at the constant rate of 48 rev/min in the direction shown. Determine the vertical components of the forces supported by the mounting brackets at *A* and *B*.

Dimensions in Millimeters

Problem 7/87

7/88 The 50-kg wheel is a solid circular disk which rolls on the horizontal plane in a circle of 600-mm radius. The wheel shaft is pivoted about the axis *O-O* and is driven by the vertical shaft at the constant rate $N = 48$ rev/min about the *Z*-axis. Determine the normal force *R* between the wheel and the horizontal surface. Neglect the weight of the horizontal shaft. *Ans.* $R = 712$ N

Problem 7/88

7/89 An experimental car is equipped with a gyro stabilizer to counteract completely the tendency of the car to tip when rounding a curve (no change in normal force between tires and road). The rotor of the gyro has a mass m_0 and a radius of gyration *k*, and is mounted in fixed bearings on a shaft which is parallel to the rear axle of the car. The center of mass of the car is a distance *h* above the road, and the car is rounding an unbanked level turn at a speed *v*. At what speed *p* should the rotor turn and in what direction to counteract completely the tendency of the car to overturn for either a right or a left turn? The combined mass of car and rotor is *m*.

7/90 A solid right-circular cone of mass *m*, base radius $r = 100$ mm, and altitude $H = 3r = 300$ mm is suspended freely at its vertex *O* and is spinning about its geometric axis at the rate $p = 2700$ rev/min. Determine the period for steady slow precession of the cone about the vertical. Does the result depend upon the angle θ made by the cone axis with the vertical? *Ans.* $\tau = 2.41$ s

Problem 7/90

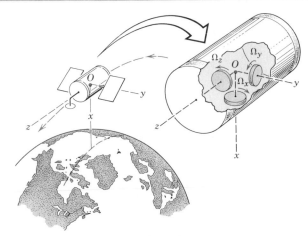

Problem 7/91

7/91 The earth-scanning satellite is in a circular orbit of period τ. The angular velocity of the satellite about its y- or pitch-axis is $\omega = 2\pi/\tau$, and the angular rates about the x- and z-axes are zero. Thus the x-axis of the satellite always points to the center of the earth. The satellite has a reaction-wheel attitude-control system consisting of the three wheels shown, each of which may be variably torqued by its individual motor. The angular rate Ω_z of the z-wheel relative to the satellite is Ω_0 at time $t = 0$, and the x- and y-wheels are at rest relative to the satellite at $t = 0$. Determine the axial torques M_x, M_y, and M_z which must be exerted by the motors on the shafts of their respective wheels in order that the angular velocity ω of the satellite will remain constant. The moment of inertia of each reaction wheel about its axis is J. The x and z reaction-wheel speeds are harmonic functions of the time with a period equal to that of the orbit. Plot the variations of the torques and the relative wheel speeds Ω_x, Ω_y, and Ω_z as functions of the time during one orbit period. (*Hint:* The torque to accelerate the x-wheel equals the reaction of the gyroscopic moment on the z-wheel, and vice versa.)

$$\textit{Ans.}\ M_x = -J\omega\Omega_0 \cos \omega t$$
$$M_y = 0$$
$$M_z = -J\omega\Omega_0 \sin \omega t$$

7/92 Show that the rate of steady precession for an axially symmetrical rotor under zero moment about its mass center G is given by $\dot{\psi} = I\omega_z/(I_0 \cos \theta)$ as well as by Eq. 7/30.

Problem 7/93

7/93 The cylindrical shell is rotating in space about its geometric axis. If the axis has a slight wobble, for what ratios of l/r will the motion be direct or retrograde precession?

$$\textit{Ans.}\ \text{Direct}\ l/r > \sqrt{6},\ \text{retrograde}\ l/r < \sqrt{6}$$

7/94 The two identical circular disks each of mass m and radius r are spinning as a rigid unit about their common axis. Determine the value of b for which no precessional motion can take place if the unit is free to move in space.

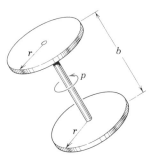

Problem 7/94

7/95 The two solid cones, each of mass m, radius r, and altitude h, are rigidly connected at their vertices and are spinning in space about their common axis. For what ratio h/r will precession be impossible? *Ans.* $h/r = \frac{1}{2}$

Problem 7/95

7/96 The thin ring is projected into the air with a spin velocity of 300 rev/min. If its geometric axis is observed to have a very slight precessional wobble, determine the frequency f of the wobble.

Problem 7/96

7/97 A boy throws a thin circular disk (like a Frisbee) with a spin rate of 300 rev/min. The plane of the disk is seen to wobble through a total angle of 10°. Calculate the period τ of the wobble and indicate whether the precession is direct or retrograde.
Ans. $\tau = 0.0996$ s, retrograde

Problem 7/97

7/98 The figure shows a football in three common in-flight configurations. Case (a) is a perfectly thrown spiral pass with a spin rate of 120 rev/min. Case (b) is a wobbly spiral pass again with a spin rate of 120 rev/min about its own axis but with the axis wobbling through a total angle of 20°. Case (c) is an end-over-end place kick with a rotational rate of 120 rev/min. For each case specify the values of p, θ, β, and $\dot{\psi}$ as defined in this article. The moment of inertia about the ball's long axis is 0.3 of that about the transverse axis of symmetry.

Problem 7/98

Problem 7/99

7/99 The spacecraft shown is symmetrical about its z-axis and has a radius of gyration of 720 mm about this axis. The radii of gyration about the x- and y-axes through the mass center are both equal to 540 mm. When moving in space the z-axis is observed to generate a cone with a total vertex angle of 4° as it precesses about the axis of total angular momentum. If the spacecraft has a spin velocity $\dot{\phi}$ about its z-axis of 1.5 rad/s, compute the period τ of each full precession. Is the spin vector in the positive or negative z-direction?

Ans. $\tau = 1.83$ s, spin vector in negative z-direction

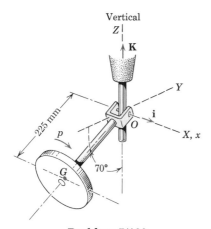

Problem 7/100

7/100 The 4-kg rotor with radius of gyration of 75 mm rotates on ball bearings at a speed of 3000 rev/min about its shaft OG. The shaft is free to pivot about the X-axis as well as to rotate about the Z-axis. Calculate the vector $\boldsymbol{\Omega}$ for precession about the Z-axis. Neglect the mass of shaft OG and compute the gyroscopic couple $\mathbf{M}$ exerted by the shaft on the rotor at G.

Ans. $\boldsymbol{\Omega} = -1.25\mathbf{K}$ rad/s, $\mathbf{M} = 8.30\mathbf{i}$ N·m

7/101 The two-bladed airplane propeller has a constant angular velocity $\dot{\phi}$ about its shaft z-axis. The airplane is turning to the pilot's left so that the airplane and propeller have a constant angular velocity $\dot{\psi}$ (precession) about a vertical axis. Determine the x-, y-, and z-components of the moment exerted on the propeller by its shaft as functions of the angular position ϕ of the propeller. The moments of inertia about the x-, y-, and z-axes are I_{xx}, 0, I_{zz}, respectively. (*Suggestion:* With axes x-y-z attached to the propeller make use of Eqs. 7/20.)

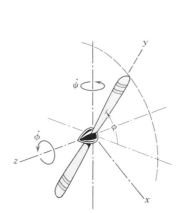

Problem 7/101

7/102 The uniform slender bar of mass m and length l is centrally mounted on the shaft A-A, about which it rotates with a constant speed $\dot{\phi} = p$. Simultaneously, the yoke is forced to rotate about the x-axis with a constant speed ω_0. As a function of ϕ determine the magnitude of the torque M required to maintain the constant speed ω_0. (*Hint*: Apply Eq. 7/19 to obtain the x-component of M.)

Problem 7/102

7/103 Let the rotor of Fig. 7/18 have a spin velocity of 3600 rev/min and execute steady precession about the vertical Z-axis at the constant angle $\theta = 30°$. Furthermore, $\bar{r} = 200$ mm, $k = 150$ mm, and the moment of inertia about the x-axis is five times that about the spin axis. Calculate the precession $\dot{\psi} = \Omega$ using Eq. 7/25 and obtain a close approximation to the percentage error e in this value.

Ans. $\dot{\psi} = 0.231$ rad/s, $e = 0.21\%$ too low

▶**7/104** The steady-state precession of an axisymmetric body is described by Eq. 7/27, which is a quadratic equation in $\dot{\psi}$. Let $\Sigma M_x = M$ and determine the two roots $\dot{\psi}_1$ and $\dot{\psi}_2$ for large values of $\dot{\phi}$. First express the solution in the general form $\dot{\psi} = [\ ](1 \pm \sqrt{1 - \{\ \}})$. Then expand the radical into a convergent series retaining only the first two terms. Compare $\dot{\psi}_1 = \dot{\psi}_{(-)}$ with the precession derived for the rotor of Fig. 7/18 in the earlier part of the article. The root $\dot{\psi}_2 = \dot{\psi}_{(+)}$ represents a fast precession which occurs at a higher energy level difficult to achieve experimentally.

Ans. $\dot{\psi}_1 = \dfrac{M}{Ip \sin \theta}$, $\dot{\psi}_2 = \dfrac{Ip}{(I_0 - I) \cos \theta}$

where $p^2 > \dfrac{4}{I^2} M(I_0 - I) \operatorname{ctn} \theta$

▶**7/105** The solid cylindrical rotor has a mass of 30 kg and is mounted in bearings A and B of the frame which rotates about the vertical Z-axis. If the rotor spins at the constant rate $p = 50$ rad/s relative to the frame and if the frame itself rotates at the constant rate $\Omega = 30$ rad/s, compute the bending moment $\mathbf{M}$ in the shaft at C which the lower portion of the shaft exerts on the upper portion. Also compute the kinetic energy T of the rotor. Neglect the mass of the frame.

Ans. $\mathbf{M} = 132.1\mathbf{i}$ N·m, $T = 99.7$ J

Problem 7/105

Problem 7/106

Problem 7/107

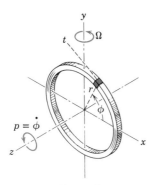

Problem 7/108

▶7/106 The cylindrical rotor B and attached shaft turn about the z-axis in a bearing which is a part of the clevis at O. The rotor and shaft have a mass of 12.5 kg with inertia properties closely approximated by those of the solid circular cylinder. The vertical shaft rotates at the constant rate $N_1 = 60$ rev/min. Rotor B rolls without slipping on the surface of the vertical cylinder A, which rotates at the constant rate $N_2 = 240$ rev/min independently of the vertical shaft and clevis. Neglect any vertical friction force between B and A and determine the contact force N between the two rotors and the forces supported by the clevis pin at O.

Ans. $N = 18.90$ N

▶7/107 A projectile moving through the atmosphere with a velocity $\bar{\mathbf{v}}$ which makes a small angle θ with its geometric axis is subjected to a resultant aerodynamic force $\mathbf{R}$ essentially opposite in direction to $\bar{\mathbf{v}}$ as shown. If $\mathbf{R}$ passes through a point C slightly ahead of the mass center G, determine the expression for the minimum spin velocity p for which the projectile will be spin-stabilized with $\dot{\theta} = 0$. The moment of inertia about the spin axis is I and that about a transverse axis through G is I_0. (*Hint:* Determine M_x and substitute into Eq. 7/27. Express the result as a quadratic equation in $\dot{\psi}$ and determine the minimum value of p for which the expression under the radical is positive.)

▶7/108 Derive Eq. 7/24 by relating the forces to the accelerations for a differential element of the thin ring of mass m. The ring has a constant angular velocity p about the z-axis and is given an additional constant angular velocity Ω about the y-axis by the application of an external moment M (not shown). (*Hint:* The acceleration of the element in the z-direction is due to (*a*) the change in magnitude of its velocity component in this direction resulting from Ω and (*b*) the change in direction of its x-component of velocity.)

▶ 7/109 The solid circular disk of mass m and small thickness is spinning freely on its shaft at the rate p. If the assembly is released in the vertical position at $\theta = 0$ with $\dot{\theta} = 0$, determine the horizontal components of the forces A and B exerted by the respective bearings on the horizontal shaft as the position $\theta = \pi/2$ is passed. Neglect the mass of the two shafts compared with m, and neglect all friction. Solve by using the appropriate moment equations.

$$Ans.\ A_z = -\frac{m\dot{\theta}}{2}\left(\frac{r^2}{2b}p + l\dot{\theta}\right)$$

$$B_z = \frac{m\dot{\theta}}{2}\left(\frac{r^2}{2b}p - l\dot{\theta}\right)$$

$$\text{where } \dot{\theta} = 2\sqrt{\frac{2gl}{r^2 + 4l^2}}$$

Problem 7/109

▶ 7/110 The elements of a gyrocompass are shown in the figure where the rotor, at north latitude γ on the surface of the earth, is mounted in a single gimbal ring which is free to rotate about the fixed vertical y-axis. The rotor axis is, then, able to rotate in the horizontal x-z plane as measured by the angle β from the north direction. Assume that the gyro spins at the rate p and that it has mass moments of inertia about the spin axis and transverse axis through G of I and I_0, respectively. Show that the gyro axis oscillates about the north direction according to the equation $\ddot{\beta} + K^2\beta = 0$ where $K^2 = I\omega_0 p \cos\gamma/I_0$, and that the period of oscillation about the north direction for small values of β is $\tau = 2\pi\sqrt{I_0/(I\omega_0 p \cos\gamma)}$. (*Hint:* The components of the angular velocity Ω of the axes in terms of the angular velocity ω_0 of the earth are

$$\Omega_x = -\omega_0 \cos\gamma \sin\beta$$

$$\Omega_y = \omega_0 \sin\gamma + \dot{\beta}$$

$$\Omega_z = \omega_0 \cos\gamma \cos\beta$$

which may be used in the y-component of the moment equations, Eqs. 7/19, to determine β as a function of time. Note that the square of the angular velocity ω_0 of the earth is small and may be neglected compared with the product $\omega_0 p$.)

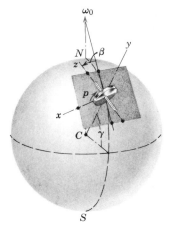

Problem 7/110

VIBRATION AND TIME RESPONSE

8

8/1 INTRODUCTION

An important and special class of problems in dynamics deals with the linear and angular motions of bodies which oscillate or respond to applied disturbances in the presence of restoring forces. A few examples of this class of dynamics problems are the response of an engineering structure to earthquakes, the vibration of an unbalanced rotating machine, the time response of the plucked string of a musical instrument, the wind-induced vibration of power lines, and the flutter of aircraft wings. In many cases, excessive vibration levels must be reduced due to material limitations or human factors.

In the analysis of every engineering problem, the system under scrutiny must be represented by a physical model. It is often permissible to represent a *continuous* or *distributed-parameter system* (one in which the mass and spring elements are continuously spread over space) by a *discrete* or *lumped-parameter model* (one in which the mass and spring elements are separate and concentrated). Such a modeling scheme is especially desirable when some portions of a continuous system are relatively massive in comparison to other portions. For example, the physical model of a ship propeller shaft is often assumed to be a massless but twistable rod with a disk rigidly attached to each end—one disk representing the turbine and the other representing the propeller. As a second example, we observe that the mass of springs may often be neglected in comparison to that of attached bodies. It should be noted that not every system is reducible to a discrete model. For example, the transverse vibration of a diving board after the departure of the diver is a somewhat difficult problem of distributed-parameter vibration. In this chapter, we shall begin the study of discrete systems, limiting our discussion to those whose configurations may be described with one displacement variable. Such systems are said to possess *one degree of freedom*. For a more detailed study which includes the treatment of two or more degrees of freedom and continuous systems, the student should consult one of the many textbooks devoted solely to the subject of vibrations.

Chapter 8 is divided into four subsequent sections: Article 8/2 treats the free vibration of particles and Art. 8/3 introduces the forced vibration of particles. Each of these two articles is subdivided into undamped- and damped-motion categories. In Art. 8/4 we discuss the vibration of rigid bodies. Lastly, an energy approach to the solution of vibration problems is presented in Art. 8/5. The study of this chapter is greatly facilitated if one observes that the topic of vibrations is a direct application of the principles of kinetics as developed in Chapters 3 and 6. In particular, a complete free-body diagram *drawn for an arbitrary positive value of the displacement variable,* followed by application of the appropriate governing equations of dynamics, will yield the equation of motion. From this equation of motion, which is a second-order ordinary differential equation, one may obtain all information of interest, such as the motion frequency, period, or the motion itself as a function of time.

8/2 FREE VIBRATION OF PARTICLES

When a spring-mounted body is disturbed from its equilibrium position, its ensuing motion in the absence of any imposed external forces is termed *free vibration.* In every actual case of free vibration, there exists some retarding or damping force which tends to diminish the motion. Common damping forces are those due to mechanical and fluid friction. In part (*a*) we consider the ideal case where the damping forces are small enough to be neglected. In part (*b*) we treat the case where the damping is appreciable and must be accounted for.

(*a*) **Undamped free vibration.**   We begin by considering the horizontal vibration of the simple frictionless spring-mass system of Fig. 8/1*a*. Note that the variable x denotes the displacement of the mass from the equilibrium position, which, for this system, is also the position of zero spring deflection. Figure 8/1*b* shows a plot of the force F_s necessary to deflect the spring versus the corresponding spring deflection for various types of springs. Although nonlinear hard and soft springs are useful in some applications, we will restrict our attention to the linear spring. Such a spring exerts a restoring force $-kx$ on the mass—that is, when the mass is displaced to the right, the spring force is to the left, and vice versa. We must be careful to distinguish between the forces of magnitude F_s which must be applied at both ends of the massless spring to cause tension or compression and the force $F = -kx$ of equal magnitude which the spring exerts on the mass.

The constant of proportionality k is known as the *spring constant, modulus,* or *stiffness* and has the units N/m.

The equation of motion for the body of Fig. 8/1a is obtained by first drawing its free-body diagram. Applying Newton's second law in the form $\Sigma F_x = m\ddot{x}$ gives

$$-kx = m\ddot{x} \qquad \text{or} \qquad m\ddot{x} + kx = 0 \qquad (8/1)$$

The oscillation of a mass subjected to a linear restoring force as described by this equation is called *simple harmonic motion* and is characterized by acceleration that is proportional to the displacement but of opposite sign. Equation 8/1 is normally written as

$$\ddot{x} + \omega_n^2 x = 0 \qquad (8/2)$$

where

$$\omega_n = \sqrt{k/m} \qquad (8/3)$$

is a convenient substitution whose physical significance will be clarified shortly.

Because we anticipate an oscillatory motion, we look for a solution which gives x as a periodic function of time. Thus a logical choice is

$$x = A \cos \omega_n t + B \sin \omega_n t \qquad (8/4)$$

or, alternatively,

$$x = C \sin (\omega_n t + \psi) \qquad (8/5)$$

Direct substitution of these expressions into Eq. 8/2 verifies that each expression is a valid solution to the equation of motion. The constants A and B, or C and ψ, are determined from knowledge of the initial displacement x_0 and initial velocity $\dot{x}_0$ of the mass. For example, if we work with the solution form of Eq. 8/4 and evaluate x and $\dot{x}$ at time $t = 0$, we obtain

$$x_0 = A \qquad \text{and} \qquad \dot{x}_0 = B\omega_n$$

Substitution of these values of A and B into Eq. 8/4 yields

$$x = x_0 \cos \omega_n t + \frac{\dot{x}_0}{\omega_n} \sin \omega_n t \qquad (8/6)$$

The constants C and ψ of Eq. 8/5 can be determined in terms of given initial conditions in a similar manner. Evaluation of Eq. 8/5 and its first time derivative at $t = 0$ gives

$$x_0 = C \sin \psi \qquad \text{and} \qquad \dot{x}_0 = C\omega_n \cos \psi$$

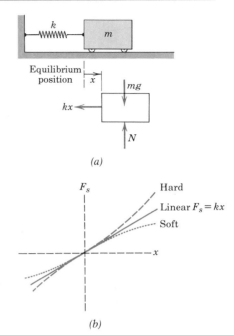

(a)

(b)

Figure 8/1

Solving for C and ψ yields

$$C = \sqrt{x_0^2 + (\dot{x}_0/\omega_n)^2} \qquad \psi = \tan^{-1}(x_0\omega_n/\dot{x}_0)$$

Substitution of these values into Eq. 8/5 gives

$$x = \sqrt{x_0^2 + (\dot{x}_0/\omega_n)^2}\,\sin\left[\omega_n t + \tan^{-1}(x_0\omega_n/\dot{x}_0)\right] \qquad (8/7)$$

Equations 8/6 and 8/7 represent two different mathematical expressions for the same time-dependent motion. We observe that $C = \sqrt{A^2 + B^2}$ and $\psi = \tan^{-1}(A/B)$.

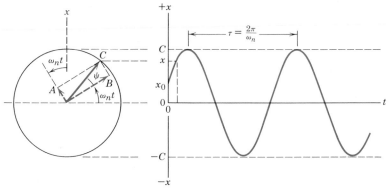

Figure 8/2

The motion may be represented graphically, Fig. 8/2, where x is seen to be the projection onto a vertical axis of the rotating vector of length C. The vector rotates at the constant angular velocity $\omega_n = \sqrt{k/m}$ which is called the *natural circular frequency* and has the units radians per second. The number of complete cycles per unit time is the *natural frequency* $f_n = \omega_n/2\pi$ and is expressed in hertz (1 hertz (Hz) = 1 cycle per second). The time required for one complete motion cycle (one rotation of the reference vector) is the *period* of the motion and is given by $\tau = 1/f_n = 2\pi/\omega_n$. We also see from the figure that x is the sum of the projections onto the vertical axis of two perpendicular vectors whose magnitudes are A and B and whose vector sum C is the *amplitude*. Vectors A, B, and C rotate together with the constant angular velocity ω_n. Thus, as we have already seen, $C = \sqrt{A^2 + B^2}$ and $\psi = \tan^{-1}(A/B)$.

As a further note on the free undamped vibration of particles, we see that, if the system of Fig. 8/1a is rotated 90° clockwise to obtain the system of Fig. 8/3 where the motion is vertical rather than horizontal, the equation of motion (and therefore all system properties) is unchanged if we continue to define x as the displacement from the equilibrium position. The equilibrium position now involves a nonzero spring deflection δ_{st}. From the free-body diagram of Fig. 8/3 Newton's second law gives

$$-k(\delta_{st} + x) + mg = m\ddot{x}$$

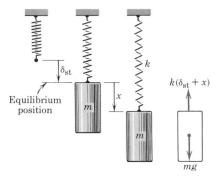

Figure 8/3

At the equilibrium position $x = 0$, the force sum must be zero, so that

$$-k\delta_{st} + mg = 0$$

Thus we see that the pair of forces $-k\delta_{st}$ and mg on the left side of the motion equation cancel, giving

$$m\ddot{x} + kx = 0$$

which is identical to Eq. 8/1. The lesson here is that by defining the displacement variable to be zero at equilibrium rather than at the position of zero spring deflection, we may ignore the equal and opposite forces associated with equilibrium.*

(b) **Damped free vibration.** Every mechanical system possesses some inherent degree of friction, which acts as a consumer of mechanical energy. Precise mathematical models of the dissipative friction forces are usually complex. The dashpot or viscous damper is a device intentionally added to systems for the purpose of limiting or retarding vibration. It consists of a cylinder filled with a viscous fluid and a piston with holes or other passages by which the fluid can flow from one side of the piston to the other. Simple dashpots arranged as shown schematically in Fig. 8/4a exert a force F_d whose magnitude is proportional to the velocity of the mass as depicted in Fig. 8/4b. The constant of proportionality c is known as the *viscous damping coefficient* and has units of N·s/m. The direction of the damping force as applied to the mass is opposite to that of the velocity $\dot{x}$. Hence the force on the mass is $-c\dot{x}$.

Complex dashpots with internal flow-rate-dependent one-way valves can produce a different damping coefficient in extension than in compression; nonlinear characteristics are also possible. We shall restrict our attention to the simple linear dashpot.

The equation of motion for the body with damping is determined from the free-body diagram as shown in Fig. 8/4a. Newton's second law gives

$$-kx - c\dot{x} = m\ddot{x} \qquad \text{or} \qquad m\ddot{x} + c\dot{x} + kx = 0 \qquad (8/8)$$

In addition to the variable substitution $\omega_n = \sqrt{k/m}$, it is convenient, for reasons which will shortly become evident, to introduce the combination of constants $\zeta = c/(2m\omega_n)$. The quantity ζ (zeta) is called the *viscous damping factor* or *damping ratio* and is a measure of the severity of the damping. The student should verify

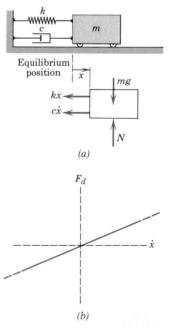

(a)

(b)

Figure 8/4

---

* For nonlinear systems, all forces, including the static forces associated with equilibrium, should be included in the analysis.

that ζ is nondimensional. Equation 8/8 may now be written as

$$\boxed{\ddot{x} + 2\zeta\omega_n\dot{x} + \omega_n^2x = 0}\qquad(8/9)$$

In order to solve the equation of motion, Eq. 8/9, we assume solutions of the form

$$x = Ae^{\lambda t}$$

Substitution into Eq. 8/9 yields the characteristic equation

$$\lambda^2 + 2\zeta\omega_n\lambda + \omega_n^2 = 0$$

whose roots are

$$\lambda_1 = \omega_n[-\zeta + \sqrt{\zeta^2 - 1}]\qquad\lambda_2 = \omega_n[-\zeta - \sqrt{\zeta^2 - 1}]$$

By superposition, the general solution is

$$\begin{aligned}x &= A_1e^{\lambda_1t} + A_2e^{\lambda_2t}\\ &= A_1e^{(-\zeta+\sqrt{\zeta^2-1})\omega_nt} + A_2e^{(-\zeta-\sqrt{\zeta^2-1})\omega_nt}\end{aligned}\qquad(8/10)$$

Since $0 \leq \zeta \leq \infty$, the radicand $(\zeta^2 - 1)$ may be positive, negative, or even zero, giving rise to the following three categories of damped motion:

I. $\zeta > 1$ (*overdamped*). The roots λ_1 and λ_2 are distinct real negative numbers. The motion as given by Eq. 8/10 decays so that x approaches zero for large values of time t. There is no oscillation and therefore no period associated with the motion.

II. $\zeta = 1$ (*critically damped*). The roots λ_1 and λ_2 are equal real negative numbers ($\lambda_1 = \lambda_2 = -\omega_n$) and the solution to the differential equation for the special case of equal roots is given by

$$x = (A_1 + A_2t)e^{-\omega_nt}$$

Again, the motion decays with x approaching zero for large time, and the motion is nonperiodic. A critically damped system, when excited with an initial velocity or displacement (or both), will approach equilibrium faster than will an overdamped system. Figure 8/5 depicts an actual response for both an overdamped and a critically damped system to an initial displacement x_0 with no initial velocity $\dot{x}_0$.

III. $\zeta < 1$ (*underdamped*). Noting that the radicand $(\zeta^2 - 1)$ is negative and recalling that $e^{(a+b)} = e^ae^b$ enables us to rewrite Eq. 8/10 as

$$x = \{A_1e^{i\sqrt{1-\zeta^2}\omega_nt} + A_2e^{-i\sqrt{1-\zeta^2}\omega_nt}\}e^{-\zeta\omega_nt}$$

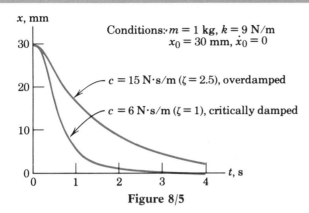

Figure 8/5

where $i = \sqrt{-1}$. It is convenient to let a new variable ω_d represent the combination $\omega_n\sqrt{1 - \zeta^2}$. Thus

$$x = \{A_1 e^{i\omega_d t} + A_2 e^{-i\omega_d t}\}e^{-\zeta\omega_n t}$$

Use of the Euler formula $e^{\pm ix} = \cos x \pm i \sin x$ allows the previous equation to be written as

$$x = \{A_1(\cos \omega_d t + i \sin \omega_d t) + A_2(\cos \omega_d t - i \sin \omega_d t)\}e^{-\zeta\omega_n t}$$

$$= \{(A_1 + A_2) \cos \omega_d t + i(A_1 - A_2) \sin \omega_d t\}e^{-\zeta\omega_n t}$$

$$= \{A_3 \cos \omega_d t + A_4 \sin \omega_d t\}e^{-\zeta\omega_n t} \qquad (8/11)$$

where $A_3 = (A_1 + A_2)$ and $A_4 = i(A_1 - A_2)$. We have shown with Eqs. 8/4 and 8/5 that the sum of two harmonics, such as those in the braces of Eq. 8/11, can be replaced by a single trigonometric function which involves a phase angle. Thus Eq. 8/11 can be written as

$$x = \{C \sin (\omega_d t + \psi)\}e^{-\zeta\omega_n t}$$

or

$$x = Ce^{-\zeta\omega_n t} \sin (\omega_d t + \psi) \qquad (8/12)$$

Equation 8/12 represents an exponentially decreasing harmonic function as shown in Fig. 8/6 for specific numerical values. The frequency $\omega_d = \omega_n\sqrt{1 - \zeta^2}$ is called the *damped natural frequency*. The *damped period* is given by $\tau_d = 2\pi/\omega_d = 2\pi/(\omega_n\sqrt{1 - \zeta^2})$.

It is important to note that the expressions developed for the constants C and ψ in terms of initial conditions for the case of no damping in part (*a*) are not valid for the case of damping in part (*b*). To find C and ψ if damping is present, one must begin anew, setting the general displacement expression of Eq. 8/12 and its first time derivative, both evaluated at time $t = 0$, equal to the initial displacement x_0 and initial velocity $\dot{x}_0$, respectively.

Conditions: $m = 1$ kg, $k = 36$ N/m
$c = 1$ N·s/m $(\zeta = 0.083)$
$x_0 = 30$ mm, $\dot{x}_0 = 0$

Figure 8/6

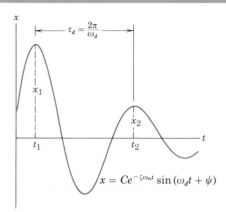

Figure 8/7

It is frequently desirable to experimentally determine the value of the damping ratio ζ for an underdamped system. The usual reason is that the value of the viscous damping coefficient c is not otherwise well known. The system is excited by initial conditions and a plot of the displacement x versus time t, such as that shown schematically in Fig. 8/7, is generated. Two successive amplitudes x_1 and x_2 are measured and their ratio

$$\frac{x_1}{x_2} = \frac{Ce^{-\zeta\omega_n t_1}}{Ce^{-\zeta\omega_n(t_1 + \tau_d)}} = e^{\zeta\omega_n \tau_d}$$

is formed. The *logarithmic decrement* δ is defined as

$$\delta = \ln\left(\frac{x_1}{x_2}\right) = \zeta\omega_n \tau_d = \zeta\omega_n \frac{2\pi}{\omega_n\sqrt{1 - \zeta^2}} = \frac{2\pi\zeta}{\sqrt{1 - \zeta^2}}$$

From this equation, we may solve for ζ and obtain

$$\zeta = \frac{\delta}{\sqrt{(2\pi)^2 + \delta^2}}$$

For a small damping ratio $x_1 \cong x_2$ and $\delta \ll 1$, so that $\zeta \cong \delta/2\pi$. If x_1 and x_2 are so close in value that experimental distinction between them is impractical, the above analysis may be modified by using two observed amplitudes which are n cycles apart.

Sample Problem 8/1

A 10-kg body is suspended from a spring of constant $k = 2.5$ kN/m. At time $t = 0$ it has a downward velocity of 0.5 m/s as it passes through the position of static equilibrium. Determine

(a) the static spring deflection δ_{st}
(b) the natural frequency of the system in both rad/s (ω_n) and cycles/s (f_n)
(c) the system period τ
(d) the displacement x as a function of time, where x is measured from the position of static equilibrium
(e) the maximum velocity v_{max} attained by the mass
(f) the maximum acceleration a_{max} attained by the mass.

$k = 2.5$ kN/m

$m = 10$ kg

Solution. (a) From the spring relationship $F_s = kx$, we see that at equilibrium,

①
$$mg = k\delta_{st} \qquad \delta_{st} = \frac{mg}{k} = \frac{10(9.81)}{2500} = 0.0392 \text{ m or } 39.2 \text{ mm} \qquad Ans.$$

(b)
$$\omega_n = \sqrt{\frac{k}{m}} = \sqrt{\frac{2500}{10}} = 15.81 \text{ rad/s} \qquad Ans.$$

$$f_n = (15.81)\left(\frac{1}{2\pi}\right) = 2.52 \text{ cycles/s (Hz)} \qquad Ans.$$

(c)
$$\tau = \frac{1}{f_n} = \frac{1}{2.52} = 0.397 \text{ s} \qquad Ans.$$

② (d) From Eq. 8/6:

$$x = x_0 \cos \omega_n t + \frac{\dot{x}_0}{\omega_n} \sin \omega_n t$$

$$= (0)\cos 15.81t + \frac{0.5}{15.81} \sin 15.81t$$

$$= 0.0316 \sin 15.81t \text{ m} \quad \text{or} \quad 31.6 \sin 15.81t \text{ mm} \qquad Ans.$$

As an exercise, let us determine x from the alternative Eq. (8/7):

$$x = \sqrt{x_0^2 + (\dot{x}_0/\omega_n)^2} \, \sin\left[\omega_n t + \tan^{-1}\left(\frac{x_0\omega_n}{\dot{x}_0}\right)\right]$$

$$= \sqrt{0^2 + \left(\frac{0.5}{15.81}\right)^2} \, \sin\left[15.81t + \tan^{-1}\left(\frac{(0)(15.81)}{0.5}\right)\right]$$

$$= 0.0316 \sin 15.81t \text{ m}$$

(e) The velocity is

$$\dot{x} = 15.81(0.0316) \cos 15.81t = 0.5 \cos 15.81t \text{ m/s}.$$

Since the cosine function can be no greater than 1 or less than -1, the maximum velocity v_{max} is 0.5 m/s, which, in this case, is the initial velocity. *Ans.*

(f) The acceleration is

$$\ddot{x} = -15.81(0.5) \sin 15.81t = -7.91 \sin 15.81t \text{ m/s}^2$$

The maximum acceleration a_{max} is 7.91 m/s^2. *Ans.*

$F_s = k\delta_{st}$

Equilibrium position

$k(\delta_{st} + x)$ kx

mg

mg

① The student should always exercise extreme caution in the matter of units. In the subject of vibrations, it is quite easy to commit errors due to mixing of feet and inches, cycles and radians, and other pairs which frequently enter the calculations.

② Recall that when we reference the motion to the position of static equilibrium, the equation of motion, and therefore its solution, for the present system is identical to that for the horizontally vibrating system.

Sample Problem 8/2

The 8-kg body is moved 0.2 m to the right of the equilibrium position and released from rest at time $t = 0$. Determine its displacement at time $t = 2$ seconds. The viscous damping coefficient c is 20 N·s/m, and the spring stiffness k is 32 N/m.

Solution. We must first determine whether the system is underdamped, critically damped, or overdamped. For that purpose, we compute the damping ratio ζ.

$$\omega_n = \sqrt{k/m} = \sqrt{32/8} = 2 \text{ rad/s} \qquad \zeta = \frac{c}{2m\omega_n} = \frac{20}{2(8)(2)} = 0.625$$

Since $\zeta < 1$, the system is underdamped. The damped natural frequency is $\omega_d = \omega_n\sqrt{1 - \zeta^2} = 2\sqrt{1 - (0.625)^2} = 1.561$ rad/s. The motion is given by Eq. 8/12 and is

$$x = Ce^{-\zeta\omega_n t} \sin(\omega_d t + \psi) = Ce^{-1.25t} \sin(1.561t + \psi)$$

The velocity is then

$$\dot{x} = -1.25Ce^{-1.25t}\sin(1.561t + \psi) + 1.561Ce^{-1.25t}\cos(1.561t + \psi)$$

Evaluating the displacement and velocity at time $t = 0$ gives

$$x_0 = C\sin\psi = 0.2 \qquad \dot{x}_0 = -1.25C\sin\psi + 1.561C\cos\psi = 0$$

Solving the two equations for C and ψ yields $C = 0.256$ m and $\psi = 0.896$ rad. Therefore the displacement in meters is

$$x = 0.256e^{-1.25t}\sin(1.561t + 0.896)$$

① Evaluation for time $t = 2$ seconds gives $x_2 = -0.0162$ m.   *Ans.*

① We note that the exponential factor $e^{-1.25t}$ is 0.082 at $t = 2$ seconds. Hence $\zeta = 0.625$ represents severe damping, although the motion is still oscillatory.

Sample Problem 8/3

The two fixed counterrotating pulleys are driven at the same angular speed ω_0. A round bar is placed off center on the pulleys as shown. Determine the natural frequency of the resulting bar motion. The coefficient of kinetic friction between the bar and pulleys is μ_k.

Solution. The free-body diagram of the bar is constructed for an arbitrary displacement x from the central position as shown. The governing equations are

$$[\Sigma F_x = m\ddot{x}] \qquad \mu_k N_A - \mu_k N_B = m\ddot{x}$$

$$[\Sigma F_y = 0] \qquad N_A + N_B - mg = 0$$

① $[\Sigma M_A = 0] \qquad aN_B - \left(\frac{a}{2} + x\right)mg = 0$

Eliminating N_A and N_B from the first equation yields

② $$\ddot{x} + \frac{2\mu_k g}{a}x = 0$$

We recognize the form of this equation as that of Eq. 8/2, so that the natural frequency in radians per second is $\omega_n = \sqrt{2\mu_k g/a}$ and the natural frequency in cycles per second is

$$f_n = \frac{1}{2\pi}\sqrt{2\mu_k g/a} \qquad \textit{Ans.}$$

① Since the bar is slender and does not rotate, the use of a moment equilibrium equation is justified.

② We note that the angular speed ω_0 does not enter the equation of motion. The reason for this is our assumption that the kinetic friction force does not depend upon the relative velocity at the contacting surface.

PROBLEMS

(Unless otherwise indicated, assume that all motion variables are referenced to the equilibrium position.)

(Undamped, free vibrations)

8/1 When a 3-kg collar is placed upon the pan which is attached to the spring of unknown constant, the additional static deflection of the pan is observed to be 50 mm. Determine the spring constant k.

Ans. $k = 589$ N/m

Problem 8/1

8/2 Show that the natural frequency of a vertically-oriented spring-mass system, such as that of Prob. 8/1, may be expressed as $\omega_n = \sqrt{g/\delta_{st}}$ where δ_{st} is the static deflection.

8/3 Determine the natural frequency of the spring-mass system in both rad/s and cycles/s (Hz).

Ans. $\omega_n = 10$ rad/s, $f_n = 1.592$ Hz

Problem 8/3

8/4 For the system of Prob. 8/3, determine the displacement x of the mass as a function of time if the mass is released from rest at time $t = 0$ from a position 30 mm to the right of the equilibrium position.

8/5 For the system of Prob. 8/3, determine the displacement x of the mass as a function of time if the mass is released at time $t = 0$ from a position 30 mm to the left of the equilibrium position with an initial velocity of 120 mm/s to the right. Determine the amplitude C of the motion.

Ans. $x = 32.3 \sin(10t - 1.190)$ mm
$C = 32.3$ mm

8/6 Replace the springs in each of the two cases shown by a single spring of stiffness k (equivalent spring stiffness) which will cause each mass to vibrate with its original frequency.

Ans. (a) $k = k_1 + k_2$

(b) $\dfrac{1}{k} = \dfrac{1}{k_1} + \dfrac{1}{k_2}$

Problem 8/6

Problem 8/7

Problem 8/8

Problem 8/9

Problem 8/10

Problem 8/11

8/7 An old car being moved by a magnetic crane pickup is dropped from a short distance above the ground. Neglect any damping effects of its worn-out shock absorbers and calculate the natural frequency f_n in cycles per second (Hz) of the vertical vibration which occurs after impact with the ground. Each of the four springs on the 1000-kg car has a constant of 17.5 kN/m. Because the center of mass is located midway between the axles and the car is level when dropped, there is no rotational motion. State any assumptions.

Ans. $f_n = 1.332$ Hz

8/8 If the 100-kg mass has a downward velocity of 0.5 m/s as it passes through its equilibrium position, calculate the magnitude a_{max} of its maximum acceleration. Each of the two springs has a stiffness $k = 180$ kN/m.   *Ans.* $a_{max} = 30$ m/s^2

8/9 If both springs are unstretched when the mass is in the central position shown, determine the static deflection δ_{st} of the mass. What is the period of oscillatory motion about the position of static equilibrium?

8/10 An energy-absorbing car bumper with its springs initially undeformed has an equivalent spring constant of 525 kN/m. If the 1200-kg car approaches a massive wall with a speed of 8 km/h, determine (*a*) the velocity v of the car as a function of time during contact with the wall, where $t = 0$ is the beginning of the impact, and (*b*) the maximum deflection x_{max} of the bumper.

Ans. (*a*) $v = 2.22 \cos 20.9t$ m/s
(*b*) $x_{max} = 106.2$ mm

8/11 Explain how the values of the mass m_1 and the spring constant k may be experimentally determined if the mass m_2 is known. Develop expressions for m_1 and k in terms of specified experimental results. Note the existence of at least three ways to solve the problem.

8/12 A 55-kg woman stands in the center of an end-supported board and causes a midspan deflection of 22 mm. If she flexes her knees slightly in order to cause a vertical vibration, what is the frequency f_n of the motion? Assume elastic response of the board and neglect its relatively small mass.

Ans. $f_n = 3.36$ Hz

22 mm

Problem 8/12

8/13 A 90-kg man stands at the end of a diving board and causes a vertical oscillation which is observed to have a period of 0.6 s. What is the static deflection δ_{st} at the end of the board? Neglect the mass of the board.

δ_{st}

Problem 8/13

8/14 Calculate the natural frequency f_n of vertical oscillation of the spring-loaded cylinder when it is set into motion. Both springs are in tension at all times. *Ans.* $f_n = 3.90$ Hz

8/15 Derive the differential equation for the vertical vibration of the cylinder and determine its frequency f_n. Neglect the friction and mass of the pulley.

8/16 The cylinder of mass m is given a vertical displacement y_0 from its equilibrium position and released. Write the differential equation for the vertical vibration of the cylinder and find the period τ of its motion. Neglect the friction and mass of the pulley.

Ans. $\tau = \pi \sqrt{\dfrac{m}{k}}$

$k = 3000$ N/m

10 kg

$k = 3000$ N/m

Problem 8/14

k

m

Problem 8/15

8/17 A small particle of mass m is attached to two highly tensioned wires as shown. Determine the system natural frequency ω_n for small vertical oscillations if the tension T in both wires is assumed to be constant. Is the calculation of the small static deflection of the particle necessary?

Ans. $\omega_n = \sqrt{\dfrac{2T}{ml}}$

k

m

y

Problem 8/16

m

l l

Problem 8/17

Problem 8/18

8/18 The cylindrical buoy floats in salt water (density 1030 kg/m³) and has a mass of 800 kg with a low center of mass to keep it stable in the upright position. Determine the frequency f_n of vertical oscillation of the buoy. Assume the water level remains undisturbed adjacent to the buoy.

8/19 The time rate of change of acceleration is called "jerk" J. Specify the maximum value of the amplitude C, measured from the neutral position, for an undamped linear harmonic oscillator of mass m which will limit the jerk to a value J_0. The stiffness of the elastic supports is k.

$$Ans. \ C = \left(\frac{m}{k}\right)^{3/2} J_0$$

Problem 8/20

8/20 A 3-kg piece of putty is dropped 2 m onto the initially stationary 28-kg block which is supported by four springs, each of which has a constant $k = 800$ N/m. Determine the displacement x as a function of time during the resulting vibration, where x is measured from the initial position of the block as shown.

8/21 Determine the period of the system shown if the mass of the lever is negligible. Assume small oscillations.

$$Ans. \ \tau = 2\pi \sqrt{\frac{m}{k_1 + \dfrac{b^2}{a^2}k_2}}$$

Problem 8/21

8/22 The weighing platform has a mass m and is connected to the spring of stiffness k by the system of levers shown. Derive the differential equation for small vertical oscillations of the platform and find the period τ. Designate y as the platform displacement from the equilibrium position and neglect the mass of the levers.

Problem 8/22

8/23 Derive the differential equation of motion for the system shown in terms of the variable x_1. The mass of the linkage is negligible. State the natural frequency ω_n' in rad/s for the case $k_1 = k_2 = k$ and $m_1 = m_2 = m$. Assume small oscillations throughout.

$$Ans. \left[m_1 + \frac{b^2}{a^2} m_2 \right] \ddot{x}_1 + \left[k_1 + \frac{b^2}{a^2} k_2 \right] x_1 = 0$$

$$\omega_n' = \sqrt{k/m}$$

Problem 8/23

8/24 Calculate the natural frequency ω_n of the system shown in the figure. The mass and friction of the pulleys are negligible.

$$Ans. \ \omega_n = \sqrt{\frac{4k}{5m}}$$

Problem 8/24

(Damped, free vibrations)

8/25 Determine the value of the damping ratio ζ for the simple spring-mass-dashpot system shown.

$$Ans. \ \zeta = 1.5$$

Problem 8/25

8/26 The period τ_d of damped linear oscillation for a certain 1-kg mass is 0.32 s. If the stiffness of the supporting linear spring is 850 N/m, calculate the damping coefficient c.

8/27 Viscous damping is added to an initially undamped spring-mass system. For what value of the damping ratio ζ will the damped natural frequency ω_d be equal to 75 percent of the natural frequency of the original undamped system?

$$Ans. \ \zeta = 0.661$$

8/28 The addition of damping to an undamped spring-mass system causes its period to increase by 20 percent. Determine the damping ratio ζ.

8/29 A linear harmonic oscillator having a mass of 1.10 kg is set into motion with viscous damping. If the frequency is 10 Hz and if two successive amplitudes a full cycle apart are measured to be 4.65 mm and 4.30 mm as shown, compute the viscous damping coefficient c.

$$Ans. \ c = 1.721 \ N \cdot s/m$$

Problem 8/29

35 kN/m

45 kg

c

Problem 8/30

8/30 Determine the value of the viscous damping co-efficient c for which the system shown is critically damped.

8/31 The 2-kg mass of Prob. 8/25 is released from rest at a distance x_0 to the right of the equilibrium position. Determine the displacement x as a function of time.

$$\textit{Ans. } x = x_0(1.171e^{-2.67t} - 0.171e^{-18.33t})$$

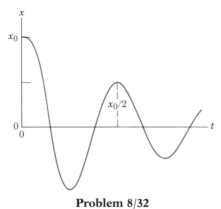

Problem 8/32

8/32 A damped spring-mass system is released from rest from a positive initial displacement x_0. If the succeeding maximum positive displacement is $x_0/2$, determine the damping ratio ζ of the system.

8/33 The system shown is released from rest from an initial position x_0. Determine the overshoot displacement x_1. $\textit{Ans. } x_1 = -0.163x_0$

8/34 Develop an expression for the logarithmic decrement δ in terms of the displacements x_1 and x_{n+1} which are n cycles apart, as shown on the plot of displacement versus time.

Problem 8/33

Problem 8/34

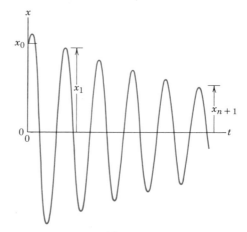

8/35 The damping constant for a spring-supported 0.4-kg mass is 3.60 N·s/m. The mass is displaced 18 mm from its equilibrium position and then released. If the frequency of the vibration is observed to be 3 cycles/s, compute the displacement at the end of the third cycle following release.

Ans. $x_3 = 0.200$ mm

8/36 The mass of a given critically damped system is released at time $t = 0$ from the position $x_0 > 0$ with a negative initial velocity. Determine the critical value $(\dot{x}_0)_c$ of the initial velocity below which the mass will pass through the equilibrium position.

8/37 The mass of the system shown is released from rest at $x_0 = 150$ mm when $t = 0$. Determine the displacement x at $t = 0.5$ s if (*a*) $c = 200$ N·s/m and (*b*) $c = 300$ N·s/m.

Ans. (*a*) $x = 110.4$ mm
(*b*) $x = 118.0$ mm

Problem 8/37

8/38 A linear oscillator with mass m, spring stiffness k, and damping coefficient c is set into motion when released from a displaced position. Derive an expression for the energy loss E in one cycle beginning at a displacement maximum x_1. Express the result in terms of x_1. (See Fig. 8/7.)

Ans. $E = \frac{1}{2}kx_1^2(1 - e^{-4\pi\zeta/\sqrt{1-\zeta^2}})$

8/39 Derive the differential equation of motion for the system shown in its equilibrium position. Neglect the mass of link AB and assume small oscillations.

Problem 8/39

8/40 Develop the equation of motion in terms of the variable x for the system shown. Determine an expression for the damping ratio ζ in terms of the given system properties. Neglect the mass of the crank AB and assume small oscillations about the equilibrium position shown.

Ans. $\ddot{x} + \frac{b^2}{a^2}\frac{c}{m}\dot{x} + \frac{k}{m}x = 0$

$\zeta = \frac{1}{2}\frac{b^2}{a^2}\frac{c}{\sqrt{km}}$

Problem 8/40

8/3 FORCED VIBRATION OF PARTICLES

Although there are many significant applications of free vibrations, the most important class of vibration problems is that where the motion is continuously excited by a disturbing force. The force may be externally applied or may be generated within the system by such means as unbalanced rotating parts. Forced vibrations may also be excited by the motion of the system foundation.

Various forms of forcing functions $F = F(t)$ and foundation displacements $x_B = x_B(t)$ are depicted in Fig. 8/8. The harmonic force shown in the *a*-part of the figure occurs frequently in engineering practice, and the understanding of the analysis associated with harmonic forces is a necessary first step in the study of more complex forms. For this reason, we will focus our attention on harmonic excitation.

We first consider the system of Fig. 8/9*a*, where the body is subjected to the external harmonic force $F = F_0 \sin \omega t$, in which F_0 is the force amplitude and ω is the driving frequency (in radians per second). The student should take care to distinguish between

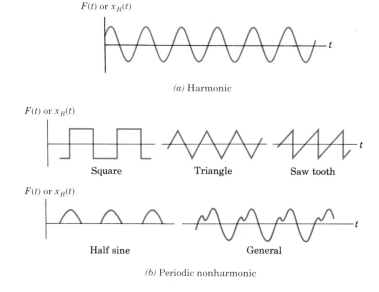

(a) Harmonic

Square Triangle Saw tooth

Half sine General

(b) Periodic nonharmonic

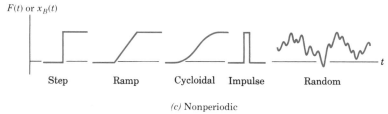

Step Ramp Cycloidal Impulse Random

(c) Nonperiodic

Figure 8/8

$\omega_n = \sqrt{k/m}$, which is a property of the system, and ω, which is a property of the force applied to the system. We also note that for a force $F = F_0 \cos \omega t$, one merely substitutes $\cos \omega t$ for $\sin \omega t$ in the results about to be developed. From the free-body diagram of Fig. 8/9a, we may apply Newton's second law to obtain

$$- kx - c\dot{x} + F_0 \sin \omega t = m\ddot{x}$$

In standard form, with the same variable substitutions made in Art. 8/2, the equation of motion becomes

$$\ddot{x} + 2\zeta\omega_n\dot{x} + \omega_n^2 x = \frac{F_0 \sin \omega t}{m} \qquad (8/13)$$

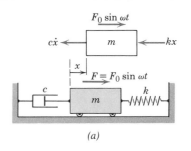

(a)

In many cases the excitation of the mass is due not to a directly applied force but to the movement of the base or foundation to which the mass is connected by springs or other compliant mountings. Examples of such applications are seismographs, vehicle suspensions, and structures shaken by earthquakes. Harmonic movement of the base is equivalent to the direct application of a harmonic force. To show this, consider the system of Fig. 8/9b where the spring is attached to the movable base. The free-body diagram shows the mass displaced a distance x from the neutral or equilibrium position it would have if the base were in its neutral position. The base, in turn, is assumed to have a harmonic movement $x_B = b \sin \omega t$. Note that the spring deflection is the difference between the inertial displacements of the mass and the base. From the free-body diagram Newton's second law gives

$$- k(x - x_B) - c\dot{x} = m\ddot{x}$$

(b)

Figure 8/9

or

$$\ddot{x} + 2\zeta\omega_n\dot{x} + \omega_n^2 x = \frac{kb \sin \omega t}{m} \qquad (8/14)$$

We see immediately that Eq. 8/14 is exactly the same as our basic equation of motion, Eq. 8/13, in that F_0 is replaced by kb. Consequently, all the results about to be developed apply to either Eq. 8/13 or 8/14.

(*a*) *Undamped forced vibration.* First we treat the case where damping is negligible ($c = 0$). Our basic equation of motion, Eq. 8/13, becomes

$$\ddot{x} + \omega_n^2 x = \frac{F_0}{m} \sin \omega t \qquad (8/15)$$

The complete solution to Eq. 8/15 is a sum of the complementary solution x_c, which is the general solution of Eq. 8/15 with the right side set to zero, and the particular solution x_p, which is *any* solution to the complete equation. Thus $x = x_c + x_p$. We developed the complementary solution in section (*a*) of Art. 8/2. A particular solution is investigated by assuming that the response to the force should resemble the form of the force term. To that end, we assume

$$x_p = X \sin \omega t \qquad (8/16)$$

where X is the magnitude (in units of length) of the particular solution. Substituting this expression into Eq. 8/15 and solving for X yield

$$X = \frac{F_0/k}{1 - (\omega/\omega_n)^2} \qquad (8/17)$$

Thus the particular solution becomes

$$x_p = \frac{F_0/k}{1 - (\omega/\omega_n)^2} \sin \omega t \qquad (8/18)$$

The complementary solution, known as the *transient solution*, is of no special interest here since, with time, it dies out with the small amount of damping which can never be completely eliminated. The particular solution x_p describes the continuing motion and is called the *steady-state solution*. Its period is $\tau = 2\pi/\omega$, the same as that of the forcing function. Of primary interest is the amplitude X of the motion. If we let δ_{st} stand for the magnitude of the static deflection of the mass under a static load F_0, then $\delta_{st} = F_0/k$, and we may form the ratio

$$M = \frac{X}{\delta_{st}} = \frac{1}{1 - (\omega/\omega_n)^2} \qquad (8/19)$$

The ratio M is known as the *amplitude ratio* or *magnification factor* and is a measure of the severity of the vibration. We especially note that *M approaches infinity* as ω approaches ω_n. Consequently, if the system possesses no damping and is excited by a harmonic force whose frequency ω approaches the natural frequency ω_n of the system, then M, and, hence X, increases without limit. Physically, this means that the motion amplitude would reach the limits of the attached spring and is a condition to be avoided. The value ω_n is known as the *resonant* or *critical frequency* of the system, and the condition of ω being close in value to ω_n with the resulting large displacement amplitude X is known as *resonance*. For $\omega < \omega_n$ the magnification factor M is positive, and the vibration is in phase with the force F. For $\omega > \omega_n$ the magnification factor is negative, and the vibration is 180° out of phase with F. Figure 8/10 shows a plot of the absolute value of M as a function of the driving-frequency ratio ω/ω_n.

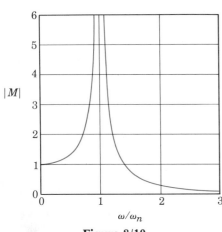

ω/ω_n

Figure 8/10

(*b*) *Damped forced vibration.* We now reintroduce damping in our expressions for forced vibration. Our basic differential equation of motion is

$$\ddot{x} + 2\zeta\omega_n\dot{x} + \omega_n{}^2x = \frac{F_0 \sin \omega t}{m} \qquad [8/13]$$

As in section (*a*) the complete solution is the sum of the complementary solution x_c, which is the general solution of Eq. 8/13 with the right side equal to zero, and the particular solution x_p, which is *any* solution to the complete equation. We have already developed the complementary solution x_c in section (*b*) of Art. 8/2. By virtue of the influence of damping, we find that a single sine or cosine expression, such as we were able to use in section (*a*), is not sufficiently general for the particular solution in the case of damping. So we try

$$x_p = X_1 \cos \omega t + X_2 \sin \omega t \qquad \text{or} \qquad x_p = X \sin (\omega t - \phi)$$

Substitution of the latter expression into Eq. 8/13, matching coefficients of $\sin \omega t$ and $\cos \omega t$, and solving the resulting two equations give

$$X = \frac{F_0/k}{\{[1 - (\omega/\omega_n)^2]^2 + [2\zeta\omega/\omega_n]^2\}^{1/2}} \qquad (8/20)$$

$$\phi = \tan^{-1}\left[\frac{2\zeta\omega/\omega_n}{1 - (\omega/\omega_n)^2}\right] \qquad (8/21)$$

The complete solution is now known, and for underdamped systems it can be written as

$$x = Ce^{-\zeta\omega_n t} \sin (\omega_d t + \psi) + X \sin (\omega t - \phi) \qquad (8/22)$$

Since the first term on the right side diminishes with time, it is known as the *transient solution*. The particular solution x_p is known as the *steady-state solution* and is the part of the solution in which we are primarily interested. All quantities on the right side of Eq. 8/22 are properties of the system and the applied force except for C and ψ (which are determinable from initial conditions) and the running time variable t.

Near resonance the magnitude X of the steady-state solution is a strong function of the damping ratio ζ and the nondimensional frequency ratio ω/ω_n. It is again convenient to form the nondimensional ratio $M = X/(F_0/k)$, which is known as the *amplitude ratio* or *magnification factor*

$$M = \frac{1}{\{[1 - (\omega/\omega_n)^2]^2 + [2\zeta\omega/\omega_n]^2\}^{1/2}} \qquad (8/23)$$

An accurate plot of the magnification factor M versus the frequency ratio ω/ω_n for various values of the damping ratio ζ is shown in Fig. 8/11. This figure reveals the most essential informa-

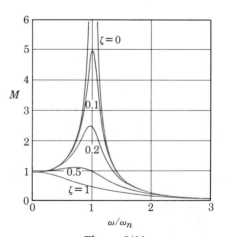

Figure 8/11

tion pertinent to the forced vibration of a single-degree-of-freedom system under harmonic excitation. It is clear from the graph that, if a motion amplitude is excessive, two possible remedies would be to (*a*) increase the damping (to obtain a larger value of ζ) or (*b*) alter the driving frequency so that ω is farther from the resonant frequency ω_n. The addition of damping is most effective near resonance. Figure 8/11 also shows that, except for $\zeta = 0$, the magnification-factor curves do not actually peak at $\omega/\omega_n = 1$. The peak for any given value of ζ can be calculated by finding the maximum value of M from Eq. 8/23.

The phase angle ϕ, given by Eq. 8/21, can vary from 0 to π, and represents the part of a cycle (and hence the time) by which the response x_p lags the forcing function F. Figure 8/12 shows how the phase angle ϕ varies with the frequency ratio for various values of the damping ratio ζ. Note that the value of ϕ at resonance is 90° for all values of ζ. To further illustrate the phase difference between the response and the forcing function, we show in Fig. 8/13 two examples of the variation of F and x_p with ωt. In the first example $\omega < \omega_n$ and ϕ is taken to be $\pi/4$. In the second example $\omega > \omega_n$ and ϕ is taken to be $3\pi/4$.

As a frequently encountered application of harmonic excitation, we consider vibration-measuring instruments such as seismometers and accelerometers. The elements of this class of

Figure 8/12

Figure 8/13

(a)

Figure 8/14

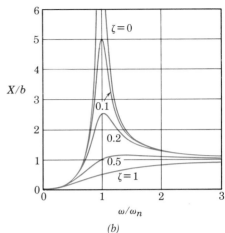

(b)

instruments are shown in Fig. 8/14a. We note that the entire system is subjected to the motion x_B of the frame. Letting x denote the position of the mass *relative* to the frame, we may apply Newton's second law and obtain

$$-c\dot{x} - kx = m\frac{d^2}{dt^2}(x + x_B) \qquad \text{or} \qquad \ddot{x} + \frac{c}{m}\dot{x} + \frac{k}{m}x = -\ddot{x}_B$$

where $(x + x_B)$ is the inertial displacement of the mass. If $x_B = b \sin \omega t$, then our equation of motion with the usual notation is

$$\ddot{x} + 2\zeta\omega_n\dot{x} + \omega_n^2 x = b\omega^2 \sin \omega t$$

which is the same as Eq. 8/13 if $b\omega^2$ is substituted for F_0/m. Again, we are interested only in the steady-state solution x_p. Thus, from Eq. 8/20, we have

$$x_p = \frac{b\,(\omega/\omega_n)^2}{\{[1 - (\omega/\omega_n)^2]^2 + [2\zeta\omega/\omega_n]^2\}^{1/2}} \sin(\omega t - \phi)$$

If X represents the amplitude of the relative response x_p, then the nondimensional ratio X/b is

$$X/b = (\omega/\omega_n)^2 M$$

where M is the magnification ratio of Eq. 8/23. A plot of X/b as a function of the driving-frequency ratio ω/ω_n is shown in the b-part of Fig. 8/14. The similarities and differences between the magnification ratios of Figs. 8/14(b) and 8/11 should be noted.

If the frequency ratio ω/ω_n is large, then $X/b \cong 1$ for all values of the damping ratio ζ. Under these conditions, the displacement of the mass relative to the frame is approximately the same as the absolute displacement of the frame, and the instrument acts as a *displacement meter*. To obtain a high value of ω/ω_n, we need a small value of $\omega_n = \sqrt{k/m}$, which means a soft spring and a large mass. With such a combination, the mass will tend to stay inertially fixed. Displacement meters generally have very light damping.

On the other hand, if the frequency ratio ω/ω_n is small, then M approaches unity (see Fig. 8/11) and $X/b \cong (\omega/\omega_n)^2$ or $X \cong b(\omega/\omega_n)^2$. But $b\omega^2$ is the maximum acceleration of the frame. Hence X is proportional to the maximum acceleration of the frame, and the instrument may be used as an *accelerometer*. The damping ratio is generally selected so that M approximates unity over the widest possible range of ω/ω_n. From Fig. 8/11, we see that a damping factor somewhere between $\zeta = 0.5$ and $\zeta = 1$ would meet this criterion.

Figure 8/15

An important analogy exists between electric circuits and mechanical spring-mass systems. Figure 8/15 shows a series circuit consisting of a voltage E which is a function of time, an inductance L, a capacitance C, and a resistance R. If we denote the charge by the symbol q, the equation which governs the charge is

$$L\ddot{q} + R\dot{q} + \frac{1}{C}q = E \tag{8/24}$$

This equation has the same form as the equation for the mechanical system. Thus by a simple interchange of symbols, the behavior of the electrical circuit may be used to predict the behavior of the mechanical system, or vice versa. The mechanical and electrical equivalents in the following table are worth noting.

MECHANICAL–ELECTRICAL EQUIVALENTS

MECHANICAL			ELECTRICAL		
QUANTITY	SYMBOL	SI UNIT	QUANTITY	SYMBOL	SI UNIT
Mass	m	kg	Inductance	L	H	henry
Spring stiffness	k	N/m	1/Capacitance	$1/C$	$1/F$	1/farad
Force	F	N	Voltage	E	V	volt
Velocity	$\dot{x}$	m/s	Current	I	A	ampere
Displacement	x	m	Charge	q	C	coulomb
Viscous damping constant	c	N·s/m	Resistance	R	Ω	ohm

Sample Problem 8/4

A 50-kg instrument is supported by four springs, each of stiffness 7500 N/m. If the instrument foundation undergoes harmonic motion given in meters by $x_B = 0.002 \cos 50t$, determine the amplitude of the steady-state motion of the instrument. Damping is negligible.

x_B

Solution. For harmonic oscillation of the base we substitute kb for F_0 in our particular-solution results, so that, from Eq. 8/17, the steady-state amplitude becomes

①
$$X = \frac{b}{1 - (\omega/\omega_n)^2}$$

The resonant frequency is $\omega_n = \sqrt{k/m} = \sqrt{4(7500)/50} = 24.5$ rad/s, and the impressed frequency $\omega = 50$ rad/s is given. Thus

② $$X = \frac{0.002}{1 - (50/24.5)^2} = -6.32(10^{-4}) \text{ m} \quad \text{or} \quad -0.632 \text{ mm} \quad \textit{Ans.}$$

Note that the frequency ratio ω/ω_n is approximately 2, so that the condition of resonance is avoided.

① Note that either $\sin 50t$ or $\cos 50t$ can be used for the forcing function with this same result.

② The minus sign indicates that the motion is 180° out of phase with the applied excitation.

Sample Problem 8/5

The spring attachment point B is given a horizontal motion $x_B = b \cos \omega t$. Determine the critical driving frequency ω_c for which the oscillations of the mass m tend to become excessively large. Neglect the friction and mass associated with the pulleys. The two springs have the same stiffness k.

Neutral position

$x_B = b \cos \omega t$

k

B

$T \quad T$

Equilibrium position

x

m

k

kx
(Dynamic forces only)

Solution. The free-body diagram is drawn for arbitrary positive displacements x and x_B. The motion variable x is measured downward from the position of static equilibrium defined as that which exists when $x_B = 0$. The additional stretch in the upper spring, beyond that
① which exists at static equilibrium, is $2x - x_B$. Therefore the *dynamic* spring force in the upper spring, and hence the *dynamic* tension T in
② the cable, is $k(2x - x_B)$. Summing forces in the x-direction gives

$[\Sigma F_x = m\ddot{x}]$ $\qquad -2k(2x - x_B) - kx = m\ddot{x}$

which becomes

$$\ddot{x} + \frac{5k}{m}x = \frac{2kb \cos \omega t}{m}$$

The natural frequency of the system is $\omega_n = \sqrt{5k/m}$. Thus

$$\omega_c = \omega_n = \sqrt{5k/m} \qquad \textit{Ans.}$$

① If a review of the kinematics of constrained motion is necessary see Art. 2/9.

② We learned from the discussion in Art. 8/2 that the equal and opposite forces associated with the position of static equilibrium may be omitted from the analysis. Our use of the terms *dynamic* spring force and *dynamic* tension stresses that only the force increments in addition to the static values need be considered.

Sample Problem 8/6

The 45-kg piston is supported by a spring of modulus $k = 35$ kN/m. A dashpot of damping coefficient $c = 1250$ N·s/m acts in parallel with the spring. A fluctuating pressure $p = 4000 \sin 30t$ in Pa acts on the piston, whose top surface area is $50(10^{-3})$ m². Determine the steady-state displacement as a function of time and the maximum force transmitted to the base.

$p = p_0 \sin \omega t$

W

k c

Solution. We begin by computing the system natural frequency and damping ratio:

$$\omega_n = \sqrt{\frac{k}{m}} = \sqrt{\frac{35(10^3)}{45}} = 27.9 \text{ rad/s}$$

$$\zeta = \frac{c}{2m\omega_n} = \frac{1250}{2(45)(27.9)} = 0.498 \text{ (underdamped)}$$

The steady-state amplitude, from Eq. 8/20, is

$$X = \frac{F_0/k}{\{[1 - (\omega/\omega_n)^2]^2 + [2\zeta\omega/\omega_n]^2\}^{1/2}}$$

$$= \frac{4000(50)(10^{-3})/[35(10^3)]}{\{[1 - (30/27.9)^2]^2 + [2(0.498)(30/27.9)]^2\}^{1/2}}$$

$$= 0.00528 \text{ m or } 5.28 \text{ mm}$$

The phase angle, from Eq. 8/21, is

$$\phi = \tan^{-1}\left[\frac{2\zeta\omega/\omega_n}{1 - (\omega/\omega_n)^2}\right]$$

$$= \tan^{-1}\left[\frac{2(0.498)(30/27.9)}{1 - (30/27.9)^2}\right]$$

$$= 1.716 \text{ rad}$$

The steady-state motion is then given by the second term on the right side of Eq. 8/22:

$$x_p = X \sin(\omega t - \phi) = 5.28 \sin(30t - 1.716) \text{ mm} \qquad \textit{Ans.}$$

The force F_{tr} transmitted to the base is the sum of the spring and damper forces, or

$$F_{\text{tr}} = kx_p + c\dot{x}_p = kX \sin(\omega t - \phi) + c\omega X \cos(\omega t - \phi)$$

The maximum value of F_{tr} is

$$(F_{\text{tr}})_{\max} = \sqrt{(kX)^2 + (c\omega X)^2} = X\sqrt{k^2 + c^2\omega^2}$$

$$= 0.00528\sqrt{(35\,000)^2 + (1250)^2(30)^2}$$

$$= 271 \text{ N} \qquad \textit{Ans.}$$

Equilibrium position

$F = pA \sin \omega t$

x

kx $c\dot{x}$

(Dynamic forces only)

① The student is encouraged to repeat these calculations with the damping coefficient c set to zero so as to observe the influence of the relatively large amount of damping present.

② Note that the argument of the inverse tangent expression for ϕ has a positive numerator and a negative denominator for the case at hand, thus placing ϕ in the second quadrant. Recall that the defined range of ϕ is $0 \le \phi \le \pi$.

PROBLEMS

(Unless otherwise indicated, assume that the damping is light to moderate so that the amplitude of the forced response is a maximum at $\omega/\omega_n \cong 1$.)

8/41 A viscously damped spring-mass system is excited by a harmonic force of constant amplitude F_0 but varying frequency ω. If the amplitude of the steady-state motion is observed to decrease by a factor of 10 as the frequency ratio ω/ω_n is varied from 1 to 2, determine the damping ratio ζ of the system. *Ans.* $\zeta = 0.1531$

8/42 Determine the amplitude of the steady-state motion if (*a*) $c = 1000$ N·s/m and (*b*) $c = 0$.

$k = 100$ kN/m c

$m = 10$ kg

$F = 1000 \cos 120t$ N

Problem 8/42

8/43 A viscously damped spring-mass system is forced harmonically at the undamped natural frequency ($\omega/\omega_n = 1$). If the damping ratio ζ is doubled from 0.1 to 0.2, compute the percentage reduction R_1 in the steady-state amplitude. Compare with the result R_2 of a similar calculation for the condition $\omega/\omega_n = 2$. Verify your results by inspecting Fig. 8/11. *Ans.* $R_1 = 50\%$, $R_2 = 2.52\%$

8/44 Each 0.5-kg ball is attached to the end of the light elastic rod and deflects 4 mm when a 2-N force is statically applied to the ball. If the central collar is given a vertical harmonic movement with a frequency of 4 Hz and an amplitude of 3 mm, find the amplitude y_0 of vertical vibration of each ball. *Ans.* $y_0 = 8.15$ mm

0.5 kg   0.5 kg

Problem 8/44

$\delta = \delta_0 \sin \omega t$

Problem 8/46

8/45 It was noted in the text that the maxima of the curves for the magnification factor M are not located at $\omega/\omega_n = 1$. Determine an expression in terms of the damping ratio ζ for the frequency ratio at which the maxima occur.

8/46 The collar A is given a harmonic oscillation along the fixed horizontal shaft with a frequency of 3 Hz. The ball is attached to the collar by the elastic rod of negligible mass and deflects 60 mm when a horizontal force of 40 N is applied statically to the ball with the collar clamped in place. If the double amplitude $2x_0$ of the ball is twice the double amplitude $2\delta_0$ of the collar with the two motions 180° out of phase, compute the mass m of the ball.
Ans. m = 2.81 kg

Problem 8/47

8/47 Attachment B is given a horizontal motion $x_B = b \cos \omega t$. Derive the equation of motion for the mass m and state the critical frequency ω_c for which the oscillations of the mass become excessively large.

Problem 8/48

8/48 Attachment B is given a horizontal motion $x_B = b \cos \omega t$. Derive the equation of motion for the mass m and state the critical frequency ω_c for which the oscillations of the mass become excessively large. What is the damping ratio ζ for the system?

$$Ans. \ m\ddot{x} + (c_1 + c_2)\dot{x} + kx = -c_2 b\omega \sin \omega t$$
$$\omega_c = \sqrt{k/m}$$
$$\zeta = \frac{c_1 + c_2}{2\sqrt{km}}$$

8/49 A spring-mounted machine with a mass of 24 kg is observed to vibrate harmonically in the vertical direction with an amplitude of 0.30 mm under the action of a vertical force which varies harmonically between F_0 and $-F_0$ with a frequency of 4 Hz. Damping is negligible. If a static force of magnitude F_0 causes a deflection of 0.60 mm, calculate the equivalent spring constant k for the springs which support the machine.

8/50 A device to produce vibrations consists of the two counter-rotating wheels, each carrying an eccentric mass $m_0 = 1$ kg with a center of mass at a distance $e = 12$ mm from its axis of rotation. The wheels are synchronized so that the vertical positions of the unbalanced masses are always identical. The total mass of the device is 10 kg. Determine the two possible values of the equivalent spring constant k for the mounting which will permit the amplitude of the periodic force transmitted to the fixed mounting to be 1500 N due to the unbalance of the rotors at a speed of 1800 rev/min. Neglect damping.

           Ans. $k = 823$ kN/m or 227 kN/m

Problem 8/50

8/51 The instrument shown has a mass of 43 kg and is spring-mounted to the horizontal base. If the amplitude of vertical vibration of the base is 0.10 mm, calculate the range of frequencies f_n of the base vibration which must be prohibited if the amplitude of vertical vibration of the instrument is not to exceed 0.15 mm. Each of the four identical springs has a stiffness of 7.2 kN/m.

           Ans. Prohibited range $2.38 < f_n < 5.32$ Hz

Problem 8/51

8/52 The circular disk of mass m is secured to an elastic shaft which is mounted in a rigid bearing at A. With the disk at rest a lateral force P applied to the disk produces a lateral deflection Δ, so that the equivalent spring constant is $k = P/\Delta$. If the center of mass of the disk is off center by a small amount e from the shaft centerline, determine the expression for the lateral deflection δ of the shaft due to unbalance at a shaft speed ω in terms of the natural frequency $\omega_n = \sqrt{k/m}$ of lateral vibration of the shaft. At what critical speed ω_c would the deflection tend to become large? Neglect damping.

Problem 8/52

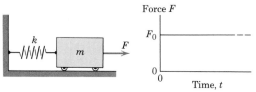

Problem 8/53

8/53 Derive and solve the equation of motion for the mass m in terms of the variable x for the system shown. Neglect the mass of the lever AC and assume small oscillations.

Problem 8/54

8/54 Derive and solve the equation of motion for the mass which is subjected to the suddenly applied force F which remains constant after application. The displacement and velocity of the mass are both zero at time $t = 0$. Plot x versus t for several motion cycles.

$$Ans. \; x = \frac{F_0}{k}(1 - \cos \omega_n t)$$

Problem 8/55

▶ **8/55** Determine the amplitude of vertical vibration of the spring-mounted trailer as it travels at a velocity of 25 km/h over the corduroy road whose contour may be expressed by a sine or cosine term. The mass of the trailer is 500 kg and that of the wheels alone may be neglected. During the loading each 75 kg added to the load caused the trailer to sag 3 mm on its springs. Assume that the wheels are in contact with the road at all times and neglect damping. At what critical speed v_c is the vibration of the trailer greatest?

$$Ans. \; X = 14.75 \text{ mm}, \; v_c = 15.23 \text{ km/h}$$

8/4 VIBRATION OF RIGID BODIES

The subject of planar rigid-body vibrations is entirely analogous to that of particle vibrations. In particle vibrations the variable of interest is one of translation (x), while in rigid-body vibrations the variable of primary concern is one of rotation (θ). Thus the principles of rotational dynamics play a central role in the development of the equation of motion. We shall see that the equation of motion for rotational vibration of rigid bodies has a mathematical form identical to that developed in Arts. 8/2 and 8/3 for translational vibration of particles. As was the case with particles, it is convenient to draw the free-body diagram for an arbitrary positive value of the displacement variable, because a negative displacement value easily leads to sign errors in the equation of motion. The practice of measuring the displacement from the position of static equilibrium rather than from the position of zero spring deflection continues to simplify the formulation for linear systems because the equal and opposite forces and moments associated with the static equilibrium position cancel from the analysis.

Rather than individually treating the cases of (a) free vibration, undamped and damped, and (b) forced vibrations, undamped and damped, as was done with particles in Arts. 8/2 and 8/3, we shall go directly to the damped, forced problem.

As an illustrative example, consider the rotational vibration of the uniform slender bar of Fig. 8/16a. Figure 8/16b depicts the free-body diagram associated with the horizontal position of static equilibrium. Equating to zero the moment sum about O yields

$$-P\left(\frac{l}{2}+\frac{l}{6}\right)+mg\left(\frac{l}{6}\right)=0 \qquad P=\frac{mg}{4}$$

where P is the magnitude of the static spring force.

Figure 8/16c depicts the free-body diagram associated with an arbitrary positive angular displacement θ. Using the rotational equation of motion $\Sigma M_O = I_O\ddot{\theta}$ as developed in Chapter 6, we may write

$$(mg)\left(\frac{l}{6}\cos\theta\right)-\left(\frac{cl}{3}\dot{\theta}\cos\theta\right)\left(\frac{l}{3}\cos\theta\right)-\left(P+k\frac{2l}{3}\sin\theta\right)\left(\frac{2l}{3}\cos\theta\right)$$

$$+\,(F_0\cos\omega t)\left(\frac{l}{3}\cos\theta\right)=\frac{1}{9}ml^2\ddot{\theta}$$

where $I_O = \bar{I}+md^2 = ml^2/12 + m(l/6)^2 = ml^2/9$ is obtained from the parallel-axis theorem for mass moments of inertia. For small angular deflections, the approximations $\sin\theta \cong \theta$ and $\cos\theta \cong 1$ may be used. With $P = mg/4$, the equation of motion, upon rearrangement and simplification, becomes

$$\ddot{\theta}+\frac{c}{m}\dot{\theta}+4\,\frac{k}{m}\,\theta=\frac{(F_0 l/3)\cos\omega t}{ml^2/9} \qquad\qquad (8/25)$$

(a)

(b)

(c)

Figure 8/16

The right side has been left unsimplified in the form $M_0(\cos \omega t)/I_O$, where $M_0 = F_0 l/3$ is the magnitude of the moment about point O of the externally applied force. Note that the two equal and opposite moments associated with static equilibrium forces canceled on the left side of the rotational equation of motion. Hence, it is not necessary to include the static-equilibrium forces and moments in the analysis.

At this point, we observe that Eq. 8/25 is identical in form to Eq. 8/13 for the linear case, so we may write

$$\ddot{\theta} + 2\zeta\omega_n\dot{\theta} + \omega_n{}^2\theta = \frac{M_0 \cos \omega t}{I_O} \tag{8/26}$$

Thus we may use all of the relations developed in Arts. 8/2 and 8/3 merely by replacing the linear quantities with their rotational counterparts. The following table shows the results of this procedure as applied to the rotating bar of Fig. 8/16.

LINEAR	ANGULAR (for current problem)
$\ddot{x} + \dfrac{c}{m}\dot{x} + \dfrac{k}{m}x = \dfrac{F_0 \sin \omega t}{m}$	$\ddot{\theta} + \dfrac{c}{m}\dot{\theta} + \dfrac{4k}{m}\theta = \dfrac{M_0 \sin \omega t}{I_O}$
$\omega_n = \sqrt{k/m}$	$\omega_n = \sqrt{4k/m} = 2\sqrt{k/m}$
$\zeta = \dfrac{c}{2m\omega_n} = \dfrac{c}{2\sqrt{km}}$	$\zeta = \dfrac{c}{2m\omega_n} = \dfrac{c}{4\sqrt{km}}$
$\omega_d = \omega_n\sqrt{1-\zeta^2} = \dfrac{1}{2m}\sqrt{4km - c^2}$	$\omega_d = \omega_n\sqrt{1-\zeta^2} = \dfrac{1}{2m}\sqrt{16km - c^2}$
$x_c = Ce^{-\zeta\omega_n t}\sin(\omega_d t + \psi)$	$\theta_c = Ce^{-\zeta\omega_n t}\sin(\omega_d t + \psi)$
$x_p = X\sin(\omega t - \phi)$	$\theta_p = \Theta\sin(\omega t - \phi)$
$X = M\left(\dfrac{F_0}{k}\right)$	$\Theta = M\left(\dfrac{M_0}{k_\theta}\right) = M\dfrac{F_0(l/3)}{\frac{4}{9}kl^2} = \dfrac{3F_0}{4kl}$

In the above table, the variable k_θ represents the equivalent torsional spring constant of the system of Fig. 8/16 and is determined by writing the restoring moment of the spring. For a small angle θ, this moment about O is

$$M_k = -[k(2l/3)\sin\theta][(2l/3)\cos\theta] \cong -(\tfrac{4}{9}kl^2)\theta$$

Thus $k_\theta = \frac{4}{9}kl^2$. Note that the quantity M_0/k_θ is the static angular deflection that would be produced by a constant external moment M_0.

We conclude that there exists an exact analogy between the linear vibration of particles and the small angular vibration of rigid bodies. Furthermore, the utilization of this analogy can save the labor of complete rederivation of the governing relationships for a given problem of general rigid-body vibration.

Sample Problem 8/7

A simplified version of a pendulum used in impact tests is shown in the figure. Derive the equation of motion and determine the period for small oscillations about the pivot. The mass center G is located a distance $\bar{r} = 0.9$ m from O, and the radius of gyration about O is $k_O = 0.95$ m. The friction of the bearing is negligible.

Solution. We draw the free-body diagram for an arbitrary, positive value of the angular-displacement variable θ, which is measured counterclockwise for the coordinate system chosen. Next we apply the governing equation of motion to obtain

① $[\Sigma M_O = I_O \ddot{\theta}]$ $\qquad -mg\bar{r}\sin\theta = mk_O^2\ddot{\theta}$

or $\qquad\qquad \ddot{\theta} + \dfrac{g\bar{r}}{k_O^2}\sin\theta = 0$ $\qquad$ *Ans.*

Note that the governing equation is independent of the mass. When θ is small, $\sin\theta \cong \theta$, and our equation of motion may be written as

$$\ddot{\theta} + \frac{g\bar{r}}{k_O^2}\theta = 0$$

② The frequency in cycles per second and the period in seconds are

$$f_n = \frac{1}{2\pi}\sqrt{\frac{g\bar{r}}{k_O^2}} \qquad \tau = \frac{1}{f_n} = 2\pi\sqrt{\frac{k_O^2}{g\bar{r}}} \qquad \text{\textit{Ans.}}$$

For the given properties: $\qquad \tau = 2\pi\sqrt{\dfrac{(0.95)^2}{(9.81)(0.9)}} = 2.01$ s $\qquad$ *Ans.*

① With our choice of point O as the moment center, the bearing reactions O_x and O_y never enter the equation of motion.

② For large angles of oscillation, determining the period for the pendulum requires the evaluation of an elliptic integral.

Sample Problem 8/8

The uniform bar of mass m and length l is pivoted at its center. The spring of constant k at the left end is attached to a stationary surface, but the right-end spring, also of constant k, is attached to a support which undergoes a harmonic motion given by $y_B = b\sin\omega t$. Determine the driving frequency ω_c which causes resonance.

① As previously, we consider only the changes in the forces due to a movement away from the equilibrium position.

Solution. We use the moment equation of motion about the fixed point O to obtain

① $-\left(k\dfrac{l}{2}\sin\theta\right)\dfrac{l}{2}\cos\theta - k\left(\dfrac{l}{2}\sin\theta - y_B\right)\dfrac{l}{2}\cos\theta = \dfrac{1}{12}ml^2\ddot{\theta}$

Assuming small deflections and simplifying give us

$$\ddot{\theta} + \frac{6k}{m}\theta = \frac{6kb}{ml}\sin\omega t$$

② The natural frequency should be recognized from the now-familiar form of the equation to be

$$\omega_n = \sqrt{6k/m}$$

Hence $\omega_c = \omega_n = \sqrt{6k/m}$ will result in resonance (as well as violation of our small-angle assumption!). $\qquad$ *Ans.*

② The standard form here is $\ddot{\theta} + \omega_n^2\theta = \dfrac{M_0\sin\omega t}{I_O}$, where $M_0 = \dfrac{klb}{2}$ and $I_O = \frac{1}{12}ml^2$. The natural frequency ω_n of a system does not depend upon the external disturbance.

Sample Problem 8/9

Derive the equation of motion for the homogeneous circular cylinder which rolls without slipping. If the cylinder mass is 50 kg, the cylinder radius 0.5 m, the spring constant 75 N/m, and the damping coefficient 10 N·s/m, determine

(a) the undamped natural frequency
(b) the damping ratio
(c) the damped natural frequency
(d) the period of the damped system.

In addition, determine x as a function of time if the cylinder is released from rest at the position $x = -0.2$ m when $t = 0$.

Solution. We have a choice of motion variables in that either x or the angular displacement θ of the cylinder may be used. Since the problem statement involves x, we draw the free-body diagram for an arbitrary, positive value of x and write the two motion equations for the cylinder as

① The angle θ is taken positive clockwise to be kinematically consistent with x.

$$[\Sigma F_x = m\ddot{x}] \qquad -c\dot{x} - kx + F = m\ddot{x}$$

$$[\Sigma M_G = \bar{I}\ddot{\theta}] \qquad -Fr = \tfrac{1}{2}mr^2\ddot{\theta}$$

② The friction force F may be assumed in either direction. We will find that the *actual* direction is to the right for $x > 0$ and to the left for $x < 0$; $F = 0$ when $x = 0$.

The condition of rolling with no slip is $\ddot{x} = r\ddot{\theta}$. Substitution of this condition into the moment equation gives $F = -\tfrac{1}{2}m\ddot{x}$. Inserting this expression for the friction force into the force equation for the x-direction yields

$$-c\dot{x} - kx - \frac{1}{2}m\ddot{x} = m\ddot{x} \qquad \text{or} \qquad \ddot{x} + \frac{2}{3}\frac{c}{m}\dot{x} + \frac{2}{3}\frac{k}{m}x = 0$$

Comparing the above equation to that for the standard damped oscillator, Eq. 8/9, allows us to state directly

(a) $\omega_n{}^2 = \dfrac{2}{3}\dfrac{k}{m} \qquad \omega_n = \sqrt{\dfrac{2}{3}\dfrac{k}{m}} = \sqrt{\dfrac{2}{3}\dfrac{75}{50}} = 1$ rad/s   *Ans.*

(b) $2\zeta\omega_n = \dfrac{2}{3}\dfrac{c}{m} \qquad \zeta = \dfrac{1}{3}\dfrac{c}{m\omega_n} = \dfrac{10}{3(50)(1)} = 0.0667$ *Ans.*

Hence the damped natural frequency and the damped period are

(c) $\omega_d = \omega_n\sqrt{1 - \zeta^2} = (1)\sqrt{1 - (0.0667)^2} = 0.998$ rad/s   *Ans.*

(d) $\tau_d = 2\pi/\omega_d = 2\pi/0.998 = 6.30$ s   *Ans.*

From Eq. 8/12, the underdamped solution to the equation of motion is

$$x = Ce^{-\zeta\omega_n t}\sin(\omega_d t + \psi) = Ce^{-(0.0667)(1)t}\sin(0.998t + \psi)$$

The velocity is

$$\dot{x} = -0.0667Ce^{-0.0667t}\sin(0.998t + \psi)$$
$$+ 0.998Ce^{-0.0667t}\cos(0.998t + \psi)$$

At time $t = 0$, x and $\dot{x}$ become

$$x_0 = C\sin\psi = -0.2$$
$$\dot{x}_0 = -0.0667C\sin\psi + 0.998C\cos\psi = 0$$

The solution to the two equations in C and ψ gives

$$C = -0.2004 \text{ m} \qquad \psi = 1.504 \text{ rad}$$

Hence the motion is given by

$$x = -0.2004e^{-0.0667t}\sin(0.998t + 1.504) \text{ m} \qquad \textit{Ans.}$$

PROBLEMS

8/56 Derive the differential equation for small oscillations of the spring-loaded pendulum and find the period τ. The equilibrium position is vertical as shown. The mass of the rod is negligible.

$$Ans. \ \tau = \frac{2\pi}{\sqrt{\frac{g}{l} + \frac{2kb^2}{ml^2}}}$$

Problem 8/56

8/57 The spring-loaded pendulum of Prob. 8/56 is inverted as shown. Determine the natural frequency f_n of small oscillations about the vertical, and state the minimum value of the spring constant k for which such oscillations will occur.

Problem 8/57

8/58 A uniform rectangular plate pivots about a horizontal axis through one of its corners as shown. Determine the natural frequency ω_n of small oscillations.

$$Ans. \ \omega_n = \frac{\sqrt{3g/2}}{\sqrt[4]{a^2 + b^2}}$$

Problem 8/58

8/59 The uniform rod of length l and mass m is suspended at its midpoint by a wire of length L. The resistance of the wire to torsion in proportional to its angle of twist θ and equals $(JG/L)\theta$ where J is the polar moment of inertia of the wire cross section and G is the shear modulus of elasticity. Derive the expression for the period τ of oscillation of the bar when it is set into rotation about the axis of the wire.

Problem 8/59

Problem 8/60

8/60 The flywheel is suspended from its center by a wire from a fixed support, and a period τ_1 is measured for torsional oscillation of the flywheel about the vertical axis. Two small weights, each of mass m, are next attached to the flywheel in opposite positions at a distance r from the center. This additional mass results in a slightly longer period τ_2. Write an expression for the moment of inertia I of the flywheel in terms of the measured quantities.

$$Ans. \ I = \frac{2mr^2}{(\tau_2/\tau_1)^2 - 1}$$

Problem 8/61

8/61 Determine the natural frequency f_n for small oscillations in the vertical plane about the bearing O for the semicircular disk of radius r.

Problem 8/62

8/62 A spring-loaded homogeneous plate of mass m pivots freely about a vertical axis through point O. Determine the natural frequency f_n of small oscillations about the equilibrium position shown.

$$Ans. \ f_n = \frac{1}{4\pi} \sqrt{\frac{13k}{m}}$$

Problem 8/63

8/63 Determine the expression for the natural frequency f_n of small oscillations of the weighted arm about O. The stiffness of the spring is k and its length is adjusted so that the arm is in equilibrium in the horizontal position shown. Neglect the mass of the spring and arm compared with m.

8/64 When the motor is slowly brought up to speed, a rather large vibratory oscillation of the entire motor about *O-O* occurs at a speed of 360 rev/min which shows that this speed corresponds to the natural frequency of free oscillation of the motor. If the motor has a mass of 43 kg and a radius of gyration of 100 mm about *O-O*, determine the stiffness *k* of each of the four identical spring mounts. *Ans. k* = 3.82 kN/m

Problem 8/64

8/65 The homogeneous 250-kg rectangular block is pivoted about a horizontal axis through *O* and supported by two springs, each of stiffness *k*. The base of the block is horizontal in the equilibrium position with each spring under a compressive force of 250 N. Determine the minimum stiffness *k* of the springs which will ensure vibration about the equilibrium position.

Problem 8/65

8/66 The uniform rod of mass *m* is freely pivoted about point *O*. Assume small oscillations and determine an expression for the damping ratio ζ. For what value c_{cr} of the damping coefficient *c* will the system be critically damped?

$$Ans. \ \zeta = \frac{1}{2}\frac{a^2}{b^2} c \sqrt{\frac{3}{km}}$$

$$c_{cr} = \frac{2b^2}{a^2}\sqrt{\frac{km}{3}}$$

Problem 8/66

8/67 The system of Prob. 8/39 is repeated here. If the link *AB* now has mass m_3 and radius of gyration k_O about point *O*, determine the equation of motion in terms of the variable *x*. Assume small oscillations. The damping coefficient for the dashpot is *c*.

$$Ans. \ \left[m_1 + \frac{a^2}{b^2} m_2 + \frac{k_O^2}{b^2} m_3 \right]\ddot{x} + \left[\frac{a^2}{b^2} c \right]\dot{x} + kx = 0$$

Problem 8/67

Problem 8/68

8/68 The system of Prob. 8/40 is repeated here. If the crank AB now has mass m_2 and a radius of gyration k_O about point O, determine expressions for the undamped natural frequency ω_n and the damping ratio ζ in terms of the given system properties. Assume small oscillations. The damping coefficient for the damper is c.

Problem 8/69

8/69 The spherical bob of the pendulum shown oscillates in a fluid, which exerts a drag force of form kv^2, where k is a constant and v is the speed of the bob. If the mass of the bob is m and the mass and drag of the rod and the friction in the bearing are negligible, derive the equation of motion of the pendulum for small oscillations and comment on the form of damping present.

Problem 8/70

8/70 Two identical uniform bars are welded together at a right angle and are pivoted about a horizontal axis through point O as shown. Determine the critical driving frequency ω_c of the block B which will result in excessively large oscillations of the assembly. The mass of the welded assembly is m.

$$Ans. \quad \omega_c = \sqrt{\frac{6}{5}\left(\frac{2k}{m} + \frac{g}{l}\right)}$$

8/71 The homogeneous solid cylindrical pulley has mass m_1 and radius r. If the attachment at B undergoes the indicated harmonic displacement, determine the equation of motion of the system in terms of the variable x. The cord which connects mass m_2 to the upper spring does not slip on the pulley.

Problem 8/71

▶8/72 The elements of the "swing-axle" type of independent rear suspension for automobiles are depicted in the figure. The differential D is rigidly attached to the car frame. The half-axles are pivoted at their inboard ends (point O for the half-axle shown) and are rigidly attached to the wheels. Suspension elements not shown constrain the wheel motion to the plane of the figure. The mass of the wheel–tire assembly is $M = 45$ kg, and its mass moment of inertia about a diametral axis passing through its mass center G is 1.4 kg·m². The mass of the half-axle is negligible. The spring rate and shock-absorber damping coefficient are $k = 8.75$ kN/m and $c = 2600$ N·s/m, respectively. If a static tire imbalance is present, as represented by the additional concentrated mass $m = 0.25$ kg as shown, determine the angular velocity ω which results in the suspension system being driven at its undamped natural frequency. What would be the corresponding vehicle speed v? Determine the damping ratio ζ. Assume small angular deflections and neglect gyroscopic effects and any car frame vibration. In order to avoid the complications associated with the varying normal force exerted by the road on the tire, treat the vehicle as being on a lift with the wheels hanging free.

Ans. $\omega = 10.21$ rad/s, $v = 12.87$ km/h
$\zeta = 1.517$

Problem 8/72

▶8/73 For the automobile suspension system of Prob. 8/72, determine the amplitude X of the vertical motion of point G if the angular velocity of the tire corresponds to (*a*) the undamped natural frequency of the system and (*b*) a vehicle speed of 90 km/h. Reconcile the two results.

Ans. (*a*) $X = 0.611$ mm
(*b*) $X = 1.731$ mm

8/5 ENERGY METHODS

In Arts. 8/2 through 8/4 we derived and solved the equations of motion for vibrating bodies by applying Newton's second law of motion to the body isolated with its free-body diagram. With this approach we were able to account for the actions of all forces acting on the body, including frictional damping forces. There are many problems where the effect of damping is small and may be neglected, so that the total energy of the system is essentially conserved. For such systems we find that the principle of conservation of energy may frequently be applied to considerable advantage in establishing the equation of motion and, when the motion is simple harmonic, in determining the frequency of vibration.

To illustrate this alternative approach consider, first, the simple case of the body of mass m attached to the spring of stiffness k and vibrating in the vertical direction without damping, Fig. 8/17. As previously, we find it convenient to measure the motion variable x from the equilibrium position. With this datum the total potential energy of the system, elastic plus gravitational, becomes

$$V = V_e + V_g = \tfrac{1}{2}k(x + \delta_{\text{st}})^2 - \tfrac{1}{2}k\delta_{\text{st}}^2 - mgx$$

where $\delta_{\text{st}} = mg/k$ is the initial static displacement. Substituting $k\delta_{\text{st}} = mg$ and simplifying give

$$V = \tfrac{1}{2}kx^2$$

Thus the total energy of the system becomes

$$T + V = \tfrac{1}{2}m\dot{x}^2 + \tfrac{1}{2}kx^2$$

Because $T + V$ is constant for a conservative system, its time derivative is zero. Consequently

$$\frac{d}{dt}(T + V) = m\dot{x}\ddot{x} + kx\dot{x} = 0$$

Canceling $\dot{x}$ gives us our basic differential equation of motion

$$m\ddot{x} + kx = 0$$

which is identical to Eq. 8/1 derived in Art. 8/2 for the same system of Fig. 8/3.

Conservation of energy may also be used to determine the period or frequency of vibration for a linear conservative system without having to derive and solve the equation of motion. For a system which oscillates with simple harmonic motion about the equilibrium position, from which the displacement x is measured, the energy changes from maximum kinetic and zero potential at the equilibrium position $x = 0$ to zero kinetic and maximum potential at the position of maximum displacement $x = x_{\max}$. Thus we may write

$$T_{\max} = V_{\max}$$

Figure 8/17

The maximum kinetic energy is $\frac{1}{2}m(\dot{x}_{max})^2$, and the maximum potential energy is $\frac{1}{2}k(x_{max})^2$. For the harmonic oscillator of Fig. 8/17 we know that the displacement may be written $x = x_{max} \sin(\omega_n t + \psi)$ so that the maximum velocity is $\dot{x}_{max} = \omega_n x_{max}$. Thus we may write

$$\frac{1}{2}m(\omega_n x_{max})^2 = \frac{1}{2}k(x_{max})^2$$

where x_{max} is the maximum displacement where the potential energy is a maximum. From this energy balance we easily obtain

$$\omega_n = \sqrt{k/m}$$

This direct determination of the frequency may be used for any linear undamped vibration.

The main advantage of the energy approach for the free vibration of conservative systems is that it becomes unnecessary to dismember the system and account for all of the forces which act on each member. In Art. 3/7 of Chapter 3 and in Arts. 6/6 and 6/7 of Chapter 6 we learned for a system of interconnected bodies that an active-force diagram of the complete system enabled us to evaluate the work U of the external active forces and to equate it to the change in the total mechanical energy $T + V$ of the system. So for a conservative mechanical system of interconnected parts with a single degree of freedom where $U = 0$, we may obtain its equation of motion simply by setting the time derivative of its constant total mechanical energy to zero which is

$$\frac{d}{dt}(T + V) = 0$$

Here $V = V_e + V_g$ is the sum of the elastic and gravitational potential energies of the system. Also, for an interconnected mechanical system as for a single body, the natural frequency of vibration is obtained by equating the expression for its maximum total kinetic energy to the expression for its maximum potential energy, where the potential energy is taken to be zero at the equilibrium position. This approach to the determination of natural frequency is valid only if it can be determined that the system vibrates with simple harmonic motion.

Sample Problem   8/10

The small sphere of mass m is mounted on the light rod pivoted at O and supported at end A by the vertical spring of stiffness k. End A is displaced a small distance y_0 below the horizontal equilibrium position and released. By the energy method derive the differential equation of motion for small oscillations of the rod and determine the expression for its natural frequency ω_n of vibration. Damping is negligible.

Equilibrium position

Solution.   With the displacement y of the end of the bar measured from the equilibrium position, the potential energy in the displaced position for small values of y becomes

① $$V = V_e + V_g = \frac{1}{2} k(y + \delta_{st})^2 - \frac{1}{2} k\delta_{st}^2 - mg\left(\frac{b}{l} y\right)$$

where δ_{st} is the static deflection of the spring at equilibrium. But the force in the spring in the equilibrium position, from a zero moment sum about O, is $(b/l)mg = k\delta_{st}$. Substituting this value in the expression for V and simplifying yield

② $$V = \frac{1}{2} ky^2$$

The kinetic energy in the displaced position is

$$T = \frac{1}{2} m\left(\frac{b}{l} \dot{y}\right)^2$$

where we see that the vertical displacement of m is $(b/l)y$. Thus, with the energy sum constant, its time derivative is zero, and we have

$$\frac{d}{dt}(T + V) = \frac{d}{dt}\left[\frac{1}{2} m\left(\frac{b}{l} \dot{y}\right)^2 + \frac{1}{2} ky^2\right] = 0$$

which yields

$$\ddot{y} + \frac{l^2}{b^2}\frac{k}{m} y = 0 \qquad\qquad Ans.$$

when $\dot{y}$ is canceled. By analogy with Eq. 8/2 we may write the motion frequency directly as

$$\omega_n = \frac{l}{b}\sqrt{k/m} \qquad\qquad Ans.$$

Alternatively we can obtain the frequency by equating the maximum kinetic energy, which occurs at $y = 0$, to the maximum potential energy, which occurs at $y = y_0 = y_{max}$ where the deflection is a maximum. Thus

$$T_{max} = V_{max} \qquad gives \qquad \frac{1}{2} m\left(\frac{b}{l} \dot{y}_{max}\right)^2 = \frac{1}{2} ky_{max}^2$$

Knowing that we have a harmonic oscillation, which can be expressed as $y = y_{max} \sin \omega_n t$, we have $\dot{y}_{max} = y_{max}\omega_n$. Substituting this relation into our energy balance gives us

$$\frac{1}{2} m\left(\frac{b}{l} y_{max}\omega_n\right)^2 = \frac{1}{2} ky_{max}^2 \qquad so\ that \qquad \omega_n = \frac{l}{b}\sqrt{k/m} \quad Ans.$$

as before.

① For large values of y the circular motion of the end of the bar would cause our expression for the deflection of the spring to be in error.

② Here again we note the simplicity of the expression for potential energy when the displacement is measured from the equilibrium position.

Sample Problem   8/11

Determine the natural frequency ω_n of vertical vibration of the 3-kg collar to which are attached the two uniform 1.2-kg links, which may be treated as slender bars. The stiffness of the spring, which is attached to both the collar and the foundation, is $k = 1.5$ kN/m, and the bars are both horizontal in the equilibrium position. A small roller on end B of each link permits end A to move with the collar. Frictional retardation is negligible.

Solution.   In the equilibrium position the compression P in the spring equals the weight of the 3-kg collar plus half the weight of each link or $P = 3(9.81) + 2(\frac{1}{2})(1.2)(9.81) = 41.20$ N. The corresponding static deflection of the spring is $\delta_{st} = P/k = 41.20/1.5(10^3) = 27.47(10^{-3})$ m. With the displacement variable y measured downward from the equilibrium position, which becomes the position of zero potential energy, the potential energy for each member in the displaced position is

$$\text{(Spring)}\quad V_e = \frac{1}{2}\,k(y + \delta_{st})^2 - \frac{1}{2}\,k\delta_{st}{}^2 = \frac{1}{2}\,ky^2 + k\delta_{st}y$$

$$= \frac{1}{2}\,(1.5)(10^3)y^2 + 1.5(10^3)(27.47)(10^{-3})y$$

$$= 750y^2 + 41.20y \text{ J}$$

$$\text{(Collar)}\quad V_g = -m_c gy = -3(9.81)y = -29.43y \text{ J}$$

$$\text{(Each link)}\quad V_g = -m_1 g\frac{y}{2} = -1.2(9.81)\frac{y}{2} = -5.89y \text{ J}$$

The total potential energy of the system then becomes

$$V = 750y^2 + 41.20y - 29.43y - 2(5.89)y = 750y^2 \text{ J}$$

The maximum kinetic energy occurs at the equilibrium position where the velocity $\dot{y}$ of the collar is its maximum value. For small motion of the collar, end B of each link has essentially no motion, so that each link may be treated as though it were rotating about B as a fixed point with an angular velocity $\dot{y}/0.3$. Thus the kinetic energy of each part is

$$\text{(Collar)}\quad T = \frac{1}{2}\,m_c\dot{y}^2 = \frac{3}{2}\,\dot{y}^2 \text{ J}$$

$$\text{(Each link)}\quad T = \frac{1}{2}\,I_B\omega^2 = \frac{1}{2}\left(\frac{1}{3}\,m_1 l^2\right)(\dot{y}/l)^2 = \frac{1}{6}\,m_1\dot{y}^2$$

$$= \frac{1}{6}\,(1.2)\dot{y}^2 = 0.2\dot{y}^2$$

Hence the kinetic energy of the collar and both links is

$$T = \frac{3}{2}\,\dot{y}^2 + 2(0.2\dot{y}^2) = 1.9\dot{y}^2$$

With the harmonic motion expressed by $y = y_{max}\sin\omega_n t$, we have $\dot{y}_{max} = y_{max}\omega_n$, so that the energy balance $T_{max} = V_{max}$ with $\dot{y} = \dot{y}_{max}$ becomes

$$1.9(y_{max}\omega_n)^2 = 750y_{max}{}^2 \quad\text{or}\quad \omega_n = \sqrt{750/1.9} = 19.87 \text{ Hz}\quad Ans.$$

① Note that the mass center of each link moves down only half as far as the collar.

② We note again that measurement of the motion variable y from the equilibrium position results in the total potential energy being simply $V = \frac{1}{2}ky^2$.

③ Our knowledge of rigid-body kinematics is essential at this point

④ To appreciate the advantage of the work-energy method for this and similar problems of interconnected systems, the student is encouraged to explore the steps required for solution by the force and moment equations of motion of the separate parts.

⑤ If the oscillations were large, the links could no longer be considered as bodies rotating about fixed points. In this event we would find that the angular velocity of each link would equal $\dot{y}/\sqrt{0.09 - y^2}$ which would cause a nonlinear response no longer described by $y = y_{max}\sin\omega t$.

PROBLEMS

(Solve the following problems by the energy method of Art. 8/5.)

8/74 Derive the equation of motion for the pendulum which consists of the slender uniform rod of mass m and the bob of mass M. Assume small oscillations and neglect the radius of the bob.

$$\textit{Ans. } \ddot{\theta} + \left(\frac{3}{2}\frac{g}{l}\frac{m+2M}{m+3M}\right)\theta = 0$$

Problem 8/74

Problem 8/75

8/75 Determine the period τ for the uniform circular hoop of radius r as it oscillates with small amplitude about the horizontal knife edge.

8/76 Determine the natural frequency f_n of the inverted pendulum, repeated from Prob. 8/57. Assume small oscillations, and note any restrictions on your solution.

$$\textit{Ans. } f_n = \frac{1}{2\pi}\sqrt{\frac{2kb^2}{ml^2} - \frac{g}{l}}$$

$$k > \frac{mgl}{2b^2}$$

Problem 8/76

8/77 The rotational axis of the turntable is inclined at an angle α from the vertical. The turntable shaft pivots freely in bearings which are not shown. If a small block of mass m is placed a distance r from point O, determine the natural frequency ω_n for small rotational oscillations through the angle θ. The mass moment of inertia of the turntable about the axis of its shaft is I.

Problem 8/77

8/78 Determine the natural frequency f_n of the vibrating cylinder, repeated from Prob. 8/15.

$$\textit{Ans. } f_n = \frac{1}{4\pi}\sqrt{\frac{k}{m}}$$

Problem 8/78

8/79 The weighted arm of Prob. 8/63 is repeated here. The length of the spring is adjusted so that the equilibrium position of the arm is horizontal as shown. Neglect the mass of the spring and the arm and calculate the natural frequency f_n for small oscillations.

Problem 8/79

8/80 Derive the equation of motion in terms of the variable x for the homogeneous circular cylinder which rolls without slipping on the horizontal surface.

$$Ans.\ \ddot{x} + \frac{2k}{3m}x = 0$$

Problem 8/80

8/81 Determine the natural frequency ω_n of the system shown. The mass and friction of the pulleys are negligible.

Problem 8/81

8/82 Develop an expression for the natural frequency ω_n of the system shown, repeated from Prob. 8/24. The mass and friction of the pulleys are negligible.

$$Ans.\ \omega_n = \sqrt{\frac{4k}{5m}}$$

Problem 8/82

8/83 Determine the period τ for the homogeneous circular cylinder which rolls without slipping on the track of radius R. Assume small oscillations.

$$Ans.\ \tau = \pi\sqrt{\frac{6(R-r)}{g}}$$

Problem 8/83

Problem 8/84

8/84 The ends of the uniform bar of mass m slide freely in the vertical and horizontal slots as shown. If the bar is in static equilibrium when $\theta = 0$, determine the natural frequency ω_n of small oscillations. What condition must be imposed on the spring constant k in order that oscillations take place?

Problem 8/85

8/85 Develop an expression for the natural frequency f_n of the system shown, repeated from Prob. 8/23. All linkages are light, and the link AB is vertical in the position of static equilibrium shown. Assume small oscillations.

$$Ans.\ f_n = \frac{1}{2\pi} \sqrt{\frac{k_1 + \dfrac{b^2}{a^2} k_2}{m_1 + \dfrac{b^2}{a^2} m_2}}$$

Problem 8/86

8/86 Determine the period τ of vertical oscillations for the system composed of the 140-kg frame and two 80-kg pulleys, each of which has a radius of gyration $k_O = 400$ mm. The flexible wires do not slip on the pulleys.

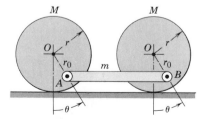

Problem 8/87

8/87 Derive the natural frequency f_n of the system composed of two homogeneous circular cylinders, each of mass M, and the connecting link AB of mass m. Assume small oscillations.

$$Ans.\ f_n = \frac{1}{2\pi} \sqrt{\frac{mgr_0}{3Mr^2 + m(r - r_0)^2}}$$

8/88 The 12-kg block is supported by the two 5-kg links with two torsion springs, each of constant $K = 500$ N·m/rad, arranged as shown. The springs are sufficiently stiff so that stable equilibrium is established in the position shown. Determine the natural frequency f_n for small oscillations about this equilibrium position.

Ans. $f_n = 1.496$ Hz

Problem 8/88

8/89 Each of the two slider blocks A has a mass m and is constrained to move in one of the smooth radial slots of the flywheel, which is driven at a constant angular speed ω. Each of the four springs has a stiffness k. Is it correct to state that the system composed of the flywheel, blocks, and springs possesses a constant energy? Explain your answer.

Problem 8/89

8/90 Determine the period τ of the system if the link AB is horizontal in the static equilibrium position shown. The radius of gyration about point O of the link AB is k_O and its center of mass is point G. Assume small oscillations.

$$Ans. \ \tau = 2\pi \sqrt{\frac{m_1 + \dfrac{k_O^2}{a^2} m_2}{k_1 + \dfrac{b^2}{a^2} k_2}}$$

Problem 8/90

8/91 Two small balls each of mass m are fastened to a rod of negligible mass, which is supported in the horizontal plane by a bifilar suspension. The rod and balls are set into small oscillation about the vertical axis through O. Derive the expression for the period τ. (*Hint:* From the auxiliary sketch note that the balls rise a distance h corresponding to an angle of twist θ. Also $l\beta \cong b\theta$ for small angles, and $\cos \beta$ may be replaced by the first two terms in its series expansion.)

$$Ans. \ \tau = 2\pi \frac{r}{b} \sqrt{\frac{l}{g}}$$

Problem 8/91

Problem 8/92

8/92 The front-end suspension of an automobile is shown. Each of the coil springs has a stiffness of 46.8 kN/m. If the mass of the front-end frame and equivalent portion of the body attached to the front end is 800 kg, determine the natural frequency f_n of vertical oscillation of the frame and body in the absence of shock absorbers. (*Hint:* To relate the spring deflection to the deflection of the frame and body, consider the frame fixed and let the ground and wheels move vertically.)

Ans. $f_n = 1.15$ Hz

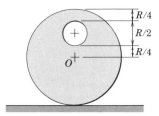

Problem 8/93

8/93 The semicircular cylindrical shell of radius r with small but uniform wall thickness is set into small rocking oscillation on the horizontal surface. If no slipping occurs, determine the expression for the period τ of each complete oscillation.

$$Ans.\ \tau = 2\pi\sqrt{\frac{r(\pi-2)}{g}}$$

Problem 8/94

▶ **8/94** A hole of radius $R/4$ is drilled through a cylinder of radius R to form a body of mass m as shown. If the body rolls on the horizontal surface without slipping, determine the period τ for small oscillations.

$$Ans.\ \tau = 41.4\sqrt{\frac{R}{g}}$$

8/6 *PROBLEM FORMULATION AND REVIEW*

In studying the vibrations of particles and rigid bodies in Chapter 8, we have observed that the subject is simply a direct application of the fundamental principles of dynamics as derived in Chapters 3 and 6. In these previous chapters we established the dynamic behavior of a body at an instant of time and also found the changes in motion resulting from finite intervals of displacement or time. Chapter 8, on the other hand, has dealt with the solution of the defining differential equations of motion, so that the linear or angular displacement can be expressed as a function of time. This information is of critical importance in many engineering problems.

Our study of time response of particles has been divided into the two categories of free and forced motion, with the further subdivisions of negligible and significant damping. It has been seen that the damping ratio ζ is a convenient variable for determining the nature of unforced but viscously damped vibrations. The prime lesson associated with harmonic forcing is that driving a lightly damped system with a force whose frequency is near the natural frequency can cause motion of excessively large amplitude— a condition called resonance which usually must be carefully avoided.

In our study of rigid-body vibrations, we observed that the equation of motion has a form identical to that for particle vibrations. Whereas particle vibrations may be described completely by the equations governing linear motion, rigid-body vibrations usually require the equations of rotational dynamics.

In the final article of Chapter 8, we learned how the energy method can facilitate the solution of free vibration problems where damping may be neglected. Here the total mechanical energy of the system is assumed to be constant. Setting its first time derivative to zero leads directly to the differential equation of motion for the system. The energy approach also permits the analysis of a conservative system of interconnected parts without dismembering the system. Where mechanical energy dissipation cannot be neglected, we must use the laws of motion in terms of force, mass, and acceleration.

Throughout the chapter, we have restricted our attention to systems having one degree of freedom, where the position of the system can be specified by a single variable. One motion variable suffices for many engineering problems. If a system possesses n degrees of freedom, it has n natural frequencies. Thus, if a harmonic force is applied to such a system which is lightly damped, there are n driving frequencies which can cause motion of large amplitude. By a process called modal analysis, a complex system with n degrees of freedom can be reduced to n single-degree-of-freedom systems. For this reason, the thorough understanding of the material of this chapter is vital for the further study of vibrations.

REVIEW PROBLEMS

Problem 8/95

8/95 Determine the natural frequency f_n of vertical oscillations of the cylinder of mass m. The mass and friction of the stepped drum are negligible.

$$Ans.\ f_n = \frac{1}{\pi}\sqrt{\frac{k}{m}}$$

Problem 8/96

8/96 Determine the value of the damping coefficient c for which the system is critically damped if $k = 70$ kN/m and $m = 90$ kg.

Problem 8/97

Labels on figure: O, $45°$, $45°$, 20 m

8/97 A 20-m I-beam is being hoisted by the cable arrangement shown. Determine the period τ of small oscillations about the junction O, which is assumed to remain fixed and about which the cables pivot freely. Treat the beam as a slender rod.

$$Ans.\ \tau = 7.33\ \text{s}$$

Problem 8/98

Labels on figure: 600 mm, 15 kg, 25 kg

8/98 Determine the natural frequency f_n of vertical vibration of the 25-kg mass after it is released from a displaced position. The pulley has a mass of 15 kg with a centroidal radius of gyration of 450 mm, and the spring has a stiffness of 540 N/m.

$$Ans.\ f_n = 1.063\ \text{Hz}$$

8/99 Determine the natural frequency ω_n of small oscillations for the system shown. Treat the links as slender rods each of mass m.

$$Ans. \ \omega_n = \sqrt{\frac{g}{l}\left(\frac{M+m}{M+\frac{2}{3}m}\right)}$$

Problem 8/99

8/100 A slender rod is shaped into the semicircle of radius r as shown. Determine the natural frequency f_n for small oscillations of the rod when it is pivoted on the horizontal knife edge at the middle of its length.

Problem 8/100

8/101 Determine the period τ for small oscillations of the assembly composed of two light bars and two particles, each of mass m. Investigate your expression as the angle α approaches values of 0 and 180°.

$$Ans. \ \tau = 2\pi \sqrt{\frac{l}{g \cos \frac{\alpha}{2}}}$$

Problem 8/101

8/102 The collar of mass m is shown in its position of static equilibrium. Find the natural frequency ω_n of small-amplitude oscillations. (*Caution:* Be certain to retain all first-order terms.)

$$Ans. \ \omega_n = \sqrt{\frac{k}{m} + \frac{g}{2d}}$$

Problem 8/102

Problem 8/103

Problem 8/104

Problem 8/105

Problem 8/106

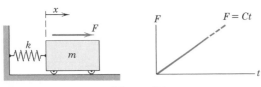

Problem 8/107

8/103 Calculate the damping ratio ζ of the system shown if the mass and radius of gyration of the stepped cylinder are $m = 9$ kg and $\bar{k} = 140$ mm, the spring constant is $k = 2.60$ kN/m, and the damping coefficient of the hydraulic cylinder is $c = 30$ N·s/m The cylinder rolls without slipping on the radius $r = 150$ mm and the spring can support tension as well as compression.

8/104 The cylinder A of radius r, mass m, and radius of gyration $\bar{k}$ is driven by a cable-spring system attached to the drive cylinder B, which oscillates as indicated. If the cables do not slip on the cylinders, and if both springs are stretched to the degree that they do not go slack during a motion cycle, determine an expression for the amplitude θ_{max} of the steady-state oscillation of cylinder A.

$$Ans. \ \theta_{max} = \phi_0 \frac{r_0/r}{1 - (\omega/\omega_n)^2} \text{ where } \omega_n = \frac{r}{\bar{k}}\sqrt{\frac{2k}{m}}$$

8/105 The uniform bar of mass M and length l has a small roller of mass m with negligible bearing friction at each end. Determine the period τ of the system for small oscillations on the curved track.

8/106 The seismic instrument shown is secured to a ship's deck near the stern where propeller-induced vibration is most pronounced. The ship has a single 3-bladed propeller which turns at 180 rev/min and operates partly out of water, thus causing a shock as each blade breaks the surface. The damping ratio of the instrument is $\zeta = 0.5$, and its undamped natural frequency is 3 Hz. If the measured amplitude of A relative to its frame is 0.75 mm, compute the amplitude δ_0 of the vertical vibration of the deck. 　　　　*Ans.* $\delta_0 = 0.712$ mm

8/107 Determine the response x of the undamped spring-mass system to the force F indicated. The initial conditions are $x_0 = \dot{x}_0 = 0$ at time $t = 0$.

▶**8/108** A 200-kg machine rests on four floor mounts each of which has an effective spring constant $k = 250$ kN/m and an effective viscous damping coefficient $c = 1000$ N·s/m. The floor is known to vibrate vertically with a frequency of 24 Hz. What would be the effect on the amplitude of the absolute machine oscillation if the mounts were replaced with new ones which have the same effective spring constant but twice the effective damping coefficient?

Ans. Amplitude increases by 28.9%!

Computer-oriented problems

***8/109** The mass of a critically damped system having a natural frequency $\omega_n = 4$ rad/s is released from rest at an initial displacement x_0. Determine the time t required for the mass to reach the position $x = 0.1x_0$.   *Ans.* $t = 0.972$ s

***8/110** The mass of the system shown is released with the initial conditions $x_0 = 0.1$ m and $\dot{x}_0 = -5$ m/s at $t = 0$. Plot the response of the system and determine the time(s) (if any) at which $x = -0.05$ m.

Ans. $t_1 = 0.0544$ s, $t_2 = 0.442$ s

Problem 8/110

***8/111** Plot the response x of the 20-kg body over the time interval $0 \le t \le 1$ second. Determine the maximum and minimum values of x and their respective times. The initial conditions are $x_0 = 0$ and $\dot{x}_0 = 2$ m/s.

Ans. $x_{max} = 62.5$ mm at $t = 0.0445$ s
$x_{min} = -20.7$ mm at $t = 0.1207$ s

Problem 8/111

***8/112** Shown in the figure are the elements of a displacement meter used to study the motion $y_B = b \sin \omega t$ of the base. The motion of the mass relative to the frame is recorded on the rotating drum. If $l_1 = 360$ mm, $l_2 = 480$ mm, $l_3 = 600$ mm, $m = 0.9$ kg, $c = 1.4$ N·s/m, and $\omega = 10$ rad/s, determine the range of the spring constant k over which the magnitude of the recorded relative displacement is less than $1.5b$. It is assumed that the ratio ω/ω_n must remain greater than unity.

Ans. $0 < k < 27.4$ N/m

Problem 8/112

AREA MOMENTS OF INERTIA

See Appendix A of *Vol. 1 Statics* for a treatment of the theory and calculation of area moments of inertia. Because this quantity plays an important role in the design of structures, especially those dealt with in statics, we present only a brief definition in this *Dynamics* volume so that the student can appreciate the basic differences between area and mass moments of inertia.

The moments of inertia of a plane area A about the x- and y-axes in its plane and about the z-axis normal to its plane, Fig. A/1, are defined by

Figure A/1

$$I_x = \int y^2 \, dA \qquad I_y = \int x^2 \, dA \qquad J_z = \int r^2 \, dA$$

where dA is the differential element of area and $r^2 = x^2 + y^2$. Clearly the polar moment of inertia J_z equals the sum $I_x + I_y$ of the rectangular moments of inertia. When dealing with thin flat plates, the area moment of inertia is useful in calculating the mass moment of inertia, as explained in Appendix B.

The area moment of inertia is a measure of the distribution of area about the axis in question, and for that axis is a constant property of the area. The dimensions of area moment of inertia are (distance)^4 expressed in m^4 or mm^4 in SI units. In contrast, mass moment of inertia is a measure of the distribution of mass about the axis in question, and its dimensions are (mass) × (distance)^2 which are expressed in kg·m^2 in SI units.

MASS MOMENTS OF INERTIA

B

B/1 MASS MOMENTS OF INERTIA ABOUT AN AXIS

The rotational equation of motion about an axis normal to the plane of motion for a rigid body in plane motion contains an integral which depends on the distribution of mass with respect to the moment axis. This integral occurs whenever a rigid body has an angular acceleration about its axis of rotation.

Consider a body of mass m, Fig. B/1, rotating about an axis O-O with an angular acceleration α. All particles of the body move in parallel planes which are normal to the rotation axis O-O. We may choose any one of the planes as the plane of motion, although the one containing the center of mass is usually the one so designated. An element of mass dm has a component of acceleration tangent to its circular path equal to $r\alpha$, and by Newton's second law of motion the resultant tangential force on this element equals $r\alpha\,dm$. The moment of this force about the axis O-O is $r^2\alpha\,dm$, and the sum of the moments of these forces for all elements is $\int r^2\alpha\,dm$. For a rigid body α is the same for all radial lines in the body and we may take it outside the integral sign. The remaining integral is known as the moment of inertia I of the mass m about the axis O-O and is

$$\boxed{I = \int r^2\,dm} \qquad\text{(B/1)}$$

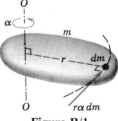

Figure B/1

This integral represents an important property of a body and is involved in the analysis of any body that has rotational acceleration about a given axis. Just as the mass m of a body is a measure of the resistance to translational acceleration, the moment of inertia is a measure of resistance to rotational acceleration of the body.

The moment-of-inertia integral may be expressed alternatively as

$$I = \Sigma r_i^2 m_i \qquad (B/1a)$$

where r_i is the radial distance from the inertia axis to the representative particle of mass m_i and where the summation is taken over all particles of the body.

If the density ρ is constant throughout the body, the moment of inertia becomes

$$I = \rho \int r^2 \, dV$$

where dV is the element of volume. In this case the integral by itself defines a purely geometrical property of the body. When the density is not constant but is expressed as a function of the coordinates of the body, it must be left within the integral sign and its effect accounted for in the integration process.

In general the coordinates which best fit the boundaries of the body should be used in the integration. It is particularly important that we make a good choice of the element of volume dV. To simplify the integration an element of lowest possible order should be chosen, and the correct expression for the moment of inertia of the element about the axis involved should be used. For example, in finding the moment of inertia of a solid right-circular cone about its central axis, we may choose an element in the form of a circular slice of infinitesimal thickness, Fig. B/2a. The differ-

(a)   (b)

Figure B/2

ential moment of inertia for this element is the expression for the moment of inertia of a circular cylinder of infinitesimal altitude about its central axis. (This expression will be obtained in Sample Problem B/1.) Alternatively we could choose an element in the form of a cylindrical shell of infinitesimal thickness as shown in Fig. B/2b. Since all of the mass of the element is at the same distance r from the inertia axis, the differential moment of inertia for this element is merely $r^2 \, dm$ where dm is the differential mass of the elemental shell.

From the definition of mass moment of inertia its dimensions are (mass)(distance)^2 and are expressed in the units kg·m^2 in SI units.

(*a*) **Radius of gyration.** The radius of gyration k of a mass m about an axis for which the moment of inertia is I is defined as

$$k = \sqrt{\frac{I}{m}} \quad \text{or} \quad I = k^2 m \qquad \textbf{(B/2)}$$

Thus k is a measure of the distribution of mass of a given body about the axis in question, and its definition is analogous to the definition of the radius of gyration for area moments of inertia. If all the mass m could be concentrated at a distance k from the axis, the correct moment of inertia would be $k^2 m$. The moment of inertia of a body about a particular axis is frequently indicated by specifying the mass of the body and the radius of gyration of the body about the axis. The moment of inertia is then calculated from Eq. B/2.

(*b*) **Transfer of axes.** If the moment of inertia of a body is known about a centroidal axis, it may be determined easily about any parallel axis. To prove this statement consider the two parallel axes in Fig. B/3, one being a centroidal axis through the mass center G and the other a parallel axis through some other point C. The radial distances from the two axes to any element of mass dm are r_0 and r, and the separation of the axes is d. Substituting the law of cosines $r^2 = r_0{}^2 + d^2 + 2r_0 d \cos \theta$ into the definition for the moment of inertia about the noncentroidal axis through C gives

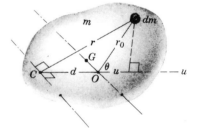

Figure B/3

$$I = \int r^2 \, dm = \int (r_0{}^2 + d^2 + 2r_0 d \cos \theta) \, dm$$

$$= \int r_0{}^2 \, dm + d^2 \int dm + 2d \int u \, dm$$

The first integral is the moment of inertia $\bar{I}$ about the mass-center axis, the second term is md^2, and the third integral equals zero, since the u-coordinate of the mass center with respect to the axis through G is zero. Thus the parallel-axis theorem is

$$I = \bar{I} + md^2 \qquad \textbf{(B/3)}$$

It must be remembered that the transfer cannot be made unless one axis passes through the center of mass and unless the axes are parallel. When the expressions for the radii of gyration are

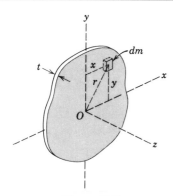

Figure B/4

substituted in Eq. B/3, there results

$$k^2 = \bar{k}^2 + d^2 \qquad (\text{B}/3a)$$

which is the parallel-axis theorem for obtaining the radius of gyration k about an axis a distance d from a parallel centroidal axis for which the radius of gyration is $\bar{k}$.

For plane-motion problems where rotation occurs about an axis normal to the plane of motion, a single subscript for I is sufficient to designate the inertia axis. Thus, if the plate of Fig. B/4 has plane motion in the x-y plane, the moment of inertia of the plate about the z-axis through O is designated I_O. For three-dimensional motion, however, where components of rotation may occur about more than one axis, we use a double subscript to preserve notational symmetry with product-of-inertia terms, which are described in Art. B/2. Thus the moments of inertia about the x-, y-, and z-axes are labeled I_{xx}, I_{yy}, and I_{zz}, respectively, and from Fig. B/5 we see that they become

$$I_{xx} = \int r_x{}^2 \, dm = \int (y^2 + z^2) \, dm$$

$$I_{yy} = \int r_y{}^2 \, dm = \int (z^2 + x^2) \, dm \qquad (\text{B}/4)$$

$$I_{zz} = \int r_z{}^2 \, dm = \int (x^2 + y^2) \, dm$$

These integrals are cited in Eqs. 7/10 of Art. 7/7 on angular momentum in three-dimensional rotation.

The similarity between the defining expressions for mass moments of inertia and area moments of inertia is easily observed. An exact relationship between the two moment-of-inertia expressions exists in the case of flat plates. Consider the flat plate of uniform thickness in Fig. B/4. If the constant thickness is t and the density is ρ, the mass moment of inertia I_{zz} of the plate about

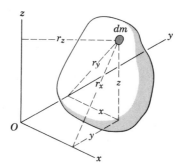

Figure B/5

the z-axis normal to the plate is

$$I_{zz} = \int r^2 \, dm = \rho t \int r^2 \, dA = \rho t J_z \qquad \text{(B/5)}$$

Thus the mass moment of inertia about the z-axis equals the mass per unit area, ρt, times the polar moment of inertia J_z of the plate area about the z-axis. If t is small compared with the dimensions of the plate in its plane, the mass moments of inertia I_{xx} and I_{yy} of the plate about the x- and y-axes are closely approximated by

$$I_{xx} = \int y^2 \, dm = \rho t \int y^2 \, dA = \rho t I_x$$

$$\text{(B/6)}$$

$$I_{yy} = \int x^2 \, dm = \rho t \int x^2 \, dA = \rho t I_y$$

Hence the mass moments of inertia equal the mass per unit area ρt times the corresponding area moments of inertia. The double subscripts for mass moments of inertia distinguish these quantities from area moments of inertia.

Inasmuch as $J_z = I_x + I_y$ for area moments of inertia, we have

$$I_{zz} = I_{xx} + I_{yy} \qquad \text{(B/7)}$$

which holds *only* for a thin flat plate. This restriction is observed from Eqs. B/6, which do not hold true unless the thickness t or z-coordinate of the element is negligible compared with the distance of the element from the corresponding x- or y-axis. Equation B/7 is very useful when dealing with a differential mass element taken as a flat slice of differential thickness, say, dz. In this case Eq. B/7 holds exactly and becomes

$$dI_{zz} = dI_{xx} + dI_{yy} \qquad \text{(B/7}a)$$

(*c*) *Composite bodies.* The defining integral, Eq. B/1, involves the square of the distance from the axis to the element and so is always positive. Thus, as in the case of area moments of inertia, the mass moment of inertia of a composite body is the sum of the moments of inertia of the individual parts about the same axis. It is often convenient to consider a composite body as defined by positive volumes and negative volumes. The moment of inertia of a negative element, such as the material removed to form a hole, must be considered a negative quantity.

Thorough familiarity with the calculation of mass moments of inertia for rigid bodies is an absolute necessity for the study of the dynamics of rotation.

A summary of some of the more useful formulas for mass moments of inertia of various masses of common shape is given in Table D/4, Appendix D.

Sample Problem B/1

Determine the moment of inertia and radius of gyration of a homogeneous right-circular cylinder of mass m and radius r about its central axis O-O.

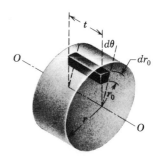

Solution. An element of mass in cylindrical coordinates is $dm = \rho\,dV = \rho t r_0\,dr_0\,d\theta$ where ρ is the density of the cylinder. The moment of inertia about the axis of the cylinder is

$$I = \int r_0{}^2\,dm = \rho t \int_0^{2\pi}\int_0^r r_0{}^3\,dr_0\,d\theta = \rho t\,\frac{\pi r^4}{2} = \tfrac{1}{2}mr^2 \qquad Ans.$$

The radius of gyration is

$$k = \sqrt{\frac{I}{m}} = \frac{r}{\sqrt{2}} \qquad Ans.$$

① If we had started with a cylindrical shell of radius r_0 and axial length t as our mass element dm, then $dI = r_0{}^2\,dm$ directly. The student should evaluate the integral.

② The result $I = \tfrac{1}{2}mr^2$ applies *only* to a solid homogeneous circular cylinder and cannot be used for any other wheel of circular periphery.

Sample Problem B/2

Determine the moment of inertia and radius of gyration of a homogeneous solid sphere of mass m and radius r about a diameter.

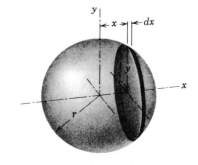

Solution. A circular slice of radius y and thickness dx is chosen as the volume element. From the results of Sample Problem B/1 the moment of inertia about the x-axis of the elemental cylinder is

$$dI_{xx} = \tfrac{1}{2}(dm)y^2 = \tfrac{1}{2}(\pi \rho y^2\,dx)y^2 = \frac{\pi\rho}{2}(r^2 - x^2)^2\,dx$$

where ρ is the constant density of the sphere. The total moment of inertia about the x-axis is

$$I_{xx} = \frac{\pi\rho}{2}\int_{-r}^{r}(r^2 - x^2)^2\,dx = \tfrac{8}{15}\pi\rho r^5 = \tfrac{2}{5}mr^2 \qquad Ans.$$

The radius of gyration is

$$k = \sqrt{\frac{I}{m}} = \sqrt{\frac{2}{5}}\,r \qquad Ans.$$

① Here is an example where we utilize a previous result to express the moment of inertia of the chosen element, which in this case is a right-circular cylinder of differential axial length dx. It would be foolish to start with a third-order element, such as $\rho\,dx\,dy\,dz$, when we can easily solve the problem with a first-order element.

Sample Problem B/3

Determine the moments of inertia of the homogeneous rectangular parallelepiped of mass m about the centroidal x_0- and z-axes and about the x-axis through one end.

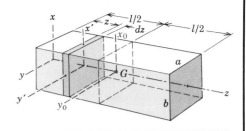

Solution. A transverse slice of thickness dz is selected as the element of volume. The moment of inertia of this slice of infinitesimal thickness equals the moment of inertia of the area of the section times the mass per unit area $\rho\, dz$. Thus the moment of inertia of the transverse slice about the y'-axis is

$$dI_{y'y'} = (\rho\, dz)(\tfrac{1}{12} ab^3)$$

and that about the x'-axis is

①

$$dI_{x'x'} = (\rho\, dz)(\tfrac{1}{12} a^3 b)$$

As long as the element is a plate of differential thickness, the principle given by Eq. B/7a may be applied to give

$$dI_{zz} = dI_{x'x'} + dI_{y'y'} = (\rho\, dz)\frac{ab}{12}(a^2 + b^2)$$

These expressions may now be integrated to obtain the desired results. The moment of inertia about the z-axis is

$$I_{zz} = \int dI_{zz} = \frac{\rho ab}{12}(a^2 + b^2)\int_0^l dz = \tfrac{1}{12}m(a^2 + b^2) \qquad \textit{Ans.}$$

where m is the mass of the block. By interchanging symbols the moment of inertia about the x_0-axis is

$$I_{x_0 x_0} = \tfrac{1}{12}m(a^2 + l^2) \qquad \textit{Ans.}$$

The moment of inertia about the x-axis may be found by the parallel-axis theorem, Eq. B/3. Thus

$$I_{xx} = I_{x_0 x_0} + m\left(\frac{l}{2}\right)^2 = \tfrac{1}{12}m(a^2 + 4l^2) \qquad \textit{Ans.}$$

This last result may be obtained by expressing the moment of inertia of the elemental slice about the x-axis and integrating the expression over the length of the bar. Again by the parallel-axis theorem

$$dI_{xx} = dI_{x'x'} + z^2\, dm = (\rho\, dz)(\tfrac{1}{12} a^3 b) + z^2 \rho ab\, dz = \rho ab\left(\frac{a^2}{12} + z^2\right) dz$$

Integrating gives the result obtained previously,

$$I_{xx} = \rho ab \int_0^l \left(\frac{a^2}{12} + z^2\right) dz = \frac{\rho abl}{3}\left(l^2 + \frac{a^2}{4}\right) = \tfrac{1}{12}m(a^2 + 4l^2)$$

The expression for I_{xx} may be simplified for a long prismatical bar or slender rod whose transverse dimensions are small compared with the length. In this case a^2 may be neglected compared with $4l^2$, and the moment of inertia of such a slender bar about an axis through one end normal to the bar becomes $I = \tfrac{1}{3}ml^2$. By the same approximation the moment of inertia about a centroidal axis normal to the bar is $I = \tfrac{1}{12}ml^2$.

① Refer to Eqs. B/6 and recall the expression for the area moment of inertia of a rectangle about an axis through its center parallel to its base.

PROBLEMS

Problem B/1

B/1 Derive expressions for the moment of inertia and the radius of gyration of the homogeneous rod of mass m and length l about axis O-O perpendicular to the rod and about the parallel mass-center axis. The cross section of the rod is an irregular shape but is constant over the length. Also, the dimensions of the cross section are small compared with the length, thus making it a "slender" rod.

$$Ans.\ I_O = \tfrac{1}{3}ml^2,\ k_O = l/\sqrt{3}$$
$$\bar{I} = \tfrac{1}{12}ml^2,\ \bar{k} = l/(2\sqrt{3})$$

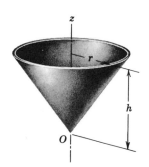

Problem B/2

B/2 The moment of inertia of a solid homogeneous cylinder of radius r about an axis parallel to the central axis of the cylinder may be obtained approximately by multiplying the mass of the cylinder by the square of the distance d between the two axes. What percentage error e results if (a) $d = 10r$, (b) $d = 2r$?

$$Ans.\ (a)\ e = 0.498\%$$
$$(b)\ e = 11.1\%$$

Problem B/3

B/3 State without calculation the moment of inertia about the z-axis of the thin conical shell of mass m and radius r from the results of Sample Problem B/1 applied to a circular disk. Observe the radial distribution of mass by viewing the cone along the z-axis.

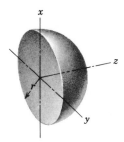

Problem B/4

B/4 From the results of Sample Problem B/2 state without computation the moments of inertia of the solid homogeneous hemisphere of mass m about the x- and z-axes.

B/5 Without integrating determine from the results of Sample Problem B/2 and Prob. B/20 the moments of inertia about the z-axis for (a) the spherical wedge and (b) the conical wedge. Each wedge has a mass m.   *Ans.* (a) $I_{zz} = \frac{2}{5}ma^2$, (b) $I_{zz} = \frac{3}{10}mr^2$

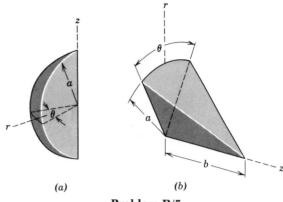

(*a*)   (*b*)

Problem B/5

B/6 The wedge-shaped segment of mass m is cut from a uniform cylindrical shell of radius r and with a thickness which is small compared with r. By inspection write the moment of inertia of the segment about the central axis O-O.

Problem B/6

B/7 A pattern is made from thin steel plate which has a mass of 31.3 kg per square meter of area. The area moments of inertia of the face of the plate are $I_x = 25.2(10^6)$ mm^4 and $I_y = 76.6(10^6)$ mm^4. Calculate the mass moment of inertia of the plate about the z-axis.   *Ans.* $I_{zz} = 3.19(10^{-3})$ kg·m^2

Problem B/7

B/8 Determine the moment of inertia of the uniform slender rod about the x-axis. Make use of your result to write the moment of inertia about the y-axis by inspection. Use these two results to determine the moment of inertia about the z-axis and check your result with that of Prob. B/1 and with the result for I_{xx} in Sample Problem B/3 for $a \ll l$.

Problem B/8

Problem B/9

B/9 The solid cylindrical wedge of mass m and radius r has an axial length b. From the results of Sample Problem B/1, write the moment of inertia of the wedge about axis O-O by inspection. Then derive an expression for the moment of inertia of the wedge about an axis through its mass center parallel to O-O.

Problem B/10

B/10 Calculate the moment of inertia of the steel cylinder with central hole about the x-x axis.

Ans. $I_{xx} = 0.0711$ kg·m^2

Dimensions in Millimeters
Problem B/11

B/11 The molded plastic block has a density of 1300 kg/m^3. Calculate its moment of inertia about the y-y axis. What percentage error e is introduced by using the approximate relation $\frac{1}{3}ml^2$ for I_{xx}?

Problem B/12

B/12 The semicircular disk has a mass of 2 kg, and its small thickness may be neglected compared with its 250-mm radius. Compute the moment of inertia of the disk about the x-, y-, y'-, and z-axes.

Ans. $I_{xx} = I_{yy} = 0.0312$ kg·m^2
$I_{zz} = 0.0625$ kg·m^2
$I_{y'y'} = 0.0501$ kg·m^2

B/13 Calculate the moment of inertia of the steel control wheel, shown in section, about its central axis. There are eight spokes, each of which has a constant cross-sectional area of 200 mm².

Dimensions in Millimeters

Problem B/13

B/14 Replace the rectangular bar of Sample Problem B/3 by the uniform circular cylinder of mass m, radius r, and length l and derive the expression for its moment of inertia about the end axis x-x.

$$\text{Ans. } I_{xx} = m\left(\frac{r^2}{4} + \frac{l^2}{3}\right)$$

Problem B/14

B/15 Use the results cited for Prob. B/14 and derive an expression for the percentage error e in calculating the moment of inertia of the cylinder about the end axis x by neglecting the term $mr^2/4$. Consider the range $0 \le r/l \le 1$. Plot your results and cite the error for $r/l = 0.2$.　　　Ans. $|e| = 2.91\%$

B/16 For what length l of the solid homogeneous cylinder are the moments of inertia about the three coordinate axes through the mass center G equal? Refer to the results of Prob. B/14 as necessary

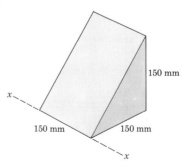

Problem B/16

B/17 A 150-mm steel cube is cut along its diagonal plane. Calculate the moment of inertia of the resulting prism about the edge x-x.

$$\text{Ans. } I_{xx} = 0.1982 \text{ kg·m}^2$$

150 mm

150 mm　　150 mm

Problem B/17

Problem B/18

B/18 Determine the moment of inertia about the tangent x-x axis for the full ring of mass m_1 and the half-ring of mass m_2.

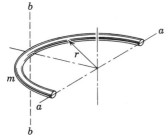

Problem B/19

B/19 Determine the moment of inertia of the half-ring of mass m about its diametral axis a-a and about axis b-b through the midpoint of the arc normal to the plane of the ring. The radius of the circular cross section is small compared with r.

$$Ans.\ I_{aa} = \frac{1}{2}mr^2,\ I_{bb} = 2mr^2\left(1 - \frac{2}{\pi}\right)$$

Problem B/20

B/20 Calculate the moment of inertia of the homogeneous right-circular cone of mass m, base radius r, and altitude h about the cone axis x and about the y-axis through its vertex.

$$Ans.\ I_{xx} = \frac{3}{10}\,mr^2$$
$$I_{yy} = \frac{3}{5}\,m\left(\frac{r^2}{4} + h^2\right)$$

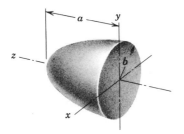

Problem B/21

B/21 Determine the moment of inertia about the x-axis of the homogeneous solid semiellipsoid of revolution having mass m. $Ans.\ I_{xx} = \frac{1}{5}m(a^2 + b^2)$

B/22 A homogeneous solid of mass m is formed by revolving the 45° right triangle about the z-axis. Determine the moment of inertia of the solid about the z-axis.

Problem B/22

B/23 In the study of high-speed reentry into the earth's atmosphere small solid cones are fired at high velocities into low-density gas. A condition of critical stability occurs when the moment of inertia of the cone about its axis of generation a-a equals that about a transverse axis b-b through the mass center. Determine the critical value of the cone angle α for this condition.   *Ans.* $\alpha = 26.6°$

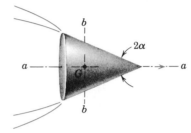

Problem B/23

B/24 A preliminary model for a spacecraft consists of a cylindrical shell and two flat panels as shown. The shell and the panels have the same thickness and density. In order for the spacecraft to have a stable spin about axis 1-1, it may be shown that the moment of inertia about axis 1-1 must be less than the moment of inertia about axis 2-2. Determine the critical value of l which must be exceeded to ensure stable spin about axis 1-1.

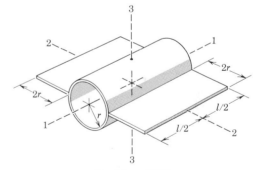

Problem B/24

B/25 A structural member in the shape shown is machined from titanium. Calculate the moment of inertia about the z-axis.
          Ans. $I_{zz} = 0.0510 \text{ kg} \cdot \text{m}^2$

Problem B/25

Dimensions in Millimeters

Problem B/26

B/26 The welded assembly shown is made from steel rod which has a mass of 0.993 kg per meter of length. Calculate the moment of inertia of the assembly about the *x-x* axis.

Problem B/27

B/27 Determine the moments of inertia of the half-spherical shell with respect to the *x*- and *z*-axes. The mass of the shell is *m*, and its thickness is negligible compared with the radius *r*.

$$\text{Ans. } I_{xx} = I_{zz} = \tfrac{2}{3}mr^2$$

Problem B/28

B/28 Determine by integration the moment of inertia of the half-cylindrical shell of mass *m* about the axis *a-a*. The thickness of the shell is small compared with *r*.

Problem B/29

Problem B/30

B/29 Determine the moment of inertia about the generating axis of a complete ring of mass *m* having a circular section (torus) with the dimensions shown in the sectional view.

$$\text{Ans. } I = m(R^2 + \tfrac{3}{4}a^2)$$

B/30 Determine the moment of inertia of the one-quarter-cylindrical shell of mass *m* about the *x-x* axis. The thickness of the shell is small compared with *r*.

$$\text{Ans. } I_{xx} = \tfrac{1}{2}m(r^2 + b^2/6)$$

▶**B/31** The volume of a homogeneous die casting of density ρ is generated by revolving the shaded area about its z-axis base through the angle θ. Determine the moment of inertia of the body about the z-axis.

$$Ans. \ I_{zz} = \frac{1}{4}\left(\frac{83}{15} - \frac{7\pi}{4}\right)\rho r^5\theta = 8.887(10^{-3})\rho r^5\theta$$

Problem B/31

▶**B/32** A shell of mass m is obtained by revolving the circular section about the z-axis. If the thickness of the shell is small compared with a and if $r = a/3$, determine the radius of gyration of the shell about the z-axis.　　　　$Ans. \ k_z = 0.890a$

Problem B/32

▶**B/33** A structural model of a spacecraft consists of a shell structure of the shape shown and is made from material with a mass of 17.5 kg/m². Both ends are closed and are made from the same material. Calculate the moment of inertia I of the model about its geometric axis.

$$Ans. \ I = 1.594 \ \text{kg} \cdot \text{m}^2$$

600 mm　　　　300 mm

◄300 mm►◄300 mm►

Problem B/33

▶**B/34** The cube with semicircular grooves in two opposite faces is cast of lead. Calculate the moment of inertia of the solid about the a-a axis.

$$Ans. \ I_{aa} = 0.367 \ \text{kg} \cdot \text{m}^2$$

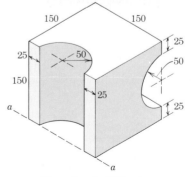

150　　　150

25

25

50

50

150

25

25

a

a

Dimensions in Millimeters

Problem B/34

B/2 PRODUCTS OF INERTIA

For problems in the rotation of three-dimensional rigid bodies the expression for angular momentum contains, in addition to the moment-of-inertia terms, *product-of-inertia* terms defined as

$$
\begin{aligned}
I_{xy} = I_{yx} = \int xy \, dm \\[2mm]
I_{xz} = I_{zx} = \int xz \, dm \\[2mm]
I_{yz} = I_{zy} = \int yz \, dm
\end{aligned}
\tag{B/8}
$$

These expressions were cited in Eqs. 7/10 in the expansion of the expression for angular momentum, Eq. 7/9.

The calculation of products of inertia involves the same basic procedure which we have followed in calculating moments of inertia and in evaluating other volume integrals insofar as the choice of element and the limits of integration are concerned. The only special precaution we need to observe is to be doubly watchful of the algebraic signs in the expressions. Whereas moments of inertia are always positive, products of inertia may be either positive or negative. The units of products of inertia are the same as those of moments of inertia.

We have seen that the calculation of moments of inertia is often simplified by using the transfer-of-axis theorem. A similar theorem exists for transferring products of inertia, and we prove it easily as follows. In Fig. B/6 is shown the x-y view of a rigid body with parallel axes x_0-y_0 passing through the mass center G and located from the x-y axes by the distances d_x and d_y. The product of inertia about the x-y axes by definition is

$$
\begin{aligned}
I_{xy} = \int xy \, dm &= \int (x_0 + d_x)(y_0 + d_y) \, dm \\[2mm]
&= \int x_0 y_0 \, dm + d_x d_y \int dm + d_x \int y_0 \, dm + d_y \int x_0 \, dm \\[2mm]
&= I_{x_0 y_0} + m d_x d_y
\end{aligned}
$$

The last two integrals vanish since the first moments of mass about the mass center are necessarily zero. Similar relations exist for the remaining two product-of-inertia terms. Dropping the zero subscripts and using the bar to designate the mass-center quantity gives

Figure B/6

$$
\begin{aligned}
I_{xy} = \bar{I}_{xy} + m d_x d_y \\
I_{xz} = \bar{I}_{xz} + m d_x d_z \\
I_{yz} = \bar{I}_{yz} + m d_y d_z
\end{aligned}
\tag{B/9}
$$

These transfer-of-axis relations are valid *only* for transfer to or from *parallel axes* through the *mass center*.

With the aid of the product-of-inertia terms we can calculate the moment of inertia of a rigid body about any prescribed axis through the origin of coordinates. For the rigid body of Fig. B/7 let it be required to determine the moment of inertia about axis *OM*. The direction cosines of *OM* are *l*, *m*, *n*, and a unit vector λ along *OM* may be written $\lambda = l\mathbf{i} + m\mathbf{j} + n\mathbf{k}$. The moment of inertia about *OM* is

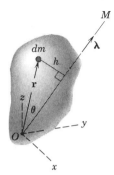

$$I_M = \int h^2 \, dm = \int (\mathbf{r} \times \lambda) \cdot (\mathbf{r} \times \lambda) \, dm$$

Figure B/7

where $|\mathbf{r} \times \lambda| = r \sin \theta = h$. The cross product is

$$(\mathbf{r} \times \lambda) = (yn - zm)\mathbf{i} + (zl - xn)\mathbf{j} + (xm - yl)\mathbf{k}$$

and, after collecting terms, the dot-product expansion gives

$$(\mathbf{r} \times \lambda) \cdot (\mathbf{r} \times \lambda) = h^2 = (y^2 + z^2)l^2 + (x^2 + z^2)m^2 + (x^2 + y^2)n^2$$
$$- 2xylm - 2xzln - 2yzmn$$

Thus, with the substitution of the expressions of Eqs. B/4 and B/8, we have

$$\boxed{I_M = I_{xx}l^2 + I_{yy}m^2 + I_{zz}n^2 - 2I_{xy}lm - 2I_{xz}ln - 2I_{yz}mn} \quad \text{(B/10)}$$

This expression gives the moment of inertia about any axis *OM* in terms of the direction cosines of the axis and the moments and products of inertia about the coordinate directions.

(a) *Principal axes of inertia.* Expansion of the angular momentum expression, Eq. 7/11, for a rigid body with attached axes yields the array

$$\begin{bmatrix} I_{xx} & -I_{xy} & -I_{xz} \\ -I_{yx} & I_{yy} & -I_{yz} \\ -I_{zx} & -I_{zy} & I_{zz} \end{bmatrix}$$

which is known as the *inertia matrix* or *inertia tensor*. If we examine the moment- and product-of-inertia terms for all possible orientations of the axes with respect to the body for a given origin, we will find in the general case one unique orientation *x-y-z* for which the product-of-inertia terms vanish and the array takes the diagonalized form

$$\begin{bmatrix} I_{xx} & 0 & 0 \\ 0 & I_{yy} & 0 \\ 0 & 0 & I_{zz} \end{bmatrix}$$

Axes *x-y-z* are called the *principal axes of inertia*, and I_{xx}, I_{yy}, and

I_{zz} are called the *principal moments of inertia* and represent the maximum, minimum, and an intermediate value for the moments of inertia for the particular origin chosen.

It may be shown* that for any given orientation of axes x-y-z the solution of the determinant equation

$$\begin{vmatrix} I_{xx} - I & -I_{xy} & -I_{xz} \\ -I_{yx} & I_{yy} - I & -I_{yz} \\ -I_{zx} & -I_{zy} & I_{zz} - I \end{vmatrix} = 0 \qquad (B/11)$$

for I yields three roots I_1, I_2, and I_3 of the resulting cubic equation which are the three principal moments of inertia. Also, the direction cosines l, m, and n of a principal inertia axis are given by

$$(I_{xx} - I)l - I_{xy}m - I_{xz}n = 0$$
$$-I_{yx}l + (I_{yy} - I)m - I_{yz}n = 0 \qquad (B/12)$$
$$-I_{zx}l - I_{zy}m + (I_{zz} - I)n = 0$$

These equations along with $l^2 + m^2 + n^2 = 1$ will enable a solution for the direction cosines to be made for each of the three I's separately.

To assist with the visualization of these conclusions, consider the rectangular block, Fig. B/8, with an arbitrary orientation with respect to the x-y-z axes. For simplicity the mass center G is located at the origin of the coordinates. If the moments and products of inertia for the block about the x-y-z axes are known, then solution of Eq. B/11 would give the three roots, I_1, I_2, and I_3, which are the principal moments of inertia. Solution of Eq. B/12 for each of the three I's, in turn, along with $l^2 + m^2 + n^2 = 1$ would give the direction cosines l, m, and n for each respective principal axis. From the proportions of the block as drawn, we see that I_1 is the maximum moment of inertia, I_2 is the intermediate value, and I_3 is the minimum moment of inertia.

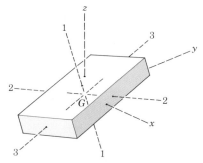

Figure B/8

---

* See, for example, the senior author's *Dynamics, SI Version*, 1975, John Wiley & Sons, Art. 41.

Sample Problem B/4

The bent plate has a uniform thickness t which is negligible compared with its other dimensions. The density of the plate material is ρ. Determine the products of inertia of the plate with respect to the axes as chosen.

Solution. Each of the two parts is analyzed separately.

① **Rectangular part.** In the separate view of this part we introduce parallel axes x_0-y_0 through the mass center G and use the transfer-of-axis theorem. By symmetry we see that $\bar{I}_{xy} = I_{x_0y_0} = 0$ so that

$$[I_{xy} = \bar{I}_{xy} + md_xd_y] \qquad I_{xy} = 0 + \rho tab\left(-\frac{a}{2}\right)\left(\frac{b}{2}\right) = -\frac{1}{4}\rho ta^2b^2$$

Since the z-coordinate of all elements of the plate is zero, it follows that $I_{xz} = I_{yz} = 0$.

Triangular part. In the separate view of this part we locate the mass center G and construct x_0-, y_0-, and z_0-axes through G. Since the x_0-coordinate of all elements is zero, it follows that $\bar{I}_{xy} = I_{x_0y_0} = 0$ and $\bar{I}_{xz} = I_{x_0z_0} = 0$. The transfer-of-axis theorems then give us

$$[I_{xy} = \bar{I}_{xy} + md_xd_y] \qquad I_{xy} = 0 + \rho t\frac{b}{2}c(-a)\left(\frac{2b}{3}\right) = -\frac{1}{3}\rho tab^2c$$

$$[I_{xz} = \bar{I}_{xz} + md_xd_z] \qquad I_{xz} = 0 + \rho t\frac{b}{2}c(-a)\left(\frac{c}{3}\right) = -\frac{1}{6}\rho tabc^2$$

We obtain I_{yz} by direct integration, noting that the distance a of the plane of the triangle from the y-z plane in no way affects the y- and z-coordinates. With the mass element $dm = \rho t\, dy\, dz$ we have

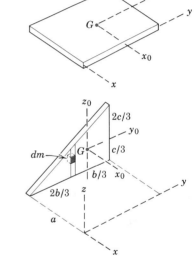

$$② \quad [I_{yz} = \int yz\, dm] \qquad I_{yz} = \rho t\int_0^b \int_0^{cy/b} yz\, dz\, dy = \rho t\int_0^b y\left[\frac{z^2}{2}\right]_0^{cy/b} dy$$

$$= \frac{\rho tc^2}{2b^2}\int_0^b y^3\, dy = \frac{1}{8}\rho tb^2c^2$$

Adding the expressions for the two parts gives

$$I_{xy} = -\tfrac{1}{4}\rho ta^2b^2 - \tfrac{1}{3}\rho tab^2c = -\tfrac{1}{12}\rho tab^2(3a + 4c) \qquad Ans.$$

$$I_{xz} = \quad 0 \quad - \tfrac{1}{6}\rho tabc^2 = -\tfrac{1}{6}\rho tabc^2 \qquad\qquad Ans.$$

$$I_{yz} = \quad 0 \quad + \tfrac{1}{8}\rho tb^2c^2 = +\tfrac{1}{8}\rho tb^2c^2 \qquad\qquad Ans.$$

① We must be careful to preserve the same sense of the coordinates. Thus plus x_0 and y_0 must agree with plus x and y.

② We choose to integrate with respect to z first, where the upper limit is the variable height $z = cy/b$. If we were to integrate first with respect to y, the limits of the first integral would be from the variable $y = bz/c$ to b.

Sample Problem   B/5

The angle bracket is made from aluminum plate with a mass of 13.45 kg per square meter. Calculate the principal moments of inertia about the origin O and the direction cosines of the principal axes of inertia. The thickness of the plate is small compared with the other dimensions.

Dimensions in Millimeters

Solution.   The masses of the three parts are

$$m_1 = 13.45(0.21)(0.1) = 0.282 \text{ kg}$$
$$m_2 = -13.45\pi(0.035)^2 = -0.0518 \text{ kg}$$
$$m_3 = 13.45(0.12)(0.11) = 0.1775 \text{ kg}$$

① Note that the mass of the hole is treated as a negative number.

Part 1

$$I_{xx} = \tfrac{1}{3}mb^2 = \tfrac{1}{3}(0.282)(0.1)^2 = 9.42(10^{-4}) \text{ kg} \cdot \text{m}^2$$

$$I_{yy} = \tfrac{1}{3}m(a^2 + b^2) = \tfrac{1}{3}(0.282)[(0.21)^2 + (0.1)^2] = 50.9(10^{-4}) \text{ kg} \cdot \text{m}^2$$

$$I_{zz} = \tfrac{1}{3}ma^2 = \tfrac{1}{3}(0.282)(0.21)^2 = 41.5(10^{-4}) \text{ kg} \cdot \text{m}^2$$

$$I_{xy} = 0 \qquad I_{yz} = 0$$

$$I_{xz} = 0 + m\frac{a}{2}\frac{b}{2} = 0.282(0.105)(0.05) = 14.83(10^{-4}) \text{ kg} \cdot \text{m}^2$$

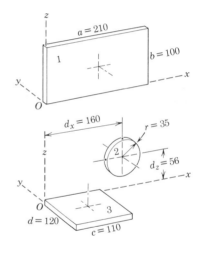

Part 2

$$I_{xx} = \tfrac{1}{4}mr^2 + md_z^2 = -0.0518\left[\frac{(0.035)^2}{4} + (0.050)^2\right]$$
$$= -1.45(10^{-4}) \text{ kg} \cdot \text{m}^2$$

$$I_{yy} = \tfrac{1}{2}mr^2 + m(d_x^2 + d_z^2) = -0.0518\left[\frac{(0.035)^2}{2} + (0.16)^2 + (0.05)^2\right]$$
$$= -14.86(10^{-4}) \text{ kg} \cdot \text{m}^2$$

$$I_{zz} = \tfrac{1}{4}mr^2 + md_x^2 = -0.0518\left[\frac{(0.035)^2}{4} + (0.16)^2\right]$$
$$= -13.41(10^{-4}) \text{ kg} \cdot \text{m}^2$$

$$I_{xy} = 0 \qquad I_{yz} = 0$$

$$I_{xz} = 0 + md_x d_z = -0.0518(0.16)(0.05) = -4.14(10^{-4}) \text{ kg} \cdot \text{m}^2$$

② You can easily derive this formula. Also check Table D/4.

Sample Problem B/5 (Continued)

Part 3

$$I_{xx} = \tfrac{1}{3}md^2 = \tfrac{1}{3}(0.1775)(0.12)^2 = 8.52(10^{-4}) \text{ kg} \cdot \text{m}^2$$

$$I_{yy} = \tfrac{1}{3}mc^2 = \tfrac{1}{3}(0.1775)(0.11)^2 = 7.16(10^{-4}) \text{ kg} \cdot \text{m}^2$$

$$I_{zz} = \tfrac{1}{3}m(c^2 + d^2) = \tfrac{1}{3}(0.1775)[(0.11)^2 + (0.12)^2]$$

$$= 15.68(10^{-4}) \text{ kg} \cdot \text{m}^2$$

$$I_{xy} = m\frac{c}{2}\left(\frac{-d}{2}\right) = 0.1775(0.055)(-0.06) = -5.86(10^{-4}) \text{ kg} \cdot \text{m}^2$$

$$I_{yz} = 0 \qquad I_{xz} = 0$$

Totals

$$I_{xx} = 16.48(10^{-4}) \text{ kg} \cdot \text{m}^2 \qquad I_{xy} = -5.86(10^{-4}) \text{ kg} \cdot \text{m}^2$$

$$I_{yy} = 43.23(10^{-4}) \text{ kg} \cdot \text{m}^2 \qquad I_{yz} = 0$$

$$I_{zz} = 43.79(10^{-4}) \text{ kg} \cdot \text{m}^2 \qquad I_{xz} = 10.69(10^{-4}) \text{ kg} \cdot \text{m}^2$$

Substitution into Eq. B/11, expansion of the determinant, and simplification yield

$$I^3 - 103.51(10^{-4})I^2 + 3179(10^{-8})I - 24\,769(10^{-12}) = 0$$

③ Solution of this cubic equation yields the following roots which are the principal moments of inertia:

$$I_1 = 48.4(10^{-4}) \text{ kg} \cdot \text{m}^2$$

$$I_2 = 11.82(10^{-4}) \text{ kg} \cdot \text{m}^2$$

$$I_3 = 43.3(10^{-4}) \text{ kg} \cdot \text{m}^2$$

The direction cosines of each principal axis are obtained by substituting each root in turn into Eq. B/12 and using $l^2 + m^2 + n^2 = 1$. The results are

$l_1 = 0.363$	$l_2 = 0.934$	$l_3 = 0.0183$
$m_1 = 0.409$	$m_2 = -0.174$	$m_3 = 0.895$
$n_1 = -0.837$	$n_2 = 0.312$	$n_3 = 0.445$

The bottom figure shows a pictorial view of the bracket and the orientation of its principal axes of inertia.

③ A computer program for the solution of a cubic equation may be used, or an algebraic solution using the formula cited in item 4 of Art. C/4, Appendix C, may be employed.

PROBLEMS

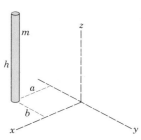

Problem B/35

B/35 Determine the products of inertia of the uniform slender rod of mass m about the coordinate axes shown.

Dimensions in Millimeters
Problem B/36

B/36 Determine the products of inertia about the coordinate axes for the uniform slender rod, which has a mass of 0.60 kg.

$$Ans.\ I_{xy} = 4.05(10^{-3})\ \text{kg} \cdot \text{m}^2$$
$$I_{yz} = 8.10(10^{-3})\ \text{kg} \cdot \text{m}^2$$
$$I_{xz} = 2.43(10^{-3})\ \text{kg} \cdot \text{m}^2$$

Problem B/37

B/37 The uniform rectangular block has a mass of 25 kg. Calculate its products of inertia about the coordinate axes shown.

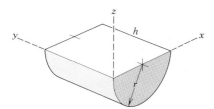

Problem B/38

B/38 Determine the products of inertia of the solid homogeneous half-cylinder of mass m for the axes shown.

$$Ans.\ I_{xy} = \tfrac{1}{2}mrh$$
$$I_{yz} = -\frac{2}{3\pi}\ mrh$$
$$I_{xz} = -\frac{4}{3\pi}\ mr^2$$

B/39 Prove that the moments of inertia about all axes passing through the center of a homogeneous cube are identical.

B/40 Prove that the moment of inertia of the rigid assembly of three identical balls each of mass m and radius r has the same value for all axes through O. Neglect the mass of the connecting rods.

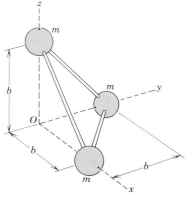

Problem B/40

B/41 Determine the product of inertia I_{xz} of the uniform slender bar of mass m and length L for the orientation given.   *Ans.* $I_{xz} = \frac{1}{6}mL^2 \sin 2\theta \cos \beta$

Problem B/41

B/42 Determine the three products of inertia with respect to the given axes for the uniform rectangular plate of mass m.

Problem B/42

B/43 The slender rod of mass m is formed into a quarter-circular arc of radius r. Determine the products of inertia of the rod with respect to the given axes.

$$\text{Ans. } I_{xy} = \frac{1}{4}mr^2, \ I_{yz} = I_{xz} = \frac{1}{\pi\sqrt{2}} \, mr^2$$

Problem B/43

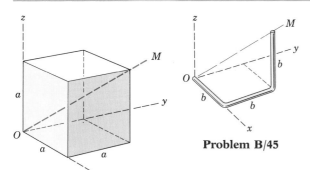

Problem B/44

Problem B/45

B/44 Determine the moment of inertia I_M of the cube of mass m about its diagonal axis OM.

B/45 The bent rod has a mass m. Determine its moment of inertia about the diagonal axis OM.

Ans. $I_M = \frac{1}{3}mb^2$

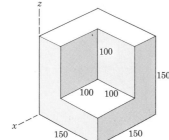

Dimensions in Millimeters
Problem B/46

B/46 The aluminum casting consists of a 150-mm cube with a 100-mm cubical recess. Calculate the products of inertia of the casting about the axes shown.

Ans. $I_{xy} = -0.0376$ kg·m^2
$I_{yz} = 0.0242$ kg·m^2
$I_{xz} = -0.0376$ kg·m^2

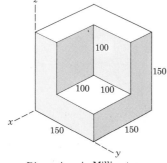

Dimensions in Millimeters
Problem B/47

B/47 The rectangular block with the central hole is machined from steel. Compute the products of inertia of the block with respect to the axes shown.

Ans. $I_{xy} = -0.0756$ kg·m^2
$I_{yz} = 0.0189$ kg·m^2
$I_{xz} = -0.0236$ kg·m^2

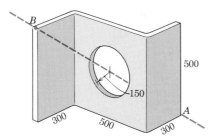

Dimensions in Millimeters
Problem B/48

B/48 The steel plate with two right-angle bends and central hole has a thickness of 15 mm. Calculate its moment of inertia about the diagonal axis through the corners A and B.   *Ans.* $I_{AB} = 2.58$ kg·m^2

Computer-oriented problems

*B/49 Each of the spheres of mass m has a diameter which is small compared with the dimension b. Neglect the mass of the connecting struts and determine the principal moments of inertia of the assembly with respect to the coordinates shown. Determine also the direction cosines of the axis of maximum moment of inertia.

$$Ans.\ I_1 = 7.525mb^2,\quad l_1 = 0.521$$
$$I_2 = 6.631mb^2,\ m_1 = -0.756$$
$$I_3 = 1.844mb^2,\ n_1 = 0.397$$

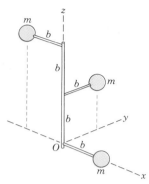

Problem B/49

*B/50 Determine the principal moments of inertia for the bent rod, which has a mass ρ per unit length.

$$Ans.\ I_1 = 3.899\rho b^3$$
$$I_2 = 0.435\rho b^3$$
$$I_3 = 3.667\rho b^3$$

Problem B/50

*B/51 Determine the moment of inertia I about axis OM for the uniform slender rod bent into the shape shown. Plot I versus θ from $\theta = 0$ to $\theta = 90°$ and determine the minimum value of I and the angle α which its axis makes with the x-direction. (*Note:* Because the analysis does not involve the z-coordinate, the expressions developed for area moments of inertia, Eqs. A/9, A/10, and A/11 in Appendix A of Vol. 1, may be utilized for this problem in place of the three-dimensional relations of Appendix B.) The rod has a mass ρ per unit length.

$$Ans.\ I_{min} = 0.187\rho r^3,\ \alpha = 38.6°$$

Problem B/51

*B/52 The thin plate has a mass ρ per unit area and is formed into the shape shown. Determine the principal moments of inertia of the plate about axes through O.

$$Ans.\ I_1 = 3.779\rho b^4$$
$$I_2 = 0.612\rho b^4$$
$$I_3 = 3.609\rho b^4$$

Problem B/52

SELECTED TOPICS OF MATHEMATICS

C

C/1 INTRODUCTION

Appendix C contains an abbreviated summary and reminder of selected topics in basic mathematics which find frequent use in mechanics. The relationships are cited without proof. The student of mechanics will have frequent occasion to use many of these relations, and he or she will be handicapped if they are not well in hand. Other topics not listed will also be needed from time to time.

As the reader reviews and applies his mathematics, he should bear in mind that mechanics is an applied science descriptive of real bodies and actual motions. Therefore the geometric and physical interpretation of the applicable mathematics should be kept clearly in mind during the development of theory and the formulation and solution of problems.

C/2 PLANE GEOMETRY

1. When two intersecting lines are, respectively, perpendicular to two other lines, the angles formed by each pair are equal.

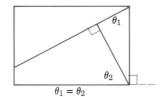

$\theta_1 = \theta_2$

2. Similar triangles

$$\frac{x}{b} = \frac{h - y}{h}$$

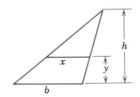

3. Any triangle

Area $= \frac{1}{2}bh$

4. Circle

Circumference $= 2\pi r$
Area $= \pi r^2$
Arc length $s = r\theta$
Sector area $= \frac{1}{2}r^2\theta$

5. Every triangle inscribed within a semicircle is a right triangle.

$\theta_1 + \theta_2 = \pi/2$

6. Angles of a triangle

$\theta_1 + \theta_2 + \theta_3 = 180°$
$\theta_4 = \theta_1 + \theta_2$

C/3 SOLID GEOMETRY

1. Sphere

Volume $= \frac{4}{3}\pi r^3$
Surface area $= 4\pi r^2$

2. Spherical wedge

Volume $= \frac{2}{3}r^3\theta$

3. Right-circular cone

Volume $= \frac{1}{3}\pi r^2 h$
Lateral area $= \pi r L$
$L = \sqrt{r^2 + h^2}$

4. Any pyramid or cone

Volume $= \frac{1}{3}Bh$
where $B =$ area of base

C/4 ALGEBRA

1. Quadratic equation

$$ax^2 + bx + c = 0$$

$$x = \frac{-b \pm \sqrt{b^2 - 4ac}}{2a}, \ b^2 \geqslant 4ac \text{ for real roots}$$

2. Logarithms

$$b^x = y, \ x = \log_b y$$

Natural logarithms

$$b = e = 2.718\ 282$$
$$e^x = y, \ x = \log_e y = \ln y$$
$$\log(ab) = \log a + \log b$$
$$\log(a/b) = \log a - \log b$$
$$\log(1/n) = -\log n$$
$$\log a^n = n \log a$$
$$\log 1 = 0$$
$$\log_{10} x = 0.4343 \ln x$$

3. Determinants
   2nd order

$$\begin{vmatrix} a_1 & b_1 \\ a_2 & b_2 \end{vmatrix} = a_1 b_2 - a_2 b_1$$

   3rd order

$$\begin{vmatrix} a_1 & b_1 & c_1 \\ a_2 & b_2 & c_2 \\ a_3 & b_3 & c_3 \end{vmatrix} = \begin{aligned} &+ a_1 b_2 c_3 + a_2 b_3 c_1 + a_3 b_1 c_2 \\ &- a_3 b_2 c_1 - a_2 b_1 c_3 - a_1 b_3 c_2 \end{aligned}$$

4. Cubic equation

$$x^3 = Ax + B$$

Let $p = A/3, \ q = B/2$

Case I: $q^2 - p^3$ negative (three roots real and distinct)

$$\cos u = q/(p\sqrt{p}), \ 0 < u < 180°$$
$$x_1 = 2\sqrt{p} \cos(u/3)$$
$$x_2 = 2\sqrt{p} \cos(u/3 + 120°)$$
$$x_3 = 2\sqrt{p} \cos(u/3 + 240°)$$

Case II: $q^2 - p^3$ positive (one root real, two roots imaginary)

$$x_1 = (q + \sqrt{q^2 - p^3})^{1/3} + (q - \sqrt{q^2 - p^3})^{1/3}$$

Case III: $q^2 - p^3 = 0$ (three roots real, two roots equal)

$$x_1 = 2q^{1/3}, \ x_2 = x_3 = -q^{1/3}$$

For general cubic equation

$$x^3 + ax^2 + bx + c = 0$$

Substitute $x = x_0 - a/3$ and get $x_0{}^3 = Ax_0 + B$. Then proceed as above to find values of x_0 from which $x = x_0 - a/3$.

C/5 ANALYTIC GEOMETRY

1. Straight line

$$y = a + mx$$

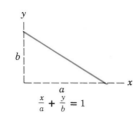

$$\frac{x}{a} + \frac{y}{b} = 1$$

2. Circle

$$x^2 + y^2 = r^2$$

$$(x - a)^2 + (y - b)^2 = r^2$$

3. Parabola

$$y = b\,\frac{x^2}{a^2}$$

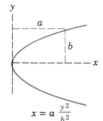

$$x = a\,\frac{y^2}{b^2}$$

4. Ellipse

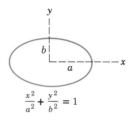

$$\frac{x^2}{a^2} + \frac{y^2}{b^2} = 1$$

5. Hyperbola

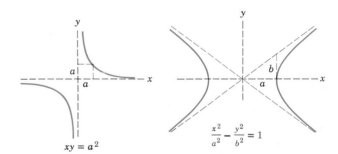

$$xy = a^2$$

$$\frac{x^2}{a^2} - \frac{y^2}{b^2} = 1$$

C/6 TRIGONOMETRY

1. Definitions

$$\sin \theta = a/c \quad \csc \theta = c/a$$
$$\cos \theta = b/c \quad \sec \theta = c/b$$
$$\tan \theta = a/b \quad \mathrm{ctn}\, \theta = b/a$$

2. Signs in the four quadrants

	I	II	III	IV
$\sin \theta$	+	+	−	−
$\cos \theta$	+	−	−	+
$\tan \theta$	+	−	+	−
$\csc \theta$	+	+	−	−
$\sec \theta$	+	−	−	+
$\mathrm{ctn}\, \theta$	+	−	+	−

3. Miscellaneous relations

$$\sin^2 \theta + \cos^2 \theta = 1$$
$$1 + \tan^2 \theta = \sec^2 \theta$$
$$1 + \text{ctn}^2 \theta = \csc^2 \theta$$

$$\sin \frac{\theta}{2} = \sqrt{\tfrac{1}{2}(1 - \cos \theta)}$$

$$\cos \frac{\theta}{2} = \sqrt{\tfrac{1}{2}(1 + \cos \theta)}$$

$$\sin 2\theta = 2 \sin \theta \cos \theta$$
$$\cos 2\theta = \cos^2 \theta - \sin^2 \theta$$
$$\sin (a \pm b) = \sin a \cos b \pm \cos a \sin b$$
$$\cos (a \pm b) = \cos a \cos b \mp \sin a \sin b$$

4. Law of sines

$$\frac{a}{b} = \frac{\sin A}{\sin B}$$

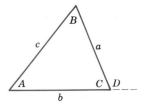

5. Law of cosines

$$c^2 = a^2 + b^2 - 2ab \cos C$$
$$c^2 = a^2 + b^2 + 2ab \cos D$$

C/7 VECTOR OPERATIONS

1. **Notation.** Vector quantities are printed in boldface type, and scalar quantities appear in lightface italic type. Thus, the vector quantity **V** has a scalar magnitude V. In longhand work vector quantities should always be consistently indicated by a symbol such as $\underline{V}$ or $\vec{V}$ to distinguish them from scalar quantities.

2. **Addition**

Triangle addition $\mathbf{P} + \mathbf{Q} = \mathbf{R}$

Parallelogram addition $\mathbf{P} + \mathbf{Q} = \mathbf{R}$

Commutative law $\mathbf{P} + \mathbf{Q} = \mathbf{Q} + \mathbf{P}$

Associative law $\mathbf{P} + (\mathbf{Q} + \mathbf{R}) = (\mathbf{P} + \mathbf{Q}) + \mathbf{R}$

3. **Subtraction**

$$\mathbf{P} - \mathbf{Q} = \mathbf{P} + (-\mathbf{Q})$$

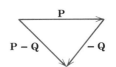

4. **Unit vectors i, j, k**

$$\mathbf{V} = V_x \mathbf{i} + V_y \mathbf{j} + V_z \mathbf{k}$$

where

$$|\mathbf{V}| = V = \sqrt{V_x^2 + V_y^2 + V_z^2}$$

5. **Direction cosines** l, m, n are the cosines of the angles between **V** and the x-, y-, z-axes. Thus

$$l = V_x/V \qquad m = V_y/V \qquad n = V_z/V$$

so that

$$\mathbf{V} = V(l\mathbf{i} + m\mathbf{j} + n\mathbf{k})$$

and

$$l^2 + m^2 + n^2 = 1$$

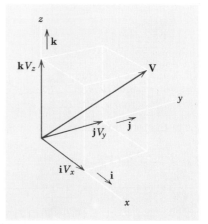

6. *Dot or scalar product*

$$\mathbf{P} \cdot \mathbf{Q} = PQ \cos \theta$$

This product may be viewed as the magnitude of **P** multiplied by the component $Q \cos \theta$ of **Q** in the direction of **P**, or as the magnitude of **Q** multiplied by the component $P \cos \theta$ of **P** in the direction of **Q**.

Commutative law $\mathbf{P} \cdot \mathbf{Q} = \mathbf{Q} \cdot \mathbf{P}$

From the definition of the dot product

$$\mathbf{i} \cdot \mathbf{i} = \mathbf{j} \cdot \mathbf{j} = \mathbf{k} \cdot \mathbf{k} = 1$$

$$\mathbf{i} \cdot \mathbf{j} = \mathbf{j} \cdot \mathbf{i} = \mathbf{i} \cdot \mathbf{k} = \mathbf{k} \cdot \mathbf{i} = \mathbf{j} \cdot \mathbf{k} = \mathbf{k} \cdot \mathbf{j} = 0$$

$$\mathbf{P} \cdot \mathbf{Q} = (P_x \mathbf{i} + P_y \mathbf{j} + P_z \mathbf{k}) \cdot (Q_x \mathbf{i} + Q_y \mathbf{j} + Q_z \mathbf{k})$$
$$= P_x Q_x + P_y Q_y + P_z Q_z$$

$$\mathbf{P} \cdot \mathbf{P} = P_x{}^2 + P_y{}^2 + P_z{}^2$$

It follows from the definition of the dot product that two vectors **P** and **Q** are perpendicular when their dot product vanishes, $\mathbf{P} \cdot \mathbf{Q} = 0$.

The angle θ between two vectors $\mathbf{P}_1$ and $\mathbf{P}_2$ may be found from their dot product expression $\mathbf{P}_1 \cdot \mathbf{P}_2 = P_1 P_2 \cos \theta$, which gives

$$\cos \theta = \frac{\mathbf{P}_1 \cdot \mathbf{P}_2}{P_1 P_2} = \frac{P_{1x} P_{2x} + P_{1y} P_{2y} + P_{1z} P_{2z}}{P_1 P_2} = l_1 l_2 + m_1 m_2 + n_1 n_2$$

where l, m, n stand for the respective direction cosines of the vectors. It is also observed that two vectors are perpendicular when their direction cosines obey the relation $l_1 l_2 + m_1 m_2 + n_1 n_2 = 0$.

Distributive law $\mathbf{P} \cdot (\mathbf{Q} + \mathbf{R}) = \mathbf{P} \cdot \mathbf{Q} + \mathbf{P} \cdot \mathbf{R}$

7. *Cross or vector product.*   The cross product $\mathbf{P} \times \mathbf{Q}$ of the two vectors **P** and **Q** is defined as a vector with a magnitude

$$|\mathbf{P} \times \mathbf{Q}| = PQ \sin \theta$$

and a direction specified by the right-hand rule as shown. Reversing the vector order and using the right-hand rule gives $\mathbf{Q} \times \mathbf{P} = -\mathbf{P} \times \mathbf{Q}$.

Distributive law $\mathbf{P} \times (\mathbf{Q} + \mathbf{R}) = \mathbf{P} \times \mathbf{Q} + \mathbf{P} \times \mathbf{R}$

From the definition of the cross product, using a *right-handed coordinate system*

$$\mathbf{i} \times \mathbf{j} = \mathbf{k} \qquad \mathbf{j} \times \mathbf{k} = \mathbf{i} \qquad \mathbf{k} \times \mathbf{i} = \mathbf{j}$$

$$\mathbf{j} \times \mathbf{i} = -\mathbf{k} \qquad \mathbf{k} \times \mathbf{j} = -\mathbf{i} \qquad \mathbf{i} \times \mathbf{k} = -\mathbf{j}$$

$$\mathbf{i} \times \mathbf{i} = \mathbf{j} \times \mathbf{j} = \mathbf{k} \times \mathbf{k} = 0$$

With the aid of these identities and the distributive law the vector product may be written

$$\mathbf{P} \times \mathbf{Q} = (P_x \mathbf{i} + P_y \mathbf{j} + P_z \mathbf{k}) \times (Q_x \mathbf{i} + Q_y \mathbf{j} + Q_z \mathbf{k})$$
$$= (P_y Q_z - P_z Q_y)\mathbf{i} + (P_z Q_x - P_x Q_z)\mathbf{j} + (P_x Q_y - P_y Q_x)\mathbf{k}$$

$\mathbf{P} \times \mathbf{Q}$

$\mathbf{Q} \times \mathbf{P} = -\mathbf{P} \times \mathbf{Q}$

The cross product may also be expressed by the determinant

$$\mathbf{P} \times \mathbf{Q} = \begin{vmatrix} \mathbf{i} & \mathbf{j} & \mathbf{k} \\ P_x & P_y & P_z \\ Q_x & Q_y & Q_z \end{vmatrix}$$

8. *Additional relations*

   Triple scalar product $(\mathbf{P} \times \mathbf{Q}) \cdot \mathbf{R} = \mathbf{R} \cdot (\mathbf{P} \times \mathbf{Q})$. The dot and cross may be interchanged as long as the order of the vectors is maintained. Parentheses are unnecessary since $\mathbf{P} \times (\mathbf{Q} \cdot \mathbf{R})$ is meaningless since a vector $\mathbf{P}$ cannot be crossed into a scalar $\mathbf{Q} \cdot \mathbf{R}$. Thus the expression may be written

   $$\mathbf{P} \times \mathbf{Q} \cdot \mathbf{R} = \mathbf{P} \cdot \mathbf{Q} \times \mathbf{R}$$

   The triple scalar product has the determinant expansion

   $$\mathbf{P} \times \mathbf{Q} \cdot \mathbf{R} = \begin{vmatrix} P_x & P_y & P_z \\ Q_x & Q_y & Q_z \\ R_x & R_y & R_z \end{vmatrix}$$

   Triple vector product $(\mathbf{P} \times \mathbf{Q}) \times \mathbf{R} = -\mathbf{R} \times (\mathbf{P} \times \mathbf{Q})$
   $$= \mathbf{R} \times (\mathbf{Q} \times \mathbf{P})$$

   Here the parentheses must be used since an expression $\mathbf{P} \times \mathbf{Q} \times \mathbf{R}$ would be ambiguous because it would not identify the vector to be crossed. It may be shown that the triple vector product is equivalent to

   $$(\mathbf{P} \times \mathbf{Q}) \times \mathbf{R} = \mathbf{R} \cdot \mathbf{P} \mathbf{Q} - \mathbf{R} \cdot \mathbf{Q} \mathbf{P}$$

   or

   $$\mathbf{P} \times (\mathbf{Q} \times \mathbf{R}) = \mathbf{P} \cdot \mathbf{R} \mathbf{Q} - \mathbf{P} \cdot \mathbf{Q} \mathbf{R}$$

   The first term in the first expression, for example, is the dot product $\mathbf{R} \cdot \mathbf{P}$, a scalar, multiplied by the vector $\mathbf{Q}$.

9. *Derivatives of vectors* obey the same rules as they do for scalars.

   $$\frac{d\mathbf{P}}{dt} = \dot{\mathbf{P}} = \dot{P}_x \mathbf{i} + \dot{P}_y \mathbf{j} + \dot{P}_z \mathbf{k}$$

   $$\frac{d(\mathbf{P}u)}{dt} = \mathbf{P}\dot{u} + \dot{\mathbf{P}}u$$

   $$\frac{d(\mathbf{P} \cdot \mathbf{Q})}{dt} = \mathbf{P} \cdot \dot{\mathbf{Q}} + \dot{\mathbf{P}} \cdot \mathbf{Q}$$

   $$\frac{d(\mathbf{P} \times \mathbf{Q})}{dt} = \mathbf{P} \times \dot{\mathbf{Q}} + \dot{\mathbf{P}} \times \mathbf{Q}$$

10. *Integration of vectors.* If $\mathbf{V}$ is a function of x, y, and z and an element of volume is $d\tau = dx\,dy\,dz$, the integral of $\mathbf{V}$ over the volume may be written as the vector sum of the three integrals of its components. Thus

    $$\int \mathbf{V}\,d\tau = \mathbf{i} \int V_x\,d\tau + \mathbf{j} \int V_y\,d\tau + \mathbf{k} \int V_z\,d\tau$$

C/8 SERIES

(Expression in brackets following series indicates range of convergence)

$$(1 \pm x)^n = 1 \pm nx + \frac{n(n-1)}{2!} x^2 \pm \frac{n(n-1)(n-2)}{3!} x^3 + \cdots \qquad [x^2 < 1]$$

$$\sin x = x - \frac{x^3}{3!} + \frac{x^5}{5!} - \frac{x^7}{7!} + \cdots \qquad [x^2 < \infty]$$

$$\cos x = 1 - \frac{x^2}{2!} + \frac{x^4}{4!} - \frac{x^6}{6!} + \cdots \qquad [x^2 < \infty]$$

$$\sinh x = \frac{e^x - e^{-x}}{2} = x + \frac{x^3}{3!} + \frac{x^5}{5!} + \frac{x^7}{7!} + \cdots \qquad [x^2 < \infty]$$

$$\cosh x = \frac{e^x + e^{-x}}{2} = 1 + \frac{x^2}{2!} + \frac{x^4}{4!} + \frac{x^6}{6!} + \cdots \qquad [x^2 < \infty]$$

$$f(x) = \frac{a_0}{2} + \sum_{n=1}^{\infty} a_n \cos \frac{n\pi x}{l} + \sum_{n=1}^{\infty} b_n \sin \frac{n\pi x}{l}$$

$$\text{where } a_n = \frac{1}{l} \int_{-l}^{l} f(x) \cos \frac{n\pi x}{l} dx, \quad b_n = \frac{1}{l} \int_{-l}^{l} f(x) \sin \frac{n\pi x}{l} dx$$

[Fourier expansion for $-l < x < l$]

C/9 DERIVATIVES

$$\frac{dx^n}{dx} = nx^{n-1}, \quad \frac{d(uv)}{dx} = u\frac{dv}{dx} + v\frac{du}{dx}, \quad \frac{d\left(\frac{u}{v}\right)}{dx} = \frac{v\frac{du}{dx} - u\frac{dv}{dx}}{v^2}$$

$$\lim_{\Delta x \to 0} \sin \Delta x = \sin dx = \tan dx = dx$$

$$\lim_{\Delta x \to 0} \cos \Delta x = \cos dx = 1$$

$$\frac{d \sin x}{dx} = \cos x, \quad \frac{d \cos x}{dx} = -\sin x, \quad \frac{d \tan x}{dx} = \sec^2 x$$

$$\frac{d \sinh x}{dx} = \cosh x, \quad \frac{d \cosh x}{dx} = \sinh x, \quad \frac{d \tanh x}{dx} = \operatorname{sech}^2 x$$

C/10 INTEGRALS

$$\int x^n \, dx = \frac{x^{n+1}}{n+1}$$

$$\int \frac{dx}{x} = \ln x$$

$$\int \sqrt{a + bx} \, dx = \frac{2}{3b} \sqrt{(a+bx)^3}$$

$$\int x\sqrt{a + bx} \, dx = \frac{2}{15b^2} (3bx - 2a)\sqrt{(a+bx)^3}$$

$$\int \frac{dx}{\sqrt{a + bx}} = \frac{2\sqrt{a+bx}}{b}$$

$$\int \frac{\sqrt{a+x}}{\sqrt{b-x}}\,dx = -\sqrt{a+x}\sqrt{b-x} + (a+b)\sin^{-1}\sqrt{\frac{a+x}{a+b}}$$

$$\int \frac{x\,dx}{a+bx} = \frac{1}{b^2}[a+bx - a\ln(a+bx)]$$

$$\int \frac{x\,dx}{(a+bx)^n} = \frac{(a+bx)^{1-n}}{b^2}\left(\frac{a+bx}{2-n} - \frac{a}{1-n}\right)$$

$$\int \frac{dx}{a+bx^2} = \frac{1}{\sqrt{ab}}\tan^{-1}\frac{x\sqrt{ab}}{a} \quad\text{or}\quad \frac{1}{\sqrt{-ab}}\tanh^{-1}\frac{x\sqrt{-ab}}{a}$$

$$\int \frac{x\,dx}{a+bx^2} = \frac{1}{2b}\ln(a+bx^2)$$

$$\int \sqrt{x^2 \pm a^2}\,dx = \tfrac{1}{2}[x\sqrt{x^2 \pm a^2} \pm a^2 \ln(x+\sqrt{x^2 \pm a^2})]$$

$$\int \sqrt{a^2 - x^2}\,dx = \tfrac{1}{2}\left(x\sqrt{a^2 - x^2} + a^2 \sin^{-1}\frac{x}{a}\right)$$

$$\int x\sqrt{a^2 - x^2}\,dx = -\tfrac{1}{3}\sqrt{(a^2 - x^2)^3}$$

$$\int x^2\sqrt{a^2 - x^2}\,dx = -\frac{x}{4}\sqrt{(a^2 - x^2)^3} + \frac{a^2}{8}\left(x\sqrt{a^2 - x^2} + a^2 \sin^{-1}\frac{x}{a}\right)$$

$$\int x^3\sqrt{a^2 - x^2}\,dx = -\tfrac{1}{5}(x^2 + \tfrac{2}{3}a^2)\sqrt{(a^2 - x^2)^3}$$

$$\int \frac{dx}{\sqrt{a+bx+cx^2}} = \frac{1}{\sqrt{c}}\ln\left(\sqrt{a+bx+cx^2} + x\sqrt{c} + \frac{b}{2\sqrt{c}}\right) \quad\text{or}\quad \frac{-1}{\sqrt{-c}}\sin^{-1}\left(\frac{b+2cx}{\sqrt{b^2 - 4ac}}\right)$$

$$\int \frac{dx}{\sqrt{x^2 \pm a^2}} = \ln(x+\sqrt{x^2 \pm a^2})$$

$$\int \frac{dx}{\sqrt{a^2 - x^2}} = \sin^{-1}\frac{x}{a}$$

$$\int \frac{x\,dx}{\sqrt{x^2 - a^2}} = \sqrt{x^2 - a^2}$$

$$\int \frac{x\,dx}{\sqrt{a^2 \pm x^2}} = \pm\sqrt{a^2 \pm x^2}$$

$$\int x\sqrt{x^2 \pm a^2}\,dx = \tfrac{1}{3}\sqrt{(x^2 \pm a^2)^3}$$

$$\int x^2\sqrt{x^2 \pm a^2}\,dx = \frac{x}{4}\sqrt{(x^2 \pm a^2)^3} \mp \frac{a^2}{8} x\sqrt{x^2 \pm a^2} - \frac{a^4}{8}\ln(x+\sqrt{x^2 \pm a^2})$$

$$\int \sin x\,dx = -\cos x$$

$$\int \cos x\,dx = \sin x$$

$$\int \sec x \, dx = \frac{1}{2} \ln \frac{1 + \sin x}{1 - \sin x}$$

$$\int \sin^2 x \, dx = \frac{x}{2} - \frac{\sin 2x}{4}$$

$$\int \cos^2 x \, dx = \frac{x}{2} + \frac{\sin 2x}{4}$$

$$\int \sin x \cos x \, dx = \frac{\sin^2 x}{2}$$

$$\int \sinh x \, dx = \cosh x$$

$$\int \cosh x \, dx = \sinh x$$

$$\int \tanh x \, dx = \ln \cosh x$$

$$\int \ln x \, dx = x \ln x - x$$

$$\int e^{ax} \, dx = \frac{e^{ax}}{a}$$

$$\int xe^{ax} \, dx = \frac{e^{ax}}{a^2} (ax - 1)$$

$$\int e^{ax} \sin px \, dx = \frac{e^{ax}(a \sin px - p \cos px)}{a^2 + p^2}$$

$$\int e^{ax} \cos px \, dx = \frac{e^{ax}(a \cos px + p \sin px)}{a^2 + p^2}$$

$$\int e^{ax} \sin^2 x \, dx = \frac{e^{ax}}{4 + a^2} \left(a \sin^2 x - \sin 2x + \frac{2}{a} \right)$$

$$\int e^{ax} \cos^2 x \, dx = \frac{e^{ax}}{4 + a^2} \left(a \cos^2 x + \sin 2x + \frac{2}{a} \right)$$

$$\int e^{ax} \sin x \cos x \, dx = \frac{e^{ax}}{4 + a^2} \left(\frac{a}{2} \sin 2x - \cos 2x \right)$$

$$\int \sin^3 x \, dx = -\frac{\cos x}{3} (2 + \sin^2 x)$$

$$\int \cos^3 x \, dx = \frac{\sin x}{3} (2 + \cos^2 x)$$

$$\int \cos^5 x \, dx = \sin x - \tfrac{2}{3} \sin^3 x + \tfrac{1}{5} \sin^5 x$$

$$\int x \sin x \, dx = \sin x - x \cos x$$

$$\int x \cos x \, dx = \cos x + x \sin x$$

$$\int x^2 \sin x \, dx = 2x \sin x - (x^2 - 2) \cos x$$

$$\int x^2 \cos x \, dx = 2x \cos x + (x^2 - 2) \sin x$$

Radius of curvature
$$\begin{cases} \rho_{xy} = \dfrac{\left[1 + \left(\dfrac{dy}{dx} \right)^2 \right]^{3/2}}{\dfrac{d^2y}{dx^2}} \\[6ex] \rho_{r\theta} = \dfrac{\left[r^2 + \left(\dfrac{dr}{d\theta} \right)^2 \right]^{3/2}}{r^2 + 2 \left(\dfrac{dr}{d\theta} \right)^2 - r \dfrac{d^2r}{d\theta^2}} \end{cases}$$

C/11 NEWTON'S METHOD FOR SOLVING INTRACTABLE EQUATIONS

Frequently, the application of the fundamental principles of mechanics leads to an algebraic or transcendental equation which is not solvable (or easily solvable) in closed form. In such cases, an iterative technique, such as Newton's method, can be a powerful tool for obtaining a good estimate to the root or roots of the equation.

Let us place the equation to be solved in the form $f(x) = 0$. The *a*-part of the accompanying figure depicts an arbitrary function $f(x)$ for values of x in the vicinity of the desired root x_r. Note that x_r is merely the value of x at which the function crosses the x-axis. Suppose that we have available (perhaps via a hand-drawn plot) a rough estimate x_1 of this root. Provided that x_1 does not closely correspond to a maximum or minimum value of the function $f(x)$, we may obtain a better estimate

(a)

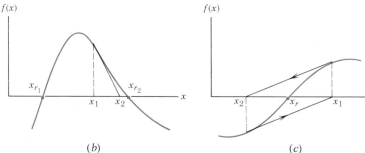

(b)   (c)

of the root x_r by projecting the tangent to $f(x)$ at x_1 such that it intersects the x-axis at x_2. From the geometry of the figure, we may write

$$\tan \theta = f'(x_1) = \frac{f(x_1)}{x_1 - x_2}$$

where $f'(x_1)$ denotes the derivative of $f(x)$ with respect to x evaluated at $x = x_1$. Solving the above equation for x_2 results in

$$x_2 = x_1 - \frac{f(x_1)}{f'(x_1)}$$

The term $-f(x_1)/f'(x_1)$ is the correction to the initial root estimate x_1. Once x_2 is calculated, we may repeat the process to obtain x_3, and so forth.

Thus we generalize the above equation to

$$x_{k+1} = x_k - \frac{f(x_k)}{f'(x_k)}$$

where

x_{k+1} = the $(k + 1)$th estimate to the desired root x_r

x_k = the kth estimate to the desired root x_r

$f(x_k)$ = the function $f(x)$ evaluated at $x = x_k$

$f'(x_k)$ = the function derivative evaluated at $x = x_k$

This equation is repetitively applied until $f(x_{k+1})$ is sufficiently close to zero and $x_{k+1} \cong x_k$. The student should verify that the equation is valid for all possible sign combinations of x_k, $f(x_k)$, and $f'(x_k)$.

Several cautionary notes are in order:

(1) Clearly, $f'(x_k)$ must not be zero or close to zero. This would mean, as restricted above, that x_k exactly or approximately corresponds to a minimum or maximum of $f(x)$. If the slope $f'(x_k)$ is zero, then the slope projection never intersects the x-axis. If the slope $f'(x_k)$ is small, then the correction to x_k may be so large that x_{k+1} is a worse root estimate than x_k. For this reason, experienced engineers usually limit the size of the correction term; that is, if the absolute value of $f(x_k)/f'(x_k)$ is larger than a preselected maximum value, the maximum value is used.

(2) If there are several roots of the equation $f(x) = 0$, we must be in the vicinity of the desired root x_r in order that the algorithm actually converges to that root. The b-part of the figure depicts the condition that the initial estimate x_1 will result in convergence to x_{r_2} rather than x_{r_1}.

(3) Oscillation from one side of the root to the other can occur if, for example, the function is antisymmetric about a root which is an inflection point. The use of one-half of the correction will usually prevent this behavior, which is depicted in the c-part of the accompanying figure.

Example: Beginning with an initial estimate of $x_1 = 5$, estimate the single root of the equation $e^x - 10 \cos x - 100 = 0$.

The table below summarizes the application of Newton's method to the given equation. The iterative process was terminated when the absolute value of the correction $-f(x_k)/f'(x_k)$ became less than 10^{-6}.

k	x_k	$f(x_k)$	$f'(x_k)$	$x_{k+1} - x_k = -\dfrac{f(x_k)}{f'(x_k)}$
1	5.000 000	45.576 537	138.823 916	$-0.328\ 305$
2	4.671 695	7.285 610	96.887 065	$-0.075\ 197$
3	4.596 498	0.292 886	89.203 650	$-0.003\ 283$
4	4.593 215	0.000 527	88.882 536	$-0.000\ 006$
5	4.593 209	$-2(10^{-8})$	88.881 956	$2.25(10^{-10})$

USEFUL TABLES

D

Table D/1   Properties

Density, ρ

	kg/m³	lbm/ft³		kg/m³	lbm/ft³
Air*	1.2062	0.07530			
Aluminum	2 690	168	Iron (cast)	7 210	450
Concrete (av.)	2 400	150	Lead	11 370	710
Copper	8 910	556	Mercury	13 570	847
Earth (wet, av.)	1 760	110	Oil (av.)	900	56
(dry, av.)	1 280	80	Steel	7 830	489
Glass	2 590	162	Titanium	3 080	192
Gold	19 300	1205	Water (fresh)	1 000	62.4
Ice	900	56	(salt)	1 030	64
* At 20°C (68°F) and atmospheric			Wood (soft pine)	480	30
pressure			(hard oak)	800	50

Coefficients of Friction, μ

(The coefficients in the following table represent typical values under normal working conditions. Actual coefficients for a given situation will depend on the exact nature of the contacting surfaces. A variation of 25 to 100 per cent or more from these values could be expected in an actual application, depending on prevailing conditions of cleanliness, surface finish, pressure, lubrication, and velocity.)

CONTACTING SURFACE	TYPICAL VALUES OF COEFFICIENT OF FRICTION, μ	
	STATIC	KINETIC
Steel on steel (dry)	0.6	0.4
Steel on steel (greasy)	0.1	0.05
Teflon on steel	0.04	0.04
Steel on babbitt (dry)	0.4	0.3
Steel on babbitt (greasy)	0.1	0.07
Brass on steel (dry)	0.5	0.4
Brake lining on cast iron	0.4	0.3
Rubber tires on smooth pavement (dry)	0.9	0.8
Wire rope on iron pulley (dry)	0.2	0.15
Hemp rope on metal	0.3	0.2
Metal on ice		0.02

Table D/2   Solar System Constants

Universal gravitational constant	$G = 6.673(10^{-11})$ m³/(kg·s²)
	$= 3.439(10^{-8})$ ft⁴/(lbf-s⁴)
Mass of Earth	$m_e = 5.976(10^{24})$ kg
	$= 4.095(10^{23})$ lbf-s²/ft
Period of Earth's rotation (1 sidereal day)	$= 23$ h 56 min 4 s
	$= 23.9344$ h
Angular velocity of Earth	$\omega = 0.7292(10^{-4})$ rad/s
Mean angular velocity of Earth–Sun line	$\omega' = 0.1991(10^{-6})$ rad/s
Mean velocity of Earth's center about Sun	$= 107\,200$ km/h
	$= 66{,}610$ mi/h

BODY	MEAN DISTANCE TO SUN km (mi)	ECCENTRICITY OF ORBIT e	PERIOD OF ORBIT solar days	MEAN DIAMETER km (mi)	MASS RELATIVE TO EARTH	SURFACE GRAVITATIONAL ACCELERATION m/s² (ft/s²)	ESCAPE VELOCITY km/s (mi/s)
Sun	—	—	—	1 392 000 (865 000)	333 000	274 (898)	616 (383)
Moon	384 398* (238 854)*	0.055	27.32	3 476 (2 160)	0.0123	1.62 (5.32)	2.37 (1.47)
Mercury	57.3 × 10⁶ (35.6 × 10⁶)	0.206	87.97	5 000 (3 100)	0.054	3.47 (11.4)	4.17 (2.59)
Venus	108 × 10⁶ (67.2 × 10⁶)	0.0068	224.70	12 400 (7 700)	0.815	8.44 (27.7)	10.24 (6.36)
Earth	149.6 × 10⁶ (92.96 × 10⁶)	0.0167	365.26	12 742† (7 918)†	1.000	9.821‡ (32.22)‡	11.18 (6.95)
Mars	227.9 × 10⁶ (141.6 × 10⁶)	0.093	686.98	6 788 (4 218)	0.107	3.73 (12.3)	5.03 (3.13)

* Mean distance to Earth (center-to-center)

† Diameter of sphere of equal volume, based upon a spheroid Earth with a polar diameter of 12 714 km (7900 mi) and an equatorial diameter of 12 756 km (7926 mi)

‡ For nonrotating spherical Earth, equivalent to absolute value at sea level and latitude 37.5°

Table D/3   Properties of Plane Figures

FIGURE	CENTROID	AREA MOMENTS OF INERTIA
Arc Segment	$\bar{r} = \dfrac{r \sin \alpha}{\alpha}$	—
Quarter and Semicircular Arcs	$\bar{y} = \dfrac{2r}{\pi}$	—
Triangular Area	$\bar{x} = \dfrac{a + b}{3}$ $\bar{y} = \dfrac{h}{3}$	$I_x = \dfrac{bh^3}{12}$ $\bar{I}_x = \dfrac{bh^3}{36}$ $I_{x_1} = \dfrac{bh^3}{4}$
Rectangular Area	—	$I_x = \dfrac{bh^3}{3}$ $\bar{I}_x = \dfrac{bh^3}{12}$ $\bar{J} = \dfrac{bh}{12}(b^2 + h^2)$
Area of Circular Sector	$\bar{x} = \dfrac{2}{3}\dfrac{r \sin \alpha}{\alpha}$	$I_x = \dfrac{r^4}{4}(\alpha - \tfrac{1}{2}\sin 2\alpha)$ $I_y = \dfrac{r^4}{4}(\alpha + \tfrac{1}{2}\sin 2\alpha)$ $J = \tfrac{1}{2}r^4\alpha$
Quarter-Circular Area	$\bar{x} = \bar{y} = \dfrac{4r}{3\pi}$	$I_x = I_y = \dfrac{\pi r^4}{16}$ $\bar{I}_x = \bar{I}_y = \left(\dfrac{\pi}{16} - \dfrac{4}{9\pi}\right)r^4$ $J = \dfrac{\pi r^4}{8}$
Area of Elliptical Quadrant Area $A = \dfrac{\pi ab}{4}$	$\bar{x} = \dfrac{4a}{3\pi}$ $\bar{y} = \dfrac{4b}{3\pi}$	$I_x = \dfrac{\pi ab^3}{16}, \quad \bar{I}_x = \left(\dfrac{\pi}{16} - \dfrac{4}{9\pi}\right)ab^3$ $I_y = \dfrac{\pi a^3 b}{16}, \quad \bar{I}_y = \left(\dfrac{\pi}{16} - \dfrac{4}{9\pi}\right)a^3 b$ $J = \dfrac{\pi ab}{16}(a^2 + b^2)$

Table D/4 Properties of Homogeneous Solids
(m = mass of body shown)

BODY	MASS CENTER	MASS MOMENTS OF INERTIA
Circular Cylindrical Shell	—	$I_{xx} = \frac{1}{2}mr^2 + \frac{1}{12}ml^2$ $I_{x_1x_1} = \frac{1}{2}mr^2 + \frac{1}{3}ml^2$ $I_{zz} = mr^2$
Half Cylinder Shell	$\bar{x} = \frac{2r}{\pi}$	$I_{xx} = I_{yy}$ $= \frac{1}{2}mr^2 + \frac{1}{12}ml^2$ $I_{x_1x_1} = I_{y_1y_1}$ $= \frac{1}{2}mr^2 + \frac{1}{3}ml^2$ $I_{zz} = mr^2$ $\bar{I}_{zz} = \left(1 - \frac{4}{\pi^2}\right)mr^2$
Circular Cylinder	—	$I_{xx} = \frac{1}{4}mr^2 + \frac{1}{12}ml^2$ $I_{x_1x_1} = \frac{1}{4}mr^2 + \frac{1}{3}ml^2$ $I_{zz} = \frac{1}{2}mr^2$
Semicylinder	$\bar{x} = \frac{4r}{3\pi}$	$I_{xx} = I_{yy}$ $= \frac{1}{4}mr^2 + \frac{1}{12}ml^2$ $I_{x_1x_1} = I_{y_1y_1}$ $= \frac{1}{4}mr^2 + \frac{1}{3}ml^2$ $I_{zz} = \frac{1}{2}mr^2$ $\bar{I}_{zz} = \left(\frac{1}{2} - \frac{16}{9\pi^2}\right)mr^2$
Rectangular Parallelepiped	—	$I_{xx} = \frac{1}{12}m(a^2 + l^2)$ $I_{yy} = \frac{1}{12}m(b^2 + l^2)$ $I_{zz} = \frac{1}{12}m(a^2 + b^2)$ $I_{y_1y_1} = \frac{1}{12}mb^2 + \frac{1}{3}ml^2$ $I_{y_2y_2} = \frac{1}{3}m(b^2 + l^2)$

Table D/4 *Continued*

(m = mass of body shown)

BODY	MASS CENTER	MASS MOMENTS OF INERTIA
Spherical Shell	—	$I_{zz} = \frac{2}{3}mr^2$
Hemispherical Shell	$\bar{x} = \frac{r}{2}$	$I_{xx} = I_{yy} = I_{zz} = \frac{2}{3}mr^2$ $\bar{I}_{yy} = \bar{I}_{zz} = \frac{5}{12}mr^2$
Sphere	—	$I_{zz} = \frac{2}{5}mr^2$
Hemisphere	$\bar{x} = \frac{3r}{8}$	$I_{xx} = I_{yy} = I_{zz} = \frac{2}{5}mr^2$ $\bar{I}_{yy} = \bar{I}_{zz} = \frac{83}{320}mr^2$
Uniform Slender Rod	—	$I_{yy} = \frac{1}{12}ml^2$ $I_{y_1y_1} = \frac{1}{3}ml^2$

Table D/4 *Continued*

(m = mass of body shown)

BODY	MASS CENTER	MASS MOMENTS OF INERTIA
Quarter-Circular Rod	$\bar{x} = \bar{y}$ $= \dfrac{2r}{\pi}$	$I_{xx} = I_{yy} = \frac{1}{2}mr^2$ $I_{zz} = mr^2$
Elliptical Cylinder		$I_{xx} = \frac{1}{4}ma^2 + \frac{1}{12}ml^2$ $I_{yy} = \frac{1}{4}mb^2 + \frac{1}{12}ml^2$ $I_{zz} = \frac{1}{4}m(a^2 + b^2)$ $I_{y_1y_1} = \frac{1}{4}mb^2 + \frac{1}{3}ml^2$
Conical Shell	$\bar{z} = \dfrac{2h}{3}$	$I_{yy} = \frac{1}{4}mr^2 + \frac{1}{2}mh^2$ $I_{y_1y_1} = \frac{1}{4}mr^2 + \frac{1}{6}mh^2$ $I_{zz} = \frac{1}{2}mr^2$ $\bar{I}_{yy} = \frac{1}{4}mr^2 + \frac{1}{18}mh^2$
Half Conical Shell	$\bar{x} = \dfrac{4r}{3\pi}$ $\bar{z} = \dfrac{2h}{3}$	$I_{xx} = I_{yy}$ $\quad = \frac{1}{4}mr^2 + \frac{1}{2}mh^2$ $I_{x_1x_1} = I_{y_1y_1}$ $\quad = \frac{1}{4}mr^2 + \frac{1}{6}mh^2$ $I_{zz} = \frac{1}{2}mr^2$ $\bar{I}_{zz} = \left(\frac{1}{2} - \frac{16}{9\pi^2}\right)mr^2$
Right-Circular Cone	$\bar{z} = \dfrac{3h}{4}$	$I_{yy} = \frac{3}{20}mr^2 + \frac{3}{5}mh^2$ $I_{y_1y_1} = \frac{3}{20}mr^2 + \frac{1}{10}mh^2$ $I_{zz} = \frac{3}{10}mr^2$ $\bar{I}_{yy} = \frac{3}{20}mr^2 + \frac{3}{80}mh^2$

Table D/4 *Continued*

(m = mass of body shown)

BODY	MASS CENTER	MASS MOMENTS OF INERTIA
Half Cone	$\bar{x} = \dfrac{r}{\pi}$ $\bar{z} = \dfrac{3h}{4}$	$I_{xx} = I_{yy}$ $\quad = \frac{3}{20}mr^2 + \frac{3}{5}mh^2$ $I_{x_1 x_1} = I_{y_1 y_1}$ $\quad = \frac{3}{20}mr^2 + \frac{1}{10}mh^2$ $I_{zz} = \frac{3}{10}mr^2$ $\bar{I}_{zz} = \left(\dfrac{3}{10} - \dfrac{1}{\pi^2}\right)mr^2$
$\dfrac{x^2}{a^2} + \dfrac{y^2}{b^2} + \dfrac{z^2}{c^2} = 1$ Semiellipsoid	$\bar{z} = \dfrac{3c}{8}$	$I_{xx} = \frac{1}{5}m(b^2 + c^2)$ $I_{yy} = \frac{1}{5}m(a^2 + c^2)$ $I_{zz} = \frac{1}{5}m(a^2 + b^2)$ $\bar{I}_{xx} = \frac{1}{5}m(b^2 + \frac{19}{64}c^2)$ $\bar{I}_{yy} = \frac{1}{5}m(a^2 + \frac{19}{64}c^2)$
$\dfrac{x^2}{a^2} + \dfrac{y^2}{b^2} = \dfrac{z}{c}$ Elliptic Paraboloid	$\bar{z} = \dfrac{2c}{3}$	$I_{xx} = \frac{1}{6}mb^2 + \frac{1}{2}mc^2$ $I_{yy} = \frac{1}{6}ma^2 + \frac{1}{2}mc^2$ $I_{zz} = \frac{1}{6}m(a^2 + b^2)$ $\bar{I}_{xx} = \frac{1}{6}m(b^2 + \frac{1}{3}c^2)$ $\bar{I}_{yy} = \frac{1}{6}m(a^2 + \frac{1}{3}c^2)$
Rectangular Tetrahedron	$\bar{x} = \dfrac{a}{4}$ $\bar{y} = \dfrac{b}{4}$ $\bar{z} = \dfrac{c}{4}$	$I_{xx} = \frac{1}{10}m(b^2 + c^2)$ $I_{yy} = \frac{1}{10}m(a^2 + c^2)$ $I_{zz} = \frac{1}{10}m(a^2 + b^2)$ $\bar{I}_{xx} = \frac{3}{80}m(b^2 + c^2)$ $\bar{I}_{yy} = \frac{3}{80}m(a^2 + c^2)$ $\bar{I}_{zz} = \frac{3}{80}m(a^2 + b^2)$
Half Torus	$\bar{x} = \dfrac{a^2 + 4R^2}{2\pi R}$	$I_{xx} = I_{yy} = \frac{1}{2}mR^2 + \frac{5}{8}ma^2$ $I_{zz} = mR^2 + \frac{3}{4}ma^2$

INDEX

CONVERSION CHARTS BETWEEN SI AND U.S. CUSTOMARY UNITS

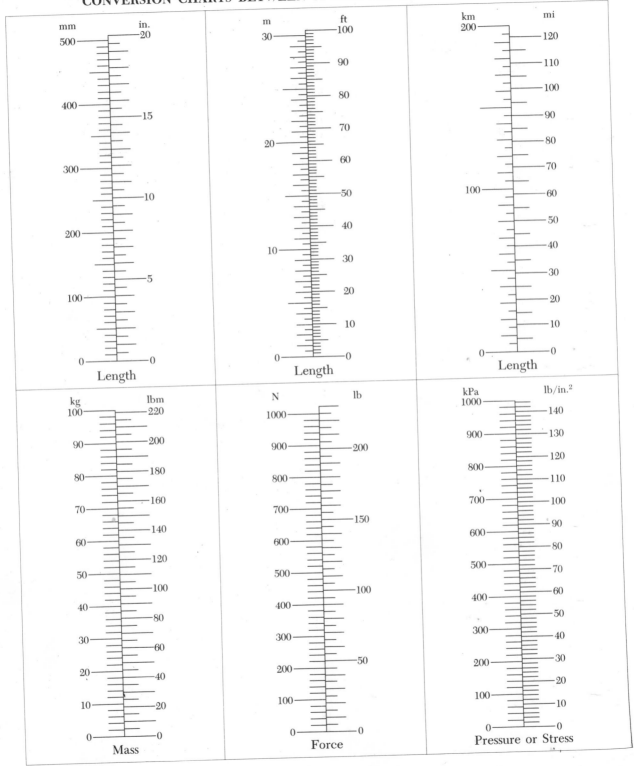

Length

Length

Length

Mass

Force

Pressure or Stress